How Fermented Foods Feed a Healthy Gut Microbiota

M. Andrea Azcarate-Peril • Roland R. Arnold
José M. Bruno-Bárcena
Editors

How Fermented Foods Feed a Healthy Gut Microbiota

A Nutrition Continuum

Editors
M. Andrea Azcarate-Peril
School of Medicine
University of North Carolina at Chapel Hill
Chapel Hill, NC, USA

Roland R. Arnold
School of Dentistry
University of North Carolina at Chapel Hill
Chapel Hill, NC, USA

José M. Bruno-Bárcena
Department of Plant and Microbial Biology
North Carolina State University
Raleigh, NC, USA

ISBN 978-3-030-28739-9   ISBN 978-3-030-28737-5   (eBook)
https://doi.org/10.1007/978-3-030-28737-5

© Springer Nature Switzerland AG 2019

This work is subject to copyright. All rights are reserved by the Publisher, whether the whole or part of the material is concerned, specifically the rights of translation, reprinting, reuse of illustrations, recitation, broadcasting, reproduction on microfilms or in any other physical way, and transmission or information storage and retrieval, electronic adaptation, computer software, or by similar or dissimilar methodology now known or hereafter developed.

The use of general descriptive names, registered names, trademarks, service marks, etc. in this publication does not imply, even in the absence of a specific statement, that such names are exempt from the relevant protective laws and regulations and therefore free for general use.

The publisher, the authors, and the editors are safe to assume that the advice and information in this book are believed to be true and accurate at the date of publication. Neither the publisher nor the authors or the editors give a warranty, express or implied, with respect to the material contained herein or for any errors or omissions that may have been made. The publisher remains neutral with regard to jurisdictional claims in published maps and institutional affiliations.

This Springer imprint is published by the registered company Springer Nature Switzerland AG
The registered company address is: Gewerbestrasse 11, 6330 Cham, Switzerland

Preface

Extensive research in recent decades has helped decode the impact of diet on the gut microbiome, together with the microbial responses to nutritional components in our diet, which converge in a delicate and balanced choreography (when healthy) or disarray (when unbalanced or unhealthy). With over 20 years of research expertise in fermented food, probiotics, and the microbiome, we felt like the pieces of the nutritional-microbial puzzle had started to fit together: could it be possible that our gut microbiota needs "reseeding" and that reseeding is done by the foods we eat? Can food have a restorative role as a microbe provider for the gut microbiota?

Traditionally, fermented foods have been consumed by humans for millennia. This method of food preservation was discovered most probably by accident as a means to prevent spoilage and, unsuspectingly to the consumer, provided our ancestors with beneficial bacteria that repopulated the gut microbiota upon consumption. However, novel methods of production and conservation of food have severed the ties between the food we consume and the gut microbiota. As a consequence, there is a documented increase in the prevalence of autoimmune diseases and obesity, which has been correlated to a decreased diversity of gut microbes, while infectious disorders have decreased in the past decades.

This book offers an introduction dedicated to the environmental microbiome as the piece that completes the circle of life or the "nutrition continuum." After the introduction, we have structured the book in three parts. The first one focuses on food and its associated microbes. From breast milk to fermented food, it is clear that food is responsible not only for providing nutrition to the host but also for "seeding" the gut-associated microbiota. The chapters in this part provide an overview of what is currently known about the microbes associated with breast milk and fermented food and recount traditional forms of food preparation with current industrial techniques in terms of the potential loss of microbial diversity associated with industrialization.

We dedicated the second part of this book to the mouth and its associated microbes. The mouth environment and the oral microbiota are clearly the gatekeepers and main microbial contributors to the gut microbiota through food and shedding of microbes. Moreover, the lack of a clear demarcation between the external environment and the oral cavity has resulted in a unique environment, which hosts a complex and diverse microbiota. In addition, the technical accessibility of the oral microbiota to sampling has allowed significant advances in microbiome research that reveal clear relationships between oral and gut health.

The final segment of our book looks into the gut microbiota across ages, starting with the lifelong consequences of the infant's initial colonization period, following with chapters on the gut microbiota during adulthood and aging. The next chapters attempt to answer questions regarding how we can beneficially modulate the microbiome with probiotics and prebiotics to rebalance a gut microbiota skewed by the modern practices of our Western society.

We are extremely grateful to the contributors of this book. Their outstanding intellectual contributions and scientific expertise made this project possible. We are also thankful to our colleagues, Apoena Ribeiro, Jason Arnold, Sue Dagher, and Hunter Whittington, for their critical reading of all or parts of the manuscript.

We live in a time where relevant scientific questions can be readily answered, thanks to available technologies. However, too often, discoveries and rational scientific conclusions are not translated into policies that benefit humankind but instead relegated by political and economic decisions with costly consequences. We hope that this book conveys the importance of making good personal and collective decisions in terms of nutritional behaviors that greatly affect human health.

Chapel Hill, NC, USA   M. Andrea Azcarate-Peril
Chapel Hill, NC, USA   Roland R. Arnold
Raleigh, NC, USA   José M. Bruno-Bárcena

Introduction: The Ancient Symbioses

I started off my career focused on social insects and the species that live with them. At the age of 19, I worked with James Danoff-Burg, then a graduate student at the University of Kansas, to study a group of beetles that live inside the colonies of *Linepithema* ants in the deserts of the southwest. Then the next year, I accepted a job working with Samantha Messier, then a graduate student at the University of Colorado, Boulder. Sam was studying a species of termite common in the forests of the Neotropics, *Nasutitermes corniger*. These termites are amazing for a variety of reasons, but one of them is the fact that the soldiers of *Nasutitermes corniger* have, in place of ordinary insect weaponry (spines, spines, stings, or mandibles), long noses. Out of these noses (nasutes), the soldiers shoot a mix of terpenes that is gooey and smells a bit like turpentine. This mix helps to defend the termites against ants and also anteaters. To me, as a then 19-year-old, that such elaborate chemical weaponry was to be found amid nonhuman societies was mind blowing. But spraying turpentine is not the best trick of which *Nasutitermes* termites are capable, not hardly.

The *Nasutitermes* soldiers, although endowed with great biochemical power, also have a weakness. They lack functional, chewing mouthparts. They cannot eat on their own and so rely on their brothers and sisters to feed them. But even after being fed by their brothers and sisters, the soldiers are still wanting in some nutrients, particularly nitrogen. The soldiers cope with this problem by playing host to bacteria that fix nitrogen from the air—"From the air!" This was really too much for me to believe, a bit of nature inspired by magical realism, and yet it was true.[1]

What I would go on to study with Sam was whether the termite soldiers (and workers) rely more on their nitrogen fixing microbes when the wood they eat is low in nitrogen and when they are actively at war with anteaters. Here then was my actual job—I was to be the anteater. My role for three months at the La Selva Biological Station in Costa Rica was to attack the colonies of this very sophisticated

---

[1] Prestwich, Glenn D., and Barbara L. Bentley. "Nitrogen fixation by intact colonies of the termite *Nasutitermes corniger*." *Oecologia* 49, no. 2 (1981): 249–251.

termite with a very unsophisticated machete and see if, once attacked, they would produce more soldiers (they did), which, in turn, would fix more nitrogen (they did), thanks to their gut microbes.[2]

At that time, the idea that termites rely on gut microbes was not novel. The very first study of the role of gut microbes in nutrition was that of Joseph Leidy on the guts of termites, which was published in 1881.[3] In the years since, it had been established that the guts of termites are full of many kinds of microbes, be they bacteria, protists, symbioses between bacteria and protists or even archaea. These relationships are, it had been shown by then, relatively fine-tuned to the life history of the termites. Grass-feeding termite species have different gut microbes from wood-feeding termite species, which, in turn, have different microbes from soil-feeding termite species.[4] The bodies of termites have slightly diversified over the last hundred and fifty million years, but their guts and the microbes in them have diversified greatly, including, in the case of *Nasutitermes*, the origin, evolution, and elaboration of relationships with nitrogen-fixing bacteria.

As a 19-year-old, it seemed obvious to me that if termites hosted microbes on which they depended, then other insects in the forests around me probably did as well (why wouldn't they?) It also seemed reasonable to me, and I suspect to anyone studying termites and their microbes, that so too did the birds and mammals I was not alone. Amid beer drinking and banjo playing at the research station where I studied termites, scientists speculated about whether the world was, if viewed through a microbial lens, one in which microbes carried out basically every process and sometimes, when they needed to, made animals carry them from place to place or, figuratively speaking, hold their sandwich.[5] But none of these ideas, although in many ways obvious, were terribly mainstream. Microbial symbioses were being studied by insect biologists or animal scientists in agricultural colleges and land grant institutions (like the one in which I now work).[6] Such studies, we know in retrospect, would prove central to understanding the biological world. But they were not central to the fields of ecology and evolutionary biology. Nor were they especially central to the story of human health. Theses, great thick theses, need to be written about why symbioses between microbes and their hosts, and the study of those symbioses, stayed in the margins for so long. Regardless of its cause the marginal status of the study of symbioses would eventually change. The role of microbes associated with animals

---

[2] Messier, Samantha Hope. "Ecology and division of labor in *Nasutitermes corniger*: The effect of environmental variation on caste ratios." (1997): 2298–2298.

[3] Leidy, Joseph. *Parasites of the termites*. Collins, Printer, 1881.

[4] Ohkuma, Moriya, and Andreas Brune. "Diversity, structure, and evolution of the termite gut microbial community." In *Biology of termites: a modern synthesis*, pp. 413–438. Springer, Dordrecht, 2010.

[5] Dyer, Betsey Dexter. "Symbiosis and organismal boundaries." *American Zoologist* 29, no. 3 (1989): 1085–1093.

[6] RE Hungate's important work spanned both of these fields. He began with termites and moved on to ruminants. Hungate, R. E. "The symbiotic utilization of cellulose." *Journal of the Elisha Mitchell Scientific Society* 62, no. 1 (1946): 9–24.

Introduction: The Ancient Symbioses

and plants would rise in prominence. This rise began in part due to the arrival of novel approaches to the study and the identification of microbes and sequencing-based approaches.

The availability of novel sequencing approaches (approaches made possible by using enzymes from microbes themselves, like those of *Thermus aquaticus*) changed everything. With new barcoding and metagenomics tools, it became possible to relatively quickly do an inventory of the kinds of life in a given place. In some subfields, these new approaches could be combined with insights from earlier work on insects or domestic animals to jump-start discoveries. In other cases, the new wave of research began as if from scratch (even where antecedents). As tens and then hundreds and then thousands of studies accumulated, it became ever clearer that humans are covered with microbes, that humans are filled with microbes, and that human food too, much of it anyway, is rich with microbes. Of course, these insights were not really new. They were instead newly appreciated and, thanks to the new sequencing approaches (and the new waves of funding that they would usher in), were no longer marginal. The microbe was out of the bag (or out of the gut, as the case might be). Humans are, just like all the species, filled with species on which we depend. We are like the termites, just bigger and gassier. The study of insects, agriculture, ecology, evolution, and medicine began to acquire a new holism, a holism made possible by the ubiquitous importance of microbes.

But there was a problem. In the years between the first studies of the gut microbes in termites and the re-recognition of the value of the microbes on and in our bodies and in our food, we had, collectively as humans, made a mistake—a very big mistake, a no good, stinking, terrible mistake. The mistake we made was simplifying the microbial communities present in our lives, dramatically. We overused antibiotics for diseases that were not bacterial. We used antimicrobials in settings where soap would have been used. We closed our windows and sealed out plant- and soil-associated microbes. We also increasingly shifted to processed food, in which wild microbes were rare, as were those associated with fermentation, and in their place were an abundance of microbes adapted to do well in freezers and refrigerators. To use the language of ecologists, these changes changed our Western baseline, they changed the condition of the body and daily life of the average person, they changed it so much that it sometimes made it difficult for us to tell exactly what is wrong. Rare chronic diseases, diseases related in one way or another to changes in the microbes in our lives, became common. Crohn's disease, inflammatory bowel disease, multiple sclerosis, allergies, asthma, and maybe even autism all fall within this circus of terrible modern maladies.

Some features of the future of the study of our bodies, our microbes, and our new chronic maladies seem predictable. It seems predictable, for instance, that the microbes we need in our environment, on our food, and in and on our bodies will be shown to depend both on our genes and on our lifestyles. It seems predictable that there will not be one kind of "healthy microbiome" but many instead. This, after all, is what was found with the termites. When the termites had less access to wood with lots of nitrogen, they needed (and typically hosted) more nitrogen-fixing bacteria. When they went to war, whether against real anteaters or against my machete, they needed more nitrogen-fixing bacteria. We should not expect our own bodies to be any simpler, and, indeed, given the great diversity of human genetic backgrounds and experiences, we are likely to be much, much more complex.

This book attempts, in light of our broad human story, and in light of a particular moment in the history of science, to consider what we know right now about microbes, health, and nutrition. It considers the microbes of the mouth, microbes in food, gut microbes, and our wellness. The authors in this book do not all agree with each other about just which set of microbes benefit us and when. They do not all agree with each other about many features of our bodies, microbes, and wellness—as it should be. We do not yet understand enough about ourselves to understand the simple answers, and so instead what we have begun to develop are kinds of regularities, things that seem to be mostly, but not always, true. Fermented food, when still alive, seems to offer health benefits—often but not always. The Western diet, rich in sugar, seems to lead to microbes in mouths that are less healthy and more likely to cause cavities—often but maybe not always.

The book is, as far as I know, the first of its kind, the first to think about what it would look like to eat food that benefits us and our beneficial microbes. Or rather, it is the first of its kind for humans. In this way, one of the most beautiful things about what this book offers is that it draws humans back into the rest of life. It reminds us that we are not only connected to and dependent upon the microbes that we ingest or fail to ingest and rub upon but also that this condition unites us with every other species of animal that has ever lived. We humans are special in our consciousness, in our ability to think about problems, and we actively make decisions about the change we would like to make. But in a microbial context, we are also unusual in that we need this book. For three hundred million years, animals acquired the microbes they needed without problem, from their food and their environments, without need to figure out how to do so. We are uniquely the species that has altered our environment and our diet so completely that we need to study how to do what other species do without thinking. This book then is the beginning of the grounding we need in order to remember how to be like the other species and remember, in other words, how to be microbially whole.

Natural History Museum of Denmark   Rob R. Dunn
Copenhagen, Denmark
Department of Applied Ecology
North Carolina State University
Raleigh, NC, USA

Contents

Part I   The Seed Source Matters: Foods and Their Associated Microbes

Baby's First Microbes: The Microbiome of Human Milk.............. 3
M. Carmen Collado, Miguel Gueimonde, Lorena Ruiz, Marina Aparicio,
Irma Castro, and Juan M. Rodríguez

Fermented Dairy Products.................................... 35
C. Peláez, M. C. Martínez-Cuesta, and T. Requena

Meat and Meat Products 57
Wim Geeraerts, Despoina Angeliki Stavropoulou, Luc De Vuyst,
and Frédéric Leroy

**Fermented Vegetables as Vectors for Relocation of Microbial
Diversity from the Environment to the Human Gut** 91
Ilenys M. Pérez-Díaz

**Production and Conservation of Starter Cultures: From
"Backslopping" to Controlled Fermentations** 125
Hunter D. Whittington, Suzanne F. Dagher, and José M. Bruno-Bárcena

Part II   The Oral Microbiota: The Seeder and Gatekeeper

Introduction to the Oral Cavity................................. 141
Roland R. Arnold and Apoena A. Ribeiro

Defining the Healthy Oral Microbiome 155
G. M. S. Soares and M. Faveri

Dysbiosis of the Oral Microbiome.............................. 171
Apoena A. Ribeiro and Roland R. Arnold

**Microbial Manipulation of Dysbiosis: Prebiotics and Probiotics
for the Treatment of Oral Diseases** 193
Eduardo Montero, Margarita Iniesta, Silvia Roldán, Mariano Sanz,
and David Herrera

Part III The Recipe for Happiness: A Balanced Gut Microbiota

Early Gut Microbiome: A Good Start in Nutrition and Growth May Have Lifelong Lasting Consequences 239
Amanda L. Thompson

"We Are What We Eat": How Diet Impacts the Gut Microbiota in Adulthood 259
Taojun Wang, Dominique I. M. Roest, Hauke Smidt, and Erwin G. Zoetendal

The Aging Gut Microbiota 285
Erin S. Keebaugh, Leslie D. Williams, and William W. Ja

Beneficial Modulation of the Gut Microbiome: Probiotics and Prebiotics 309
M. Andrea Azcarate-Peril

The Disappearing Microbiota: Diseases of the Western Civilization 325
Emiliano Salvucci

Conclusions: What Is Next for the Healthy Human-Microbe "Holobiome"? 349
M. Andrea Azcarate-Peril

Index 357

Part I
The Seed Source Matters: Foods and Their Associated Microbes

Baby's First Microbes: The Microbiome of Human Milk

M. Carmen Collado, Miguel Gueimonde, Lorena Ruiz, Marina Aparicio, Irma Castro, and Juan M. Rodríguez

Abstract At the beginning of the twenty-first century, microbiological studies on human milk started to describe the existence of its own microbiota. Hygienically collected milk samples from healthy women contain a relatively low bacterial load consisting mostly of *Staphylococcus*, *Streptococcus*, and other Gram-positive bacteria (*Corynebacterium*, *Propionibacterium*, *Lactobacillus and Bifidobacterium*). DNA from strict anaerobic bacteria is also detected in human milk samples. Colostrum and milk bacteria may play a key role in driving the development of the infant gut microbiota, the correct maturation of the infant immune system and the improvement of tolerance mechanisms. A well-balanced human milk microbiota is also relevant for maternal breast health. The origin of human milk bacteria still remains largely unknown. Infant's oral cavity and maternal skin may contaminate milk. Additionally, selected bacteria of the maternal digestive microbiota may access the mammary glands through oral- and entero-mammary pathways by involving mononuclear cells for their transport. These pathways would provide new opportunities for manipulating maternal-fetal microbiota, reducing the risk of preterm birth or infant diseases.

Keywords Human milk · Human milk oligosaccharides · Infant nutrition · Probiotics · Prebiotics

---

M. C. Collado
Institute of Agrochemistry and Food Technology-National Research Council (IATA-CSIC), Valencia, Spain

M. Gueimonde · L. Ruiz
Department of Microbiology and Biochemistry of Dairy Products, Instituto de Productos Lácteos de Asturias, Consejo Superior de Investigaciones Científicas (IPLA-CSIC), Villaviciosa, Spain

M. Aparicio · I. Castro · J. M. Rodríguez (✉)
Department of Nutrition and Food Science, Complutense University of Madrid, Madrid, Spain
e-mail: jmrodrig@vet.ucm.es

© Springer Nature Switzerland AG 2019
M. A. Azcarate-Peril et al. (eds.), *How Fermented Foods Feed a Healthy Gut Microbiota*, https://doi.org/10.1007/978-3-030-28737-5_1

Introduction

Human milk is the gold standard for infant nutrition during the first months of life since it is perfectly adapted to the nutritional requirements of the baby; it contains a plethora of biologically active components, including immunoglobulins, chemokines, growth factors, cytokines, bioactive lipids, oligosaccharides, microRNAs, hormones, immune cells and microorganisms, among others (Hennet and Borsig 2016). Human milk composition varies among individuals and it is contingent on several factors, such as mother's genotype, geographical location, gestational age, maternal health status, diet and time of lactation (Cabrera-Rubio et al. 2012; Andreas et al. 2015; Feng et al. 2016; Kumar et al. 2016; Gomez-Gallego et al. 2016; Ruiz et al. 2017). Therefore, the composition of human milk adapts dynamically to the variable needs of the baby along the first months of life. Globally, these complex and dynamic compositions promote a healthy growth and development of the infants (Mosca and Giannì 2017). This recognition has prompted all national and international organizations focused on health, infancy, pediatrics, nutrition or epidemiology to recommend exclusive breast-feeding during at least the first 6 months of life; thereafter, infants should receive nutritionally adequate and safe complementary foods while breastfeeding continues for up to 2 years of age or beyond (World Health Organization 2003).

The short- and long-term health-promoting effects of breastfeeding have been known for decades and apply both to developing and developed countries. Historically, these effects were partly attributed to the presence of the so-called *"bifidogenic factors"*, leading to the predominance of microorganisms of the genus *Bifidobacterium* in the gut of breast-fed babies. However, the role of human milk as a complex ecological niche and as a relevant source of bacteria to seed the infant gut had remained unstudied until recently.

Microbial Diversity in Human Milk

Culture-Based Studies

The first culture-based studies testing breast milk were carried out during the second half of the past century and were mainly focused in the detection of "contaminants" and/or potentially harmful microbes and their role in infant infection (Rantasalo and Kauppinen 1959; Foster and Harris 1960; Kenny 1977; Williamson et al. 1978; Eidelman and Szilagyi 1979). Although some cases of infant infections and sepsis have been linked to the transmission of pathogens from human milk (Qutaishat et al. 2003; Kayıran et al. 2014; Weems et al. 2015; Zimmermann et al. 2017) the presence of contaminant microorganisms in human milk are not valid predictors of infection risk (Boer and Anido 1981; Schanler et al. 2011; Zimmermann et al. 2017).

During the last 15 years, several studies have described the presence of viable commensal, mutualistic, or potentially probiotic bacteria in human milk from healthy individuals [reviewed in (Fernández et al. 2013)], leading to an increasing interest in the assessment of its microbiota and microbiome, the potential mother-to-infant bacterial transfer through breastfeeding, and their role in the maternal and/or infant health. It also stimulated the search for new bacterial strains to be used as probiotics for the mother-infant dyad.

The cultivable bacteria usually found in human milk are dominated by Gram-positive belonging to the genera *Staphylococcus*, *Streptococcus*, *Corynebacterium* and *Propionibacterium* (Jiménez et al. 2008b; Solís et al. 2010; Schanler et al. 2011). At a lower extend, lactic acid bacteria (*Lactobacillus, Lactococcus, Leuconostoc, Weissella, Enterococcus*, among others) and bifidobacteria are also commonly isolated from human milk (Martín et al. 2003, 2009; Abrahamsson et al. 2009; Solís et al. 2010; Arboleya et al. 2011; Murphy et al. 2017). Among them, *Lactobacillus* (*L. salivarius, L. gasseri, L. fermentum, L. reuteri,* among others) and *Bifidobacterium* (*B. longum* and *B. breve*) species have been the subject of the highest interest because of their potential application as probiotics. Noticeably, isolates from these two bacterial genera seem to be more abundant in human milk samples from locations with a low use of antibiotics (Soto et al. 2014). Even though globally, more than 200 different bacterial species, belonging to approximately 50 different genera have been isolated from human milk up to the present (Fernández et al. 2013), including new bacterial species, such as *Streptococcus lactarius* (Martín et al. 2011).

The microbial load in human milk may range from 10^1 to 10^6 CFU/mL, depending on the health status of the mother (e.g., mastitis) and, also, on the milk collection method. As an example, the use of milk pumps may result in high concentrations of contaminating Gram-negative bacteria (*Enterobacteria, Pseudomonas, Stenotrophomonas,* among others) and yeasts appearing from rinsing water and/or poor hygienic manipulation practices (Jiménez et al. 2017).

Limitations of culture-dependent methods may rely in the inability to assess the presence of viable but non-cultivable organisms but, in contrast, they enable the isolation, preservation and characterization of bacterial strains (Lara-Villoslada et al. 2007b; Jiménez et al. 2008b; Delgado et al. 2009, 2011; Arboleya et al. 2011; Langa et al. 2012; Cárdenas et al. 2014). The availability of bacterial strains isolated from human milk, together with the novel genetic tools, is allowing the sequencing and annotation of its genomes (Jiménez et al. 2010a, b; Martín et al. 2012b, 2013; Langa et al. 2012; Gueimonde et al. 2012; Cárdenas et al. 2015), which will facilitate further functional studies and future applications.

From pioneer human milk studies to the most recent culture-based analysis aimed at isolating human-milk strains for potential probiotic applications, the use of culture-methods has unveiled human milk as a complex ecological niche and potential source of probiotics. In addition, we should not forget that human milk might contain yeasts, moulds and viruses (Daudi et al. 2012; Liu et al. 2015; Dupont-Rouzeyrol et al. 2016; Mutschlechner et al. 2016). The transmissions of three specific viruses (CMV, HIV, and HTLV-I) to the infants through breastfeeding are of particular concern, and are taken into consideration during management of human milk banks. In addition, human milk may contain bacteriophages, which might play a role in modulating the human milk microbiota (Jiménez et al. 2015; Duranti et al. 2017). Moreover, the human milk ecosystem also contains a complex population of human cells (Fan et al. 2010; Hassiotou and Geddes 2015; Witkowska-Zimny and Kaminska-El-Hassan 2017), which may interact with the microorganisms, both in human milk and in the infant gut.

Culture-Independent Studies

Cultivable microorganisms may represent a fraction of the natural microbial communities inhabiting a specific ecological niche. Therefore, the application of culture-independent molecular techniques, including quantitative PCR, denaturing gradient gel electrophoresis (DGGE), temperature gradient gel electrophoresis (TGGE), and Next Generation Sequencing (NGS) approaches, from metataxonomics (16S rRNA amplicon analysis) to metagenomics (total DNA sequencing), has provided a complementary assessment of the microbiome in human milk (Jeurink et al. 2013; McGuire and McGuire 2015). It is important to point out that such techniques detect nucleic acids and not living microbial cells, which means that bacterial DNA may belong to either live or dead organisms. Other limitations and bias that molecular techniques may introduce in the assessment of complex microbial communities include an over- or underestimation of some microbial groups with difficult to break cell wall, outer membranes or plasmatic membrane composition, DNA extraction methods, number of copies of the targeted gene, the specificity of the selected primers to 16S rRNA region(s), and current limitations inherent to the bioinformatics analysis (McGuire and McGuire 2015, 2017; Gomez-Gallego et al. 2016). Future studies considering the bacterial cell wall integrity coupled with 16S rRNA sequencing as well as RNA-based (metatranscriptomics) methodologies will provide novel information about the microbiota present in the milk at a functional level (Gosalbes et al. 2012).

Globally, culture-independent studies have confirmed the presence of DNA from bacterial groups previously identified with culture-dependent techniques, such as *Staphylococcus, Streptococcus, Corynebacterium, Propionibacterium. Lactococcus, Leuconostoc, Weissella, Bifidobacterium* and/or *Lactobacillus* spp. (Gueimonde et al. 2007; Martín et al. 2007a, b; Delgado et al. 2008; Collado et al. 2009; Hunt et al. 2011; Cabrera-Rubio et al. 2012, 2016; Ward et al. 2013; Jost et al. 2013, 2014; Jiménez et al. 2015; Boix-Amorós et al. 2016; Fitzstevens et al. 2017). In addition, some studies have also reported the presence of DNA from strictly anaerobic gut-associated microbes (*Bacteroides, Blautia, Clostridium, Collinsella, Coprococcus, Eubacterium, Faecalibacterium, Roseburia, Ruminococcus, Veillonella,* among others), which are either non-cultivable or very difficult to culture in the laboratory and, therefore, may not be detected using culture-based methods (Cabrera-Rubio et al. 2012; Jost et al. 2013, 2014; Jiménez et al. 2015; Gomez-Gallego et al. 2016).

Microbiome studies focused on human milk or breast tissue have also revealed the presence of DNA belonging to a third group of soil- and water-associated bacterial genera, including *Acinetobacter*, *Bradyrhizobium*, *Methylobacterium*, *Microbacterium*, *Novosphingobium*, *Pseudomonas*, *Ralstonia*, *Sphingopyxis*, *Sphingobium*, *Sphingomonas*, *Stenotrophomonas* and *Xanthomonas* (Hunt et al. 2011; Cabrera-Rubio et al. 2012; Urbaniak et al. 2014a, b; Xuan et al. 2014; Cacho et al. 2017). In some works, the amplified sequences related to such microorganisms were so frequent and abundant among individuals that some of them were considered members of the "core microbiome" of human milk. However, it has been pointed out that the molecular techniques used to study low abundance microbiomes (such as that of human milk from healthy women) have a high susceptibility to false positives because of contamination with DNA sequences from the water- and soil-associated bacterial genera cited above (Lauder et al. 2016; Perez-Muñoz et al. 2017). Presence of contaminating DNA in PCR reagents, DNA extraction kits and molecular biology grade water (Grahn et al. 2003; Mühl et al. 2010; Salter et al. 2014) area particularly relevant challenge when working with samples containing low microbial load since, upon amplification, the low amount of starting material may be widely overcome by the contaminating DNA and lead to inaccurate results and conclusions (Laurence et al. 2014; Lauder et al. 2016).

The possible impact of the presence contaminating DNA on 16S rRNA gene-based profiling and shotgun metagenomics analyses from typically low biomass samples has often not been taken into consideration among microbiome researchers (Perez-Muñoz et al. 2017). In fact, most DNA sequence-based studies describing microbial communities in low-biomass environments neither report sequencing of negative controls, nor describe their contaminant removal procedures. Recommendations to reduce the impact of contaminants in sequence-based, low-biomass microbiota studies have already been provided (Salter et al. 2014), and future studies will require having them taken into consideration when concluding which observations are actually genuine.

In relation to this issue, a non-critical analysis of sequences obtained in a human milk metagenomic study suggested that the bacterial core microbiome was composed of the genera: *Pseudomonas*, *Staphylococcus*, *Streptococcus*, *Bacteroides*, *Faecalibacterium*, *Ruminococcus*, *Lactobacillus*, *Propionibacterium*, *Sphingomonas*, *Novosphingobium*, *Sphingopyxis, Sphingobium* and *Burkholderia*, since their sequences could be detected in the samples obtained from most or all recruited women (Jiménez et al. 2015). In the same study, no bacteria belonging to the genera *Pseudomonas*, *Methylobacterium*, *Sphingomonas*, *Novosphingobium*, *Sphingopyxis* or *Sphingobium* could be isolated from any of the tested samples despite providing culture media and conditions suitable for their growth. Considering the DNA contamination problem and the fact that soil and water-associated Gram-negative bacteria have been seldom isolated from human milk despite many of them, including *Pseudomonas* and closely-related bacteria, grow well in standard laboratory conditions. The authors concluded that it was highly probable that the core bacteriome of the analyzed samples was actually constituted by *Staphylococcus*, *Streptococcus*, *Lactobacillus*, *Propionibacterium* and gut-associated obligate anaerobes (*Bifidobacterium, Bacteroides, Roseburia, Eubacterium, Faecalibacterium, Ruminococcus*). A previous metagenomic study of 10 pooled human milk samples reported more than 360 prokaryotic genera being Proteobacteria (65%) and Firmicutes (34%) the predominant phyla, and *Pseudomonas, Staphylococcus* and *Streptococcus* the predominant genera (Ward et al. 2013).

The fact that sequences from lactobacilli, bifidobacteria and strict anaerobes can be detected in some studies and are scarce or absent in others may also be attributable to genetic, environmental, medical or dietary differences among subjects. Additionally, differences in the high-throughput sequencing techniques used may be responsible for these conflicting findings (Lagier et al. 2012). In this context, a metagenomic analysis of samples from birth to adulthood to study the development of the infant gut microbiota found relatively low frequency and low abundance of bifidobacteria in feces (Palmer et al. 2007). In contrast, an assessment, by pyrosequencing and analysis of PCR amplicons, of the complexity of the infant bifidobacterial population in the gut suggested a predominance of bifidobacteria in the infant gut as well as co-occurrence of bifidobacterial species (Turroni et al. 2012).

Factors Influencing Microbiota/Microbiome Composition in the Human Milk

It is known that the quantitative and/or qualitative composition of many components of human milk (peptides, proteins, lipids, immunological compounds, oligosaccharides, etc.) may be influenced by several factors, including genetic background, geographical location, maternal nutrition, part of the feeding (foremilk, hindmilk), gestational age, circadian rhythm, lactation stage, and others (Quinn et al. 2014; Nishimura et al. 2014; Atiya Ali et al. 2014; Collado et al. 2015; Ares Segura et al. 2016; Hoashi et al. 2016; Kumar et al. 2016; Munblit et al. 2016; Sprenger et al. 2017; Kunz et al. 2017; McGuire et al. 2017; Ruiz et al. 2017; Bardanzellu et al. 2017; Toscano et al. 2017). However, little is known on the interaction and impact of these and other factors on microbial communities composition in the human milk (Fernández et al. 2014; Gomez-Gallego et al. 2016) (Fig. 1).

Human milk oligosaccharides (HMOs) seem to stimulate the growth of specific bacterial groups frequently found either in breast milk or in the feces of breast-fed infants, such as *Staphylococcus epidermidis* or *Bifidobacterium* spp. (Hunt et al. 2012; Thongaram et al. 2017). Associations between microbes found in milk and the HMOs profile, human milk cells and/or macronutrients have been described recently (Boix-Amorós et al. 2016; Williams et al. 2017a, b; Aakko et al. 2017). Interestingly, the HMOs profile has been described to influence infant gut microbial colonization (Wang et al. 2015). Infants fed by non-secretor mothers, with a lower presence of 2FL (2'-fucosyllactose), exhibit delayed and lower *Bifidobacterium* colonization when compared to infants receiving human milk from secretor mothers (Lewis et al. 2015).

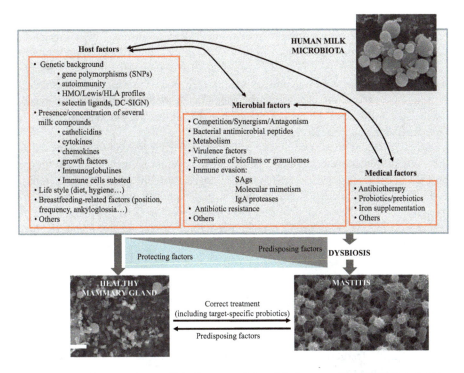

Fig. 1 Factors that may play a role in the composition of the human milk microbiota in healthy women and, also, in protecting or predisposing to mastitis

Some studies have investigated the impact of a variety of factors in the human milk microbiota/microbiome composition. Such factors include gestational age, postpartum period geographical location, mode of delivery, maternal diet, maternal health status (healthy, mastitis, metabolic syndrome, obesity, allergy, celiac disease, HIV-positive women), medical treatments (antibiotics, chemotherapy) and use of pumps and other devices for sampling (Grönlund et al. 2007; Albesharat et al. 2011; Hunt et al. 2011; Collado et al. 2012; Cabrera-Rubio et al. 2012, 2016; González et al. 2013; Urbaniak et al. 2014b; Soto et al. 2014; Khodayar-Pardo et al. 2014; Olivares et al. 2015; Boix-Amorós et al. 2016; Davé et al. 2016; Hoashi et al. 2016; Sakwinska et al. 2016; Williams et al. 2017a; Gómez-Gallego et al. 2017; Li et al. 2017; Jiménez et al. 2017). Some of the studies found significant differences between the compared groups while others did not. Conflicting and controversial results have also been obtained when different research groups have compared the effect of the same factor on the human milk microbiome. So, while it is becoming evident that human milk microbiome may be influenced by several factors and, also, that microbiota found in human milk may exert a strong influence on other milk components and, globally, on maternal/infant health. The exact triggers or drivers of differences in the composition of the human milk microbiota/microbiome need to be elucidated. Conflicting results between studies can be explained, at least partially, by host factors, environmental factors, perinatal factors, differences in milk collection and storage procedures, growth media and conditions, DNA extraction and amplification protocols, DNA sequencing methods, and bioinformatics analysis, among other factors (Gomez-Gallego et al. 2016). International and collaborative research, sharing common protocols from recruitment criteria to bioinformatics, is required in order to enable the comparison of results between groups and to evaluate the actual impact of the factors cited above (McGuire and McGuire 2015, 2017; Gomez-Gallego et al. 2016).

Anyway, recent studies analyzing milk samples from different locations using the same protocols have reported distinct HMO, immunological and microbiome profiles in different locations (Kumar et al. 2016; Munblit et al. 2016; McGuire et al. 2017; Ruiz et al. 2017). Such results would suggest the relevance of environmental factors on these components. Furthermore, specific factors (such as mode of delivery) affect in a different manner depending on the environment (Kumar et al. 2016).

Origin of Bacteria in Human Milk

The origin of the bacteria present in human milk has been reviewed recently (Mira 2016) (Fig. 2). Traditionally, it was believed that any prokaryote found in human milk was just the result of contamination from the infant's oral cavity or the mother's skin. However, the detection of live bacterial cells and/or DNA from anaerobic species that are usually related to gut environments and that cannot survive in aerobic locations has fuelled a scientific debate on the origin of milk-associated bacteria. These findings suggest that at least some of the bacteria present in the maternal gut could reach the mammary gland through an endogenous route, involving complex interactions between bacteria, epithelial cells and immune cells (Martín et al. 2004). Although the pathway and mechanisms that some bacteria could exploit to transit from the oral and/or intestinal epithelium to reach the mammary gland and other locations has not been elucidated yet, some works have offered a plausible scientific basis (Vazquez-Torres et al. 1999; Rescigno et al. 2001; Perez et al. 2007; Rodríguez 2014).

An increased bacterial translocation from the gut to mesenteric lymph nodes and mammary glands in pregnant and lactating mice has been described previously (Perez et al. 2007). Bacteria could be observed histologically in the subepithelial dome and interfollicular regions of Peyer's patches, in the *lamina propria* of the small bowel, and associated with cells in the glandular tissue of the mammary gland. In the same study, acridine orange staining of human milk and blood cyto-preparations identified bacterial cells in association with maternal mononuclear cells. In addition, other studies have reported that oral administration of *L. reuteri*, *L. gasseri*, *L. fermentum* and *L. salivarius* strains isolated from human milk to lactating women led to their presence in human milk (Jiménez et al. 2008a, b, c;

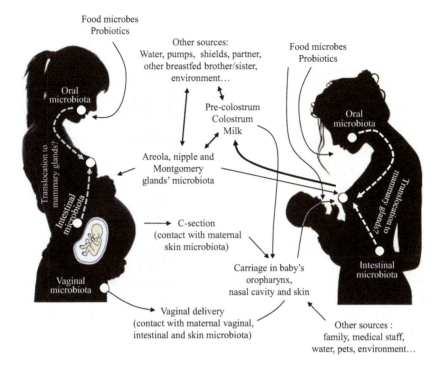

Fig. 2 Potential sources of the bacteria present in human milk

Abrahamsson et al. 2009; Arroyo et al. 2010). A previous study showed that oral administration of a *Lactobacillus* strain to women during pregnancy resulted in colonization of their intestine and, subsequently, of their respective breastfed infants' gut, even if the infants had been born by cesarean section (Schultz et al. 2004). Unfortunately, the presence of bacteria in the milk bacteria as the potential source of the strain was not investigated in that study. The microbiome of the different human body locations constitutes a dynamic network of interrelated communities. Therefore, the fact that the infant's mouth or the maternal skin may provide some bacteria to the milk is not incompatible with the role of human milk as a source of bacteria to the infant's mouth, the maternal skin and other infant/mother locations (Fig. 2).

A couple of works have also described the presence of bacteria in breast tissue biopsies (even from non-lactating women), which could represent, eventually, an additional source of microorganisms to human milk (Xuan et al. 2014; Urbaniak et al. 2014a). However, it is unclear if such findings are actually valid or a reflection of the technical bias resulting from the application of molecular-based techniques to low abundance microbiomes. Therefore, well-controlled studies are required either to confirm or to refute the presence of bacteria in breast tissue.

Mother-to-Infant Transfer of Bacteria Through Human Milk

After birth, the bacterial colonization process represents the first massive contact with microbes; different human studies have established a link between early gut microbiota composition and the risk of disease later in life (Kalliomäki et al. 2001, 2008; Fujimura et al. 2016), underlining the important role of the microbiota-host interactions in the neonatal period.

By providing a supply of live microorganisms, together with different bioactive substances including carbohydrates to be selectively fermented, human milk seems to play a key role in the proper establishment and further development of the infant microbiota and, as a consequence, on important host functions such as nutrient absorption, formation of host barriers against pathogens, or maturation of the immune and nervous systems (Jost et al. 2015). This also applies to preterm infants (Gregory et al. 2016).

Bacteria present in human milk are among the first colonizers of the infant gut and, therefore, may play a key role in driving the establishment of a healthy microbiota (Fernández et al. 2013). Several studies have reported a mother-to-infant transfer of microorganisms through human milk (at the species and strain level), using both culture-dependent (Martín et al. 2003, 2006, 2009; Solís et al. 2010; Makino et al. 2011, 2015; Murphy et al. 2017), and culture-independent techniques (Milani et al. 2015; Asnicar et al. 2017; Murphy et al. 2017). In fact, the initial microbiota of healthy breastfed babies resemble closely that found in the mother's milk, and it has been estimated that ~20% of the bacterial community present during the first month of life derives from that contained in human milk and 10% from areola skin (Pannaraj et al. 2017). The networks established between the intestinal microorganisms and the host in breast-fed babies are different from those found in formula-fed infants (Harmsen et al. 2000; Martin et al. 2016), leading to differences in the host transcriptome when the two feeding types are compared (Praveen et al. 2015).

Until recently, it was thought that the development of a more diverse gut microbiota in breastfed infants started at the weaning period (Favier et al. 2002). However, a recent work showed that stopping breastfeeding—rather than introducing solids—drives maturation of the infant gut microbiota (Bäckhed et al. 2015). These researchers found more adult-like taxa in the microbiomes of babies who stopped breastfeeding earlier, while the microbiota of babies breastfed for longer periods were dominated by bacteria present in breast milk.

Human Milk: A Source of Probiotic Bacteria to the Infant Gut

As stated above, human milk contains a vast array of bioactive compounds, which may act synergistically in order to preserve infants' health. Therefore, it may be difficult to delineate the specific functions of a given milk component (such as the human milk microbiota) without taking in account its potential interactions with other human milk ingredients.

Nonetheless, bacteria isolated from human milk of healthy women are particularly attractive organisms since they fulfill some of the criteria generally recommended for human probiotics, such as human origin, a history of safe prolonged intake by a sensitive population (from preterms to infants), and adaptation to both mucosal and dairy substrates (Lara-Villoslada et al. 2007a; Fernández et al. 2013). Because of their origin, they seem to be uniquely adapted to reside in the human digestive tract and to interact with us in symbiosis from the time we are born (Jeurink et al. 2013). Among the bacterial species isolated from human milk, some of them (*L. gasseri*, *L. salivarius*, *L. rhamnosus*, *L. plantarum*, *L. fermentum*, *L. reuteri*, *B. breve*, *B. longum...*) are considered among the potentially probiotic bacteria and enjoy the GRAS (Generally Recognised As Safe) and the QPS (Qualified Presumption of Safety) status conceded by the Food and Drug Administration (FDA, USA) and the European Food Safety Authority (EFSA), respectively.

Bacteria found in human milk may play several roles in the infant gut, including an important role in reducing the incidence and severity of infections in the breast-fed infant. In fact, some of the lactic acid bacteria strains isolated from this biological fluid have the ability to inhibit (both *in vitro* and *in vivo*) the growth of a wide spectrum of pathogenic bacteria, including *E. coli*, *Salmonella* and *Listeria monocytogenes*, by competitive exclusion or through the production of antimicrobial compounds, such as organic acids, hydrogen peroxide, reuterin or bacteriocins (Heikkilä and Saris 2003; Beasley and Saris 2004; Martín et al. 2005, 2006). It has also been shown that some bacteria present in human milk improve the intestinal barrier function by increasing mucin production and reducing intestinal permeability (Olivares et al.2006c). In addition, some lactobacilli found in human milk inhibit the adhesion of *Salmonella* to mucins and increase the survival of mice infected with this pathogen (Olivares et al. 2006c).

More recently, a randomized double-blinded controlled study that included infants at the age of 6 months, was conducted to examine the effects of a follow-on formula containing *Lactobacillus fermentum* CECT5716 plus galactooligosaccharide (experimental group), or the same formula supplemented with only galactooligosaccharide (control group), on the incidence of infections in infants between the ages of 6 and 12 months (Maldonado et al. 2012). The experimental group showed a significant 46% reduction in the incidence rate of gastrointestinal infections, 27% reduction in the incidence of upper respiratory tract infections, and 30% reduction in the total number of infections, at the end of the study period compared with the control group.

The ability of some lactic acid bacteria strains isolated from human milk to inhibit infection of HIV-1 *in vitro* has also been demonstrated (Martín et al. 2010). HIV-inhibitory activity has been associated with both the killed bacteria and the conditioned cell-free supernatant from the bacterial cultures, suggesting that more than one mechanism may be used by LAB to block viral infectivity. The highest levels of HIV-inhibitory activity were associated with the bacterial strains

L. salivarius VM5, *L. gasseri* VM22, *L. lactis* VM17, and *Streptococcus salivarius* VM18, with distinct inhibitory effects against R5- and X4-tropic HIV-1. More recently, it was observed that the genome of *L. salivarius* CECT 5713 encodes a protein (1230) that contains a motif for recognition of high mannose N-linked oligosaccharides present in a variety of pathogen antigens, including HIV gp120. Consequently, it might have the potential to block gp120 from binding to target cells and, therefore, to inhibit HIV infectivity (Langa et al. 2012). To test such predictions, the ability of *L. salivarius* CECT 5713 to inhibit the *in vitro* infectivity of HIV-1 was assessed. Presence of heat-killed *L. salivarius* CECT 5713 cells led to reductions in the infectivity of R5 (HIV-1_{BaL}), CXCR4 (HIV-1_{HC4}) and R5/X4 (HIV-$1_{C7/86}$) viruses by 42.3%, 58.9%, and 49.8%, respectively (Langa et al. 2012).

Such results have particular relevance in HIV-1 breast milk transmission, where breastfeeding infants are exposed daily to the virus found in maternal breast milk. Transmission of HIV-1 to the breastfeeding infant presumably occurs across mucosal surfaces in the infant oral and gastrointestinal tissues, sites that are also abundantly colonized with bacteria within the first days of birth. As some of the strains that colonize the infant have their origins in the mother's breast milk, this suggests a unique mechanism for conferring protection to the infant against mucosal pathogens. Despite clear evidence that HIV-1 is transmitted to infants during breastfeeding, most breastfed infants remain uninfected even after repeated exposure of their oral and gastrointestinal mucosal surfaces to high amounts of both cell-free HIV-1 and cell-associated virus in the milk (Kourtis et al. 2003). It is known that infants who are exclusively breastfed for the first months of life have a significantly lower risk of being HIV-infected when compared to infants that are mixed-fed (Coutsoudis et al. 1999). Exclusive breastfeeding may help prevent damage to the gut mucosa and/or stimulate growth and colonization of bacterial species harbored in maternal breast milk.

Bacteria found in human milk may also play key roles in the correct maturation of the infant immune system and their function seems to be flexible depending on the gut conditions. As an example, *L. salivarius* CECT5713 and *L. fermentum* CECT5716 enhance macrophage production of Th1 cytokines, such as IL-2 and IL-12 and the inflammatory mediator TNF-*a*, in the absence of an inflammatory stimulus (Díaz-Ropero et al. 2007). However, both probiotics lead to a reduction of Th1 cytokines when cells are incubated in the presence of lipopolysaccharide. This regulatory mechanism is probably based on the induction of the synthesis of IL-10, an immunosuppressive cytokine, by these strains (Díaz-Ropero et al. 2007).

The immunomodulatory effects of probiotics have also been reported in animal models of pathologies where the immune system is involved. Different probiotic strains isolated from human milk have been reported to enhance the immune defence of mice, increasing both natural and acquired immune responses (Díaz-Ropero et al. 2007). This immune-stimulating activity could be also involved in the anti-infective role previously mentioned for these bacteria in an animal model of *Salmonella* infection (Olivares et al. 2006c). In addition, the human milk strain *L. gasseri* CECT5714 in combination with *L. coryniformis* CECT5711 reduces the incidence and severity of the allergic response in an animal model of cow's milk protein allergy (Olivares et al. 2005) while *L. fermentum* CECT5716 showed a beneficial effect in an animal model of intestinal inflammation, reducing the inflammatory response and the intestinal damage (Peran et al. 2007).

Strains isolated from human milk have also been reported to modulate the immune response of healthy humans, as shown by a study reporting an increase in phagocytic activity, in the number of natural killer cells and in the plasma concentration of IgA in healthy humans consuming human milk-isolated probiotics daily for 3 months (Olivares et al. 2006b). The consumption of *L. fermentum* CECT5716 enhances the response to influenza vaccination in healthy volunteers aged 26–40 and reduces the incidence of influenza-like illness (Olivares et al. 2007).

A recent study has confirmed that *L. fermentum* CECT5716 and *L. salivarius* CECT5713 have a broad array of effects on the immune system (Pérez-Cano et al. 2010). They are potent activators of NK cells and moderate activators of CD4+ and CD8+ T cells and regulatory T cells. Thus, they have an impact on both innate and acquired immunity. They strongly induce a wide range of pro- and anti-inflammatory cytokines and chemokines. The authors compare these strains with others belonging to the same species but isolated from sources different to breast milk and found some milk strain-specific effects, such as a higher induction of IL-10 and IL-1 production.

Finally, there is increasing interest in the manipulation of intestinal microbiota with the aim of improving gastrointestinal function and nutrient absorption. Probiotics isolated from human milk colonise the intestine and increase faecal lactobacilli counts, thus modifying intestinal microbiome both in rodents (Peran et al. 2005) and humans (Olivares et al. 2006a), including infants (Maldonado et al. 2010). Strains obtained from human milk are metabolically active in the human gut, modulating the production of functional metabolites such as butyrate, which is the main energy source for colonocytes and plays a key role in the modulation of intestinal function. As a result, they lead to a better intestinal habit, with an increase in faecal moisture, and in stool frequency and volume. A recent study evaluated the impact of L. fermentum CECT 5716 on stress-induced intestinal epithelial barrier dysfunction, systemic immune response and exploratory behavior in rat pups (Vanhaecke et al. 2017). The results showed that the *L.* fermentum strain prevented such stress-induced dysfunction *in vivo*, reduced permeability to both fluorescein sulfonic acid and horseradish peroxidase in the small intestine, and increased expression of zonula occludens-1 (ZO-1) and prevented stress-induced ZO-1 disorganization in ileal epithelial cells. In addition, this strain also significantly reduced stress-induced increase in plasma corticosteronemia and enhanced IFNγ secretion while preventing IL-4 secretion from activated splenocytes.

Streptococci (mainly *S. mitis* and *S. salivarius* groups) and coagulase-negative staphylococci (CNS) have received marginal attention regarding their role in the human mammary gland and during the early colonization of the infant gut despite being the dominant bacteria in human milk (Jiménez et al. 2008b; Hunt et al. 2012; Martín et al. 2012a; Cacho et al. 2017). Interestingly, an abundant presence of *S. epidermidis* in the infant gut seems to be a differential feature of the feces of breast-fed infants when compared to those of formula-fed infants (Lundequist et al.

1985; Sakata et al. 1985; Balmer and Wharton 1989; Adlerberth et al. 2006; Jiménez et al. 2008b).

Indeed, CNS and mitis/salivarius streptococci provided by human milk can be particularly useful in reducing the acquisition of undesired pathogens by infants [including preterm neonates) exposed to hospital environments. It has been proposed that *S. epidermidis* and other CNS may have a probiotic function by preventing colonization of the host by more severe pathogens, such as *S. aureus* (Otto 2009). In fact, some *S. epidermidis* strains that inhibit *in vivo* colonization by *S. aureus* have been postulated as a future strategy to eradicate such pathogens from the mucosal surfaces (Iwase et al. 2010; Park et al. 2011). Similarly, it has been shown that viridans streptococci inhibit oral colonization by methicillin-resistant *S. aureus* in high-risk newborns exposed to hospital environments (Uehara et al. 2001). In addition, the presence of viridans streptococci seems to be a feature of the healthy infant gut in contrast with the atopic infant gut (Kirjavainen et al. 2001). Therefore, at least some staphylococcal and streptococcal strains present in human milk may play important empirical probiotic roles in the breast and in breastfed infants.

Future sequencing of the genomes of a wide variety of isolates from human milk and an accurate functional analysis of the human milk microbiome will provide additional clues on the safety and potential probiotic properties of microbes found in human milk.

From Physiology to Pathology: Lactational Mastitis

In practice, it is often difficult to cope with the WHO recommendations in relation to the duration of breastfeeding. From the medical point of view, mastitis represent the first cause of undesired premature weaning, with an incidence among lactating women as high as 35% when any clinical mastitis case is considered. Since a history of breastfeeding is associated with a reduced risk of many diseases in infants and mothers, both in developed and developing countries (U.S. Department of Health and Human Services 2011; American Academy of Pediatrics 2012; Renfrew et al. 2012), any reason for hampering such feeding options should be considered a relevant Public Health issue, instead of receiving marginal attention from the medical community, as it is the case of mastitis.

The process of lactation has been remarkably successful since the earliest mammals, allowing thousands of species to occupy a vast range of ecological niches. However, mastitis remains as a common feeding complication among most, if not all, mammalian species (Michie et al. 2003). Literally, mastitis means the inflammation of any part of a mammary gland, including not only intramammary tissues but also nipples and mammary areolas in the species that harbor such structures. However, in practice, the term mastitis is generally used to define an infectious process of the mammary gland characterized by a variety of local and, in some cases, systemic symptoms (Lawrence and Lawrence 2005). The infectious nature of lactational mastitis usually serves to differentiate this condition from other inflammatory processes of the mammary gland, such as those associated to different types of breast cancer and from Raynaud's disease, a painful vasoconstriction of the nipple during human breastfeeding.

As described in previous sections of this chapter, the lactating mammary gland ecosystem is hospitable to many microorganisms, including bacterial groups that have the potential to cause mastitis (Fernández et al. 2013; Jeurink et al. 2013); however, upon disturbance of this balanced state, infection can occur and, in fact, recent studies suggest that mastitis is a process characterized by a mammary bacterial dysbiosis (Delgado et al. 2008; Fernández et al. 2014) (Fig. 1).

In this context, microbiological analysis of milk is the only method that allows an etiological diagnosis of mastitis. It may seem simple but is not an easy issue, partly due to the absence of uniform or standard protocols for the collection of this biological fluid, the doubts that often arise for the interpretation of the results and, in humans, the lack of tradition in milk microbiological analysis. The collection of a representative sample for microbial analysis is of outmost importance in order to get a correct diagnosis since there are many sampling-related factors that may affect the result (Arroyo et al. 2010). As explained before, the use of milk pumps to collect the samples is associated with a high concentration of some contaminant bacteria, that arise from the rinsing water and other sources but are not related to the particular mastitis case (Jiménez et al. 2017). Other relevant factors that may be considered in making an etiological diagnosis include a reliable identification of the organism(s) detected on culture, its/their concentration(s), antibiogram, concurrent evidence of inflammation and, if so, at what degree. The introduction of molecular microbiology techniques to mastitis diagnosis has been extremely useful. Matrix-assisted laser desorption/ionization time-of-flight mass spectrometry (MALDI-TOF MS) is also spreading as a bacterial identification tool with high confidence and speed (Marín et al. 2017).

Etiopathogenesis of Lactational Mastitis

S. aureus is the main etiological agent of acute mastitis. Once in the mammary gland, it can proliferate and produce toxins that lead to a strong inflammation of the mammary tissue; as a consequence, intense local symptoms (breast redness, heat, pain...) usually arise (Fig. 3). Since the mammary gland is highly vascularized throughout the lactation period, toxins are rapidly absorbed and reach the bloodstream causing an alteration in the host cytokine patterns and, eventually, leading to systemic flu-like symptoms, which may include fever, muscular and articular pain, and general physical discomfort (Fig. 3). Acute mastitis constitutes a small fraction of human mastitis cases but due to the evident local and systemic signs, tends to be the only type of mastitis that is correctly diagnosed.

As explained above, CNS and viridans streptococci are normal inhabitants of the mammary ecosystem during lactation. However, different factors (that will be

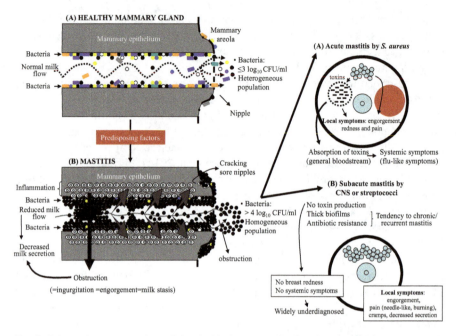

Fig. 3 Schematic representation of the dysbiosis process leading from a healthy human milk microbiota to acute and subacute mastitis. Mammary epithelium in physiological conditions (**a**), and during mastitis (**b**). Gray arrows indicate the excessive pressure of milk through an inflamed mammary epithelium, leading to cramps and typical burning and/or needle-like pain. *CNS* coagulase-negative staphylococci. (Adapted from Fernández et al. (2014))

discussed later) may favor an overgrowth of such bacterial species, leading to subacute or subclinical mastitis. Since CNS and viridans streptococci do not produce the toxins responsible for acute mastitis, there are no systemic flu-like symptoms and, generally, local breast symptoms are milder and do not include breast redness. However, in such circumstances, they can form thick biofilms inside the ducts, inflaming the mammary epithelium and forcing milk to pass through an increasingly narrower lumen. The increasing milk pressure on an inflamed epithelium results in a characteristic needle- or prick-like pain, often accompanied by breast cramps and a burning feeling. Eventually, bacterial biofilms may fill up some ducts, obstructing or blocking the milk flow and leading to breast engorgement (Fig. 3).

Among CNS, *S. epidermidis* is the species most commonly associated with lactational mastitis in women (Thomsen et al. 1985; Delgado et al. 2008, 2009). *Streptococcus* species associated with mastitis seem to be host-specific. *Streptococcus agalactiae, Streptococcus uberis* or *Streptococcus dysgalactiae* are frequent agents of mastitis in cattle (Keefe 1997) but are not (or rarely) implicated in human mastitis. In contrast, the most common streptococcal species affecting humans include *Streptococcus mitis* and *Streptococcus salivarius* (Mediano et al. 2017). It is important to note that streptococci have been submitted for important taxonomical rearrangements and that many novel streptococcal species and closely related genera have only been described in the last few years; therefore, the implication of streptococci in mastitis should be carefully reevaluated.

Some corynebacteria, including *Corynebacterium kroppenstedtii, C. amycolatum* and *C. tuberculostearicum*, are involved in human granulomatous lobular mastitis, a chronic inflammatory disease that was previously considered of unknown etiology (Renshaw et al. 2011). Patients usually present an enlarged, mildly tender breast lump, which is sometimes associated with local inflammation, tenderness, and sinus formation and can become chronic and disfiguring in a large number of patients. Since corynebacteria stain poorly using the Gram technique, often fail to grow on routine media, and are found forming granulomatous structures deep in the breast tissue, it is probable that corynebacteria may have been overlooked as mastitis-causing agents (Paviour et al. 2002; Bercot et al. 2009). At present, *C. kroppenstedtii* is considered to be the main cause of granulomatous mastitis. It is a lipophilic species and this feature seems particularly relevant in the pathogenesis of this condition. Such a property allows a firm attachment to fat globules, providing easy access to an abundant source of fatty acids. Therefore, it is not unusual that histological preparations always show the bacterial cells within a central lipid-containing vacuole (Renshaw et al. 2011).

Mastitis Predisposing Factors

The discovery of the existence of a site-specific microbiota in the mammary ecosystem during late pregnancy and throughout the lactation period suggests that, similarly to other body locations, breast health during such life stages may depend on the balance between the state of the host and its mammary-associated microbiome. Although the exact causal events leading to the transition from colonization to infection are still ill-defined *in vivo*, different host, microbial and medical factors may play important roles in the protection against or predisposition to mastitis (Fernández et al. 2014) (Fig. 1). The composition of the microbiota in breast milk is host-dependent (Martín et al. 2007b; Hunt et al. 2011; Cabrera-Rubio et al. 2012; Jost et al. 2013), and strain-specific traits of some of its staphylococcal or streptococcal members, such as molecular mimetism mechanisms to evade the immune system response, presence of virulence factors or resistance to antibiotics, may be essential in determining whether a woman will suffer from mastitis or not.

The host's genetic background can also play a key role in the lactation outcome since differences in selectin, Lewis antigens and human milk oligosaccharides (HMO) gene determinants also predispose or protect against mastitis by altering neutrophils' activation and production of reactive oxygen species (Bode et al. 2004). HMOs present in human milk are able to modulate the microbiome of breastfed infants and, therefore, it is highly probable that they can modulate the bacterial communities in the mammary gland, too (Bode 2012). In other words, it is possible that the susceptibility of suffering from mastitis is determined not only by the bacterial composition of the human milk but, also, by the HMOs composition the milk, which is, in turn, related to secretor and Lewis blood group systems (Thurl et al. 2010; Albrecht et al. 2011).

It is known that human milk contains a wide spectrum of other biologically active substances, including eukaryotic antimicrobial peptides, such as cathelicidin LL-37. This peptide is expressed in the mammary gland and secreted in milk, and displays a relevant antimicrobial activity against potential mastitis-causing agents (Murakami et al. 2005), including a strong anti-biofilm effect even at subinhibitory concentrations (Jacobsen and Jenssen 2012). Polymorphisms or variations in the copy number or in the expression of genes encoding the biosynthesis of such anti-microbial peptides may be linked to mastitis susceptibility (Rivas-Santiago et al. 2009). The existence of a genetic basis for host responses to bacterial intramam-mary infections has been widely documented in ruminants while human granuloma-tous mastitis due to corynebacterial infection has been associated to a single nucleotide polymorphism (SNP) within the *NOD2* gene, impairing the neutrophil responses to Nod2 agonists (Bercot et al. 2009).

Peripartum antibiotherapy, frequently related to Caesarean section, premature or prolonged rupture of membranes and GBS intrapartum prophylaxis, has emerged as a strong risk factor for human mastitis because of the selection of antibiotic-resistant staphylococci in the mammary environment and the elimination of potential natural competitors (Delgado et al. 2009; Contreras and Rodríguez 2011; Willing et al. 2011; Nogacka et al. 2017).

Human Mastitis: A Target for Probiotics?

Since resistance to antibiotics and ability to drive the formation of biofilms are com-mon properties among mastitis-causing bacteria, many cases are refractory to anti-biotic therapy (Fernández et al. 2014). As a consequence, alternative strategies are required to improve mastitis healing rates while reducing the use of antibiotics. In this context, the development of new strategies for mastitis management based on human milk probiotics, as an alternative or complement to antibiotic therapy, is particularly appealing.

Initially, a pilot trial highlighted the potential of *L. salivarius* CECT 5713 and *L. gasseri* CECT 5714, two strains isolated from breast milk, for the treatment of staphylococcal mastitis (Jiménez et al. 2008c). In the study, 20 women with staphy-lococcal mastitis were randomly divided in two groups. Those in the probiotic group daily ingested 10 $\log_{10}$ cfu of *Lactobacillus salivarius* CECT5713 and the same quantity of *Lactobacillus gasseri* CECT5714 for 4 weeks while those in the placebo group only ingested the excipient. On day 0, the mean staphylococcal counts in the probiotic and placebo groups were similar (4.74 and 4.81 $\log_{10}$ cfu/mL, respec-tively) but lactobacilli could not be detected. By day 14, no clinical signs of mastitis were observed in the women assigned to this group but persisted throughout the study period in placebo group women. On day 30, the mean staphylococcal count in the probiotic group (2.96 $\log_{10}$ cfu/mL) was significantly lower than that of the placebo group (4.79 $\log_{10}$ cfu/mL). These results revealed that *L. salivarius* CECT5713 and *L. gasseri* CECT5714 were an efficient alternative for the treatment of lactational mastitis.

Later, the efficacy of *L. fermentum* CECT 5716 or *L. salivarius* CECT 5713, two lactobacilli strains isolated from breast milk, to treat lactational mastitis when administered orally was evaluated and compared to antibiotic therapy (Arroyo et al. 2010). A total of 352 women with infectious mastitis were randomly divided in three groups. Those in groups A (n = 124) and B (n = 127) ingested daily 9 $\log_{10}$ CFU of *L. fermentum* CECT 5716 or *L. salivarius* CECT 5713, respectively, for 3 weeks while those in group C (n = 101) were submitted to antibiotic therapy prescribed in their respective Primary Care Centres. On day 0, the mean staphylococcal and/or streptococcal counts in milk samples of the three groups were similar (4.35–4.47 $\log_{10}$ CFU/mL) and lactobacilli could not be detected. On day 21, the mean staphylococcal and/or streptococcal counts in the probiotic groups (2.61 and 2.33 $\log_{10}$ CFU/mL) were lower than that of the control group (3.28 $\log_{10}$ CFU/mL). The probiotic treatment led to a significant reduction (1.7–2.1 $\log_{10}$ CFU/mL) in the milk bacterial count and to a rapid improvement of the condition. The final staphylococcal and/or streptococcal count was approximately 2.5 $\log_{10}$ CFU/mL, an acceptable bacterial load in milk of healthy women. On the basis of the bacterial counts, pain scores and clinical evolution, women ascribed to any of the probiotic groups improved significantly more than those ascribed to the antibiotic group. In addition, mothers who used the probiotics strains avoided suffering from side effects often associated with antibiotic treatment such as vaginal infections and recurrent mastitis episodes. More recently, a randomized controlled trial showed that *L. fermentum* CECT 5716 was able to reduce the staphylococcal load in the milk of lactating mothers suffering breast pain (Maldonado-Lobón et al. 2015).

A subsequent study was aimed toward finding microbiological, biochemical and/or immunological biomarkers of the probiotic effect. Women with and without symptoms of mastitis received three daily doses (10^9 CFU) of *L. salivarius* PS2 for 21 days. Samples of milk, blood and urine were collected before and after the probiotic intervention, and screened for a wide spectrum of microbiological, biochemical and immunological parameters. In the mastitis group, *L. salivarius* PS2 intake led to a reduction in milk bacterial counts, milk and blood leukocyte counts and IL-8 level in milk, an increase in those of IgE, IgG3, EGF and IL-7, a modification of the milk electrolyte profile, and a reduction of some oxidative stress biomarkers (Espinosa-Martos et al. 2016). In the same cohort, the NMR characterization of the urine metabolic profile of the lactating women with mastitis at the beginning of a probiotic intervention showed increased energy metabolism (lactate, citrate, formate, acetate, malonate) and decreased branched-chain amino acid catabolism (isocaproate and isovalerate) when compared to that after probiotic intake (Vázquez-Fresno et al. 2014). Probiotic supplementation led to a normalization of breast permeability. Changes in the levels of acetate and 2-phenylpropionate after probiotic intake suggested an immunomodulatory while increased level of malonate indicated an important antagonistic strategy of *L. salivarius* PS2 since this catabolite is a well-known repressor of the tricarboxylic acid (TCA) cycle, which may alter staphylococcal and streptococcal metabolism and negatively affect their survival, virulence and ability for biofilm formation.

Finally, a recent clinical trial evaluated the potential of *L. salivarius* PS2 to prevent mastitis when orally administered during late pregnancy to women that had suffered infectious mastitis after a previous pregnancy (Fernández et al. 2016). A total of 108 pregnant women were randomly divided in two groups. Those in the probiotic group (n = 55) ingested daily 9 $\log_{10}$ cfu of *L. salivarius* PS2 from ~30 weeks of pregnancy until delivery while those in the control group (n = 53) received a placebo. The occurrence of mastitis was evaluated during the first 3 months after delivery. Globally, 44 out of 108 women (59.26%) suffered mastitis; however, the percentage of women suffering mastitis in the probiotic group (25.45%, n = 14) was significantly lower than in the control group (56.50%, n = 30). When mastitis occurred, the milk bacterial counts in the probiotic group were significantly lower than in the placebo one. As a conclusion, the oral administration of *L. salivarius* PS2 during late pregnancy appears to be an efficient method to prevent infectious mastitis in a susceptible population.

The potential mechanisms by which some lactobacilli strains are able to control mastitis-causing agents in the breast after oral administration have been reviewed recently (Fernández et al. 2014). Ingestion of probiotic strains during late pregnancy and/or breastfeeding increases IgA and TGF-β2 levels in breast milk (Rautava et al. 2002; Prescott et al. 2008; Nikniaz et al. 2013), which may control the local growth of mastitis-causing bacteria while limiting their ability to access or to damage the mammary epithelium.

In addition, local competitive exclusion and production of antimicrobials may also explain the control of mastitis-causing bacteria by certain lactobacilli strains (Beasley and Saris 2004; Martín et al. 2005; Olivares et al. 2006c). This would imply that a lactobacilli strain must be able to reach the mammary gland upon ingestion, and, as explained above, it has been suggested that the origin at least in part, of the live bacteria found in human milk could be from the maternal gut through an endogenous route (the so-called entero-mammary pathway), involving complex interactions with immune cells (Martín et al. 2004; Perez et al. 2007; Rodríguez 2014).

Conclusions

Human milk has been traditionally considered sterile; however, studies carried out in the last 15 years by using both culture-dependent and -independent techniques have shown that it represents a continuous supply of commensal, mutualistic and/or potentially probiotic bacteria to the infant gut. Once in the infant gut, these bacteria may play several roles, contributing—among others—to the protection against infections and the maturation of the immune system functions. Bacteria found in human milk may have different origins such as maternal skin and infant mouth. Other studies suggest that some bacteria present in the maternal digestive tract could reach the mammary gland during late pregnancy and lactation through a mechanism involving gut monocytes. The microbiota composition in human milk has implications not only on the infant but also on the mammary health. Mammary dysbiosis may lead to acute, subacute or subclinical mastitis, a frequently under-rated and underdiagnosed condition that represents the first medical cause for unde-sired weaning. Since breastfeeding provides short-, and long-term benefits to the mother-infant pair, lactational mastitis should be considered as a relevant Public Health issue. It seems clear that many host, microbial, medical and environmental factors may predispose to or protect against mastitis development. In the future, a better knowledge of the microbiota found in human milk and the influencing factors could be used to design novel means to improve it or to develop probiotics derived from human milk able to achieve better maternal and infant health.

References

Aakko, J., Kumar, H., Rautava, S., et al. (2017). Human milk oligosaccharide categories define the microbiota composition in human colostrum. *Beneficial Microbes, 8*, 563–567. https://doi.org/10.3920/BM2016.0185.

Abrahamsson, T. R., Sinkiewicz, G., Jakobsson, T., et al. (2009). Probiotic lactobacilli in breast milk and infant stool in relation to oral intake during the first year of life. *Journal of Pediatric Gastroenterology and Nutrition, 49*, 349–354. https://doi.org/10.1097/MPG.0b013e31818f091b.

Adlerberth, I., Lindberg, E., Åberg, N., et al. (2006). Reduced enterobacterial and increased staphylococcal colonization of the infantile bowel: An effect of hygienic lifestyle? *Pediatric Research, 59*, 96–101. https://doi.org/10.1203/01.pdr.0000191137.12774.b2.

Albesharat, R., Ehrmann, M. A., Korakli, M., et al. (2011). Phenotypic and genotypic analy-ses of lactic acid bacteria in local fermented food, breast milk and faeces of mothers and their babies. *Systematic and Applied Microbiology, 34*, 148–155. https://doi.org/10.1016/j.syapm.2010.12.001.

Albrecht, S., Schols, H. A., van den Heuvel, E. G. H. M., et al. (2011). Occurrence of oligosaccha-rides in feces of breast-fed babies in their first six months of life and the corresponding breast milk. *Carbohydrate Research, 346*, 2540–2550. https://doi.org/10.1016/j.carres.2011.08.009.

American Academy of Pediatrics. (2012). Breastfeeding and the use of human milk. *Pediatrics, 129*, 827–841.

Andreas, N. J., Kampmann, B., & Mehring Le-Doare, K. (2015). Human breast milk: A review on its composition and bioactivity. *Early Human Development, 91*, 629–635. https://doi.org/10.1016/J.EARLHUMDEV.2015.08.013.

Arboleya, S., Ruas-Madiedo, P., Margolles, A., et al. (2011). Characterization and *in vitro* proper-ties of potentially probiotic *Bifidobacterium* strains isolated from breast-milk. *International Journal of Food Microbiology, 149*, 28–36. https://doi.org/10.1016/j.ijfoodmicro.2010.10.036.

Ares Segura, S., Arena Ansótegui, J., Díaz-Gómez, N. M., en representación del Comité de Lactancia Materna de la Asociación Española de Pediatría. (2016). La importancia de la nutrición materna durante la lactancia, ¿necesitan las madres lactantes suplementos nutriciona-les? *An Pediatría, 84*, 347.e1–347.e7. https://doi.org/10.1016/j.anpedi.2015.07.024

Arroyo, R., Martín, V., Maldonado, A., et al. (2010). Treatment of infectious mastitis during lacta-tion: Antibiotics versus oral administration of Lactobacilli isolated from breast milk. *Clinical Infectious Diseases, 50*, 1551–1558. https://doi.org/10.1086/652763.

Asnicar, F., Manara, S., Zolfo, M., et al. (2017). Studying vertical microbiome transmission from mothers to infants by strain-level metagenomic profiling. *mSystems, 2*(1), e00164-16. https://doi.org/10.1128/mSystems.00164-16.

Atiya Ali, M., Strandvik, B., Sabel, K.-G., et al. (2014). Polyamine levels in breast milk are associated with mothers' dietary intake and are higher in preterm than full-term human milk and formulas. *Journal of Human Nutrition and Dietetics, 27*, 459–467. https://doi.org/10.1111/jhn.12156.

Bäckhed, F., Roswall, J., Peng, Y., et al. (2015). Dynamics and stabilization of the human gut microbiome during the first year of life. *Cell Host & Microbe, 17*, 690–703.

Balmer, S. E., & Wharton, B. A. (1989). Diet and faecal flora in the newborn: Breast milk and infant formula. *Archives of Disease in Childhood, 64*, 1672–1677.

Bardanzellu, F., Fanos, V., & Reali, A. (2017). "Omics" in human colostrum and mature milk: Looking to old data with new eyes. *Nutrients, 9*(8), pii: E843. https://doi.org/10.3390/nu9080843.

Beasley, S. S., & Saris, P. E. J. (2004). Nisin-producing *Lactococcus lactis* strains isolated from human milk. *Applied and Environmental Microbiology, 70*, 5051–5053. https://doi.org/10.1128/AEM.70.8.5051-5053.2004.

Bercot, B., Kannengiesser, C., Oudin, C., et al. (2009). First description of NOD2 variant associated with defective neutrophil responses in a woman with granulomatous mastitis related to corynebacteria. *Journal of Clinical Microbiology, 47*, 3034–3037. https://doi.org/10.1128/JCM.00561-09.

Bode, L. (2012). Human milk oligosaccharides: every baby needs a sugar mama. *Glycobiology, 22*, 1147–1162. https://doi.org/10.1093/glycob/cws074.

Bode, L., Rudloff, S., Kunz, C., et al. (2004). Human milk oligosaccharides reduce platelet-neutrophil complex formation leading to a decrease in neutrophil beta 2 integrin expression. *Journal of Leukocyte Biology, 76*, 820–826. https://doi.org/10.1189/jlb.0304198.

Boer, H. R., & Anido, G. M. N. (1981). Bacterial colonization of human milk. *Southern Medical Journal, 74*, 716–718.

Boix-Amorós, A., Collado, M. C., & Mira, A. (2016). Relationship between milk microbiota, bacterial load, macronutrients, and human cells during lactation. *Frontiers in Microbiology, 7*, 492. https://doi.org/10.3389/fmicb.2016.00492.

Cabrera-Rubio, R., Collado, M. C., Laitinen, K., et al. (2012). The human milk microbiome changes over lactation and is shaped by maternal weight and mode of delivery. *The American Journal of Clinical Nutrition, 96*, 544–551. https://doi.org/10.3945/ajcn.112.037382.

Cabrera-Rubio, R., Mira-Pascual, L., Mira, A., & Collado, M. C. (2016). Impact of mode of delivery on the milk microbiota composition of healthy women. *Journal of Developmental Origins of Health and Disease, 7*, 54–60. https://doi.org/10.1017/S2040174415001397.

Cacho, N. T., Harrison, N. A., Parker, L. A., et al. (2017). Personalization of the microbiota of donor human milk with mother's own milk. *Frontiers in Microbiology, 8*, 1470. https://doi.org/10.3389/fmicb.2017.01470.

Cárdenas, N., Martín, V., Delgado, S., Rodríguez, J. M., & Fernández, L. (2014). Characterization of *Lactobacillus gastricus* strains isolated from human milk. *International Dairy Journal, 39*, 167–177.

Cárdenas, N., Laiño, J. E., Delgado, S., et al. (2015). Relationships between the genome and some phenotypical properties of *Lactobacillus fermentum* CECT 5716, a probiotic strain isolated from human milk. *Applied Microbiology and Biotechnology, 99*, 4343–4353. https://doi.org/10.1007/s00253-015-6429-0.

Collado, M. C., Delgado, S., Maldonado, A., & Rodríguez, J. M. (2009). Assessment of the bacterial diversity of breast milk of healthy women by quantitative real-time PCR. *Letters in Applied Microbiology, 48*, 523–528. https://doi.org/10.1111/j.1472-765X.2009.02567.x.

Collado, M. C., Laitinen, K., Salminen, S., & Isolauri, E. (2012). Maternal weight and excessive weight gain during pregnancy modify the immunomodulatory potential of breast milk. *Pediatric Research, 72*, 77–85. https://doi.org/10.1038/pr.2012.42.

Collado, M. C., Santaella, M., Mira-Pascual, L., et al. (2015). Longitudinal study of cytokine expression, lipid profile and neuronal growth factors in human breast milk from term and preterm deliveries. *Nutrients, 7*, 8577–8591. https://doi.org/10.3390/nu7105415.

Contreras, G. A., & Rodríguez, J. M. (2011). Mastitis: Comparative etiology and epidemiology. *Journal of Mammary Gland Biology and Neoplasia, 16*, 339–356.

Coutsoudis, A., Pillay, K., Spooner, E., et al. (1999). Influence of infant-feeding patterns on early mother-to-child transmission of HIV-1 in Durban, South Africa: A prospective cohort study. South African Vitamin A Study Group. *Lancet (London, England), 354*, 471–476.

Daudi, N., Shouval, D., Stein-Zamir, C., & Ackerman, Z. (2012). Breastmilk hepatitis A virus RNA in nursing mothers with acute hepatitis A virus infection. *Breastfeeding Medicine, 7*, 313–315. https://doi.org/10.1089/bfm.2011.0084.

Davé, V., Street, K., Francis, S., et al. (2016). Bacterial microbiome of breast milk and child saliva from low-income Mexican-American women and children. *Pediatric Research, 79*, 846–854. https://doi.org/10.1038/pr.2016.9.

Delgado, S., Arroyo, R., Martín, R., & Rodríguez, J. M. (2008). PCR-DGGE assessment of the bacterial diversity of breast milk in women with lactational infectious mastitis. *BMC Infectious Diseases, 8*, 51. https://doi.org/10.1186/1471-2334-8-51.

Delgado, S., Arroyo, R., Jiménez, E., et al. (2009). *Staphylococcus epidermidis* strains isolated from breast milk of women suffering infectious mastitis: potential virulence traits and resistance to antibiotics. *BMC Microbiology, 9*, 82. https://doi.org/10.1186/1471-2180-9-82.

Delgado, S., García, P., Fernández, L., et al. (2011). Characterization of *Staphylococcus aureus* strains involved in human and bovine mastitis. *FEMS Immunology and Medical Microbiology, 62*, 225–235. https://doi.org/10.1111/j.1574-695X.2011.00806.x.

Díaz-Ropero, M. P., Martín, R., Sierra, S., et al. (2007). Two *Lactobacillus* strains, isolated from breast milk, differently modulate the immune response. *Journal of Applied Microbiology, 102*, 337–343. https://doi.org/10.1111/j.1365-2672.2006.03102.x.

Dupont-Rouzeyrol, M., Biron, A., O'Connor, O., et al. (2016). Infectious Zika viral particles in breastmilk. *Lancet (London, England), 387*, 1051. https://doi.org/10.1016/S0140-6736(16)00624-3.

Duranti, S., Lugli, G. A., Mancabelli, L., et al. (2017). Maternal inheritance of bifidobacterial communities and bifidophages in infants through vertical transmission. *Microbiome, 5*, 66. https://doi.org/10.1186/s40168-017-0282-6.

Eidelman, A. I., & Szilagyi, G. (1979). Patterns of bacterial colonization of human milk. *Obstetrics and Gynecology, 53*, 550–552.

Espinosa-Martos, I., Jiménez, E., de Andrés, J., et al. (2016). Milk and blood biomarkers associated to the clinical efficacy of a probiotic for the treatment of infectious mastitis. *Beneficial Microbes, 7*, 305–318. https://doi.org/10.3920/BM2015.0134.

Fan, Y., Chong, Y. S., Choolani, M. A., et al. (2010). Unravelling the mystery of stem/progenitor cells in human breast milk. *PLoS One, 5*, e14421. https://doi.org/10.1371/journal.pone.0014421.

Favier, C. F., Vaughan, E. E., De Vos, W. M., & Akkermans, A. D. L. (2002). Molecular monitoring of succession of bacterial communities in human neonates. *Applied and Environmental Microbiology, 68*, 219–226.

Feng, P., Gao, M., Burgher, A., et al. (2016). A nine-country study of the protein content and amino acid composition of mature human milk. *Food & Nutrition Research, 60*, 31042. https://doi.org/10.3402/fnr.v60.31042.

Fernández, L., Langa, S., Martín, V., et al. (2013). The human milk microbiota: origin and potential roles in health and disease. *Pharmacological Research, 69*, 1–10. https://doi.org/10.1016/j.phrs.2012.09.001.

Fernández, L., Arroyo, R., Espinosa, I., et al. (2014). Probiotics for human lactational mastitis. *Beneficial Microbes, 5*, 169–183. https://doi.org/10.3920/BM2013.0036.

Fernández, L., Cárdenas, N., Arroyo, R., et al. (2016). Prevention ofinfectious mastitis by oral administration of *Lactobacillus salivarius* PS2 during late pregnancy. *Clinical Infectious Diseases, 62*, 568–573. https://doi.org/10.1093/cid/civ974.

Fitzstevens, J. L., Smith, K. C., Hagadorn, J. I., et al. (2017). Systematic review of the human milk microbiota. *Nutrition in Clinical Practice, 32*, 354–364. https://doi.org/10.1177/0884533616670150.

Foster, W. D., & Harris, R. E. (1960). The incidence of *Staphylococcus pyogenes* in normal human breast milk. *The Journal of Obstetrics and Gynaecology of the British Empire, 67*, 463–464.

Fujimura, K. E., Sitarik, A. R., Havstad, S., et al. (2016). Neonatal gut microbiota associates with childhood multisensitized atopy and T cell differentiation. *Nature Medicine, 22*, 1187–1191. https://doi.org/10.1038/nm.4176.

Gomez-Gallego, C., Garcia-Mantrana, I., Salminen, S., & Collado, M. C. (2016). The human milk microbiome and factors influencing its composition and activity. *Seminars in Fetal & Neonatal Medicine, 21*, 400–405. https://doi.org/10.1016/j.siny.2016.05.003.

Gómez-Gallego, C., Kumar, H., García-Mantrana, I., et al. (2017). Breast milk polyamines and microbiota interactions: Impact of mode of delivery and geographical location. *Annals of Nutrition & Metabolism, 70*, 184–190. https://doi.org/10.1159/000457134.

González, R., Maldonado, A., Martín, V., et al. (2013). Breast milk and gut microbiota in African mothers and infants from an area of high HIV prevalence. *PLoS One, 8*, e80299. https://doi.org/10.1371/journal.pone.0080299.

Gosalbes, M. J., Abellan, J. J., Durban, A., et al. (2012). Metagenomics of human microbiome: beyond 16s rDNA. *Clinical Microbiology and Infection, 18*, 47–49. https://doi.org/10.1111/j.1469-0691.2012.03865.x.

Grahn, N., Olofsson, M., Ellnebo-Svedlund, K., et al. (2003). Identification of mixed bacterial DNA contamination in broad-range PCR amplification of 16S rDNA V1 and V3 variable regions by pyrosequencing of cloned amplicons. *FEMS Microbiology Letters, 219*, 87–91.

Gregory, K. E., Samuel, B. S., Houghteling, P., et al. (2016). Influence of maternal breast milk ingestion on acquisition of the intestinal microbiome in preterm infants. *Microbiome, 4*, 68. https://doi.org/10.1186/s40168-016-0214-x.

Grönlund, M.-M., Gueimonde, M., Laitinen, K., et al. (2007). Maternal breast-milk and intestinal bifidobacteria guide the compositional development of the Bifidobacterium microbiota in infants at risk of allergic disease. *Clinical & Experimental Allergy, 37*, 1764–1772. https://doi.org/10.1111/j.1365-2222.2007.02849.x.

Gueimonde, M., Laitinen, K., Salminen, S., & Isolauri, E. (2007). Breast milk: A source of bifidobacteria for infant gut development and maturation? *Neonatology, 92*, 64–66. https://doi.org/10.1159/000100088.

Gueimonde, M., Bottacini, F., van Sinderen, D., et al. (2012). Genome sequence of *Parascardovia denticolens* IPLA 20019, isolated from human breast milk. *Journal of Bacteriology, 194*, 4776–4777. https://doi.org/10.1128/JB.01035-12.

Harmsen, H. J., Wildeboer-Veloo, A. C., Raangs, G. C., et al. (2000). Analysis of intestinal flora development in breast-fed and formula-fed infants by using molecular identification and detection methods. *Journal of Pediatric Gastroenterology and Nutrition, 30*, 61–67.

Hassiotou, F., & Geddes, D. T. (2015). Immune cell-mediated protection of the mammary gland and the infant during breastfeeding. *Advances in Nutrition, 6*, 267–275. https://doi.org/10.3945/an.114.007377.

Heikkilä, M. P., & Saris, P. E. J. (2003). Inhibition of *Staphylococcus aureus* by the commensal bacteria of human milk. *Journal of Applied Microbiology, 95*, 471–478.

Hennet, T., & Borsig, L. (2016). Breastfed at Tiffany's. *Trends in Biochemical Sciences, 41*, 508–518. https://doi.org/10.1016/j.tibs.2016.02.008.

Hoashi, M., Meche, L., Mahal, L. K., et al. (2016). Human milk bacterial and glycosylation patterns differ by delivery mode. *Reproductive Sciences, 23*, 902–907. https://doi.org/10.1177/1933719115623645.

Hunt, K. M., Foster, J. A., Forney, L. J., et al. (2011). Characterization of the diversity and temporal stability of bacterial communities in human milk. *PLoS One, 6*, e21313. https://doi.org/10.1371/journal.pone.0021313.

Hunt, K. M., Preuss, J., Nissan, C., et al. (2012). Human milk oligosaccharides promote the growth of staphylococci. *Applied and Environmental Microbiology, 78*, 4763–4770. https://doi.org/10.1128/AEM.00477-12.

Iwase, T., Uehara, Y., Shinji, H., et al. (2010). *Staphylococcus epidermidis*ssp inhibits *Staphylococcus aureus* biofilm formation and nasal colonization. *Nature, 465*, 346–349. https://doi.org/10.1038/nature09074.

Jacobsen, A. S., & Jenssen, H. (2012). Human cathelicidin LL-37 prevents bacterial biofilm formation. *Future Medicinal Chemistry, 4*, 1587–1599. https://doi.org/10.4155/fmc.12.97.

Jeurink, P. V., van Bergenhenegouwen, J., Jiménez, E., et al. (2013). Human milk: A source of more life than we imagine. *Beneficial Microbes, 4*, 17–30. https://doi.org/10.3920/BM2012.0040.

Jiménez, E., Delgado, S., Fernández, L., et al. (2008a). Assessment of the bacterial diversity of human colostrum and screening of staphylococcal and enterococcal populations for potential virulence factors. *Research in Microbiology, 159*, 595–601. https://doi.org/10.1016/j.resmic.2008.09.001.

Jiménez, E., Delgado, S., Maldonado, A., et al. (2008b). *Staphylococcus epidermidis*: A differential trait of the fecal microbiota of breast-fed infants. *BMC Microbiology, 8*, 143. https://doi.org/10.1186/1471-2180-8-143.

Jiménez, E., Fernandez, L., Maldonado, A., et al. (2008c). Oral administration of *Lactobacillus*strains isolated from breast milk as an alternative for the treatment of infectious mastitis during lactation. *Applied and Environmental Microbiology, 74*, 4650–4655. https://doi.org/10.1128/AEM.02599-07.

Jiménez, E., Langa, S., Martín, V., et al. (2010a). Complete genome sequence of *Lactobacillus fermentum* CECT 5716, a probiotic strain isolated from human milk. *Journal of Bacteriology, 192*, 4800. https://doi.org/10.1128/JB.00702-10.

Jiménez, E., Martin, R., Maldonado, A., et al. (2010b). Complete genome sequence of *Lactobacillus salivarius* CECT 5713, a probiotic strain isolated from human milk and infant feces. *Journal of Bacteriology, 192*, 5266–5267. https://doi.org/10.1128/JB.00703-10.

Jiménez, E., de Andrés, J., Manrique, M., et al. (2015). Metagenomic analysis of milk of healthy and mastitis-suffering women. *Journal of Human Lactation, 31*, 406–415. https://doi.org/10.1177/0890334415585078.

Jiménez, E., Arroyo, R., Cárdenas, N., et al. (2017). Mammary candidiasis: A medical condition without scientific evidence? *PLoS One, 12*, e0181071. https://doi.org/10.1371/journal.pone.0181071.

Jost, T., Lacroix, C., Braegger, C., & Chassard, C. (2013). Assessment of bacterial diversity in breast milk using culture-dependent and culture-independent approaches. *The British Journal of Nutrition, 110*, 1253–1262. https://doi.org/10.1017/S0007114513000597.

Jost, T., Lacroix, C., Braegger, C. P., et al. (2014). Vertical mother-neonate transfer of maternal gut bacteria via breastfeeding. *Environmental Microbiology, 16*, 2891–2904. https://doi.org/10.1111/1462-2920.12238.

Jost, T., Lacroix, C., Braegger, C., & Chassard, C. (2015). Impact of human milk bacteria and oligosaccharides on neonatal gut microbiota establishment and gut health. *Nutrition Reviews, 73*, 426–437. https://doi.org/10.1093/nutrit/nuu016.

Kalliomäki, M., Kirjavainen, P., Eerola, E., et al. (2001). Distinct patterns of neonatal gut microflora in infants in whom atopy was and was not developing. *The Journal of Allergy and Clinical Immunology, 107*, 129–134. https://doi.org/10.1067/mai.2001.111237.

Kalliomäki, M., Collado, M. C., Salminen, S., & Isolauri, E. (2008). Early differences in fecal microbiota composition in children may predict overweight. *The American Journal of Clinical Nutrition, 87*, 534–538. https://doi.org/10.1093/ajcn/87.3.534.

Kayıran, P. G., Can, F., Kayıran, S. M., et al. (2014). Transmission of methicillin-sensitive *Staphylococcus aureus* to a preterm infant through breast milk. *The Journal of Maternal-Fetal & Neonatal Medicine, 27*, 527–529. https://doi.org/10.3109/14767058.2013.819332.

Keefe, G. (1997). *Streptococcus agalactiae* mastitis: A review. *The Canadian Veterinary Journal, 38*, 429–437.

Kenny, J. F. (1977). Recurrent group B streptococcal disease in an infant associated with the ingestion of infected mother's milk. *The Journal of Pediatrics, 91*, 158–159.

Khodayar-Pardo, P., Mira-Pascual, L., Collado, M. C., & Martínez-Costa, C. (2014). Impact of lactation stage, gestational age and mode of delivery on breast milk microbiota. *Journal of Perinatology, 34*, 599–605. https://doi.org/10.1038/jp.2014.47.

Kirjavainen, P. V., Apostolou, E., Arvola, T., et al. (2001). Characterizing the composition of intestinal microflora as a prospective treatment target in infant allergic disease. *FEMS Immunology and Medical Microbiology, 32*, 1–7.

Kourtis, A. P., Butera, S., Ibegbu, C., et al. (2003). Breast milk and HIV-1: Vector of transmission or vehicle of protection? *The Lancet Infectious Diseases, 3*, 786–793.

Kumar, H., du Toit, E., Kulkarni, A., et al. (2016). Distinct patterns in human milk microbiota and fatty acid profiles across specific geographic locations. *Frontiers in Microbiology, 7*, 1619. https://doi.org/10.3389/fmicb.2016.01619.

Kunz, C., Meyer, C., Collado, M. C., et al. (2017). Influence of gestational age, secretor, and Lewis blood group status on the oligosaccharide content of human milk. *Journal of Pediatric Gastroenterology and Nutrition, 64*, 789–798. https://doi.org/10.1097/MPG.0000000000001402.

Lagier, J.-C., Million, M., Hugon, P., et al. (2012). Human gut microbiota: Repertoire and variations. *Frontiers in Cellular and Infection Microbiology, 2*. https://doi.org/10.3389/fcimb.2012.00136.

Langa, S., Maldonado-Barragán, A., Delgado, S., et al. (2012). Characterization of *Lactobacillus salivarius* CECT 5713, a strain isolated from human milk: From genotype to phenotype. *Applied Microbiology and Biotechnology, 94*, 1279–1287. https://doi.org/10.1007/s00253-012-4032-1.

Lara-Villoslada, F., Olivares, M., Sierra, S., et al. (2007a). Beneficial effects of probiotic bacteria isolated from breast milk. *The British Journal of Nutrition, 98*, S96–S100. https://doi.org/10.1017/S0007114507832910.

Lara-Villoslada, F., Sierra, S., Martín, R., et al. (2007b). Safety assessment of two probiotic strains, *Lactobacillus coryniformis* CECT5711 and *Lactobacillus gasseri* CECT5714. *Journal of Applied Microbiology, 103*, 175–184. https://doi.org/10.1111/j.1365-2672.2006.03225.x.

Lauder, A. P., Roche, A. M., Sherrill-Mix, S., et al. (2016). Comparison of placenta samples with contamination controls does not provide evidence for a distinct placenta microbiota. *Microbiome, 4*, 29. https://doi.org/10.1186/s40168-016-0172-3.

Laurence, M., Hatzis, C., & Brash, D. E. (2014). Common contaminants in next-generation sequencing that hinder discovery of low-abundance microbes. *PLoS One, 9*, e97876. https://doi.org/10.1371/journal.pone.0097876.

Lawrence, R. A., & Lawrence, R. M. (2005). *Breastfeeding: A guide for the medical profession* (6th ed.). Philadelphia: Mosby.

Lewis, Z. T., Totten, S. M., Smilowitz, J. T., et al. (2015). Maternal fucosyltransferase 2 status affects the gut bifidobacterial communities of breastfed infants. *Microbiome, 3*, 13. https://doi.org/10.1186/s40168-015-0071-z.

Li, S.-W., Watanabe, K., Hsu, C.-C., et al. (2017). Bacterial composition and diversity in breast milk samples from mothers living in Taiwan and mainland China. *Frontiers in Microbiology, 8*, 965. https://doi.org/10.3389/fmicb.2017.00965.

Liu, A. Y., Lohman-Payne, B., Chung, M. H., et al. (2015). Maternal plasma and breastmilk viral loads are associated with HIV-1-specific cellular immune responses among HIV-1-exposed, uninfected infants in Kenya. *Clinical and Experimental Immunology, 180*, 509–519. https://doi.org/10.1111/cei.12599.

Lundequist, B., Nord, C. E., & Winberg, J. (1985). The composition of the faecal microflora in breastfed and bottle fed infants from birth to eight weeks. *Acta Paediatrica Scandinavica, 74*, 45–51.

Makino, H., Kushiro, A., Ishikawa, E., et al. (2011). Transmission of intestinal *Bifidobacterium longum* subsp. *longum* strains from mother to infant, determined by multilocus sequencing typing and amplified fragment length polymorphism. *Applied and Environmental Microbiology, 77*, 6788–6793. https://doi.org/10.1128/AEM.05346-11.

Makino, H., Martin, R., Ishikawa, E., et al. (2015). Multilocus sequence typing of bifidobacterial strains from infant's faeces and human milk: Are bifidobacteria being sustainably shared during breastfeeding? *Beneficial Microbes, 6*, 563–572. https://doi.org/10.3920/BM2014.0082.

Maldonado, J., Lara-Villoslada, F., Sierra, S., et al. (2010). Safety and tolerance of the human milk probiotic strain *Lactobacillus salivarius* CECT5713 in 6-month-old children. *Nutrition, 26*, 1082–1087. https://doi.org/10.1016/j.nut.2009.08.023.

Maldonado, J., Cañabate, F., Sempere, L., et al. (2012). Human milk probiotic *Lactobacillus fermentum* CECT5716 reduces the incidence of gastrointestinal and upper respiratory tract infections in infants. *Journal of Pediatric Gastroenterology and Nutrition, 54*, 55–61. https://doi.org/10.1097/MPG.0b013e3182333f18.

Maldonado-Lobón, J. A., Díaz-López, M. A., Carputo, R., et al. (2015). *Lactobacillus fermentum* CECT 5716 reduces *Staphylococcus*load in the breastmilk of lactating mothers suffering breast pain: A randomized ccntrolled trial. *Breastfeeding Medicine, 10*, 425–432. https://doi.org/10.1089/bfm.2015.0070.

Marín, M., Arroyo, R., Espinosa-Martos, I., et al. (2017). Identification of emerging human mastitis pathogens by MALDI-TOF and assessment of their antibiotic resistance patterns. *Frontiers in Microbiology, 8*, 1258. https://doi.org/10.3389/fmicb.2017.01258.

Martín, R., Langa, S., Reviriego, C., et al. (2003). Human milk is a source of lactic acid bacteria for the infant gut. *The Journal of Pediatrics, 143*, 754–758. https://doi.org/10.1016/j.jpeds.2003.09.028.

Martín, R., Langa, S., Reviriego, C., Jiménez, E., Marín, M. L., Olivares, M., Boza, J., Jiménez, J., Fernández, L., Xaus, J., & Rodríguez, J. M. (2004). The commensal microflora of human milk: New perspectives for food bacteriotherapy and probiotics. *Trends in Food Science and Technology, 15*, 121–127.

Martín, R., Olivares, M., Marín, M. L., et al. (2005). Probiotic potential of 3 lactobacilli strains isolated from breast milk. *Journal of Human Lactation, 21*, 8–17. https://doi.org/10.1177/0890334404272393.

Martín, R., Jiménez, E., Olivares, M., et al. (2006). *Lactobacillus salivarius* CECT 5713, a potential probiotic strain isolated from infant feces and breast milk of a mother-child pair. *International Journal of Food Microbiology, 112*, 35–43. https://doi.org/10.1016/j.ijfoodmicro.2006.06.011.

Martín, R., Heilig, G. H. J., Zoetendal, E. G., et al. (2007a). Diversity of the *Lactobacillus* group in breast milk and vagina of healthy women and potential role in the colonization of the infant gut. *Journal of Applied Microbiology, 103*, 2638–2644. https://doi.org/10.1111/j.1365-2672.2007.03497.x.

Martín, R., Heilig, H. G. H. J., Zoetendal, E. G., et al. (2007b). Cultivation-independent assessment of the bacterial diversity of breast milk among healthy women. *Research in Microbiology, 158*, 31–37. https://doi.org/10.1016/j.resmic.2006.11.004.

Martín, R., Jiménez, E., Heilig, H., et al. (2009). Isolation of bifidobacteria from breast milk and assessment of the bifidobacterial population by PCR-denaturing gradient gel electrophoresis and quantitative real-time PCR. *Applied and Environmental Microbiology, 75*, 965–969. https://doi.org/10.1128/AEM.02063-08.

Martín, V., Maldonado, A., Fernández, L., et al. (2010). Inhibition of human immunodeficiency virus type 1 by lactic acid bacteria from human breastmilk. *Breastfeeding Medicine, 5*, 153–158. https://doi.org/10.1089/bfm.2010.0001.

Martín, V., Mañes-Lázaro, R., Rodríguez, J. M., & Maldonado-Barragán, A. (2011). *Streptococcus lactarius* sp. nov., isolated from breast milk of healthy women. *International Journal of Systematic and Evolutionary Microbiology, 61*, 1048–1052. https://doi.org/10.1099/ijs.0.021642-0.

Martín, V., Maldonado-Barragán, A., Jiménez, E., et al. (2012a). Complete genome sequence of *Streptococcus salivarius* PS4, a strain isolated from human milk. *Journal of Bacteriology, 194*, 4466–4467. https://doi.org/10.1128/JB.00896-12.

Martín, V., Maldonado-Barragán, A., Moles, L., et al. (2012b). Sharing of bacterial strains between breast milk and infant feces. *Journal of Human Lactation, 28*, 36–44. https://doi.org/10.1177/0890334411424729.

Martín, V., Cárdenas, N., Jiménez, E., et al. (2013). Genome sequence of *Lactobacillus gastricus* PS3, a strain isolated from human milk. *Genome Announc, 1*(4), e00489-13. https://doi.org/10.1128/genomeA.00489-13.

Martin, R., Makino, H., Cetinyurek Yavuz, A., et al. (2016). Early-life events, including mode of delivery and type of feeding, siblings and gender, shape the developing gut microbiota. *PLoS One, 11*, e0158498. https://doi.org/10.1371/journal.pone.0158498.

McGuire, M. K., & McGuire, M. A. (2015). Human milk: Mother nature's prototypical probiotic food? *Advances in Nutrition, 6*, 112–123. https://doi.org/10.3945/an.114.007435.

McGuire, M. K., & McGuire, M. A. (2017). Got bacteria? The astounding, yet not-so-surprising, microbiome of human milk. *Current Opinion in Biotechnology, 44*, 63–68. https://doi.org/10.1016/J.COPBIO.2016.11.013.

McGuire, M. K., Meehan, C. L., McGuire, M. A., et al. (2017). What's normal? Oligosaccharide concentrations and profiles in milk produced by healthy women vary geographically. *The American Journal of Clinical Nutrition, 105*, 1086–1100. https://doi.org/10.3945/ajcn.116.139980.

Mediano, P., Fernández, L., Jiménez, E., et al. (2017). Microbial diversity in milk of women with mastitis: Potential role of coagulase-negative staphylococci, viridans group streptococci, and corynebacteria. *Journal of Human Lactation, 33*, 309–318. https://doi.org/10.1177/0890334417692968.

Michie, C., Lockie, F., & Lynn, W. (2003). The challenge of mastitis. *Archives of Disease in Childhood, 88*, 818–821.

Milani, C., Mancabelli, L., Lugli, G. A., et al. (2015). Exploring vertical transmission of bifidobacteria from mother to child. *Applied and Environmental Microbiology, 81*, 7078–7087. https://doi.org/10.1128/AEM.02037-15.

Mira, A. R. J. (2016). The origin of human milk bacteria. In M. McGuire & M. A. B. L. McGuire (Eds.), *Prebiotics and probiotics in human milk. Origins and functions of milk-borne oligosaccharides and bacteria* (pp. 349–364). London: Elsevier.

Mosca, F., & Giannì, M. L. (2017). Human milk: Composition and health benefits. *La Pediatria Medica e Chirurgica, 39*, 155. https://doi.org/10.4081/pmc.2017.155.

Mühl, H., Kochem, A.-J., & Disqué, C. S. S. (2010). Activity and DNA contamination of commercial polymerase chain reaction reagents for the universal 16SrDNA real-time polymerase chain reaction detection of bacterial pathogens in blood. *Diagnostic Microbiology and Infectious Disease, 66*, 41–49.

Munblit, D., Treneva, M., Peroni, D. G., et al. (2016). Colostrum and mature human milk of women from London, Moscow, and Verona: Determinants of immune composition. *Nutrients, 8*, 695. https://doi.org/10.3390/nu8110695.

Murakami, M., Dorschner, R. A., Stern, L. J., et al. (2005). Expression and secretion of cathelicidin antimicrobial peptides in murine mammary glands and human milk. *Pediatric Research, 57*, 10–15. https://doi.org/10.1203/01.PDR.0000148068.32201.50.

Murphy, K., Curley, D., O'Callaghan, T. F., et al. (2017). The composition of human milk and infant faecal microbiota over the first three months of life: A pilot study. *Scientific Reports, 7*, 40597. https://doi.org/10.1038/srep40597.

Mutschlechner, W., Karall, D., Hartmann, C., et al. (2016). Mammary candidiasis: Molecular-based detection of *Candida* species in human milk samples. *European Journal of Clinical Microbiology & Infectious Diseases, 35*, 1309–1313. https://doi.org/10.1007/s10096-016-2666-0.

Nikniaz, L., Ostadrahimi, A., Mahdavi, R., et al. (2013). Effects of synbiotic supplementation on breast milk levels of IgA, TGF-β1, and TGF-β2. *Journal of Human Lactation, 29*, 591–596. https://doi.org/10.1177/0890334413490833.

Nishimura, R. Y., Barbieiri, P., de Castro, G. S. F., et al. (2014). Dietary polyunsaturated fatty acid intake during late pregnancy affects fatty acid composition of mature breast milk. *Nutrition, 30*, 685–689. https://doi.org/10.1016/j.nut.2013.11.002.

Nogacka, A., Salazar, N., Suárez, M., et al. (2017). Impact of intrapartum antimicrobial prophylaxis upon the intestinal microbiota and the prevalence of antibiotic resistance genes in vaginally delivered full-term neonates. *Microbiome, 5*, 93. https://doi.org/10.1186/s40168-017-0313-3.

Olivares, M., Díaz-Ropero, M. P., Lara-Villoslada, F., Rodriguez, J. M., & Xaus, J. (2005). Effectiveness of probiotics in allergy: A child's game or adult affair? *Nutrafoods, 4*, 59–64.

Olivares, M., Díaz-Ropero, M. A. P., Gómez, N., et al. (2006a). Antimicrobial potential of four *Lactobacillus* strains isolated from breast milk. *Journal of Applied Microbiology, 101*, 72–79.

Olivares, M., Díaz-Ropero, M. P., Gómez, N., et al. (2006b). The consumption of two new probiotic strains, *Lactobacillus gasseri* CECT 5714 and *Lactobacillus coryniformis* CECT 5711, boosts the immune system of healthy humans. *International Microbiology, 9*, 47–52.

Olivares, M., Díaz-Ropero, M. P., Gomez, N., et al. (2006c). Oral administration of two probiotic strains, Lactobacillus gasseri CECT5714 and Lactobacillus coryniformis CECT5711, enhances the intestinal function of healthy adults. *International Journal of Food Microbiology, 107*, 104–111.

Olivares, M., Díaz-Ropero, M. P., Sierra, S., et al. (2007). Oral intake of *Lactobacillus fermentum* CECT5716 enhances the effects of influenza vaccination. *Nutrition, 23*, 254–260. https://doi. org/10.1016/j.nut.2007.01.004.

Olivares, M., Albrecht, S., De Palma, G., et al. (2015). Human milk composition differs in healthy mothers and mothers with celiac disease. *European Journal of Nutrition, 54*, 119–128. https:// doi.org/10.1007/s00394-014-0692-1.

Otto, M. (2009). Staphylococcus epidermidis—The "accidental" pathogen. *Nature Reviews. Microbiology, 7*, 555–567. https://doi.org/10.1038/nrmicro2182.

Palmer, C., Bik, E. M., DiGiulio, D. B., et al. (2007). Development of the human infant intestinal microbiota. *PLoS Biology, 5*, e177. https://doi.org/10.1371/journal.pbio.0050177.

Pannaraj, P. S., Li, F., Cerini, C., et al. (2017). Association between breast milk bacterial communities and establishment and development of the infant gut microbiome. *JAMA Pediatrics, 171*, 647–654. https://doi.org/10.1001/jamapediatrics.2017.0378.

Park, B., Iwase, T., & Liu, G. Y. (2011). Intranasal application of *S. epidermidis* prevents colonization by methicillin-resistant *Staphylococcus aureus* in mice. *PLoS One, 6*, e25880. https://doi. org/10.1371/journal.pone.0025880.

Paviour, S., Musaad, S., Roberts, S., et al. (2002). *Corynebacterium* species isolated from patients with mastitis. *Clinical Infectious Diseases, 35*, 1434–1440. https://doi.org/10.1086/344463.

Peran, L., Camuesco, D., Comalada, M., et al. (2005). Preventative effects of a probiotic, *Lactobacillus salivarius* ssp. *salivarius*, in the TNBS model of rat colitis. *World Journal of Gastroenterology, 11*, 5185–5192.

Peran, L., Sierra, S., Comalada, M., et al. (2007). A comparative study of the preventative effects exerted by two probiotics, Lactobacillus reuteri and *Lactobacillus fermentum*, in the trinitrobenzenesulfonic acid model of rat colitis. *The British Journal of Nutrition, 97*, 96–103. https:// doi.org/10.1017/S0007114507257770.

Perez, P. F., Doré, J., Leclerc, M., et al. (2007). Bacterial imprinting of the neonatal immune system: Lessons from maternal cells? *Pediatrics, 119*, e724–e732. https://doi.org/10.1542/ peds.2006-1649.

Pérez-Cano, F. J., Dong, H., & Yaqoob, P. (2010). *In vitro* immunomodulatory activity of *Lactobacillus fermentum* CECT5716 and *Lactobacillus salivarius* CECT5713: two probiotic strains isolated from human breast milk. *Immunobiology, 215*, 996–1004. https://doi. org/10.1016/j.imbio.2010.01.004.

Perez-Muñoz, M. E., Arrieta, M.-C., Ramer-Tait, A. E., & Walter, J. (2017). A critical assessment of the "sterile womb" "in utero colonization" hypotheses: Implications for research on the pioneer infant microbiome. *Microbiome, 5*, 48. https://doi.org/10.1186/s40168-017-0268-4.

Praveen, P., Jordan, F., Priami, C., & Morine, M. J. (2015). The role of breast-feeding in infant immune system: A systems perspective on the intestinal microbiome. *Microbiome, 3*, 41. https://doi.org/10.1186/s40168-015-0104-7.

Prescott, S. L., Wickens, K., Westcott, L., et al. (2008). Supplementation with *Lactobacillus rhamnosus* or *Bifidobacterium lactis* probiotics in pregnancy increases cord blood interferon-γ and breast milk transforming growth factor-β and immunoglobin A detection. *Clinical & Experimental Allergy, 38*, 1606–1614. https://doi.org/10.1111/j.1365-2222.2008.03061.x.

Quinn, R. A., Lim, Y. W., Maughan, H., et al. (2014). Biogeochemical forces shape the composition and physiology of polymicrobial communities in the cystic fibrosis lung. *MBio, 5*, e00956–e00913. https://doi.org/10.1128/mBio.00956-13.

Qutaishat, S. S., Stemper, M. E., Spencer, S. K., et al. (2003). Transmission of *Salmonella enterica* serotype *typhimurium* DT104 to infants through mother's breast milk. *Pediatrics, 111*, 1442–1446.

Rantasalo, I., & Kauppinen, M. A. (1959). The occurrence of *Staphylococcus aureus* in mother's milk. *Annales Chirurgiae et Gynaecologiae Fenniae, 48*, 246–258.

Rautava, S., Kalliomäki, M., & Isolauri, E. (2002). Probiotics during pregnancy and breast-feeding might confer immunomodulatory protection against atopic disease in the infant. *The Journal of Allergy and Clinical Immunology, 109*, 119–121.

Renfrew, M. J., Pokhrel, S., Quigley, M., McCormick, F., Fos-Rushby, J., Dodds, R., Duffy, S., & Trueman, P. W. A. (2012). *Preventing disease and saving resources: the potential contribution of increasing breastfeeding rates in the UK*. London: UNICEF UK.

Renshaw, A. A., Derhagopian, R. P., & Gould, E. W. (2011). Cystic neutrophilic granulomatous mastitis: An underappreciated pattern strongly associated with gram-positive bacilli. *American Journal of Clinical Pathology, 136*, 424–427. https://doi.org/10.1309/AJCP1W9JBRYOQSNZ.

Rescigno, M., Urbano, M., Valzasina, B., et al. (2001). Dendritic cells express tight junction proteins and penetrate gut epithelial monolayers to sample bacteria. *Nature Immunology, 2*, 361–367. https://doi.org/10.1038/86373.

Rivas-Santiago, B., Serrano, C. J., & Enciso-Moreno, J. A. (2009). Susceptibility to infectious diseases based on antimicrobial peptide production. *Infection and Immunity, 77*, 4690–4695. https://doi.org/10.1128/IAI.01515-08.

Rodríguez, J. M. (2014). The origin of human milk bacteria: Is there a bacterial entero-mammary pathway during late pregnancy and lactation? *Advances in Nutrition, 5*, 779–784. https://doi.org/10.3945/an.114.007229.

Ruiz, L., Espinosa-Martos, I., García-Carral, C., et al. (2017). What's normal? Immune profiling of human milk from healthy women living in different geographical and socioeconomic settings. *Frontiers in Immunology, 8*, 696. https://doi.org/10.3389/fimmu.2017.00696.

Sakata, H., Yoshioka, H., & Fujita, K. (1985). Development of the intestinal flora in very low birth weight infants compared to normal full-term newborns. *European Journal of Pediatrics, 144*, 186–190.

Sakwinska, O., Moine, D., Delley, M., et al. (2016). Microbiota in breast milk of Chinese lactating mothers. *PLoS One, 11*, e0160856. https://doi.org/10.1371/journal.pone.0160856.

Salter, S. J., Cox, M. J., Turek, E. M., et al. (2014). Reagent and laboratory contamination can critically impact sequence-based microbiome analyses. *BMC Biology, 12*, 87. https://doi.org/10.1186/s12915-014-0087-z.

Schanler, R. J., Fraley, J. K., Lau, C., et al. (2011). Breastmilk cultures and infection in extremely premature infants. *Journal of Perinatology, 31*, 335–338. https://doi.org/10.1038/jp.2011.13.

Schultz, M., Göttl, C., Young, R. J., et al. (2004). Administration of oral probiotic bacteria to pregnant women causes temporary infantile colonization. *Journal of Pediatric Gastroenterology and Nutrition, 38*, 293–297.

Solís, G., de Los Reyes-Gavilan, C. G., Fernández, N., et al. (2010). Establishment and development of lactic acid bacteria and bifidobacteria microbiota in breast-milk and the infant gut. *Anaerobe, 16*, 307–310. https://doi.org/10.1016/j.anaerobe.2010.02.004.

Soto, A., Martín, V., Jiménez, E., et al. (2014). Lactobacilli and bifidobacteria in human breast milk: Influence of antibiotherapy and other host and clinical factors. *Journal of Pediatric Gastroenterology and Nutrition, 59*, 78–88. https://doi.org/10.1097/MPG.0000000000000347.

Sprenger, N., Lee, L. Y., De Castro, C. A., et al. (2017). Longitudinal change of selected human milk oligosaccharides and association to infants' growth, an observatory, single center, longitudinal cohort study. *PLoS One, 12*, e0171814. https://doi.org/10.1371/journal.pone.0171814.

Thomsen, A. C., Mogensen, S. C., & Løve Jepsen, F. (1985). Experimental mastitis in mice induced by coagulase-negative staphylococci isolated from cases of mastitis in nursing women. *Acta Obstetricia et Gynecologica Scandinavica, 64*, 163–166.

Thongaram, T., Hoeflinger, J. L., Chow, J., & Miller, M. J. (2017). Human milk oligosaccharide consumption by probiotic and human-associated bifidobacteria and lactobacilli. *Journal of Dairy Science, 100*, 7825–7833. https://doi.org/10.3168/jds.2017-12753.

Thurl, S., Munzert, M., Henker, J., et al. (2010). Variation of human milk oligosaccharides in relation to milk groups and lactational periods. *The British Journal of Nutrition, 104*, 1261–1271. https://doi.org/10.1017/S0007114510002072.

Toscano, M., De Grandi, R., Peroni, D. G., et al. (2017). Impact of delivery mode on the colostrum microbiota composition. *BMC Microbiology, 17*, 205. https://doi.org/10.1186/s12866-017-1109-0.

Turroni, F., Peano, C., Pass, D. A., et al. (2012). Diversity of bifidobacteria within the infant gut microbiota. *PLoS One, 7*, e36957. https://doi.org/10.1371/journal.pone.0036957.

U.S. Department of Health and Human Services. (2011). *The surgeon general's call to action to support breastfeeding*. Washington, DC: Department of Health and Human Services, Office of the Surgeon General.

Uehara, Y., Kikuchi, K., Nakamura, T., et al. (2001). H(2)O(2) produced by viridans group streptococci may contribute to inhibition of methicillin-resistant *Staphylococcus aureus* colonization of oral cavities in newborns. *Clinical Infectious Diseases, 32*, 1408–1413. https://doi.org/10.1086/320179.

Urbaniak, C., Cummins, J., Brackstone, M., et al. (2014a). Microbiota of human breast tissue. *Applied and Environmental Microbiology, 80*, 3007–3014. https://doi.org/10.1128/AEM.00242-14.

Urbaniak, C., McMillan, A., Angelini, M., et al. (2014b). Effect of chemotherapy on the microbiota and metabolome of human milk, a case report. *Microbiome, 2*, 24. https://doi.org/10.1186/2049-2618-2-24.

Vanhaecke, T., Aubert, P., Grohard, P.-A., et al. (2017). *L. fermentum* CECT 5716 prevents stress-induced intestinal barrier dysfunction in newborn rats. *Neurogastroenterology & Motility, 29*, e13069. https://doi.org/10.1111/nmo.13069.

Vázquez-Fresno, R., Llorach, R., Marinic, J., et al. (2014). Urinary metabolomic fingerprinting after consumption of a probiotic strain in women with mastitis. *Pharmacological Research, 87*, 160–165. https://doi.org/10.1016/j.phrs.2014.05.010.

Vazquez-Torres, A., Jones-Carson, J., Bäumler, A. J., et al. (1999). Extraintestinal dissemination of *Salmonella* by CD18-expressing phagocytes. *Nature, 401*, 804–808. https://doi.org/10.1038/44593.

Wang, M., Li, M., Wu, S., et al. (2015). Fecal microbiota composition of breast-fed infants is correlated with human milk oligosaccharides consumed. *Journal of Pediatric Gastroenterology and Nutrition, 60*, 825–833. https://doi.org/10.1097/MPG.0000000000000752.

Ward, T. L., Hosid, S., Ioshikhes, I., & Altosaar, I. (2013). Human milk metagenome: A functional capacity analysis. *BMC Microbiology, 13*, 116. https://doi.org/10.1186/1471-2180-13-116.

Weems, M. F., Dereddy, N. R., & Arnold, S. R. (2015). Mother's milk as a source of *Enterobacter cloacae* sepsis in a preterm infant. *Breastfeeding Medicine, 10*, 503–504. https://doi.org/10.1089/bfm.2015.0146.

Williams, J. E., Carrothers, J. M., Lackey, K. A., et al. (2017a). Human milk microbial community structure is relatively stable and related to variations in macronutrient and micronutrient intakes in healthy lactating women. *The Journal of Nutrition, 147*, 1739–1748. https://doi.org/10.3945/jn.117.248864.

Williams, J. E., Price, W. J., Shafii, B., et al. (2017b). Relationships among microbial communities, maternal cells, oligosaccharides, and macronutrients in human milk. *Journal of Human Lactation, 33*, 540–551. https://doi.org/10.1177/0890334417709433.

Williamson, S., Finucane, E., Gamsu, H. R., & Hewitt, J. H. (1978). *Staphylococcus aureus* in raw human milk for neonates. *British Medical Journal, 1*, 1146.

Willing, B. P., Russell, S. L., & Finlay, B. B. (2011). Shifting the balance: Antibiotic effects on host–microbiota mutualism. *Nature Reviews. Microbiology, 9*, 233–243. https://doi.org/10.1038/nrmicro2536.

Witkowska-Zimny, M., & Kaminska-El-Hassan, E. (2017). Cells of human breast milk. *Cellular & Molecular Biology Letters, 22*, 11. https://doi.org/10.1186/s11658-017-0042-4.

World Health Organization. (2003). *Global strategy for infant and young child feeding*. Geneva: World Health Organization.

Xuan, C., Shamonki, J. M., Chung, A., et al. (2014). Microbial dysbiosis is associated with human breast cancer. *PLoS One, 9*, e83744. https://doi.org/10.1371/journal.pone.0083744.

Zimmermann, P., Gwee, A., & Curtis, N. (2017). The controversial role of breast milk in GBS late-onset disease. *The Journal of Infection, 74*(Suppl 1), S34–S40. https://doi.org/10.1016/S0163-4453(17)30189-5.

Fermented Dairy Products

C. Peláez, M. C. Martínez-Cuesta, and T. Requena

Abstract The microbiota of fermented dairy products contributes to the safety, flavor, and organoleptic qualities of the products. Moreover, metabolites obtained from the fermentation process enhance the milk nutritive value and digestibility, whereas dairy microorganisms could be the perfect carriers for reseeding the gut microbiota. The structural food matrix of fermented milk facilitates the delivery of viable microorganisms to the intestinal tract. Fermented dairy products may be beneficial to human health by improving lactose intolerance symptoms and for the production of bioactive compounds such as vitamins, gamma-amino butyric acid, exopolysaccharides, and bioactive peptides, among others. Also, fermented dairy products contribute to the modulation of the gut microbiota and the prevention of infections, inflammation, and cardiometabolic diseases. Furthermore, fermented dairy products constitute the hallmark of probiotics supply in the food market.

Keywords Fermented dairy · Yogurt · Kefir · Cheese · Probiotics · Bioactive compounds

Introduction

The fermented dairy products consumed today are generated through controlled microbial culturing and enzymatic conversions of major and minor milk components (see (Macori and Cotter 2018) for a recent review). Fermentation improves shelf life, increases microbiological safety, adds flavor, and enhances palatability and organoleptic qualities. The fermentation process involves a series of complex reactions carried out by microorganisms, which transform milk constituents rendering new molecules of

---

C. Peláez · M. C. Martínez-Cuesta · T. Requena (✉)
Department of Food Biotechnology and Microbiology, Institute of Food Science Research, CIAL (CSIC), Madrid, Spain
e-mail: t.requena@csic.es

© Springer Nature Switzerland AG 2019
M. A. Azcarate-Peril et al. (eds.), *How Fermented Foods Feed a Healthy Gut Microbiota*, https://doi.org/10.1007/978-3-030-28737-5_2

enhanced nutritive value and digestibility. Moreover, fermentation generates metabolites that can be major contributors of a daily healthful diet (Marco et al. 2017).

The contributions of milk components and dairy products to human health have been comprehensively reviewed (Tunick and Van Hekken 2015). These can be summarized as enhancing muscle building, lowering blood pressure, reducing low density lipoprotein cholesterol, and preventing diabetes, obesity and cancer, among others. Additionally, due to the reduced consumption of dietary fiber in Western societies and the overall decrease of microbial diversity in processed foods, fermented dairy products could be the perfect carriers for reseeding the gut microbiota. The above listed health benefits and the fact that they are viewed as natural products have placed yogurt, kefir, and cheese in the forefront of consumers' preferences. In the present chapter we describe existing data regarding the microbiota found in fermented dairy products and the advances in knowledge of microbial properties that may benefit human health.

Fermented Dairy Products

Fermentation is one of the oldest forms of milk preservation and has been used by humans since ancient times (Markowiak and Slizewska 2017). The beneficial effects of fermented dairy products was empirically known by Romans, Greeks and Egyptians. They produced different types of sour milk from buffalo, cow, or goat's milk. However, it was not until the twentieth century that the beneficial properties of lactic acid bacteria (LAB) started to be scientifically substantiated by the immunologist Élie Metchnikoff, who was awarded the Nobel Prize in 1908. He concluded that the unusual high numbers of Balkan centenarians was due to the consumption of sour milk containing large numbers of the lactic acid-producing bacterium *Lactobacillus bulgaricus bacillus*, currently classified as *Lactobacillus delbrueckii* subsp. *bulgaricus*. In the book 'The Prolongation of Life' (Metchnikoff 1908), Metchnikoff recommended the daily consumption of milk fermented with pure cultures of this *Lactobacillus* to discourage microbial putrefactive growth in the colon, setting the stage for further studies on beneficial effects of LAB in fermented products.

Yogurt

Among fermented dairy products, yogurt is consumed the most in Western societies, being a common component in the daily diet of populations from the Netherlands and Scandinavian countries. Although yogurt has been manufactured commercially for over a century, its concept has changed over time into a very segmented market. Today fermented dairy products include a broad catalog of flavored, low-fat, drinkable, probiotic, and other products marketed as health-promoting. The European

Codex Alimentarius Commission explicitly defines yogurt as the product of milk fermentation by *Streptococcus thermophilus* and *L. delbrueckii* subsp. *bulgaricus* (*L. bulgaricus*) (Codex-Alimentarius 2003). According to the Codex STAN 243-2003 these microorganisms must reach a minimum of 10^7 cfu/g, be viable, active, and abundant in the product until the set expiration date. However, differences in labeling laws allow, for example in the United Kingdom, to include any *Lactobacillus* species in fermented milks labeled as 'yogurt'. In this case, the term 'yogurt-like product' is used and defined as an alternative dairy product in which *L. bulgaricus* can be substituted by other *Lactobacillus* species for the fermentation, or yogurt containing probiotic bacteria, when probiotic or alternative organisms are added to yogurt (Guarner et al. 2005).

Yogurt is an excellent source of macro- and micronutrients like high-quality, digestible proteins and carbohydrates, minerals, and vitamins. It contributes to the growth and fitness of muscle mass and helps maintaining bone health due to their calcium and phosphorus content. The nutritional value of yogurt has been recognized by the Canadian Food Guide (Health-Canada 2011), USA Department of Agriculture (USDA 2010), and the British Nutrition Foundation (BNF 2015). These international agencies recommend the inclusion of fermented dairy products in the daily diet. Furthermore, they emphasize that food guides must state whether dairy products are fermented or non-fermented since fermented dairy products have additional health claims compared to non-fermented products (Chilton et al. 2015). As an example, studies in the Netherlands and Sweden (Keszei et al. 2010; Sonestedt et al. 2011) showed that regular consumption of fermented dairy products, but not non-fermented dairy products, significantly decrease the risk of bladder cancer and cardiovascular disease. Likewise, a significant positive effect of calcium intake on teeth health was specifically associated with dairy fermented products (Adegboye et al. 2012). This may be explained by the breakdown of milk components during microbial fermentation into more bioavailable and new metabolites of potential health benefits.

Kefir

Kefir is a traditional fermented dairy beverage produced by a complex natural microbiota from kefir grains. These grains are traditionally obtained from periodic coagulation of cow's milk with calf or sheep abomasum (forth stomach) in goatskin bags. Kefir originated in the Caucasus but gained popularity in Eastern and Central European countries starting in the second half of the nineteenth century. Kefir can also be made with milk from other sources including as goat, sheep and buffalo milk (Bourrie et al. 2016). Distinctive microorganisms in kefir are homofermentative lactobacilli (*Lactobacillus kefiranofaciens*), which produces a kefiran complex that surrounds yeasts (*Saccharomyces cerevisiae*), and other bacteria (*Lactococcus, Leuconostoc*, thermophilic and mesophilic lactobacilli and acetic acid bacteria). Microbiological, technological, as well as nutritional, and health benefits of kefir have been recently summarized (Bourrie et al. 2016; Kesenkas et al. 2017).

Cheese

Animal skins and inflated internal organs, particularly the rumen, have provided storage vessels for a range of foodstuffs since ancient times (Marciniak 2011). Hence, we can presume that cheese making was discovered accidentally when storing milk in ruminant stomachs, which resulted in milk curdling by the residual gastric rennin. Most modern cheeses are manufactured from pasteurized milk coagulated in a vat by recombinant enzymes or proteases of vegetable origin with added *Lactococcus, Lactobacillus* and *Leuconostoc* as starters. However, traditional raw milk cheeses naturally fermented by its indigenous microbiota are still produced in some Mediterranean countries. For specific cheeses like blue or soft cheeses, bacteria of the genera *Brevibacterium* and *Propionibacterium* and molds of the genus *Penicillium* are added to develop their characteristic organoleptical properties. Raw milk can contain over 400 bacterial species. This microbial biodiversity decreases in the cheese core usually dominated by few species of LAB but persists on the cheese surface with high numbers of species of bacteria, yeasts and molds (Montel et al. 2014; Orla O'Sullivan 2017). It is commonly accepted that cheese flavor develops as a result of the overall microbial metabolism beginning during clotting, progressing further during cheese ripening (Weimer 2007).

Semi-hard cheeses typically contain non-starter lactobacilli (NSLAB), which can reach up to 10^7–10^8 cfu/g for long periods of time during production and storage (Pelaez and Requena 2005). NSLAB have been shown to generate bioactive peptides and gamma-aminobutyric acid (GABA) (Settanni and Moschetti 2010). *Propionibacterium freudenreichii*, a ripening culture in Swiss-type cheese, produces conjugated linoleic acid (CLA) and may have bifidogenic and immunomodulatory properties (Thierry et al. 2011).

The Microbes in Fermented Dairy Products

Microorganisms already present in the raw milk, from the environment or added from a previously fermented product were the main actors of the traditional fermentation process. Because humans have consumed fermented foods since ancient times, the human gastrointestinal tract (GIT) adapted to a constant supply of live bacteria on a nearly daily basis. In fact, many of the microbial species found in fermented foods are either identical to or share physiological traits with species known to promote GIT health (Marco et al. 2017). However, the industrialization of food production has reduced the variety of foods that humans consume and their associated microbiota. Hygienic industrial practices, including thermal treatment, have decreased the microbial diversity of fermented foods, modern dairy fermentations have to rely on standardized starter cultures and microorganisms as well as cultivation protocols that extend shelf life, improve food safety and enhance perceived health benefits (Hill et al. 2017). The consumption of these products limits the

Fermented Dairy Products

traditional exposure of the human gut to the highly biodiverse traditional food ecosystem. In response, artisanal dairy fermentation and consumption of traditional fermented dairy products as part of Western diets have regained popularity (Prakash Tamang and Kailasapathy 2010).

In 1873, Lister (1873) first isolated *Bacterium lactis* [later renamed as *Streptococcus lactis* and more recently *Lactococcus lactis* subsp. *lactis* (Schleifer et al. 1985)]. This bacterium together with other species of the genera *Lactococcus, Lactobacillus* and *Leuconostoc* are usually included in well-defined starter cultures currently used in modern cheese fermentation. Likewise, most common starters for yogurt and fermented milks include species of *Lactobacillus, Streptococcus, Bifidobacterium* and some yeast used in the manufacture of kefir and koumiss (Carminati et al. 2016). Nevertheless, undefined starter communities composed of a complex undefined mixture of LAB strains are still in use in dairy fermentation due to advantages in reducing sensitivity to bacteriophage attack. State of the art advances in high-throughput DNA sequencing technology and the potential of predictive metabolic modelling of the multi-strain cultures, are emerging as powerful tools to investigate structure and function of these complex communities (Smid et al. 2014).

The structural food matrix of fermented milk facilitates the delivery of viable microorganisms to the intestinal tract by contributing to the microorganisms' survival during transit, enhancing their interaction with the gut microbiota, and participating in the reinforcement of the intestinal barrier. Similarly, the solid matrix of cheese and its buffering capacity may protect bacteria during the intestinal transit more efficiently than yogurt or fermented milks (Karimi et al. 2011). The fact that semi-hard ripened cheeses like Cheddar, Gouda or Manchego, maintain viable bacteria for up to 12 months, is a good reason to consider cheeses as excellent carriers of health promoting bacteria into the gastrointestinal tract (Ross et al. 2002).

Do Microbes in Fermented Dairy Products Survive Passage Through the Gastrointestinal Tract?

Upon entering the human gastrointestinal tract, fermentation-associated microorganisms must survive environmental challenges including acidity of the stomach and bile salts and enzymes in the small intestine to reach the colon. Hence, survival of fermentative bacteria within the gastrointestinal tract became an important research topic since the beginning of the last century. As early as 1920, Cheplin and Rettger (1920) were able to recover live *Lactobacillus acidophilus* but not *L. bulgaricus* from rat stools after daily consumption. Since then, there has been conflicting evidence concerning the viability of the yogurt cultures in the gastrointestinal tract. In a double-blind placebo prospective study including 114 healthy young volunteers that were usual yogurt consumers the authors did not detect *S. thermophilus* nor *L. bulgaricus* in feces by culturing or by specific PCR and DNA hybridization of total fecal DNA (del Campo et al. 2005). Detection of LAB from yogurt in total fecal DNA was consistently negative, even after repeated yogurt consumption during 15 days. In a contrasting study, out of 39 samples recovered from 13 healthy subjects over a 12-day period of fresh yogurt intake, 32 and 37 samples contained viable *S. thermophilus* and *L. bulgaricus*, respectively (Mater et al. 2005). Furthermore, in a population of 51 healthy subjects living in the Paris area, the yogurt species *L. bulgaricus* was detected in 73% of fecal samples from consumers (200–400 g yogurt/day) *vs.* 28% from non-consumers (Alvaro et al. 2007). Accordingly, Elli et al. (2006) demonstrated that yogurt bacteria survived gastrointestinal transit being recovered in feces from 20 healthy volunteers fed commercial yogurt for 1 week.

Lactose Hydrolysis by Microbes in Fermented Dairy Foods

During milk fermentation, most lactose is converted into lactic acid by fermentative microorganisms. Although the final product still contains traces of the carbohydrate, lactose-intolerant populations are able to consume yogurt without experiencing adverse symptoms. The improvement in lactose absorption was also demonstrated when healthy subjects with lactose maldigestion consumed yogurt containing live bacterial cultures in comparison with heated yogurt (Rizkalla et al. 2000). This can be attributed to the intestinal release of β-galactosidase by yogurt cultures that must be viable when ingested (Guarner et al. 2005; Savaiano 2014). In the study carried out by Alvaro et al. (2007) with 51 healthy adult subjects of yogurt and non-yogurt consumers, they found that among the nine metabolic bacterial enzyme activities investigated, the only significant difference concerned β-galactosidases.

Lactose intolerance is caused by a decreased expression of the enzyme β-galactosidase or lactase normally secreted by the epithelial cells within the villi. This enzyme is required to digest the lactose from milk and dairy products. Undigested lactose consequently enters the colon where it is fermented by gas producing microbes, resulting in symptoms including abdominal pain, bloating, diarrhea, and flatulence (Misselwitz et al. 2013). Expression of lactase decreases after weaning in most individuals and as a result become relatively lactose intolerant. It is, therefore, not surprising that as adults, as much as 75% of the world's human population is intolerant to ingested dietary lactose (Silanikove et al. 2015). Over centuries of evolution, humans have adapted to lactose ingestion by several mechanisms including genetic mutations that allow lactose digestion in a classic example of evolutionary nutrigenetics. But also, colonic microbiome adaptation and the development of fermented dairy products have contributed to lactose tolerance in humans. As a result, the intolerance to lactose occurs in the Central European population at a 5% rate, while Asia and Latin America populations observe up to 90% intolerance rates. Furthermore, some authors have proposed to modulate the colonic microbiome of lactose-intolerant individuals, increasing the abundance of lactose metabolizing bacteria that are non-gas producers (i.e. *Bifidobacterium*), by administration of short-chain galactooligosaccharides (Azcarate-Peril et al. 2017).

Fermented Dairy Products 41

The generation and release of β-galactosidases is a species-related trait in the yogurt-associated species (*L. bulgaricus* and *S. thermophilus*). The European Food Safety Authority Panel on Dietetic Products, Nutrition and Allergies in 2010 formally approved yogurt's beneficial effects on reverting lactose intolerance. They indicated that the dose of live microorganisms should be at least 10^8 cfu/g (EFSA 2010). This claim does not require survival and reproduction of the bacterial cells during intestinal transit.

Besides the improvement in lactose intolerance symptoms, the conversion of lactose into lactic acid reduces the intestinal pH and confers a protective effect against foodborne pathogens infection. The low pH also increases peristalsis, thereby indirectly removing pathogens by accelerating their transit through the intestine (Kailasapathy and Chin 2000). Moreover, LAB cultures enhance pathogens elimination by other mechanisms including competitive exclusion and production of antimicrobial metabolites like bacteriocins (Arqués et al. 2015). Alvaro et al. (2007) showed significantly lower numbers of Enterobacteriaceae in human feces of yogurt consumers versus non-consumers. Accordingly, Van der Meer and Bovee-Oudenhoven (1998) reported that lactic acid in yogurt and calcium in other dairy products inhibit the gastrointestinal survival and colonization of *Salmonella* Enteritidis.

Bioactive Compounds

One important mechanism by which the fermented dairy foods and its associated microbiota (LAB, propionibacteria, yeasts, and molds) may be beneficial to human health is through the production of bioactive compounds. Some of these bioactive compounds include vitamins, bioactive peptides, exopolysaccharides (EPS), GABA and CLA. See Fernández et al. (2015) for a recent overview of the impact of metabolites produced by microorganisms found in fermented dairy products.

Particular attention to bioactive components resulting from the fermentation of milk and their impact on health is rapidly been explored using high throughput, multi-omic approaches. Over the last decades these new technologies have allowed for more sophisticated metabolite analysis, which integrates fermented dairy products composition and functional assessments following ingestion (Zheng et al. 2015; Hagi et al. 2016). Still, we should consider that the acidic environment of the stomach and the subsequent stages of digestion can lead to early inactivation of certain bioactive compounds (Stanton et al. 2005).

Vitamins Vitamins are compounds essential for human individuals that are insufficiently or not synthesized at all by the human organism. Although they are present in foods, vitamin deficiencies still exist due to malnutrition or lack of a daily balanced diet (Arth et al. 2016). Furthermore, food processing and cooking may destroy or remove some vitamins such as vitamins of the B-group. This group of vitamins includes thiamine (B1), riboflavin (B2), niacin (B3), pyridoxine (B6), pantothenic acid (B5), biotin (B7 or H), folate (B9, B11 or M) and cobalamin (B12). Although shown beneficial, programs with synthetic vitamin B fortification of foods to correct vitamin deficiencies has not been adopted in many countries due to the potential side effects from excessive intake in populations with normal vitamin B levels (Atta et al. 2016).

Folate vitamin deficiency is associated with megaloblastic anemia as well as congenital malformations, including spina bifida and anencephaly, although only a small fraction of these diseases is actively being prevented worldwide (Arth et al. 2016). Today, it is recognized that the yeast *S. cerevisiae*, used as starter in fermented milks like kefir or koumiss, is a folate producer (Moslehi-Jenabian et al. 2010). As for LAB, production of folate still remains controversial. It has been shown that *S. thermophilus* can produce folate whereas *L. bulgaricus* is a folate consumer. Still, the concentration of the vitamin in yogurt may reach values up to 200 µg/L (Wouters et al. 2002). In addition to *S. thermophilus*, other bacteria present in dairy fermentations such as *L. lactis*, *L. acidophilus*, *L. plantarum*, *L. fermentum*, *Leuconostoc lactis*, *Bifidobacterium longum*, and some strains of *Propionibacterium*, also have the ability to produce folate (Iyer and Tomar 2009; LeBlanc et al. 2017). Therefore, the consumption of fermented dairy foods could be an attractive approach for improving the world wide nutritional deficiencies of folate (Saubade et al. 2017).

Gamma-amino butyric acid Gamma-aminobutyric acid (GABA) is a major inhibitory neurotransmitter in the adult mammalian brain. GABA also has different functions in the central and peripheral nervous systems, and in some non-neuronal tissues (Watanabe et al. 2002). In general, anti-hypertensive and antidepressant activities are the major functions of GABA or GABA-rich foods. However, the mechanisms responsible for these activities are still unknown due to scarce studies carried out on the pathway for GABA absorption (Dhakal et al. 2012).

LAB can generate GABA as end product from the decarboxylation of glutamic acid. The enzyme that converts glutamic acid into GABA is the glutamic acid decarboxylase (GAD). Genes encoding GADs are broadly distributed in *Lactobacillus brevis*, *L. plantarum*, *L. fermentum*, *L. reuteri*, *S. thermophilus*, *L. lactis* subsp. *cremoris*, *L. lactis* subsp. *lactis* and some *Bifidobacterium* species, indicating that these bacteria may be able to synthesize GABA (Wu and Shah 2017). In fact, production of GABA has been demonstrated in *L. lactis*, *S. thermophilus*, and *L. bulgaricus* isolated from milk, a trait that could be used to produce GABA-rich fermented milk products. Other high GABA producers are *L. brevis* strains, which are not usually associated with the dairy environment. Hence, to enhance GABA concentrations in fermented dairy products, strategies have been proposed that include co-culturing *L. brevis* with conventional dairy starters in dairy fermentation (Wu and Shah 2017). Enhancement of GABA production in fermented products has shown hypotensive effects on rats (Quilez and Diana 2017). Additionally, with the aim of producing GABA enriched cheeses, GABA producing LAB have been isolated from several cheeses made from raw milk such as Spanish artisanal cheeses (Diana et al. 2014) and traditional alpine Italian cheeses (Franciosi et al. 2015).

Exopolysaccharides Exopolysaccharides (EPS) are high molecular weight carbohydrate polymers loosely connected to the cell surface of microorganisms. EPS protect against food processing promoting biofilm formation, and also mediate cell-to-cell interactions in the human gut, facilitating microbial adhesion to intestinal mucosa and preventing adhesion of pathogens. Multiple strains including the yogurt starter species *L. bulgaricus* and *S. thermophilus* have been reported generate EPS. Also many probiotic strains of *Lactobacillus* and *Bifidobacterium* are being investigated for their ability to produce EPS (Salazar et al. 2016). The use of EPS producers in dairy fermentation can be beneficial not only for intestinal health but also technologically to improve texture and flavor of the final products (Ryan et al. 2015; Nampoothiri et al. 2017).

Bioactive peptides Bioactive peptides are released from animal and plant proteins by endogenous proteolytic enzymes or enzymes of microbial origin. They usually contain between two and 20 amino acids, which can be naturally resistant to gastro-intestinal digestion due to partial protection conferred by their high hydrophobicity and the usual presence of proline. The hydrolysis degree of bioactive peptides can vary depending on peptide chain length, nature of the peptide, presence of other peptides in the medium and the food texture (Fang et al. 2016).

Bioactive peptides have been the subject of intensive research due to their potential physiological effects on various human systems such as cardiovascular, digestive, endocrine, immune, and nervous systems [see recent review of (Martinez-Villaluenga et al. 2017)]. Furthermore, industrial-scale technologies suitable for the commercial production of bioactive milk peptides have been developed (Korhonen and Pihlanto 2006; Urista et al. 2011). By far, the most studied bioactive peptides are those derived from milk proteolysis (Nagpal et al. 2011; Beermann and Hartung 2013). Several peptides have been isolated from yogurt, kefir, other fermented milks and cheese and a number of them have shown to be released by the proteolytic system of the LAB from these fermented products (López-Expósito et al. 2017).

Microbial antihypertensive properties have been related to peptides with the capacity to inhibit the angiotensin converting enzyme (ACE) that suppresses angiotensin II-mediated vasoconstriction. The responsible inhibitor peptides of the antihypertensive effect derive mainly from β-casein and have been found in milk fermented by strains of *Enterococcus faecalis*, *L. lactis,* and *Bifidobacterium* (Martinez-Villaluenga et al. 2017). The most studied antihypertensive peptides are the tripeptides VPP and IPP, released from β-casein after fermentation of milk. VPP and IPP are now added to fermented sour-milk products that claim antihypertensive effects launched in Japan and Finland. The Japanese product called "Calpis" consists of milk fermented with *L. helveticus* CP790 and *S. cerevisiae* containing both peptides VPP and IPP. This fermented milk has demonstrated properties to prevent the development of hypertension (Sipola et al. 2002). The Finnish product called "Evolus" contains the tripeptide IPP and claims to have similar antihypertensive effects. It is produced by *L. helveticus* LBK-16H strain as starter in milk fermentation

(Seppo et al. 2003). A comprehensive meta-analysis of data from relevant human studies showed a modest reduction in blood pressure in individuals consuming VPP and IPP compared to antihypertensive drugs (Fekete et al. 2015) demonstrating a potential for use in cardiovascular therapy as a complement to traditional medications.

Other beneficial effects attributed to peptides derived from milk include immunomodulating, antioxidant, antimutagenic, mucin-stimulating, and opioid effects (Martinez-Villaluenga et al. 2017). Some peptides have multifunctional characteristics such as the peptide YQEPVLGPVRGPFPIIV (fragment 193–209 of β-casein) obtained after fermentation of milk with the strain *L. casei* Shirota. This peptide showed an inhibition efficiency ratio for ACE (antihypertensive activity) of 0.14%/peptide concentration (mg/mL), and a thrombin inhibition efficiency ratio of 4.6%/peptide concentration (mg/mL) (Rojas-Ronquillo et al. 2012). Some strategies employed to enhance the release of bioactive peptides during fermentation by LAB include supplementing milk with milk peptide fractions. This strategy facilitates the proteolytic activity of *L. acidophilus*, *L. helveticus*, *L. bulgaricus*, and *S. thermophilus* and enhances the production of ACE-inhibitory peptides (Gandhi and Shah 2014).

Cheese can also contain peptides with antihypertensive, antioxidant, opioid, antimicrobial, antiproliferative, mineral absorption and modulatory effects (López-Expósito et al. 2017). As example, an intensive ACE-inhibitory activity (75.7%) was detected with peptides isolated from Gouda cheese aged for 8 months (Saito et al. 2000). Dimitrov et al. (2015) evaluated the ACE inhibitory activity of 180 LAB and selected as starters several *L. helveticus*, *L. bulgaricus* and *L. casei* strains for Bulgarian cheese production, which led to increased production of bioactive peptides. Finally, addition of *L. casei* 279 or *L. casei* LAFTI® L26 as adjuncts in Cheddar cheese production increased the ACE-inhibitory activity during ripening at 4 °C, possibly due to increased proteolysis (Ong et al. 2007).

Scientific Evidence for Health Promoting Effects by Fermented Dairy Products

Although the presence of beneficial microbiota and their biological active metabolites in fermented dairy products is well documented, many of their claimed physiological actions have only been assayed *in vitro* and in animal models. Thus, there is a significant challenge in trying to extrapolate animal studies to humans. Scientific evidence from human interventions and clinical trials that include the evaluation of the fermented products and their matrices are needed to validate their functional properties.

Modulation of the gut microbiota It has been postulated that the live microorganisms in yogurt and fermented milks benefit gastrointestinal health by modulating the resident gut microbiota. However, it is worth pointing out that this modulation is based on the notion that a "normal" healthy gut microbiota exists; however, a normal healthy gut microbiota has not been defined, except perhaps as microbiota without a pathogenic bacteria overgrowth. The type and amount of microbes in the human intestine differ substantially between individuals, which means that the phylogenetic composition could be considered a subjective fingerprint (Schmidt et al. 2018). However, while our microbiota varies phylogenetically, metagenomic analyses have revealed that at the highest functional level, the functional potential of the microbiome of healthy individuals remains very similar (Heintz-Buschart and Wilmes 2018). This could be a good starting point for developing the concept of "normal" healthy gut microbiota.

To exert their benefits, transient dairy bacteria entering the human gastrointestinal tract not only must survive the hostile conditions of the stomach and the small intestine to reach the colon, they also need to survive and compete in a colon environment fully seeded with resident microorganisms (Hillman et al. 2017). Fermented food and beverages represent between 5 and 40% of the daily food intake in the world (Prakash Tamang and Kailasapathy 2010), which corresponds to 0.1–1.0% of the bacteria present in the gastrointestinal tract. The bacteria that survive the conditions of the gastrointestinal tract are an extra source of microbial metabolites being conceivable that they might alter the proportions of autochthonous bacteria, impacting diversity and functionality.

Unno et al. (2015) investigated changes in the human gut microbial community structure after consumption of fermented milk containing probiotics. The microbiota was stable at the phylum level, although the relative abundance of Bacteroidetes increased during the ingestion of the fermented milk and decreased during the non-ingestion period. This suggested that consumption of the fermented milk can temporarily alter the gut microbial community structure maintaining its stability. Nevertheless, interactions between resident and transient microorganisms are yet insufficiently clear and can be highly dependent on the colonization resistance of the autochthonous microbes. In another study, conventional and gnotobiotic rats fed fermented milk containing five strains of *Bifidobacterium animalis*, *L. lactis* subsp. *lactis*, *L. bulgaricus* and *S. thermophilus* over a 15-day period showed that the clearance kinetic of *L. lactis* was strongly dependent on the structure of the resident gut microbiota and its susceptibility to be modulated by the transient strain (Zhang et al. 2016). One group of rats promptly eliminated *L. lactis* after fermented milk ingestion, whereas another group shed the strain over an additional 24–48 h. Overall, the specific contribution of dairy fermentative bacteria to the human gut ecosystem composition and functionality remains unclear.

Prevention of infection by pathogens Cheese has been associated with the prevention of bacterial gastroenteritis caused by the foodborne pathogen *Campylobacter jejuni*. A case control study carried out in South Australia involving children aged 1–5 years diagnosed with *C. jejuni* infection, showed that frequent consumption of foods including Cheddar and soft processed cheese was associated with a lower risk of gastroenteritis symptoms (Cameron et al. 2004). This fact reinforces the body of evidence pointing to traditional fermented products such as cheese be considered and included regularly in the daily diet.

Inflammation Maturation, function and defense mechanisms of the immune system are built up over the first years of life and are greatly influenced by the established gut microbiota (Nash et al. 2017). Inflammation is part of the human host defense system for facing unwanted environmental challenges. The immune system reacts with the production of proinflammatory mediators that causes systemic inflammation.

Human cross-sectional studies have supported the premise that the consumption of yogurt can be associated with a lower inflammatory state. Meyer et al. (2007) studied the effect of daily intake of conventional yogurt containing *S. thermophilus* and *L. bulgaricus* on cytokine production in 33 healthy women aged 22–29 years. The subjects consumed 100 g of yogurt per day for 2 weeks, and then 200 g for the subsequent 2 weeks. Consumption of yogurt enhanced cellular immune function, stimulating significant production of TNF-α (63% compared with baseline). On the other hand, Olivares et al. (2006) studied the immunological effects of the dietary deprivation of fermented foods in 30 healthy adult human volunteers (15 females and 15 males) aged from 23 to 43 years. After deprivation for 2 weeks, a decrease in phagocytic activity in leukocytes was observed. The fall in immune response was counteracted after the ingestion of conventional yogurt. Nevertheless, methodological factors limit comparisons between these studies and do not allow differentiation between a beneficial or neutral impact of dairy products on inflammation. Hence, further studies specifically designed to assess inflammation-related outcomes are warranted.

Chronic intestinal inflammation has been associated with development of colorectal cancer. Several studies have shown that LAB present in fermented products may protect against cancer by binding mutagens, inhibiting bacterial enzymes that form carcinogens and reducing inflammation (Zhong et al. 2014). Accordingly, Perdigón et al. (2002) demonstrated in BALBc mice that yogurt may exert antitumor activity by decreasing the inflammatory immune response mediated by IgA(+), apoptosis induction and IL-10 release. Likewise, Pala et al. (2011) conducted a prospective study on 45,241 (14,178 men; 31,063 women) volunteers of the EPIC-Italy cohort and found that high yogurt intake can be significantly associated with decreased colorectal cancer risk in humans.

Cardiometabolic diseases Studies have concluded that yogurt may help to improve diet quality and maintain metabolic well-being as part of a healthy, energy-balanced dietary pattern. A cross-sectional study that examined whether yogurt consumption was associated with better diet quality and metabolic profile among American adults (n = 6526) participating in the Framingham Heart Study Offspring (1998–2001) and Third Generation (2002–2005) cohorts, concluded that yogurt consumption was associated with lower levels of circulating triglycerides, glucose, and lower systolic blood pressure and insulin resistance (Wang et al. 2013). Healthier insulin profile after frequent yogurt consumption was also observed with children in a cross-sectional study using data from the National Health and Nutrition Examination Survey (NHANES) in the USA (Zhu et al. 2015). Accordingly, the PREDIMED study following prospectively 3454 non-diabetic individuals concluded that consumption of yogurt was inversely associated with type 2 diabetes risk in the elderly at high cardiovascular risk (Diaz-Lopez et al. 2016). Association of high yogurt intake with a reduced risk of diabetes type 2 was also observed in a meta-analysis of 14 prospective cohorts with 459,790 participants (Chen et al. 2014).

Unlike type 2 diabetes, an inverse correlation between yogurt consumption and obesity has not been clearly established. A prospective study in an elderly population at high cardiovascular risk (PREDIMED study), concluded that consumption of whole-fat yogurt was associated with positive changes in waist circumference and higher probability for reversion of abdominal obesity (Santiago et al. 2016). Furthermore, a comprehensive literature search on MEDLINE and ISI Web of Knowledge from 1966 through June 2016 (Sayon-Orea et al. 2017) indicated that an inverse association between yogurt consumption and the risk of becoming overweight or obese was not fully consistent in prospective cohort studies although the results showed a tendency to improvement of most parameters of weight gain, risk of overweight or obesity, and risk of metabolic syndrome associated to yogurt consumption.

Finally there is evidence suggesting a moderate cholesterol-reducing action by fermented dairy products (St-Onge et al. 2000). Nevertheless, there is still a need for more prospective studies and high-quality randomized clinical trials to confirm improvement of metabolic markers apparently associated to yogurt and fermented milk consumption.

Probiotics in Fermented Dairy Foods

The International Scientific Association for Probiotics and Prebiotics has agreed to define probiotics as "live microorganisms that, when administered in adequate amounts, confer a health benefit on the host" (Hill et al. 2014). Traditionally, *Lactobacillus, Bifidobacterium* and the yeast *Saccharomyces boulardii* are added fermented milks as probiotics (Aryana and Olson 2017). However, new species recently identified as relevant to intestinal health like *Faecalibacterium prausnitzii, Akkermansia muciniphila* and *Roseburia* could be added to fermented products as probiotics (El Hage et al. 2017). These species contribute to intestinal homeostasis, mucus integrity, and generation of short chain fatty acids.

Viability and functionality of probiotics in both food matrix and the gastrointestinal tract are essential to exert their beneficial effects. Therefore fermented products containing probiotics must have a minimum 10^6 cfu/g live bacteria at the expiration date, since the minimum therapeutic dose per day is suggested to be 10^8–10^9 cfu. Nevertheless, the presence and activity of probiotics in fermented products is sometimes far from being optimal. Although specific strains are inherently resistant to the conditions of production and transit through the gastrointestinal tract, like *Lactobacillus salivarius* CECT5713 and PS2, both isolated from human milk, which were shown to survive during storage for 28 days at 4 °C with only a significant reduction in their viable counts was observed after 21 days (Cárdenas et al. 2014), some strains of bifidobacteria added to commercial fermented milks do not survive gastric transit and their numbers decline during storage (Ladero and Sanchez 2017). Consequently, technological strategies for improving viability and functionality of probiotics have been developed (Tripathi and Giri 2014) and include control of the product's final pH, the addition of oxygen scavengers, the use of probiotics producing protective EPS, microencapsulation or addition of prebiotics to the fermented milk in order to improve survival (Fernández et al. 2015). The review by Castro et al. (2015) summarizes different aspects related to the technological stability barriers encountered in the development of cheeses containing live probiotics.

Bifidobacterium and *Lactobacillus* have been exploited as probiotic beneficial microorganisms for centuries; however, the molecular mechanisms by which these bacteria exert their beneficial effects are still under investigation. Genome sequencing has provided insights into the diversity and evolution of commensal and probiotic bacteria to reveal the molecular basis for their health-promoting properties (Ventura et al. 2009). Full genome sequences of lactobacilli and bifidobacteria are publicly available and have significantly expanded our understanding of the biology of these microorganisms and how they have adapted so well to the human gastrointestinal tract despite their distant original niches. Combined with advanced postgenomic mammalian host response analyses, the molecular interactions that underlie the host-health effects observed are being elucidated (Sommer and Backhed 2016). Following metagenomics, the current metaproteomic and metatranscriptomic analyses have revealed that although there are similarities at the highest functional level among individuals, the microbiota is affected by a variety of factors, including diet, host genetics, and health status. As example, a placebo-controlled intervention trial of 16 healthy subjects that consumed *Lactobacillus rhamnosus* GG for 3 weeks found a common core of shared microbial protein functions in all subjects but no significant changes in the metaproteome attributable to the probiotic intervention (Kolmeder et al. 2016).

A new era in the probiotic field to be considered is the link between bacteria and brain activity in humans, the so called psychobiotics (Bambury et al. 2018). The link between the gut microbiota and brain was first demonstrated in germ free mice showing impaired emotional behaviors and brain biochemistry (Diaz Heijtz et al. 2011). In this model, microbial colonization of the gut initiates signaling mechanisms that affect neuronal circuits involved in motor control and anxiety behavior. A recent randomized, double-blind placebo controlled clinical trial conducted among 60 patients with Alzheimer's disease showed that patients drinking 200 mL/day of a probiotic milk containing *L. acidophilus*, *B. bifidum* and *L. fermentum* during 12 weeks were positively affected in its cognitive function (Akbari et al. 2016). Although promising, this field is still in infancy and high-quality clinical trials are needed to provide enough evidence before probiotics could be therapeutically used in neurodegenerative and emotional disorders. Furthermore, human randomized control trials and hypothesis-driven mechanistic-based experimental studies are needed to validate health claims before fermented dairy products including probiotics can be advocated for specific disease prevention.

Conclusions

Fermented dairy products are ideal carriers of live microorganisms and their metabolites allowing for specific delivery where they will exert their physiological functionality. Moreover, dairy products constitute the hallmark of probiotics supply in the food market. Although industrialized processes have reduced the microbial diversity in milk and dairy products contributing to the limited exposure to a traditionally rich and highly biodiverse microbiota in food products, recent efforts are addressing this issue by applying diversification methods to dairy starters and by enriching fermented dairy products with bioactive compounds. New omics technologies are providing a better understanding of the microbial metabolism, interaction between transient and resident microbiota, and contribution of the microbiota to the host homeostasis at a molecular level essential to increase our understanding of the beneficial effects of milk and their microbially generated products.

Acknowledgements This work was supported by the Spanish Ministry (Project AGL2016-75951-R), CDTI (INDEKA IDI-20190077) and CYTED (Project P917PTE0537/PCIN-2017-075).

References

Adegboye, A. R., Christensen, L. B., Holm-Pedersen, P., Avlund, K., Boucher, B. J., & Heitmann, B. L. (2012). Intake of dairy products in relation to periodontitis in older Danish adults. *Nutrients, 4*(9), 1219–1229.

Akbari, E., Asemi, Z., Kakhaki, R. D., Bahmani, F., Kouchaki, E., Tamtaji, O. R., Hamidi, G. A., & Salami, M. (2016). Effect of probiotic supplementation on cognitive function and metabolic status in Alzheimer's disease: A randomized, double-blind and controlled trial. *Frontiers in Aging Neuroscience, 8*, 256.

Alvaro, E., Andrieux, C., Rochet, V., Rigottier-Gois, L., Lepercq, P., Sutren, M., Galan, P., Duval, Y., Juste, C., & Dore, J. (2007). Composition and metabolism of the intestinal microbiota in consumers and non-consumers of yogurt. *The British Journal of Nutrition, 97*(1), 126–133.

Arqués, J. L., Rodríguez, E., Langa, S., Landete, J. M., & Medina, M. (2015). Antimicrobial activity of lactic acid bacteria in dairy products and gut: Effect on pathogens. *BioMed Research International, 2015*, 584183.

Arth, A., Kancherla, V., Pachon, H., Zimmerman, S., Johnson, Q., & Oakley, G. P., Jr. (2016). A 2015 global update on folic acid-preventable spina bifida and anencephaly. *Birth Defects Research. Part A, Clinical and Molecular Teratology, 106*(7), 520–529.

Aryana, K. J., & Olson, D. W. (2017). A 100-year review: Yogurt and other cultured dairy products. *Journal of Dairy Science, 100*(12), 9987–10013.

Atta, C. A., Fiest, K. M., Frolkis, A. D., Jette, N., Pringsheim, T., St Germaine-Smith, C., Rajapakse, T., Kaplan, G. G., & Metcalfe, A. (2016). Global birth prevalence of spina bifida by folic acid fortification status: A systematic review and meta-analysis. *American Journal of Public Health, 106*(1), e24–e34.

Azcarate-Peril, M. A., Ritter, A. J., Savaiano, D., Monteagudo-Mera, A., Anderson, C., Magness, S. T., & Klaenhammer, T. R. (2017). Impact of short-chain galactooligosaccharides on the gut microbiome of lactose-intolerant individuals. *Proceedings of the National Academy of Sciences of the United States of America, 114*(3), E367–E375.

Bambury, A., Sandhu, K., Cryan, J. F., & Dinan, T. G. (2018). Finding the needle in the haystack: Systematic identification of psychobiotics. *British Journal of Pharmacology, 175*, 4430–4438.

Beermann, C., & Hartung, J. (2013). Physiological properties of milk ingredients released by fermentation. *Food & Function, 4*(2), 185–199.

BNF. (2015). *Healthy eating*. Retrieved March 20, 2015, from http://www.nutrition.org.uk/healthyliving/healthyeating.html

Bourrie, B. C., Willing, B. P., & Cotter, P. D. (2016). The microbiota and health promoting characteristics of the fermented beverage kefir. *Frontiers in Microbiology, 7*, 647.

Cameron, S., Ried, K., Worsley, A., & Topping, D. (2004). Consumption of foods by young children with diagnosed campylobacter infection—A pilot case-control study. *Public Health Nutrition, 7*(1), 85–89.

Cárdenas, N., Calzada, J., Peiroten, A., Jiménez, E., Escudero, R., Rodríguez, J. M., Medina, M., & Fernández, L. (2014). Development of a potential probiotic fresh cheese using two *Lactobacillus salivarius* strains isolated from human milk. *BioMed Research International, 2014*, 801918.

Carminati, D., Giraffa, G., Zago, M., Marco, M. B., Guglielmotti, D., Binetti, A., & Reinheimer, J. (2016). Lactic acid bacteria for dairy fermentations: Specialized starter cultures to improve dairy products. In F. Mozzi, R. R. Raya, & G. M. Vignolo (Eds.), *Biotechnology of lactic acid bacteria: Novel applications* (2nd ed., pp. 191–208). Chichester, UK: Wiley.

Castro, J. M., Tornadijo, M. E., Fresno, J. M., & Sandoval, H. (2015). Biocheese: A food probiotic carrier. *BioMed Research International, 2015*, 723056.

Chen, M., Sun, Q., Giovannucci, E., Mozaffarian, D., Manson, J. E., Willett, W. C., & Hu, F. B. (2014). Dairy consumption and risk of type 2 diabetes: 3 cohorts of US adults and an updated meta-analysis. *BMC Medicine, 12*, 215.

Cheplin, H. A., & Rettger, L. F. (1920). Studies on the transformation of the intestinal flora, with special reference to the implantation of *Bacillus acidophilus*, II. Feeding experiments on man. *Proceedings of the National Academy of Sciences of the United States of America, 6*, 704–705.

Chilton, S. N., Burton, J. P., & Reid, G. (2015). Inclusion of fermented foods in food guides around the world. *Nutrients, 7*(1), 390–404.

Codex-Alimentarius. (2003). *CODEX standard for fermented milks*. Codex Stan 243–2003. Retrieved from http://www.codexalimentarius.net/download/standards/400/CXS_243e.pdf, http://www.codexalimentarius.net/download/standards/400/CXS_243e.pdf

del Campo, R., Bravo, D., Canton, R., Ruiz-Garbajosa, P., Garcia-Albiach, R., Montesi-Libois, A., Yuste, F. J., Abraira, V., & Baquero, F. (2005). Scarce evidence of yogurt lactic acid bacteria in human feces after daily yogurt consumption by healthy volunteers. *Applied and Environmental Microbiology, 71*(1), 547–549.

Dhakal, R., Bajpai, V. K., & Baek, K. H. (2012). Production of GABA (gamma-aminobutyric acid) by microorganisms: A review. *Brazilian Journal of Microbiology, 43*(4), 1230–1241.

Diana, M., Rafecas, M., Arco, C., & Quilez, J. (2014). Free amino acid profile of Spanish artisanal cheeses: Importance of gamma-aminobutyric acid (GABA) and ornithine content. *Journal of Food Composition and Analysis, 35*(2), 94–100.

Diaz Heijtz, R., Wang, S., Anuar, F., Qian, Y., Bjorkholm, B., Samuelsson, A., Hibberd, M. L., Forssberg, H., & Pettersson, S. (2011). Normal gut microbiota modulates brain development and behavior. *Proceedings of the National Academy of Sciences of the United States of America, 108*(7), 3047–3052.

Diaz-Lopez, A., Bullo, M., Martinez-Gonzalez, M. A., Corella, D., Estruch, R., Fito, M., Gomez-Gracia, E., Fiol, M., Garcia de la Corte, F. J., Ros, E., Babio, N., Serra-Majem, L., Pinto, X., Munoz, M. A., Frances, F., Buil-Cosiales, P., & Salas-Salvado, J. (2016). Dairy product consumption and risk of type 2 diabetes in an elderly Spanish Mediterranean population at high cardiovascular risk. *European Journal of Nutrition, 55*(1), 349–360.

Dimitrov, Z., Chorbadjiyska, E., Gotova, I., Pashova, K., & Ilieva, S. (2015). Selected adjunct cultures remarkably increase the content of bioactive peptides in Bulgarian white brined cheese. *Biotechnology & Biotechnological Equipment, 29*(1), 78–83.

EFSA. (2010). Scientific opinion on the substantiation of health claims related to live yoghurt cultures and improved lactose digestion. *EFSA Journal, 8*(10), 1763.

El Hage, R., Hernandez-Sanabria, E., & Van de Wiele, T. (2017). Emerging trends in "Smart probiotics": Functional consideration for the development of novel health and industrial applications. *Frontiers in Microbiology, 8*, 1889.

Elli, M., Callegari, M. L., Ferrari, S., Bessi, E., Cattivelli, D., Soldi, S., Morelli, L., Goupil Feuillerat, N., & Antoine, J. M. (2006). Survival of yogurt bacteria in the human gut. *Applied and Environmental Microbiology, 72*(7), 5113–5117.

Fang, X. X., Rioux, L. E., Labrie, S., & Turgeon, S. L. (2016). Commercial cheeses with different texture have different disintegration and protein/peptide release rates during simulated in vitro digestion. *International Dairy Journal, 56*, 169–178.

Fekete, A. A., Givens, D. I., & Lovegrove, J. A. (2015). Casein-derived lactotripeptides reduce systolic and diastolic blood pressure in a meta-analysis of randomised clinical trials. *Nutrients, 7*(1), 659–681.

Fernández, M., Hudson, J. A., Korpela, R., & de los Reyes-Gavilán, C. G. (2015). Impact on human health of microorganisms present in fermented dairy products: An overview. *BioMed Research International, 2015*, 412714.

Franciosi, E., Carafa, I., Nardin, T., Schiavon, S., Poznanski, E., Cavazza, A., Larcher, R., & Tuohy, K. M. (2015). Biodiversity and gamma-aminobutyric acid production by lactic acid bacteria isolated from traditional alpine raw cow's milk cheeses. *BioMed Research International, 2015*, 625740.

Gandhi, A., & Shah, N. P. (2014). Cell growth and proteolytic activity of *Lactobacillus acidophilus*, *Lactobacillus helveticus*, *Lactobacillus delbrueckii* ssp *bulgaricus*, and *Streptococcus thermophilus* in milk as affected by supplementation with peptide fractions. *International Journal of Food Sciences and Nutrition, 65*(8), 937–941.

Guarner, F., Perdigon, G., Corthier, G., Salminen, S., Koletzko, B., & Morelli, L. (2005). Should yoghurt cultures be considered probiotic? *The British Journal of Nutrition, 93*(6), 783–786.

Hagi, T., Kobayashi, M., & Nomura, M. (2016). Metabolome analysis of milk fermented by gamma-aminobutyric acid-producing *Lactococcus lactis*. *Journal of Dairy Science, 99*(2), 994–1001.

Health-Canada. (2011). *Eating well with Canada's food guide*. Retrieved February 26, 2018, from http://www.hc-sc.gc.ca/fn-an/food-guide-aliment/index-eng.php

Heintz-Buschart, A., & Wilmes, P. (2018). Human gut microbiome: Function matters. *Trends in Microbiology, 26*(7), 563–574.

Hill, C., Guarner, F., Reid, G., Gibson, G. R., Merenstein, D. J., Pot, B., Morelli, L., Canani, R. B., Flint, H. J., Salminen, S., Calder, P. C., & Sanders, M. E. (2014). Expert consensus document. The International Scientific Association for Probiotics and Prebiotics consensus statement on the scope and appropriate use of the term probiotic. *Nature Reviews. Gastroenterology & Hepatology, 11*(8), 506–514.

Hill, D., Sugrue, I., Arendt, E., Hill, C., Stanton, C., & Ross, R. P. (2017). Recent advances in microbial fermentation for dairy and health. *F1000Research, 6*, 751.

Hillman, E. T., Lu, H., Yao, T., & Nakatsu, C. H. (2017). Microbial ecology along the gastrointestinal tract. *Microbes and Environments, 32*(4), 300–313.

Iyer, R., & Tomar, S. K. (2009). Folate: A functional food constituent. *Journal of Food Science, 74*(9), R114–R122.

Kailasapathy, K., & Chin, J. (2000). Survival and therapeutic potential of probiotic organisms with reference to *Lactobacillus acidophilus* and *Bifidobacterium* spp. *Immunology and Cell Biology, 78*(1), 80–88.

Karimi, R., Mortazavian, A. M., & Da Cruz, A. G. (2011). Viability of probiotic microorganisms in cheese during production and storage: A review. *Dairy Science & Technology, 91*(3), 283–308.

Kesenkas, H., Gursoy, O., & Ozbas, H. (2017). Kefir. In J. Frias, C. Martinez-Villaluenga, & E. Penas (Eds.), *Fermented foods in health and disease prevention* (pp. 339–361). London: Academic Press Ltd/Elsevier Science Ltd.

Keszei, A. P., Schouten, L. J., Goldbohm, R. A., & van den Brandt, P. A. (2010). Dairy intake and the risk of bladder cancer in the Netherlands Cohort Study on Diet and Cancer. *American Journal of Epidemiology, 171*(4), 436–446.

Kolmeder, C. A., Salojarvi, J., Ritari, J., de Been, M., Raes, J., Falony, G., Vieira-Silva, S., Kekkonen, R. A., Corthals, G. L., Palva, A., Salonen, A., & de Vos, W. M. (2016). Faecal metaproteomic analysis reveals a personalized and stable functional microbiome and limited effects of a probiotic intervention in adults. *PLoS One, 11*(4), e0153294.

Korhonen, H., & Pihlanto, A. (2006). Bioactive peptides: Production and functionality. *International Dairy Journal, 16*(9), 945–960.

Ladero, V., & Sanchez, B. (2017). Molecular and technological insights into the aerotolerance of anaerobic probiotics: Examples from bifidobacteria. *Current Opinion in Food Science, 14*, 110–115.

LeBlanc, J. G., Chain, F., Martin, R., Bermudez-Humaran, L. G., Courau, S., & Langella, P. (2017). Beneficial effects on host energy metabolism of short-chain fatty acids and vitamins produced by commensal and probiotic bacteria. *Microbial Cell Factories, 16*(1), 79.

Lister, J. (1873). Further contribution to the natural history of bacteria and the germ theory of fermentative changes. *The Quarterly Journal of Microscopical Science, 13*, 380–408.

López-Expósito, I., Miralles, B., Amigo, L., & Hernandez-Ledesma, B. (2017). Health effects of cheese components with a focus on bioactive peptides. In J. Frias, C. Martinez-Villaluenga, & E. Penas (Eds.), *Fermented foods in health and disease prevention* (pp. 239–273). London: Academic Press Ltd/Elsevier Science Ltd.

Macori, G., & Cotter, P. D. (2018). Novel insights into the microbiology of fermented dairy foods. *Current Opinion in Biotechnology, 49*, 172–178.

Marciniak, A. (2011). The secondary products revolution: Empirical evidence and its current zoo-archaeological critique. *Journal of World Prehistory, 24*(2–3), 117–130.

Marco, M. L., Heeney, D., Binda, S., Cifelli, C. J., Cotter, P. D., Foligne, B., Ganzle, M., Kort, R., Pasin, G., Pihlanto, A., Smid, E. J., & Hutkins, R. (2017). Health benefits of fermented foods: Microbiota and beyond. *Current Opinion in Biotechnology, 44*, 94–102.

Markowiak, P., & Slizewska, K. (2017). Effects of probiotics, prebiotics, and synbiotics on human health. *Nutrients, 9*(9), 1021.

Martinez-Villaluenga, C., Penas, E., & Frias, J. (2017). Bioactive peptides in fermented foods: Production and evidence for health effects. In J. Frias, C. Martinez-Villaluenga, & E. Penas (Eds.), *Fermented foods in health and disease prevention* (pp. 23–47). London: Academic Press Ltd/Elsevier Science Ltd.

Mater, D. D., Bretigny, L., Firmesse, O., Flores, M. J., Mogenet, A., Bresson, J. L., & Corthier, G. (2005). *Streptococcus thermophilus* and *Lactobacillus delbrueckii* subsp. *bulgaricus* survive gastrointestinal transit of healthy volunteers consuming yogurt. *FEMS Microbiology Letters, 250*(2), 185–187.

Metchnikoff, E. (1908). *The prolongation of life. Optimistic studies*. New York: G. P. Putnam's Sons/The Knickerbocker Press.

Meyer, A. L., Elmadfa, I., Herbacek, I., & Micksche, M. (2007). Probiotic, as well as conventional yogurt, can enhance the stimulated production of proinflammatory cytokines. *Journal of Human Nutrition and Dietetics, 20*(6), 590–598.

Misselwitz, B., Pohl, D., Fruhauf, H., Fried, M., Vavricka, S. R., & Fox, M. (2013). Lactose malabsorption and intolerance: Pathogenesis, diagnosis and treatment. *United European Gastroenterology Journal, 1*(3), 151–159.

Fermented Dairy Products

Montel, M. C., Buchin, S., Mallet, A., Delbes-Paus, C., Vuitton, D. A., Desmasures, N., & Berthier, F. (2014). Traditional cheeses: Rich and diverse microbiota with associated benefits. *International Journal of Food Microbiology, 177*, 136–154.

Moslehi-Jenabian, S., Pedersen, L. L., & Jespersen, L. (2010). Beneficial effects of probiotic and food borne yeasts on human health. *Nutrients, 2*(4), 449–473.

Nagpal, R., Behare, P., Rana, R., Kumar, A., Kumar, M., Arora, S., Morotta, F., Jain, S., & Yadav, H. (2011). Bioactive peptides derived from milk proteins and their health beneficial potentials: An update. *Food & Function, 2*(1), 18–27.

Nampoothiri, K. M., Beena, D. J., Vasanthakumari, D. S., & Ismail, B. (2017). Health benefits of exopolysaccharides in fermented foods. In J. Frias, C. Martinez-Villaluenga, & E. Penas (Eds.), *Fermented foods in health and disease prevention* (pp. 49–62). London: Academic Press Ltd/ Elsevier Science Ltd.

Nash, M. J., Frank, D. N., & Friedman, J. E. (2017). Early microbes modify immune system development and metabolic homeostasis—The "Restaurant" hypothesis revisited. *Frontiers in Endocrinology, 8*, 349.

Olivares, M., Diaz-Ropero, P., Gomez, N., Sierra, S., Lara-Villoslada, F., Martin, R., Rodriguez, J. M., & Xaus, J. (2006). Dietary deprivation of fermented foods causes a fall in innate immune response. Lactic acid bacteria can counteract the immunological effect of this deprivation. *Journal of Dairy Research, 73*(4), 492–498.

Ong, L., Henriksson, A., & Shah, N. P. (2007). Angiotensin converting enzyme-inhibitory activity in Cheddar cheeses made with the addition of probiotic *Lactobacillus casei* sp. *Le Lait, 87*(2), 149–165.

Orla O'Sullivan, P. D. C. (2017). Microbiota of raw milk and raw milk cheeses. In P. L. H. McSweeney, P. F. Fox, P. D. Cotter, & D. W. W. Everett (Eds.), *Cheese: Chemistry, physics and microbiology* (pp. 301–316). London: Elsevier Academic Press.

Pala, V., Sieri, S., Berrino, F., Vineis, P., Sacerdote, C., Palli, D., Masala, G., Panico, S., Mattiello, A., Tumino, R., Giurdanella, M. C., Agnoli, C., Grioni, S., & Krogh, V. (2011). Yogurt consumption and risk of colorectal cancer in the Italian European prospective investigation into cancer and nutrition cohort. *International Journal of Cancer, 129*(11), 2712–2719.

Pelaez, C., & Requena, T. (2005). Exploiting the potential of bacteria in the cheese ecosystem. *International Dairy Journal, 15*(6–9), 831–844.

Perdigón, G., de Moreno de LeBlanc, A., Valdez, J., & Rachid, M. (2002). Role of yoghurt in the prevention of colon cancer. *European Journal of Clinical Nutrition, 56*(Suppl 3), S65–S68.

Prakash Tamang, J., & Kailasapathy, K. (2010). *Fermented foods and beverages of the world*. Boca Raton: CRC Press.

Quilez, J., & Diana, M. (2017). Gamma-aminobutyric acid-enriched fermented foods. In J. Frias, C. Martinez-Villaluenga, & E. Penas (Eds.), *Fermented foods in health and disease prevention* (pp. 85–103). London: Academic Press Ltd/Elsevier Science Ltd.

Rizkalla, S. W., Luo, J., Kabir, M., Chevalier, A., Pacher, N., & Slama, G. (2000). Chronic consumption of fresh but not heated yogurt improves breath-hydrogen status and short-chain fatty acid profiles: A controlled study in healthy men with or without lactose maldigestion. *American Journal of Clinical Nutrition, 72*(6), 1474–1479.

Rojas-Ronquillo, R., Cruz-Guerrero, A., Flores-Najera, A., Rodriguez-Serrano, G., Gomez-Ruiz, L., Reyes-Grajeda, J. P., Jimenez-Guzman, J., & Garcia-Garibay, M. (2012). Antithrombotic and angiotensin-converting enzyme inhibitory properties of peptides released from bovine casein by *Lactobacillus casei* Shirota. *International Dairy Journal, 26*(2), 147–154.

Ross, R. P., Fitzgerald, G., Collins, K., & Stanton, C. (2002). Cheese delivering biocultures— Probiotic cheese. *Australian Journal of Dairy Technology, 57*(2), 71–78.

Ryan, P. M., Ross, R. P., Fitzgerald, G. F., Caplice, N. M., & Stanton, C. (2015). Sugar-coated: Exopolysaccharide producing lactic acid bacteria for food and human health applications. *Food & Function, 6*(3), 679–693.

Saito, T., Nakamura, T., Kitazawa, H., Kawai, Y., & Itoh, T. (2000). Isolation and structural analysis of antihypertensive peptides that exist naturally in Gouda cheese. *Journal of Dairy Science, 83*(7), 1434–1440.

Salazar, N., Gueimonde, M., de los Reyes-Gavilan, C. G., & Ruas-Madiedo, P. (2016). Exopolysaccharides produced by lactic acid bacteria and bifidobacteria as fermentable substrates by the intestinal microbiota. *Critical Reviews in Food Science and Nutrition, 56*(9), 1440–1453.

Santiago, S., Sayon-Orea, C., Babio, N., Ruiz-Canela, M., Marti, A., Corella, D., Estruch, R., Fito, M., Aros, F., Ros, E., Gomez-Garcia, E., Fiol, M., Lapetra, J., Serra-Majem, L., Becerra-Tomas, N., Salas-Salvado, J., Pinto, X., Schroder, H., & Martinez, J. A. (2016). Yogurt consumption and abdominal obesity reversion in the PREDIMED study. *Nutrition, Metabolism, and Cardiovascular Diseases, 26*(6), 468–475.

Saubade, F., Hemery, Y. M., Guyot, J.-P., & Humblot, C. (2017). Lactic acid fermentation as a tool for increasing the folate content of foods. *Critical Reviews in Food Science and Nutrition, 57*(18), 3894–3910.

Savaiano, D. A. (2014). Lactose digestion from yogurt: Mechanism and relevance. *The American Journal of Clinical Nutrition, 99*(5 Suppl), 1251S–1255S.

Sayon-Orea, C., Martinez-Gonzalez, M. A., Ruiz-Canela, M., & Bes-Rastrollo, M. (2017). Associations between yogurt consumption and weight gain and risk of obesity and metabolic syndrome: A systematic review. *Advances in Nutrition, 8*(1), 146S–154S.

Schleifer, K. H., Kraus, J., Dvorak, C., Kilpperbalz, R., Collins, M. D., & Fischer, W. (1985). Transfer of *Streptococcus lactis* and related streptococci to the genus *Lactococcus* gen. nov. *Systematic and Applied Microbiology, 6*(2), 183–195.

Schmidt, T. S. B., Raes, J., & Bork, P. (2018). The human gut microbiome: From association to modulation. *Cell, 172*(6), 1198–1215.

Seppo, L., Jauhiainen, T., Poussa, T., & Korpela, R. (2003). A fermented milk high in bioactive peptides has a blood pressure-lowering effect in hypertensive subjects. *American Journal of Clinical Nutrition, 77*(2), 326–330.

Settanni, L., & Moschetti, G. (2010). Non-starter lactic acid bacteria used to improve cheese quality and provide health benefits. *Food Microbiology, 27*(6), 691–697.

Silanikove, N., Leitner, G., & Merin, U. (2015). The interrelationships between lactose intolerance and the modern dairy industry: Global perspectives in evolutional and historical backgrounds. *Nutrients, 7*(9), 7312–7331.

Sipola, M., Finckenberg, P., Korpela, R., Vapaatalo, H., & Nurminen, M. L. (2002). Effect of long-term intake of milk products on blood pressure in hypertensive rats. *Journal of Dairy Research, 69*(1), 103–111.

Smid, E. J., Erkus, O., Spus, M., Wolkers-Rooijackers, J. C., Alexeeva, S., & Kleerebezem, M. (2014). Functional implications of the microbial community structure of undefined mesophilic starter cultures. *Microbial Cell Factories, 13*(Suppl 1), S2.

Sommer, F., & Backhed, F. (2016). Know your neighbor: Microbiota and host epithelial cells interact locally to control intestinal function and physiology. *BioEssays, 38*(5), 455–464.

Sonestedt, E., Wirfalt, E., Wallstrom, P., Gullberg, B., Orho-Melander, M., & Hedblad, B. (2011). Dairy products and its association with incidence of cardiovascular disease: The Malmo diet and cancer cohort. *European Journal of Epidemiology, 26*(8), 609–618.

Stanton, C., Ross, R. P., Fitzgerald, G. F., & Van Sinderen, D. (2005). Fermented functional foods based on probiotics and their biogenic metabolites. *Current Opinion in Biotechnology, 16*(2), 198–203.

St-Onge, M. P., Farnworth, E. R., & Jones, P. J. (2000). Consumption of fermented and nonfermented dairy products: Effects on cholesterol concentrations and metabolism. *The American Journal of Clinical Nutrition, 71*(3), 674–681.

Thierry, A., Deutsch, S. M., Falentin, H., Dalmasso, M., Cousin, F. J., & Jan, G. (2011). New insights into physiology and metabolism of *Propionibacterium freudenreichii. International Journal of Food Microbiology, 149*(1), 19–27.

Tripathi, M. K., & Giri, S. K. (2014). Probiotic functional foods: Survival of probiotics during processing and storage. *Journal of Functional Foods, 9*, 225–241.

Tunick, M. H., & Van Hekken, D. L. (2015). Dairy products and health: Recent insights. *Journal of Agricultural and Food Chemistry, 63*(43), 9381–9388.

Unno, T., Choi, J. H., Hur, H. G., Sadowsky, M. J., Ahn, Y. T., Huh, C. S., Kim, G. B., & Cha, C. J. (2015). Changes in human gut microbiota influenced by probiotic fermented milk ingestion. *Journal of Dairy Science, 98*(6), 3568–3576.

Urista, C. M., Fernandez, R. A., Rodriguez, F. R., Cuenca, A. A., & Jurado, A. T. (2011). Review: Production and functionality of active peptides from milk. *Food Science and Technology International, 17*(4), 293–317.

USDA. (2010). *Dietary guidelines for Americans.* Washington, DC: U.S. Government Printing Office.

Ventura, M., O'Flaherty, S., Claesson, M. J., Turroni, F., Klaenhammer, T. R., van Sinderen, D., & O'Toole, P. W. (2009). Genome-scale analyses of health-promoting bacteria: Probiogenomics. *Nature Reviews. Microbiology, 7*(1), 61–71.

Van der Meer, R., Bovee-Oudenhoven, I. M. J. (1998). Dietary modulation of intestinal bacterial infections. International Dairy Journal, 8(5-6), 481–486.

Wang, H., Livingston, K. A., Fox, C. S., Meigs, J. B., & Jacques, P. F. (2013). Yogurt consumption is associated with better diet quality and metabolic profile in American men and women. *Nutrition Research, 33*(1), 18–26.

Watanabe, M., Maemura, K., Kanbara, K., Tamayama, T., & Hayasaki, H. (2002). GABA and GABA receptors in the central nervous system and other organs. *International Review of Cytology, 213*, 1–47.

Weimer, B. (2007). *Improving the flavour of cheese.* Boca Raton: Woodhead Publishing/CRC Press.

Wouters, J. T. M., Ayad, E. H. E., Hugenholtz, J., & Smit, G. (2002). Microbes from raw milk for fermented dairy products. *International Dairy Journal, 12*(2–3), 91–109.

Wu, Q. L., & Shah, N. P. (2017). High gamma-aminobutyric acid production from lactic acid bacteria: Emphasis on *Lactobacillus brevis* as a functional dairy starter. *Critical Reviews in Food Science and Nutrition, 57*(17), 3661–3672.

Zhang, C., Derrien, M., Levenez, F., Brazeilles, R., Ballal, S. A., Kim, J., Degivry, M. C., Quere, G., Garault, P., van Hylckama Vlieg, J. E., Garrett, W. S., Dore, J., & Veiga, P. (2016). Ecological robustness of the gut microbiota in response to ingestion of transient food-borne microbes. *The ISME Journal, 10*(9), 2235–2245.

Zheng, H., Yde, C. C., Clausen, M. R., Kristensen, M., Lorenzen, J., Astrup, A., & Bertram, H. C. (2015). Metabolomics investigation to shed light on cheese as a possible piece in the French paradox puzzle. *Journal of Agricultural and Food Chemistry, 63*(10), 2830–2839.

Zhong, L., Zhang, X., & Covasa, M. (2014). Emerging roles of lactic acid bacteria in protection against colorectal cancer. *World Journal of Gastroenterology, 20*(24), 7878–7886.

Zhu, Y., Wang, H., Hollis, J. H., & Jacques, P. F. (2015). The associations between yogurt consumption, diet quality, and metabolic profiles in children in the USA. *European Journal of Nutrition, 54*(4), 543–550.

Meat and Meat Products

Wim Geeraerts, Despoina Angeliki Stavropoulou, Luc De Vuyst, and Frédéric Leroy

Abstract Meat is an important foodstuff, both from a nutritional and economic standpoint, available under a wide variety of raw and processed variants, including cooked, dry-cured, fermented, and smoked products. This chapter outlines the microbial diversity of meat and different meat products. Considerable microbial heterogeneity is found when comparing between meat types and their derived products, which is largely to be ascribed to variability on the level of the substrates, ingredients, and recipes, the processing conditions, and the storage methods. Upon consumption, the microorganisms that are present within the meat matrix enter the human gastrointestinal system and potentially interact with the gut microbiota. Whether they thus play a role in health and disease still needs to be established.

Keywords Meat microbiome · Raw meat microbiome · Red meat microbiome · Poultry microbiome · Meat fermented products

Introduction

Meat and meat products have been valuable components of the human omnivore diet from an evolutionary perspective and remain so today (Leroy and Praet 2015; Smil 2013). They contain abundant amounts of high-quality protein and several micronutrients of interest, such as iron, zinc and vitamin B12, that are often present in lower concentrations or that do not always display the same level of bioavailability or nutritional quality in non-meat sources (De Smet and Vossen 2016). When economically affordable and societally relevant, meat is often central to the composition of meals and much appreciated for its role in culinary heritage and, especially,

---

W. Geeraerts · D. A. Stavropoulou · L. De Vuyst · F. Leroy (⊠)
Research Group of Industrial Microbiology and Food Biotechnology, Faculty of Sciences and Bioengineering Sciences, Vrije Universiteit Brussel, Brussels, Belgium
e-mail: frederic.leroy@vub.be

© Springer Nature Switzerland AG 2019
M. A. Azcarate-Peril et al. (eds.), *How Fermented Foods Feed a Healthy Gut Microbiota*, https://doi.org/10.1007/978-3-030-28737-5_3

its wide range of important cultural connotations (Leroy and Praet 2015). Moreover, meat products frequently display an elevated level of convenience, for instance as ready-to-eat foods, which makes them very popular in Western diets (Leroy and Degreef 2015). Beef, pork, and poultry are the most consumed meat types worldwide, although in some cultures relatively important amounts originating from other animals may also be eaten, such as meat from horse, sheep, goat, rabbit, wild animals (game meat and bushmeat), or dog. Although meat consumption in Western countries seems to somewhat level off or even decline, the anticipation of a further rise in global meat consumption is leading to major environmental and economic challenges (Gerber et al. 2015; Vranken et al. 2014). Additionally, despite the popularity of meat and its desirable status, controversies are gaining ground. The latter are related to the ethical aspects of its production and slaughter (Leroy and Praet 2017), its ecological impact and sustainability (Smil 2013), as well as some health concerns (Carr et al. 2016; Wang et al. 2015). As a result, the demand for meat alternatives is on the rise, including a search for high-protein foods of a vegetarian or vegan character (Leitzmann 2014), as well as the consumption of insects (Caparros Megido et al. 2014; Verbeke 2015). Also, new technologies such as the development of *in vitro* meat production systems are being investigated and developed (Sharma et al. 2015). Be that as it may, it needs to be established still if such trends are to become truly significant, whereas meat and meat products are expected to remain key elements in human diets, contributing substantially to the overall food intake. As such, they likely serve as non-negligible vehicles of microorganisms that enter the human digestive system upon ingestion. Yet, only little is known about the potential role of meat (products) as suppliers of microorganisms and how this may impact the human gut (David et al. 2014). Potentially, the meat matrix itself may even exert protective effects towards some of the beneficial microorganisms it contains (Klingberg and Budde 2006).

In this chapter, an overview is given of the microbial communities that are associated with meat and meat products and how those are affected by the processing factors and storage conditions. In addition, based on the general availability of ingestible microorganisms, potential implications for the human gut microbiome are highlighted, in particular with respect to those elements of the microbiome that have a potential impact on the health of the host. In all cases, focus will be on the dominant, non-pathogenic microbiota rather than on the potential presence of foodborne pathogens.

Meat-Associated Microorganisms

Meat as a Microbial Ecosystem

Although microorganisms are often rather ubiquitous, restricted consortia are generated in specific ecological niches by selective environmental pressure (Plé et al. 2015; Pothakos et al. 2015). Meat is to be considered as a favourable matrix for microbial growth, as it has an elevated water activity and contains an abundancy of nutrients, although it is rather low in carbohydrates and heterogeneous with respect to its biochemical composition and pH (Toldrá 2017). The microbiota present on raw meat can originate from a variety of sources, of which the initial shaping typically traces back to the slaughterhouses. At this stage, a primary contamination of the animal carcasses takes place, which is largely of intestinal or skin origin (Belluco et al. 2015; Chaillou et al. 2015; De Filippis et al. 2013; Hue et al. 2011; Mc Nulty et al. 2016). This contamination is further affected by overall hygiene practices, cleaning techniques used for the equipment, and automatization of the process (Borch and Arinder 2002; Milios et al. 2014; Yalcin et al. 2001). Manipulation of the carcasses, such as skinning, removal of feathers, deboning, and chopping, will result in additional alterations of the microbiota (Arnold 2007; Borch and Arinder 2002; Kang et al. 2001). In a next stage, processing and storage conditions have a main impact on both the total load and composition of the meat microbiome, usually narrowing down the biodiversity and selecting for specific taxonomic groups (Borilova et al. 2016; Chaillou et al. 2015; Fougy et al. 2016). This is of major importance as the quantity of microorganisms as well as the composition of the microbial communities, even on the subdominant level, will determine shelf-life and spoilage due to metabolic activity (Fougy et al. 2016; Vasilopoulos et al. 2015). The major factors that are influencing the microbiota and, thus, the risk on spoilage due to greening, off-flavour formation, gas production, or textural defects, include changes in water activity, temperature, pH, carbohydrate content, atmospheric conditions, and the use of additives (Doulgeraki et al. 2012; Fougy et al. 2016; Koutsoumanis et al. 2006). For instance, the use of non-sterilised marinades and the addition of herbs or vegetable pieces may bring in new microorganisms and lead to alterations of the microbiota that are already present (Björkroth et al. 2005; Säde et al. 2016). In contrast, process actions such as cooking, irradiation, and the use of preservatives or other inhibitory compounds will normally lead to a reduction of the microbial loads (Haugaard et al. 2014; Milios et al. 2014; Vasilopoulos et al. 2015). Cross-contamination within slaughterhouses and meat-processing plants is also not to be underestimated as a contributor to the shaping of the final microbial communities of meat and meat products that will eventually reach the consumption stage. The level of cross-contamination is enhanced by the formation of biofilms on the processing equipment, despite the use of cleaning methods (Brightwell et al. 2006; Srey et al. 2013; Vasilopoulos et al. 2015). Worryingly, many potential pathogens have the potential to form such biofilms, as has been shown for half of the *Salmonella* isolates retrieved from a poultry farm (Marin et al. 2009). Multiple-species biofilms have also been found in meat-processing plants, including communities of *Acinetobacter calcoaceticus* and *Escherichia coli* O157:H7 (Srey et al. 2013).

As will be exemplified below, it is important to highlight that temperature is a major effector in the establishment of microbial communities during the storage of meat and meat products. Hereby, the animal-derived microbiota from the slaughterhouse becomes overtaken by a core microbiota of a more environmental and cold-adapted nature, for instance originating from water reservoirs (Chaillou et al. 2015). In general, temperature control ranges from chilling (usually to be set within a range of 4–7 °C) to freezing conditions. Although storage at chilling temperatures or just below the freezing point will prolong the shelf-life of meat products, a transition in dominance will take place from mesophilic to psychrophilic and psychrotolerant microorganisms that are better adapted to the low temperatures of the cold chain (Doulgeraki et al. 2012; Ercolini et al. 2009b; Smolander et al. 2004). At freezing temperatures, bacterial growth will be halted, although moulds may still be of concern and some bacterial enzymatic activity may remain (Chipley and May 1968; D'Amico et al. 2006; Lowry and Gill 1984).

Application of a cold chain is usually not to be considered independently from the use of packaging. The latter can be done under a variety of different atmospheres, ranging from vacuum, over air, to modified-atmosphere packaging (MAP). In general, vacuum packaging and MAP cause a shift from an aerobic microbial consortium to a more fermentative one, mostly centred around lactic acid bacteria (Doulgeraki et al. 2010; Ercolini et al. 2006; La Storia et al. 2012; Schirmer et al. 2009). The use of vacuum or MAP packaging thus reduces or slows down the production of off-flavour compounds and stabilizes the sensory characteristics (La Storia et al. 2012).

Due to differences in raw materials, product formulation, processing, cleaning methods, and in-house storage conditions, relatively comparable products that originate from different production facilities may still generate rather dissimilar microbial ecologies (Geeraerts et al. 2017). Such facility-specific microbial consortia may also evolve over time (Schirmer et al. 2009). Taken together, this heterogeneity helps to explain batch variation of meat microbiota and the potential metabolic phenomena to which they are related. Important to note is that only a fraction of the initial microbiota will lead to spoilage (Pothakos et al. 2015). In this context, it has been reported that spoiled samples of ground beef, ground veal, poultry sausage, and diced bacon display lower values of operational taxonomic units (OTUs) than their fresh equivalents (Chaillou et al. 2015).

Raw Meats

General aspects Raw meats can be found in different structural forms, following a spectrum ranging from intact pieces to mince. For intact meat, bacterial growth occurs primarily on the surface, whereby the internal parts are sterile or have only low microbial loads. Grounded meat generally has a shorter shelf-life due to higher initial contamination levels caused by manipulation, the presence of meat juice, and a higher exposed surface (Cerveny et al. 2010). In butcher shops, basic packaging of raw meat is usually through air-exposed wrapping in paper, whereas more advanced methods of vacuum packaging or MAP are applied in the retailing via supermarkets (Cerveny et al. 2010). Packaged raw meats are then stored at reduced temperatures to prolong shelf-life (Jones 2004).

Depending on the type of raw meat, the microbial genera that can be encountered are mostly *Acinetobacter, Brochothrix, Flavobacterium, Micrococcus, Moraxella,*

Pseudomonas, Psychrobacter, Shewanella, and *Staphylococcus,* as well as several genera of lactic acid bacteria and *Enterobacterales* (Benson et al. 2014; Doulgeraki et al. 2011, 2012; Ercolini et al. 2009a, b; Fougy et al. 2016; Pennacchia et al. 2011; Pothakos et al. 2015). An overview of typical microorganisms associated with raw meat and poultry is given in Table 1. Below, more specificities are given with respect to the difference between red meats and poultry and the effects of formulation, packaging, and temperature.

Red Meats

Red meats, usually pork and beef cuts, are commonly stored under vacuum or high-oxygen MAP. High concentrations of oxygen simulate the formation of oxymyoglobin from the muscle myoglobin, leading to a bright red colour of the meat that is preferred by consumers (Arvanitoyannis and Stratakos 2012; Pothakos et al. 2015; Troy and Kery 2010). High-oxygen MAP red meats are colonized by several genera of lactic acid bacteria, such as *Carnobacterium, Lactobacillus, Lactococcus,* and *Leuconostoc,* that can adapt to the oxidative stress (Pothakos et al. 2015). To a lesser degree, enterobacterial species can be found on MAP products, such as *Hafnia alvei, Proteus vulgaris, Serratia liquefaciens,* and *Serratia proteamaculans* (Doulgeraki et al. 2011, 2012). When comparing pork and beef that are either packed anaerobically or under high-oxygen MAP, it seems that *Hafnia* species tend to dominate under anaerobic conditions whereas *Serratia* species are more present under high-oxygen packaging (Säde et al. 2013). With the advent of next-generation, culture-independent characterization methods for bacterial community analysis, additional species that may have been neglected previously are now being identified on meat (Stoops et al. 2015). Sometimes they may even be part of the core microbiota. As an example, analysis of MAP pork led to the detection of *Photobacterium phosphoreum,* a species that is usually associated with raw fish but nevertheless seems to be a regular member of meat microbiota too (Nieminen et al. 2016). Using next-generation sequencing, beef has also been shown to contain relatively high numbers of *Corynebacterium,* a skin-related bacterium, whereas veal is higher in *Prevotella,* suggesting that veal is more contaminated by rumen bacteria (Chaillou et al. 2015). Currently, specific attention goes to psychrotolerant and psychrophilic bacteria that emerge upon storage in the cold chain of high-oxygen MAP meat cuts. This is the case for *Leuconostoc gelidum* subsp. *gasicomitatum,* which is not only cold-adapted but also has the ability for respiration in the presence of exogenous haem and high concentrations of oxygen (Jääskeläinen et al. 2013; Johansson et al. 2011). Besides oxygen, MAP also may contain carbon dioxide, which leads to reduced microbial growth when compared to air storage and to shifts in microbial diversity (Ercolini et al. 2006; La Storia et al. 2012; Lorenzo and Gómez 2012). Yet, when present in too high concentrations, infiltrating carbon dioxide forms carbonic acid and causes texture deterioration (Sivertsvik et al. 2004). The combination of packaging and low temperatures thus improves the shelf-life of red meats. For

62 W. Geeraerts et al.

Table 1 Non-exhaustive overview of the bacterial species diversity of raw red meats and raw poultry products as well as some raw or cooked meat products based on studies between 2007 and 2017

Product type	Bacterial species diversity	Reference
Sliced cooked ham, chicken, and turkey (MAP)	*C. divergens, Lb. curvatus, Lb. sakei/fuchuensis,* and *Leuc. carnosum*	Audenaert et al. (2010)
Pork sausage	*Ac. junii, Acinetobacter* sp., *Bifidobacterium* sp., *Bu. brennerae, C. divergens, Ci. braakii, Lb. gasseri, Lb. graminis, Lb. sakei* subsp. *sakei, Lc lactis* subsp. *cremoris, Leuc. citreum, Ps. lini, Ps. psychrophila, Serratia* sp., *Streptococcus* sp., *Str. suis, Str. minor, Str. thermophilus, W. confusa,* and *Y. mollaretii* (metagenomic identification)	Benson et al. (2014)
Beef (vacuum-packed)	*Brochothrix* sp., *C. maltaromaticum, Cl. algidicarnis, Cl. putrefaciens, Enterobacterales* species, *Lactobacillus* sp., and *Pseudomonas* sp.	Brightwell et al. (2009)
Goat (MAP)	*Ba. cereus, B. thermosphacta, Eb. cloacae, Ec. durans, Ec. faecium, Ec. hirae, Ec. lactis, Er. persicina, H. alvei, Lb. sakei, Lc. curvatus, Lc. lactis, M. caseolyticus, Pa. agglomerans, Ps. fragi, Sr. liquefaciens, Sr. proteamaculans, St. equorum, St. epidermidis, St. saprophyticus,* and *St. xylosus*	Carrizosa et al. (2017)
Cooked ham (vacuum-packed)	*Lb. sakei, Leuc. carnosum, Leuc. mesenteroides,* and *W. viridescens*	Comi and Iacumin (2012)
Cooked bacon (vacuum-packed)	*Leuc. mesenteroides*	Comi et al. (2016)
Minced beef (air-stored or MAP)	*Lb. sakei* and *Leuconostoc* sp.	Doulgeraki et al. (2010)
Minced beef (air-stored or MAP)	*Ci. freudii, H. alvei, Pr. vulgaris, Serratia* sp., *Sr. liquefaciens,* and *Sr. proteomaculans*	Doulgeraki et al. (2011)
Hot smoked, non-fermented dry sausage (vacuum-packed)	*Ec. durans, Ec. Faeca:lis, Lb. fructivorans/curvatus, Lb. mucosae, Lb. plantarum, Lb. sakei, Lc. lactis, Lc. garviae, Leuc. mesenteroides, Leuc. citreum, P. pentosaceus, Str. salivarius, W. hellenica,* and *W. viridescens* (identification by MALDI-TOF)	Dušková et al. (2015)
Cooked ham (MAP)	*Lb. curvatus, Lb. sakei, Leuc. carnosum, Leuc. gelidum* subsp. *gelidum, Leuc. mesenteroides, Leuc. pseudomesenteroides,* and *W. viridescens* (identification by MALDI-TOF)	Dušková et al. (2016)
Beef (vacuum-packed)	*Ac. baumannii, B. thermosphacta, Bu. agrestis, Bu. noackiae, Carnobacterium* sp., *C. divergens, C. maltaromaticum, H. alvei, Halomonas* sp., *Lb. sakei, Leuc. gelidum, Pseudomonas* sp., *Ps. fragi, Ps. putida, Ra. aquatilis, Serratia* sp., *Sr. proteamaculans, Ste. maltophilia,* and *Str. parauberis*	Ercolini et al. (2009a)

(continued)

Table 1 (continued)

Product type	Bacterial species diversity	Reference
Beef (vacuum-packed, with and without nisin)	*B. thermosphacta, Carnobacterium* sp., *C. divergens, Lactobacillus* sp., *Leuc. mesenteroides, Pseudomonas* sp., *Ph. kishitaniiclade, Rhanella* sp., *Sr. grimesii, Sr. proteomaculans, Staphylococcus* sp., *St. xylosus*, and *Weissella* sp.	Ercolini et al. (2009a)
Beef chop (different types of packaging)	*Acinetobacter* sp., *Bradyrhizobium* sp., *Brochothrix* sp., *Carnobacterium* sp., *C. divergens, Lactobacillus* sp., *Lactococcus* sp., *Lc. piscium, Limnobacter* sp., *Lm. thiooxidans, Pseudomonas* sp., *Ralstonia* sp., *Ru. cellulosiytica*, and *Streptococcus* sp.	Ercolini et al. (2011)
Raw pork sausage (MAP or vacuum-packed)	*B. thermosphacta, C. divergens, C. maltaromaticum, Enterobacterales, Enterococcaceae, Lactobacillaceae, Lb. sakei, Lc. piscium, Leuconostocaceae*, and *S. proteamaculans* (metagenomic identification)	Fougy et al. (2016)
Sliced cooked pork products (MAP)	*B. thermosphacta, C. divergens, C. funditum, C. maltaromaticum, Lb. curvatus/graminis, Lb. sakei, Leuc. carnosum, Leuc. gelidum* subsp. *gasicomitatum*, and *Leuc. gelidum* subsp. *gelidum*	Geeraerts et al. (2017)
Chicken meat and chicken liver	*Ec. saigonensis*	Harada et al. (2016)
Skinless chicken breast (MAP)	*B. thermosphacta, Carnobacterium* sp., *E. coli, H. alvei, Janthinobacterium* sp., *Lactobacillus* sp., *Microbacterium* sp., *Pseudochrobactrum* sp., *Pseudomonas* sp., *Rhodococcus* sp., *Rothia nasimurium, Serratia* sp., *Staphylococcus* sp., *Stenotrophomonas* sp., and *Yersinia* sp. (identification by MALDI-TOF)	Höll et al. (2016)
Cooked sausage (vacuum-packed)	*Lactobacillus* sp., *Leuconostoc* sp., and *Streptococcus* sp. (metagenomic identification)	Hultman et al. (2015)
Spontaneously acidified sausage (air-stored)	*Lb. algidus, Lb. paralimentarus/mindensis/crustorum, Lb. sakei, Lactococcus* sp., *Leuc. mesenteriodes, Leuc. carnosum, Pseudomonas* sp., *St. carnosus*, and *St. saprophyticus*	Janssens et al. (2012)
Pork loin (vacuum-packed)	*Carnobacterium* sp., *C. divergens, Lb. curvatus, Lb. sakei, Lactococcus* sp., *Lc. lactis* subsp. *lactis, Lc. piscium, Leuc. mesenteroides, W. cibaria*, and *W. viridescens*	Jiang et al. (2010)
Beef steak (different types of packaging)	*B. thermosphacta, Carnobacterium* sp., *C. maltaromaticum, Pseudomonas* sp., *Ps. fragi, Ps. lundensis, R. aquatilis, Sr. proteamaculans*, and *St. saprophyticus*	La Storia et al. (2012)
Rullepølse (MAP)	*C. divergens, Lb. sakei, Leuc. carnosum*, and *Leuc. mesenteroides*	Laursen et al. (2009)
Beef carpaccio (MAP or vacuum-packed)	*B. thermosphacta, C. divergens, Lb. curvatus, Lb. fuchuensis, Lb. sakei, Leuc. carnosum, Leuc. gelidum* subsp. *gasicomitatum, Leuc. gelidum* subsp. gelidum, *Leuc. mesenteroides, St. warneri*, and *W. viridescens*	Lucquin et al. (2012)

(continued)

Table 1 (continued)

Product type	Bacterial species diversity	Reference
Goose foie gras	*Carnobacterium* sp., *Lb. buchneri/parabuchneri*, *Lb. coryniformis*, *Lb. casei/paracasei*, *Lb. curvatus*, *Lb. plantarum/pentosus*, *Lb. sakei*, and *W. viridescens*	Matamoros et al. (2010)
Spoiled marinated chicken fillet (packed with oxygen-permeable film)	*Ps. fragi*, *Ps. fluorescens*, and *Ps. lundensis*	Morales et al. (2016)
Beef and pork mixture (MAP)	*B. thermosphacta*, *C. divergens*, *C. maltaromaticum*, *Ec. raffinosus*, *Lb. algidus*, *Lb. sakei*, *Lactococcus* sp., *Leuc. carnosum*, *Leuc. gasicomitatum*, *Leuc. gelidum*, *Leuc. mesenteroides*, *Str. parauberis*, and *Weissella* sp.	Nieminen et al. (2011)
(Marinated) broiler fillet strip (MAP)	*Carnobacterium* sp., *Lactobacillus* sp., *Lactococcus* sp., *Leuconostoc* sp., and *Vagococcus* sp. (metagenomic identification)	Nieminen et al. (2012)
Pork (MAP or vacuum-packed)	*Arthrobacter* sp., *Brochothrix* sp., *Carnobacterium* sp., *Lactobacillus* sp., *Lactococcus* sp., *Leuconostoc* sp., *Photobacterium* sp., *Propionibacterium* sp., *Psychrobacter* sp., *Staphylococcus* sp., *Vagococcus* sp., and *Weissella* sp. (metagenomic identification)	Nieminen et al. (2016)
Raw chicken parts	*Pseudomonas* sp., *Ps. fragi*, *Ps. meridiana*, and *Ps. psychrophila* (metagenomic identification)	Oakley et al. (2013)
Cooked chicken (vacuum-packed)	*Bacillus* sp., *Enterobacter* sp., *Lactococcus* sp., *Ra. aquitilis*, *Sr. proteamaculans*, and *W. viridescens*	Patterson et al. (2010)
Beef steak (air-stored or vacuum-packed)	*Ae. salmonicida*, *B. thermosphacta*, *C. divergens*, *H. alvei*, *Lb. algidus*, *Lb. sakei*, *Lc. piscium*, *Leuc. mesenteroides*, *Pantoea* sp., *Pa. agglomerans*, *Pseudomonas* sp., *Rahnella* sp., *Ra. aquitilis*, *Sr. grimessi*, *Sr. marcescens*, *Sr. proteamaculans*, *Staphylococcus* sp., and *St. pasteuri*	Pennacchia et al. (2011)
Ham (vacuum-packed)	*Brochothrix* sp., *Citrobacter* sp., *Corynebacterium* sp., *Enterobacter* sp., *Enterococcus* sp., *Escherichia* sp., *Flavobacterium* sp., *Lactobacillus* sp., *Kocuria* sp., *Micrococcus* sp., *Proteus* sp., *Pseudomonas* sp., *Serratia* sp., *Streptomyces* sp., *Vagococcus* sp., and *Wautersiella* sp. (metagenomic identification)	Piotrowska-Cyplik et al. (2017)
Fresh raw meat products	*C. divergens*, *Ec. raffinosus Lb. algidus*, *Lb. fuchuensis*, *Lb. sakei*, *Lb. oligofermentans*, *Lc. piscium*, *Leuc. gelidum* subsp. *gasicomitatum*, and *Leuc. gelidum* subsp. *gelidum*	Pothakos et al. (2014)
Cooked meat products	*Lb. sakei*, *Lc. piscium*, *Leuc. carnosum*, *Leuc. gelidum* subsp. *gasicomitatum*, and *Leuc. gelidum* subsp. *gelidum*	Pothakos et al. (2014)
Chicken leg (MAP)	*B. thermosphacta* and *Pseudomonas* sp.	Rouger et al. (2017)

(continued)

Meat and Meat Products

Table 1 (continued)

Product type	Bacterial species diversity	Reference
Chicken leg (MAP)	*Ac. cyllenbergii, Ac. lwoffii, An. tetradius, B. thermosphacta, Bd. aquatica, C. divergens, C. maltaromaticum, C. pleistocenium, Fl. antarcticum, Ja. lividum, Kl. pneumoniae, Sh. baltica, Sh. profunda, Sh. xiamenensis, Ps. cedrina, Ps. extremaustralis, Ps. fragi, Psy. urativorans,* and *V. fluvialis*	Rouger (2017)
Minced, intact, or marinated beef and pork (MAP)	*Buttiauxella* sp., *Hafnia* sp., *Rahnella* sp., *Serratia* sp., and *Yersinia* sp.	Säde et al. (2013)
(Marinated) broiler and turkey (MAP)	*Hafnia* sp., *Rahnella* sp., *Serratia* sp., and *Yersinia* sp.	Säde et al. (2013)
Marinated pork steak (vacuum-packed)	*B. thermosphacta, Carnobacterium* sp., *C. divergens, C. maltaromaticum, Ec. faecalis, Enterobacter* sp., *Flavobacterium* sp., *Halomonas* sp., *Lb. algidus, Lb. sakei/curvatus, Lactococcus* sp., *Lc. lactis, Leuconostoc* sp., *Leuc. carnosum, Leuc. mesenteroides, Pseudomonas* sp., *Sr. proteamaculans, St. hominis,* and *St. pasteuri/warneri*	Schirmer et al. (2009)
Raw pork and beef	*Acinetobacter* sp., *Brochothrix* sp., *Pseudomonas* sp., *Psychrobacter* sp., and *Streptococcus* sp. (metagenomic identification)	Stellato et al. (2016)
Minced beef (MAP)	*C. divergens, Lb. algidus, Lactococcus* sp., *Leuconostoc* sp., *Photobacterium* sp., *Pr. acnes, Pseudoxanthomonas* sp., *Pseudomonas* sp., *Ps. grimontii,* and *Psy. urativorans* (metagenomic identification)	Stoops et al. (2015)
Cooked pork ham (MAP)	*B. thermosphacta, C. divergens, Enterococcus* sp., *Ec. faecalis, Lb. sakei, Lc. lactis,* and *Leuc. carnosum*	Vasilopoulos et al. (2008)
Marinated beef steak (MAP)	*Lb. algidus, Lb. sakei, Leuc. gasicomitatum, Leuc. gelidum,* and *C. divergens*	Vihavainen and Björkroth (2007)
Chicken meat	*Ae. hydrophila, Ae. media, Ae. salmonicida, Ch. shigense, Ps. fluorescens, Ps. fragi,* and *Ps. putida*	Wang et al. (2017)

Bacteria were identified using standard molecular identification methods unless otherwise specified (identification by metagenomics or MALDI-TOF). The identified genera encompassed *Ac., Acinetobacter; Ae., Aeromonas; An., Anaerococcus; Ba., Bacillus; Bf., Bifidobacterium; B., Brochothrix; Bd., Budvicia; Bu., Buttiauxella; Ca., Campylobacter; C., Carnobacterium; Ch., Chryseobacterium; Ci., Citrobacter; Cl., Clostridium; Eb., Enterobacter; Ec., Enterococcus; Er., Erwinia; E., Escherichia; Fl., Flavobacterium; H., Hafnia; Ja., Janthinobacterium; Kl., Klebsiella; Lb., Lactobacillus; Lc., Lactococcus; Leuc., Leuconostoc; Lm., Limnobacter; L., Listeria; M., Macrococcus; Pa., Pantoea; P., Pediococcus; Ph., Photobacterium; Pb., Propionibacterium; Pr., Proteus; Ps., Pseudomonas; Psy., Psychrobacter; R., Rahnella; Ru., Rudaea; Sr., Serratia; Sh., Shewanella; St., Staphylococcus; Ste., Stenotrophomonas; Str., Streptococcus; V., Vagococcus; W., Weissella;* and *Y., Yersinia*

instance, the recommended storage life of vacuum-packed beef at -1.5 °C amounts up to 84 days (Bell 2001).

Other than pork and beef, some less-consumed red meats include sheep, goat, equine, camel, and game meat. Due to their limited consumption, these meat variants have been far less studied. Spanish goat meat stored under oxygen-containing MAP, for instance, has been shown to be dominated by members of the *Enterocbacterales*, with *H. alvei* and *Sr. proteamaculans* being the most important species (Carrizosa et al. 2017). For MAP equine meats, it has been demonstrated that several lactic acid bacteria, *Enterobacterales*, and *Pseudomonas* species are present, whereby *Lactobacillus sakei* acts as the most important species on vacuum-packed meat variants (Lorenzo and Gómez 2012). Information on game meat is limited, although some basic microbiological studies have been done on certain types of African game meat and on wild boar meat (Borilova et al. 2016; Hoffman and Dicks 2011).

Processed derivatives of raw red meats, such as fresh sausage, hamburgers, and mince, are stored in a similar manner as intact meat cuts, whereby the type of packaging influences the microbial diversity. Minced beef, for instance, has been shown to display a shift from *H. alvei* and *Pr. vulgaris* to *Citrobacter freundii* when stored under air instead of MAP (Doulgeraki et al. 2011). In the case of air-stored, freshly-cut beefsteaks, *Rahnella* species, *Pseudomonas* species., and *Lb. sakei* are the dominant species, whereas MAP variants are dominated by either *Pseudomonas* species and *Lb. sakei* when stored under 60% O_2 and 40% CO_2 or by *Rahnella* species and *Lb. sakei* when stored under 20% O_2, 40% CO_2, and 40% N_2 (Ercolini et al. 2006). Metagenetic analysis of raw pork sausages has indicated a core microbial community of *Brochothrix thermosphacta*, *Carnobacterium divergens*, *Carnobacterium maltaromaticum*, *Lb. sakei*, *Lactococcus piscium*, and *Sr. proteamaculans* (Fougy et al. 2016). In addition, a subdominant fraction of enterococci, leuconostocs, and *Enterobacterales* has been found, becoming more manifest when increasing the salt concentration from 1.5 to 2.0% and applying vacuum-packaging instead of MAP, which also correlates with reduced spoilage.

Poultry

The slaughtering process of poultry differs on some points from the one for mammal species, including feather removal, the use of several water baths, and mechanical treatments that are typical for small carcasses. These factors may lead to specificities within the bacterial contamination (Rouger 2017). As for red meats, poultry is distributed either as MAP products or under air (Arvanitoyannis and Stratakos 2012). However, a main difference relies in the fact that MAP poultry does not require oxygen in the headspace of the package, as in white meat types no red colour formation based on oxymyoglobin development is expected. Nevertheless, some producers start using high-oxygen MAP (up to 80%) to inhibit specific pathogens such as *Campylobacter jejuni* (Höll et al. 2016). MAP generally leads to a reduction of aerobic bacteria and a longer shelf-life of poultry (Balamatsia et al. 2007; Chouliara et al. 2007). When comparing three types of MAP (*i.e.*, 30% CO_2/70% N_2; 60% CO_2/40% N_2, and 90% CO_2/10% N_2) with air storage, it has been found that a level of 30% CO_2 exposure prolongs shelf-life by 4 days and 60–90% CO_2 extended it to 6 days (Patsias et al. 2006). On MAP poultry, the non-pathogenic microbial communities habitually consist of *B. thermosphacta* and a variety of pseudomonads, lactic acid bacteria, and *Enterobacterales* (Remenant et al. 2015). Within the group of lactic acid bacteria, the genera *Carnobacterium*, *Lactococcus*, *Lactobacillus*, and *Leuconostoc* are the most representative ones (Björkroth et al. 2005; Morales et al. 2016; Pavelková et al. 2014). Several enterococci, such as *Enterococcus saigonensis* and *Enterococcus viikkiensis* have also been found within the lactic acid bacteria communities (Harada et al. 2016; Rahkila et al. 2011). *Enterobacterales* found on MAP poultry often belong to the genera *Buttiauxella*, *Hafnia*, *Rahnella*, *Serratia*, and *Yersinia* (Säde et al. 2013; Smolander et al. 2004). *Hafnia* (*i.e.*, *H. alvei* and *Hafnia paralvei*) has been found as the most important genus on a variety of poultry samples, followed by *Serratia* (*Sr. liquefaciens* and *Serratia quinivorans*) (Säde et al. 2013). *Enterobacterales* are frequently found in relatively high concentrations at the end of the shelf-life, potentially causing spoilage even if they do not represent the most prevalent fraction, at levels of up to 10^7 colony-forming units (CFU) per gram of meat (Nieminen et al. 2012; Säde et al. 2013; Smolander et al. 2004). In a study by Höll et al. (2016), skinless chicken breasts stored at 4 °C under high-oxygen MAP samples were dominated by *B. thermosphacta* and *Carnobacterium* species, whereas the absence of oxygen in the package gave rise to species of *Carnobacterium*, *Serratia*, and *Yersinia*. At 10 °C, the presence or absence of oxygen favoured, respectively, *Pseudomonas* species and *Serratia* species or *H. alvei*. In the presence of air, microbial consortia have been described as being based on the presence of *Aeromonas*, *Chryseobacterium*, and *Pseudomonas* (Wang et al. 2017). Although all the above-mentioned bacteria are of interest because of their potential role in spoilage, analysis of raw poultry has mostly looked at the presence of pathogenic species, in particular *Campylobacter coli*, *Ca. jejuni*, *Listeria monocytogenes*, and *Salmonella enterica* serovar Typhimurium (Ahmed et al. 2017; Kudirkienė et al. 2011; Raeisi et al. 2016; Stella et al. 2017).

Microbial diversity in poultry can be considerably affected by the process of marination, which has become a quite common practice in the poultry industry due to its popularity with consumers (Lytou et al. 2017). When this process is implemented, meat is injected with a brine and mixed with a marinade (oil, water, spices, and organic acids) (Björkroth et al. 2005). The addition of a marinade can, for instance, lead to a shift from a dominance by *Carnobacterium* and *Lactococcus* species towards communities that are led by *Leuconostoc gelidum* subsp. *gelidum* and *Leuc. gelidum* subsp. *gasicomitatum* (Nieminen et al. 2012). This effect may partially be ascribed to the higher tolerance of *Leuconostoc* species towards organic acids (Leisner et al. 2007; Nieminen et al. 2012). In addition, marination may lead to a reduction of the fractions of *B. thermosphacta*, *Enterobacterales*, and *Pseudomonas* species, but not so for the lactic acid bacteria (Lytou et al. 2017).

Poultry is sometimes used as a mince in sausage and hamburger formulations (Patsias et al. 2006). This results in an increase of the overall microbial loads. For instance, an average total count of 7 log CFU/g has been found for Spanish chicken hamburgers and sausages, as compared to an average of 6 log CFU/g for chicken thighs and wings (Álvarez-Astorga et al. 2002). Despite these higher counts, some bacterial fractions may be less present. *Campylobacter* was less retrieved from minced Italian poultry compared to intact variants, possibly due to extra exposure to oxidative stress resulting from the mixing (Stella et al. 2017).

Meat Products

General aspects Several types of meat products are produced from raw meats, whereby pork is widely used as starting material next to, among others, beef and poultry. The production of these products is often done by curing, fermentation, drying, and/or smoking, based on their historical use as empirical preservation methods for raw meat (Leroy et al. 2013, 2015; Šimko 2005). Preservative action is based on a variety of effects due to the presence of antimicrobials (*e.g.*, lactic acid from the fermentation and phenolic compounds from the smoking) and a reduced water activity due to salting and/or drying. Cooking is also used to stabilize meat, at the same time generating specific sensory properties (Vasilopoulos et al. 2015). Cooked meat products, such as cooked hams, are nevertheless still at risk of spoilage due to their high remaining water activity, and they require additional chilling as compared to the dried-in meat product variants.

Besides salting, several other preserving compounds can be added to meat products to enhance their stability (Drosinos et al. 2006; Geeraerts et al. 2017; Vasilopoulos et al. 2015). Examples include the combination of potassium lactate and sodium diacetate to inhibit the growth of *L. monocytogenes* and the addition of nitrate and/or nitrite salts to prevent the outgrowth of *Clostridium botulinum* (Hospital et al. 2016; Vasilopoulos et al. 2015). Despite their effectiveness, many consumers perceive those preservatives as unauthentic and unhealthy, so that efforts are being made to replace them by alternatives that are perceived as more natural, such as the use of antimicrobial strategies for bioprotection (Leroy et al. 2006; Sánchez Mainar et al. 2016) or the introduction of specific herbal components (Ballester-Costa et al. 2017; Zhang et al. 2009). The latter, however, may also interfere with flavour or even act as an additional contamination route when used as spices or herbs (Säde et al. 2016).

Cooked meat products The production of cooked meat products generally proceeds along a series of processing steps. In a first phase, raw meats and, sometimes, fat pieces are tumbled and mixed with ingredients, followed by an injection with brine and compression into logs. Nowadays consumers demand cooked meat products that are low in salt and contain as few chemicals and preservatives as possible, which has the potential to affect the overall microbiology in negative ways (Geeraerts et al. 2017; Vasilopoulos et al. 2015). During a second phase, the products are cooked at least once and cooled down (Vasilopoulos et al. 2015). Thermal treatment will lead to a very strong reduction or even elimination of most of the original microbial communities, although some thermotolerant bacteria may survive. Also, an elevated risk of secondary contamination from the production facility remains, mainly due to slicing and packaging (Dušková et al. 2016; Vasilopoulos et al. 2010a). Most cooked meat products are stored under MAP or under vacuum, thus disfavouring the aerobic species and selecting for lactic acid bacteria and *B. thermosphacta*. Nevertheless, the shelf-life of sliced cooked meat products in the cold chain is rather limited, for instance to 6 weeks (Leroy et al. 2009). The most frequently retrieved lactic acid bacteria mainly include species belonging to the genera *Carnobacterium*, *Lactobacillus*, *Leuconostoc*, and *Weissella*, although considerable differences can be found between different production facilities (Audenaert et al. 2010; Geeraerts et al. 2017; Pothakos et al. 2014; Vasilopoulos et al. 2008). Several studies suggest that leuconostocs are among the most prevalent bacteria on cooked pork products, mainly the species *Leuconostoc carnosum*, *Leuconostoc mesenteroides*, and *Leuc. gelidum* (Dušková et al. 2016; Geeraerts et al. 2017; Samelis et al. 2000; Vasilopoulos et al. 2008). A market survey of Belgian cooked pork products has mostly retrieved *Leuc. carnosum* (Geeraerts et al. 2017), whereas Greek products have been reported to be dominated mostly by *Leuc. mesenteroides* (Samelis et al. 2000). However, such geographical links should not be taken too strictly. Carnobacteria also form a prevalent group, in particular the species *Cb. divergens* and *Cb. maltaromaticum*. Yet, carnobacteria are sometimes overlooked due to the use of acetate-containing growth media that are inhibitory to these bacteria (Davidson and Cronin 1973; Geeraerts et al. 2017).

Dry-cured (fermented) meat products Dry-cured meat products are mostly encompassing fermented sausages and dry-cured hams (Toldrá 2014). Meat fermentation is commonly based on the stuffing of a seasoned and cured meat batter into casings, so that microaerobic conditions are obtained, followed by a phase of maturation and drying that can be very short or take up 1–2 years. In general, the worldwide variety of different recipes and products is overwhelming (Leroy et al. 2013, 2015). The main microbiota of fermented meats consists of lactic acid bacteria that cause the acidification and, thus, contribute to their shelf-life stability and the triggering of proteolytic cascades that are important for flavour development (Aquilanti et al. 2016; Leroy et al. 2006). In Europe, where fermentations are usually performed at temperatures between 18 and 28 °C, depending on the type, lactobacilli take the overhand. This is the case for both spontaneous meat fermentations and, the much more common, starter culture-induced industrial productions. In both instances, *Lb. sakei* is the most prevalent lactic acid bacterial species (Ammor et al. 2005; Aymerich et al. 2006; Janssens et al. 2012, 2013; Ravyts et al. 2012; Urso et al. 2006). It is at the same time the most often used species in meat starter cultures and the species that is most adapted to the stringent conditions that typify meat fermentations, spontaneously emerging from the background microbiota (Chaillou et al. 2005; Nyquist et al. 2011; Rimaux et al. 2011, 2012). Sometimes, *Lactobacillus curvatus*, and to an even lesser degree, other lactic acid bacterial species as *Lactobacillus plantarum* and *Lactobacillus pentosus* can be encountered (Aymerich et al. 2006; Cocolin et al. 2009; Rantsiou et al. 2005; Tremonte et al. 2017; Villani et al. 2007). Moreover, enterococci have often been isolated from these products, with *Enterococcus faecium* and *Enterococcus faecalis* being the most commonly found species, but their level depends on the product type (Aymerich et al. 2003; Chevallier et al. 2006; Lebert et al. 2007; Martín et al. 2005). In North America, however, fermentations are carried out at higher temperatures (up to 40 °C) for faster production, thereby selecting for pediococci rather than lactobacilli. The most commonly encountered species are *Pediococcus acidilactici* and *Pediococcus cerevisiae* (Cocconcelli and Fontana 2014; Hammes and Hertel 1998).

Besides lactic acid bacteria, catalase-positive cocci are commonly found in fermented meats, especially when acidification is not too harsh (Aquilanti et al. 2007; Benito et al. 2007; Coton et al. 2010; Lebert et al. 2007; Leroy et al. 2010; Ravyts et al. 2012). These bacteria are desirable as well because they contribute to the formation of flavour and colour (Sánchez Mainar et al. 2017). They comprise a variety of coagulase-negative staphylococci, besides some occasional kocuria (mostly *Kocuria varians*). Once more, these bacteria either emerge from the background of the meats or are added via the starter culture, usually in combination with the lactic acid bacteria (Leroy et al. 2006; Ravyts et al. 2012). In traditional fermented meat products, which are sometimes still produced without the addition of starter cultures, a diverse staphylococcal consortium has been described (Aquilanti et al. 2016). Although the type of microorganisms depends on the different formulations and manufacturing processes, *Staphylococcus equorum*, *Staphylococcus saprophyticus*, and *Staphylococcus xylosus* are in most cases found as the prevalent species (Leroy et al. 2010; Martín et al. 2006; Mauriello et al. 2004; Sánchez Mainar et al. 2017). Apart from these main species, a whole range of other members may be encountered with the staphylococcal consortia, including *Staphylococcus succinus*, *Staphylococcus haemolyticus*, *Staphylococcus epidermidis*, *Staphylococcus pasteuri*, *Staphylococcus sciuri*, *Staphylococcus carnosus*, and many others (Aquilanti et al. 2016; Fonseca et al. 2013; Greppi et al. 2015; Leroy et al. 2010; Martín et al. 2006; Mauriello et al. 2004; Sánchez Mainar et al. 2017). For industrially produced fermented sausages, starter cultures of catalase-positive cocci are commonly added based on selected strains of *St. carnosus* and/or *St. xylosus* (Leroy et al. 2006; Ravyts et al. 2012).

As a third group, yeasts and moulds may be present in fermented meats (Leroy et al. 2006). Concerning the yeasts, a wide diversity has been reported, including species belonging to the genera *Debaryomyces*, *Rhodotorula*, *Yarrowia*, *Candida*, *Hansenula*, and *Torulopsis* (Mendonça et al. 2013; Selgas and García 2014). *Debaryomyces hansenii* is the most frequently isolated yeast (Cocolin et al. 2006; Mendonça et al. 2013) and is of particular interest, potentially used as starter culture for reasons of flavour, colour, and bioprotection (Andrade et al. 2010; Flores et al. 2015; Núñez et al. 2015). Whereas some fermented meat products are smoked, many others are moulded. Moulds can be obtained spontaneously or by application of a surface starter culture, whereby *Penicillium nalgiovense* is a commonly applied species (Iacumin et al. 2009; Papagianni et al. 2007; Sunesen and Stahnke 2003).

External moulding is not only desirable for its traditional whitish covering of the surface, but also for its contribution to flavour and quality stabilisation (Magistà et al. 2017; Sunesen and Stahnke 2003). Moreover, moulding has an impact on the microbial diversity when compared to smoking, causing a shift from *St. saprophyticus* to *St. equorum* (Janssens et al. 2013).

Finally, dry-cured hams are also part of the larger family of dry-cured meat products. Although they are not fermented in the strict sense, they also contain microorganisms that are roughly comparable with the ones present on the fermented meat products mentioned above, albeit usually in smaller numbers. The microbiota of dry-cured hams thus contain similar species of lactic acid bacteria, catalase-positive cocci, yeasts, and moulds (Martínez-Onandi et al. 2017; Simoncini et al. 2007; Virgili et al. 2012).

Smoked meat products A wide variety of smoked meat products exists, such as certain types of bacon, sausage, and ham-like products (Dušková et al. 2015). Unlike for cooked and dry-cured meat products, only few microbiological studies are available for smoked meat products, which may be partially ascribed to the fact that these products are less prone to microbial growth and spoilage (Roseiro et al. 2011; Škaljac et al. 2014). During the smoking step, chemical compounds from the smoke enter the meat matrix and alternate the flavour profile of the product. These compounds, which can be inhibitory towards specific groups of microorganisms, are combinations of polyaromatic hydrocarbons, phenols, and different carbonyls, depending on the technique used (Djinovic et al. 2008; Lingbeck et al. 2014; Šimko 2005). Liquid smoke is sometimes applied not only to add flavour to the meat but also to inhibit the growth of several pathogenic microorganisms, including *E. coli*, *L. monocytogenes*, *S.* Typhimurium, and *Yersinia enterocolitica* (Lingbeck et al. 2014). Many smoked meat products also undergo a cooking, fermentation, or ripening phase. Whereas fermented variants were discussed above, an example of a non-fermented smoked sausage are Vysočina sausages from the Czech Republic (Dušková et al. 2015). Species found on such smoked sausages belong the genera *Enterococcus*, *Lactobacillus*, *Lactococcus*, *Streptococcus,* and *Weissella*. A dominance of *Leuc. mesenteroides* has been found on both spoiled and non-spoiled smoked bacon (Comi et al. 2016). Whereas smoked samples of pork loin and bacon are dominated mostly by *Lb. sakei*, non-smoked variants contain mixtures of *Lb. sakei*, *Leuc. carnosum*, and *Leuc. mesenteroides* (Samelis et al. 2000). This finding may be due to the fact that some strains of *Lb. sakei* are able to break down polycyclic aromatic hydrocarbons in smoked meat products (Bartkiene et al. 2017).

Some smoked meat products, especially when they are also dried, as in the case of smoked hams, contain a variety of yeasts and moulds. Among the yeasts, the genera *Debaryomyces* and *Candida* seem to be the most prevalent ones, for instance encompassing the species *D. hansenii* and *Candida zeylanoides* that have been found on smoked Norwegian meat products (Asefa et al. 2009b). Moulds retrieved from smoked Norwegian hams belong mostly to the genus *Penicillium* (Asefa et al. 2009a). The genera *Penicillium* dominates the mould microbiota, specifically consisting of the species *P. nalgiovense*, being somehow favoured by smoking as compared to the subdominant species *Penicillium solitum* and *Penicillium commune* and certain species of *Cladosporium* (Asefa et al. 2009a, 2010).

Potential Impact on the Gut Microbiota

General Aspects

The ingestion of food has the potential to affect the human gut microbiome, with latent effects on the health status of the host (Dutton and Turnbaugh 2012). As a central part of life-style, for instance with respect to differences between hunter-gatherers, rural, or Western societies, the type of food consumed seems to be of importance (Quercia et al. 2014). Even short-time dietary changes can alter the composition of the intestinal bacterial communities, for instance when comparing different degrees of omnivore dietary setup. Diets that are based on animal products, and thus contain high amounts of protein, seem to favour the genera *Alistipes*, *Bilophila*, and *Bacteroides*, whereas plant-based diets increase the presence of Firmicutes species (David et al. 2014). In addition to effects caused by the different biochemical components that are present within a food matrix, food-associated microorganisms are introduced upon ingestion (Dutton and Turnbaugh 2012). Although meats and meat products can contain substantial numbers of microorganisms, it is not clear to what degree they will end up as significant elements of the gut microbiome. The latter consists of a large variety of bacterial species and reaches densities of up to 10^{12} cells per gram of intestinal content (O'Hara and Shanahan 2006; Plé et al. 2015). Gut microorganisms are a mixture of (semi)permanent intestinal inhabitants that co-exist with transient bacteria, whereby the latter only stay within the gut for a limited time (Zhang et al. 2016). The consumption of foods with high microbial loads may thus lead to temporary changes in the bacterial species diversity of the gut (Plé et al. 2015). The latter can either reshape the microbiome in beneficial ways and improve health (Doré and Blottière 2015), change the communities minimally with a fast return to the original status, or lead to microbial dysbiosis and illness (Josephs-Spaulding et al. 2016). With respect to the latter effect, meat can contain several pathogenic bacteria that have the potential to cause disease, such as *Ca. jejuni*, *Cl. botulinum*, *Clostridium perfringens*, *E. coli*, *L. monocytogenes*, and *Staphylococcus aureus* (Buchanan et al. 2017; Guran et al. 2014; Hospital et al. 2016; Huang et al. 2015; Josephs-Spaulding et al. 2016; Mc Nulty et al. 2016; Ortega et al. 2010). Using culture-independent methods, several species of lactic acid bacteria that have been retrieved from meat have also been found in human faeces, including *Lactobacillus* spp., *Leuc. mesenteroides*, *Leuc. carnosum*, *P. acidilactici*, *Pediococcus pentosaceus* and *Weissella viridescens* (Sanz et al. 2007; Walter et al. 2001). However, a causal link between meat-eating and the composition of the gut microbiota is not necessarily present. In the sections below, the focus will be on meat-associated microorganisms other than pathogens, potentially leading to health benefits.

Meat-Associated Microorganisms of Potential Relevance for the Gut Microbiome

Meat products are a natural source of lactic acid bacteria, a group of microorganisms that is often presented as beneficial to gut health (De Vuyst et al. 2008). It has been demonstrated that their consumption may partially contribute to an improved immune system, the prevention of infections and diarrhoea, and the reduction of food allergies and inflammatory bowel conditions (Fijan 2014; Kumari et al. 2011; Zhong et al. 2014). In fermented meats, lactic acid bacteria occur in high numbers, up to 10^9 CFU per gram; they originate from the starter cultures used but may also emerge from the background microbiota (Ravyts et al. 2012). As such, fermented meats may contribute to the overall diversity of ingested beneficial microorganisms, a property that has led to suggestions that the regular consumption of fermented foods is generally to be recommended via nutritional guidelines based on the potential contributions to health (Bell et al. 2017; Chilton et al. 2015; Marco et al. 2017), and even the explicit prevention of disease (Olivares et al. 2006). As outlined above, the most encountered bacteria in fermented meat products are certain species of lactobacilli and pediococci, of which the relevance for the gut is not entirely clear. Nevertheless, their incorporation into the gut microbiota is likely to some degree. Several strains of lactic acid bacteria, for which possible albeit often very uncertain probiotic effects have been mentioned, have been applied in meat fermentations as starter cultures, encompassing strains of *P. acidilactici*, *P. pentosaceus*, and *Lb. plantarum*, besides some strains from species that are non-conventional for meat fermentation, such as *Lactobacillus brevis*, *Lactobacillus rhamnosus*, *Lactobacillus casei*, and *Lactobacillus reuteri* (Table 2; De Vuyst et al. 2008; Rouhi et al. 2013).

According to recommendations, probiotic foods should carry a population of at least 10^6 CFU per g (or mL), so that a consumption of 100 g (or mL) would lead to a dose of 10^8 CFU in the gut, which is supposed to be able to confer health benefits to the consumers (Ashraf and Shah 2011; Jayamanne and Adams 2006). Thus, survival of the probiotic strains during the production process and storage as well as the passage through the upper gastrointestinal track is of major importance (Cavalheiro et al. 2015; Shori 2015; Tripathi and Giri 2014). Interestingly, the use of fermented meat as a carrying matrix for probiotic strains seems to increase their survival rate (Klingberg and Budde 2006). As an example, strains of *Lb. rhamnosus* are able to survive the human digestive system and are sometimes mentioned in a meat fermentation context (Albano et al. 2009; Erkkilä et al. 2001; Rubio et al. 2014a, b, c). Studies have already demonstrated that during an animal-based diet, including the consumption of cured meats, *P. acidilactici* increases significantly in faecal samples, indicating that at least some bacteria from meat products can survive and reach the human gut in considerable numbers (David et al. 2014).

Even though bifidobacteria do not belong to the typical microbial communities of meat and meat products, attempts have been made to use probiotic bifidobacterial strains in fermented meat products, including strains of *Bifidobacterium animalis*, *Bifidobacterium lactis*, and *Bifidobacterium longum* (Holko et al. 2013;

Table 2 Non-exhaustive overview of the use of candidate probiotic strains in different dry-cured meat products based on studies between 2005 and 2017

Product type	(Candidate) probiotic strain(s)	Reference
Portuguese fermented sausage	*Ec. faecium* 120	Barbosa et al. (2014)
Harbin-style fermented sausage	*Lb. brevis* R4, *Lb. fermentum* R6, and *P. pentosaceus* R1	Han et al. (2017)
Scandinavian-type fermented sausage	*Lb. pentosus* MF 1300 and *Lb. plantarum* MF1291 and MD1298	Klingberg et al. (2005)
Fermented sausage	*Lb. plantarum* MF1298	Klingberg and Budde (2006)
Dry-cured pork neck	*Bf. animalis* subsp. *lactis* BB-12	Libera et al. (2015)
Fermented sausage	Strains of *Bf. lactis* and *Lb. acidophilus*	Nogueira Ruiz et al. (2014)
Spanish fermented sausage	*Lb. plantarum* 299V and *Lb. rhamnosus* GG	Rubio et al. (2013a)
Low-acid fermented sausage	*Lb. casei* CTC1677, CTC1678, and Shirota; *Lb. plantarum* 299V; and *Lb. rhamnosus* CTC1679 and GG	Rubio et al. (2014a)
Model-type fermented sausage	*Lb casei/paracasei* CTC1677 and CTC1678 and *Lb rhamnosus* CTC1679	Rubio et al. (2014b)
Low-acid fermented sausage	*Lb. rhamnosus* CTC1679	Rubio et al. (2014c)
Iberian fermented sausage	*Lb. fermentum* HL57 and *Lb. reuteri* PL519 and PL542	Ruiz-Moyano et al. (2009)
Iberian fermented sausage	*P. acidilactici* SP979	Ruiz-Moyano et al. (2010)
Iberian fermented sausage	*Lb. fermentum* HL57 and *P. acidilactici* SP979	Ruiz-Moyano et al. (2011)
Dry-cured pork neck and sausage	*Bf. animalis* subsp. *lactis* BB-12, *Lb. acidophilus* Bauer, and *Lb. rhamnosus* LOCK900	Wójciak et al. (2016)

Bacterial genera encompass *Bf.*, *Bifidobacterium*; *Ec.*, *Enterococcus*; *Lb.*, *Lactobacillus*; and *P.*, *Pediococcus*

Muthukumarasamy and Holley 2007; Nogueira Ruiz et al. 2014; Ruiz et al. 2014; Pidcock et al. 2002). Apart from lactobacilli and bifidobacteria, strains of which are the most representative microorganisms used as probiotics, some strains of other bacterial groups have been suggested as potential functional cultures in meat products. The main examples include some candidate probiotic strains of enterococci, despite some issues on the potential presence of antibiotic resistance genes and virulence factors (Barbosa et al. 2014; Foulquié Moreno et al. 2006; Franz et al. 2011; Hugas et al. 2003), as well as strains of *Bacillus coagulans* and *Bacillus subtilis* (Jafari et al. 2017).

Although fermented meat products as carriers of probiotic bacteria have attracted attention in recent years, there is still controversy over their use (De Vuyst et al. 2008). Critique relates mostly to the fact that so-called probiotic properties, such as survival of the gastrointestinal tract, are often taken for granted, whereas true probiotic effects need to be validated in clinical trials to convincingly demonstrate health effects on the host. The mere addition of strains with (often very preliminary) probiotic potential to a food product, *in casu* fermented meat, does not suffice. As fermented meats are a challenging matrix for microorganisms (high content of curing salts, low pH and water activity, competition with the background microbiota; Rouhi et al. 2013), efforts need to be made in developing robust delivery strategies for probiotics in meat products (Cavalheiro et al. 2015). Moreover, it is doubtful if such products have the right nutritional profile in view of consumer expectations and market potential as probiotic foods (De Vuyst ct al. 2008).

Except for fermented meats, high numbers of lactic acid bacteria may also be encountered in fresh meat and non-fermented meat products (*e.g.*, cooked ham), for which bioprotective cultures have been applied (Table 3). The latter are added to

Table 3 Non-exhaustive overview of the use of bioprotective strains used in different meat products based on studies between 2005 and 2017

Product type	Bioprotective strain(s)	Reference
Portuguese fermented sausage	*P. acidilactici* HA-6111-2	Albano et al. (2009)
Salami	*Lb. curvatus* MBSa2	Barbosa et al. (2014)
Smoked pork sausages	*Lb. sakei* KTU05-6, *P. acidilactici* KTU05-7, and *P. pentosaceus* KTU05-9	Bartkiene et al. (2017)
Chicken hamburger	*Lb. acidophilus* CRL1014	Bomdespacho et al. (2014)
Vacuum-packed meat	*Lb. curvatus* CRL705	Castellano and Vignolo (2006)
Vacuum-packed beef	*Lb. curvatus* CRL705	Castellano et al. (2010)
Grounded beef	*Lb. curvatus* CRL705 and *Lc. lactis* subsp. *lactis* CRL1109	Castellano et al. (2011)
Grounded beef	*Lb. sakei* CIP105422 and 23K, and strains 18, 64, 112, 156, 160x1, 332, and G3	Chaillou et al. (2014)
Cooked bacon	*Lb. sakei* B-2 Safe Pro® and *Lc. lactis* Rubis	Comi et al. (2016)
Sliced and cooked ham	*Lb. curvatus* 2711 and *Lb. sakei* 2512	Héquet et al. (2007)
Ready-to-eat cured and smoked pork product	Commercial culture containing strains of *Lb. curvatus*, *Lb. sakei* ST153, *St. xylosus*, and *Pe. acidilactici*	Jacome et al. (2014)
Fermented sausage	*D. hansenii* 253H and 226G	Núñez et al. (2015)
Ground pork meat and Iberian chorizo	Commercial culture containing strains of *Lb. curvatus*, *Lb. sakei*, *P. acidilactici*, and *St. xylosus*	Ortiz et al. (2014)
Fermented sausage	Strain of *Lb. sakei*	Urso et al. (2006)
Cooked ham	*Leuc. carnosum* 3M42	Vasilopoulos et al. (2010b)
Cooked model ham	*Lb. sakei* subsp. *carnosus* 10A and *Lb. sakei* 148	Vermeiren et al. (2006)

Bacterial genera encompass *D.*, *Debaryomyces*; *Lb.*, *Lactobacillus*; *Lc.*, *Lactococcus*; *Leuc.*, *Leuconostoc*; *P.*, *Pediococcus*; and *St.*, *Staphylococcus* counter the growth of spoilage bacteria and pathogens (Pothakos et al. 2015; Vasilopoulos et al. 2015). Examples of bioprotective cultures applied in the context of both fresh and processed meats encompass strains of *Ec. faecalis* (Sparo et al. 2008, 2013), *Ec. faecium* (Huang et al. 2016; Rubio et al. 2013b), *Lactobacillus alimentarius* (Lemay et al. 2002), *Lb. curvatus* (Castellano et al. 2010, 2011), *Lb. sakei* (Bartkiene et al. 2017; Chaillou et al. 2014; Jacome et al. 2014; Ortiz et al. 2014; Vermeiren et al. 2006), *Lactococcus lactis* (Castellano et al. 2011; Comi et al. 2016), *Leuconostoc* spp. (Budde et al. 2003; Metaxopoulos et al. 2002; Vasilopoulos et al. 2010b), *P. acidilactici* (Albano et al. 2009; Bartkiene et al. 2017), and *P. pentosaceus* (Bartkiene et al. 2017). Sometimes, strains of species with probiotic connotations have also been applied as bioprotective cultures in meat, such as *Lb. acidophilus* (Bomdespacho et al. 2014).

Besides lactic acid bacteria and the other above-mentioned bacterial groups, strains of which are sometimes associated with probiotic action, the group of meat-associated coagulase-negative staphylococci may also require further attention. It is well established that staphylococci are among the early gut colonisers of infants, as the gut microbiota is partially dominated by staphylococci during the first weeks of life (Chang et al. 2011; Jacquot et al. 2011; Jimenez et al. 2008; Salminen et al. 2015). However, little is known about their presence and abundance in the gut of adults and to which degree this may affect the host. Staphylococci can reach the human gut upon the consumption of cured meat products, including salami and prosciutto, as has been demonstrated for *St. carnosus* (David et al. 2014). Although *St. carnosus* is considered to be a non-pathogenic species (Müller et al. 2016; Rosenstein and Götz 2013), it needs to be taken into account that certain staphylococcal species may pose health concerns (Becker et al. 2014).

Conclusions

A large variety of different meats and meat products can be found worldwide. They are dominated by different communities of microorganisms that, in principle, can have either positive or negative effects on human health, provided they survive the harsh conditions of the gastrointestinal tract. These microorganisms are adapted to the specific conditions that prevail in each type of meat or meat product. The microbiota are thus being shaped by a superposition of different selective pressures caused by, among others, temperature and packaging, as well as technological interventions, such as salting, smoking, and fermentation. In some cases, for instance after meat fermentation or upon the use of bioprotective cultures, their numbers can reach 10^9 CFU or more per gram. In those cases, they mostly consist of lactic acid bacteria, of which several members have been associated with human health advantages. Although some studies have indicated that several meat-associated microorganisms are able to enter the gut, it is yet unclear what their true impact on the host would be. Further investigation of their role in health and disease is therefore needed.

Meat and Meat Products

Acknowledgements The authors acknowledge financial support of the Research Council of the Vrije Universiteit Brussel (SRP7 and IOF342 projects, and in particular the HOA21 project 'Artisan quality of fermented foods: myth, reality, perceptions, and constructions' and the Interdisciplinary Research Program IRP2 'Food quality, safety, and trust since 1950: societal controversy and biotechnological challenges'), and the Hercules Foundation (projects UABR 09/004 and UAB 13/002).

References

Ahmed, J., Mulla, M., & Arfat, Y. A. (2017). Application of high-pressure processing and polylactide/cinnamon oil packaging on chicken sample for inactivation and inhibition of *Listeria monocytogenes* and *Salmonella* Typhimurium, and post-processing film properties. *Food Control, 78*, 160–168.

Albano, H., van Reenen, C. A., Todorov, S. D., Cruz, D., Fraga, L., Hogg, T., Dicks, L. M. T., & Teixeira, P. (2009). Phenotypic and genetic heterogeneity of lactic acid bacteria isolated from "Alheira", a traditional fermented sausage produced in Portugal. *Meat Science, 82*, 389–398.

Álvarez-Astorga, M., Capita, R., Alonso-Calleja, C., & Capita, R. (2002). Microbiological quality of retail chicken by-products in Spain. *Meat Science, 62*, 45–50.

Ammor, S., Rachman, C., Chaillou, S., Prévost, H., Dousset, X., Zagorec, M., Dufour, E., & Chevallier, I. (2005). Phenotypic and genotypic identification of lactic acid bacteria isolated from a small-scale facility producing traditional dry sausages. *Food Microbiology, 22*, 373–382.

Andrade, M. J., Rodríguez, M., Casado, E., & Córdoba, J. J. (2010). Efficiency of mitochondrial DNA restriction analysis and RAPD-PCR to characterize yeasts growing on dry-cured Iberian ham at the different geographic areas of ripening. *Meat Science, 84*, 377–383.

Aquilanti, L., Garofalo, C., Osimani, A., Silvestri, G., Vignaroll, C., & Clementi, F. (2007). Isolation and molecular characterization of antibiotic-resistant lactic acid bacteria from poultry and swine meat products. *Journal of Food Protection, 70*, 557–565.

Aquilanti, L., Garofalo, C., Osimani, A., & Clementi, F. (2016). Ecology of lactic acid bacteria and coagulase-negative cocci in fermented dry sausages manufactured in Italy and other Mediterranean countries: An overview. *International Food Research Journal, 23*, 429–445.

Arnold, J. W. (2007). Bacterial contamination on rubber picker fingers before, during, and after processing. *Poultry Science, 86*, 2671–2675.

Arvanitoyannis, I. S., & Stratakos, A. C. (2012). Application of modified atmosphere packaging and active/smart technologies to red meat and poultry: A review. *Food Bioprocess Technology, 5*, 1423–1446.

Asefa, D. T., Gjerde, R. O., Sidhu, M. S., Langsrud, S., Kure, C. F., Nesbakken, T., & Skaar, I. (2009a). Moulds contaminants on Norwegian dry-cured meat products. *International Journal of Food Microbiology, 128*, 435–439.

Asefa, D. T., Møretrø, T., Gjerde, R. O., Langsrud, S., Kure, C. F., Sidhu, M. S., Nesbakken, T., & Skaar, I. (2009b). Yeast diversity and dynamics in the production processes of Norwegian dry-cured meat products. *International Journal of Food Microbiology, 133*, 135–140.

Asefa, D. T., Kure, C. F., Gjerde, R. O., Omer, M. K., Langsrud, S., Nesbakken, T., & Skaar, A. (2010). Fungal growth pattern, sources and factors of mould contamination in a dry-cured meat production facility. *International Journal of Food Microbiology, 140*, 131–135.

Ashraf, R., & Shah, N. P. (2011). Selective and differential enumerations of *Lactobacillus delbrueckii* subsp. *bulgaricus*, *Streptococcus thermophilus*, *Lactobacillus acidophilus*, *Lactobacillus casei* and *Bifidobacterium* ssp. in yogurt—A review. *International Journal of Food Microbiology, 149*, 194–208.

Audenaert, K., D'Haene, K., Messens, K., Ruyssen, T., Vandamme, P., & Huys, G. (2010). Diversity of lactic acid bacteria from modified atmosphere packaged sliced cooked meat products at sell-by date assessed by PCR-denaturing gradient gel electrophoresis. *Food Microbiology, 27,* 12–18.

Aymerich, T., Martín, B., Garriga, M., & Hugas, M. (2003). Microbial quality and direct PCR identification of lactic acid bacteria and nonpathogenic staphylococci from artisanal low-acid sausages. *Applied and Environmental Microbiology, 69,* 4583–4594.

Aymerich, T., Martín, B., Garriga, M., Vidal-Carou, M. C., Bover-Cid, S., & Hugas, M. (2006). Safety properties and molecular strain typing of lactic acid bacteria isolated from slightly fermented sausages. *Journal of Applied Microbiology, 100,* 40–49.

Balamatsia, C. C., Patsias, A., Kontominas, M. G., & Savvaidis, I. N. (2007). Possible role of volatile amines as quality-indicating metabolites in modified atmosphere-packaged chicken fillets: Correlation with microbiological and sensory attributes. *Food Chemistry, 104,* 1622–1628.

Ballester-Costa, C., Sendra, E., Fernández-López, L., Pérez-Álvarez, J. A., & Viuda-Martos, M. (2017). Assessment of antioxidant and antibacterial properties on meat homogenates of essential oils obtained from four *Thymus* species achieved from organic growth. *Foods, 6,* 59.

Barbosa, J., Borges, S., & Teixeira, P. (2014). Selection of potential probiotic *Enterococcus faecium* isolated from Portuguese fermented food. *International Journal of Food Microbiology, 191,* 144–148.

Bartkiene, E., Bartkevics, V., Mozuriene, E., Krungleviciute, V., Novoslavskij, A., Santini, A., Rozentale, I., Juodeikiene, G., & Cizeikiene, D. (2017). The impact of lactic acid bacteria with antimicrobial properties on biodegradation of polycyclic aromatic hydrocarbons and biogenic amines in cold smoked pork sausages. *Food Control, 71,* 285–292.

Becker, K., Heilmann, C., & Peters, G. (2014). Coagulase-negative staphylococci. *Clinical Microbiology Reviews, 27,* 870–926.

Bell, R. G. (2001). Meat packaging: Protection, preservation and presentation. In Y. H. Hui, W. K. Nip, R. W. Rogers, & G. A. Young (Eds.), *Meat science and applications* (pp. 463–490). New York: Marcel Dekker.

Bell, V., Ferrão, J., & Fernandes, T. (2017). Nutritional guidelines and fermented food frameworks. *Foods, 6,* 65.

Belluco, S., Barco, L., Roccato, A., & Ricci, A. (2015). Variability of *Escherichia coli* and *Enterobacteriaceae* counts on pig carcasses: A systematic review. *Food Control, 55,* 115–126.

Benito, M. J., Martín, A., Aranda, E., Pérez-Nevado, F., Ruiz-Moyano, S., & Cordoba, M. G. (2007). Characterization and selection of autochthonous lactic acid bacteria isolated from traditional Iberian dry-fermented salchichón and chorizo sausages. *Journal of Food Protection, 72,* 193–201.

Benson, A. K., David, J. R. D., Gilbreth, S. E., Smith, G., Nietfeldt, J., Legge, R., Kim, J., Sinha, R., Duncan, C. E., Ma, J., & Singh, I. (2014). Microbial successions are associated with changes in chemical profiles of a model refrigerated fresh pork sausage during an 80-day shelf life study. *Applied and Environmental Microbiology, 80,* 5178–5194.

Björkroth, J., Ristiniemi, M., Vandamme, P., & Korkeala, H. (2005). *Enterococcus* species dominating in fresh modified-atmosphere-packaged, marinated broiler legs are overgrown by *Carnobacterium* and *Lactobacillus* species during storage at 6°C. *International Journal of Food Microbiology, 97,* 267–276.

Bomdespacho, L. Q., Cavallini, D. C. U., Zavarizi, A. C. M., Pinto, R. A., & Rossi, E. A. (2014). Evaluation of the use of probiotic acid lactic bacteria in the development of chicken hamburger. *International Food Research Journal, 21,* 965–972.

Borch, E., & Arinder, P. (2002). Bacteriological safety issues in red meat and ready-to-eat meat products, as well as control measures. *Meat Science, 62,* 381–390.

Borilova, G., Hulankova, R., Svobodova, I., Jezek, F., Hutarova, Z., Vecerek, V., & Steinhauserova, I. (2016). The effect of storage conditions on the hygiene and sensory status of wild boar meat. *Meat Science, 118,* 71–77.

Brightwell, E., Boerema, J., Mills, J., Mowat, E., & Pulford, D. (2006). Identifying the bacterial community on the surface of Intralox™ belting in a meat boning room by culture-dependent and culture-independent 16S rDNA sequence analysis. *International Journal of Food Microbiology, 109,* 47–53.

Brightwell, G., Clemens, R., Adam, K., Urlich, S., & Boerema, J. (2009). Comparison of culture-dependent and independent techniques for characterisation of the microflora of peroxyacetic acid treated, vacuum-packaged beef. *Food Microbiology, 26*, 283–288.

Buchanan, R. L., Garris, L. G. M., Hayman, M. M., Jackson, T. C., & Whiting, R. C. (2017). A review of *Listeria monocytogenes*: An update on outbreaks, virulence, dose-response, ecology, and risk assessments. *Food Control, 75*, 1–13.

Budde, B. B., Hornbaek, T., Jacobsen, T., Barkholt, V., & Koch, A. G. (2003). *Leuconostoc carnosum* 4010 has the potential for use as a protective culture for vacuum-packed meats: Culture isolation, bacteriocin identification, and meat application experiments. *International Journal of Food Microbiology, 83*, 171–184.

Caparros Megido, R., Sablon, L., Geuens, M., Brostaux, Y., Alabi, T., Blecker, C., Drugmand, D., Haubruge, É., & Francis, F. (2014). Edible insects acceptance by Belgian consumers: Promising attitude for entomophagy development. *Journal of Sensory Studies, 29*, 14–20.

Carr, P. R., Walter, V., Brenner, H., & Hoffmeister, M. (2016). Meat subtypes and their association with colorectal cancer: Systematic review and meta-analysis. *International Journal of Cancer, 138*, 293–302.

Carrizosa, E., Benito, M. J., Ruiz-Moyano, S., Hernández, A., Villalobos, C., Martin, A., & Córdoba, M. G. (2017). Bacterial communities of fresh goat meat packaged in modified atmosphere. *Food Microbiology, 65*, 57–63.

Castellano, P., & Vignolo, G. (2006). Inhibition of *Listeria innocua* and *Brochothrix thermosphacta* in vacuum-packaged meat by addition of bacteriocinogenic *Lactobacillus curvatus* CRL705 and its bacteriocins. *Letters in Applied Microbiology, 43*, 194–199.

Castellano, P., González, C., Carduza, F., & Vignolo, G. (2010). Protective action of *Lactobacillus curvatus* CRL705 on vacuum-packaged raw beef. Effect on sensory and structural characteristics. *Meat Science, 85*, 394–401.

Castellano, P., Belfiore, C., & Vignolo, G. (2011). Combination of bioprotective cultures with EDTA to reduce *Escherichia coli* O157:H7 in frozen ground-beef patties. *Food Control, 22*, 1461–1465.

Cavalheiro, C. P., Ruiz-Capillas, C., Herrero, A. M., Jiménez-Colmenero, F., de Menezes, C. R., & Fries, L. L. M. (2015). Application of probiotic delivery systems in meat products. *Trends in Food Science and Technology, 46*, 120–131.

Cerveny, J., Meyer, J. D., & Hall, P. A. (2010). Microbiological spoilage of meat and poultry products. In W. H. Sperber & M. P. Doyle (Eds.), *Compendium of the microbiological spoilage of foods and beverages* (pp. 69–86). New York: Springer.

Chaillou, S., Champomier-Vergès, M. C., Cornet, M., Crutz-Le Coq, A. M., Dudez, A. M., Martin, V., Beaufils, S., Darbon-Rongère, E., Bossy, R., Loux, V., & Zagorec, M. (2005). The complete genome sequence of the meat-borne lactic acid bacterium *Lactobacillus sakei* 23k. *Nature Biotechnology, 23*, 1527–1533.

Chaillou, S., Christieans, S., Rivollier, M., Lucquin, I., Champomier-Vergès, M. C., & Zagorec, M. (2014). Quantification and efficiency of *Lactobacillus sakei* strain mixtures used as protective cultures in ground beef. *Meat Science, 97*, 332–338.

Chaillou, S., Choulot-Talmon, A., Caekebeke, H., Cardinal, M., Christieans, S., Denis, C., Desmonts, M. H., Dousset, X., Feurer, C., Hamon, E., Joffraud, J. J., La Carbona, S., Leroi, F., Leroy, S., Lorre, S., Macé, S., Pilet, M. F., Prévost, H., Rivollier, M., Roux, D., Talon, R., Zagorec, M., & Champomier-Vergès, M. C. (2015). Origin and ecological selection of core and food-specific bacterial communities associated with meat and seafood spoilage. *The ISME Journal, 9*, 1105–1118.

Chang, J. Y., Shin, S. M., Chun, J., Lee, J. H., & Seo, J. K. (2011). Pyrosequencing-based molecular monitoring of the intestinal bacterial colonization in preterm infants. *Journal of Pediatric Gastroenterology and Nutrition, 53*, 512–519.

Chevallier, I., Ammor, S., Laguet, A., Labayle, S., Castanet, V., Dufour, E., & Talon, R. (2006). Microbial ecology of a small-scale facility producing traditional dry sausage. *Food Control, 17*, 446–453.

Chilton, S. N., Burton, J. P., & Reid, G. (2015). Inclusion of fermented foods in food guides around the world. *Nutrients, 7*, 390–404.

Chipley, J. R., & May, K. N. (1968). Survival of aerobic and anaerobic bacteria in chicken meat during freeze-dehydration, rehydration and storage. *Applied Microbiology, 16*, 445–449.

Chouliara, E., Karatapanis, A., Savvaidis, I. N., & Kontominas, M. G. (2007). Combined effect of oregano essential oil and modified atmosphere packaging on shelf-life extension of fresh chicken breast meat, stored at 4°C. *Food Microbiology, 24*, 607–617.

Cocconcelli, P. S., & Fontana, C. (2014). Bacteria. In F. Toldrá (Ed.), *Handbook of fermented meat and poultry* (2nd ed., pp. 117–128). Hoboken, NJ: Wiley-Blackwell.

Cocolin, L., Urso, R., Rantsiou, K., Cantoni, C., & Comi, G. (2006). Dynamics and characterization of yeasts during natural fermentation of Italian sausages. *FEMS Yeast Research, 6*, 692–701.

Cocolin, L., Dolci, P., Rantsiou, K., Urso, R., Cantoni, C., & Comi, G. (2009). Lactic acid bacteria ecology of three traditional fermented sausages produced in the North Italy as determined by molecular methods. *Meat Science, 82*, 125–132.

Comi, G., & Iacumin, L. (2012). Identification and process origin of bacteria responsible for cavities and volatile off-flavour compounds in artisan cooked ham. *Food Science and Technology, 47*, 114–121.

Comi, G., Andyanto, D., Manzano, M., & Iacumin, L. (2016). *Lactococcus lactis* and *Lactobacillus sakei* as bio-protective culture to eliminate *Leuconostoc mesenteroides* spoilage and improve the shelf life and sensorial characteristics of commercial cooked bacon. *Food Microbiology, 58*, 16–22.

Coton, E., Desmonts, M. H., Leroy, S., Coton, M., Jamet, E., Christieans, S., Donnio, P. Y., Lebert, I., & Talon, R. (2010). Biodiversity of coagulase-negative staphylococci in French cheeses, dry fermented sausages, processing environments and clinical samples. *International Journal of Food Microbiology, 137*, 221–229.

D'Amico, S., Collins, T., Marx, J. C., Feller, G., & Gerday, C. (2006). Psychrophilic microorganisms: Challenges for life. *EMBO Reports, 7*, 385–389.

David, L. A., Maurice, C. F., Carmody, R. N., Gootenberg, D. B., Button, J. E., Wolfe, B. E., Ling, A. V., Devlin, A. S., Varma, Y., Fischbach, M. A., Biddinger, S. B., Dutton, R. J., & Turnbaugh, P. J. (2014). Diet rapidly and reproducibly alters the human gut microbiome. *Nature, 505*, 559–563.

Davidson, C. M., & Cronin, F. (1973). Medium for the selective enumeration of lactic acid bacteria from foods. *Applied and Environmental Microbiology, 26*, 439–440.

De Filippis, F., La Storia, A., Villani, F., & Ercolini, D. (2013). Exploring the sources of bacterial spoilers in beefsteaks by culture-independent high-throughput sequencing. *PLoS One, 8*, e70222.

De Smet, S., & Vossen, E. (2016). Meat: the balance between nutrition and health. A review. *Meat Science, 120*, 145–156.

De Vuyst, L., Falony, G., & Leroy, F. (2008). Probiotics in fermented sausage. *Meat Science, 80*, 75–78.

Djinovic, J., Popovic, A., & Jira, W. (2008). Polycyclic aromatic hydrocarbons (PAHs) in different types of smoked meat products from Serbia. *Meat Science, 80*, 449–456.

Doré, J., & Blottière, H. (2015). The influence of diet on the gut microbiota and its consequences for health. *Current Opinion in Biotechnology, 32*, 195–199.

Doulgeraki, A. I., Paramithiotis, S., Kagkli, D. M., & Nychas, G.-J. E. (2010). Lactic acid bacteria population dynamics during minced beef storage under aerobic or modified atmosphere packaging conditions. *Food Microbiology, 27*, 1028–1034.

Doulgeraki, A. I., Paramithiotis, S., & Nychas, G.-J. E. (2011). Characterization of the *Enterobacteriaceae* community that developed during storage of minced beef under aerobic or modified atmosphere packaging conditions. *International Journal of Food Microbiology, 145*, 77–83.

Doulgeraki, A. I., Ercolini, D., Villani, F., & Nychas, G.-J. E. (2012). Spoilage microbiota associated to the storage of raw meat in different conditions. *International Journal of Food Microbiology, 157*, 130–141.

Drosinos, E. H., Mataragas, M., Kampani, A., Kritikos, D., & Metaxopoulos, I. (2006). Inhibitory effect of organic acid salts on spoilage flora in culture medium and cured cooked meat products under commercial manufacturing conditions. *Meat Science, 73*, 75–81.

Dušková, M., Kameník, J., Šedo, O., Zdráhal, Z., Salàkovà, A., Karpíšková, R., & Lačanin, I. (2015). Survival and growth of lactic acid bacteria in hot smoked dry sausages (non-fermented salami) with and without sensory deviations. *Food Control, 50*, 804–808.

Dušková, M., Kameník, J., Lačanin, I., Šedo, O., & Zdráhal, Z. (2016). Lactic acid bacteria in cooked hams as sources of contamination and chances of survival in the product. *Food Science and Technology, 61*, 492–495.

Dutton, R. J., & Turnbaugh, P. J. (2012). Taking a metagenomic view of human nutrition. *Current Opinion in Clinical and Nutrition and Metabolic Care, 15*, 448–454.

Ercolini, D., Russo, F., Torrieri, E., Masi, P., & Villani, F. (2006). Changes in the spoilage-related microbiota of beef during refrigerated storage under different packaging conditions. *Applied and Environmental Microbiology, 70*, 4663–4671.

Ercolini, D., Ferrocino, I., La Storia, A., Mauriello, G., Gigli, S., Masi, P., & Villani, F. (2009a). Development of spoilage microbiota in beef stored in nisin activated packaging. *Food Microbiology, 27*, 137–143.

Ercolini, D., Russo, F., Nasi, A., Ferranti, P., & Villani, F. (2009b). Mesophilic and psychrotrophic bacteria from meat and their spoilage potential in vitro and in beef. *Applied and Environmental Microbiology, 75*, 1990–2001.

Ercolini, D., Ferrocino, I., Nasi, A., Ndagijimana, M., Vernocchi, P., La Storia, A., Laghi, L., Mauriello, M., Guerzoni, M. E., & Villani, F. (2011). Monitoring of microbial metabolites and bacterial diversity in beef stored under different packaging conditions. *Applied and Environmental Microbiology, 77*, 7372–7381.

Erkkilä, S., Petäjä, E., Eerola, S., Lilleberg, L., Mattila-Sandholm, T., & Suihko, M. L. (2001). Flavour profiles of dry sausages fermented by selected novel meat starter cultures. *Meat Science, 58*, 111–116.

Fijan, S. (2014). Microorganisms with claimed probiotic properties: An overview of recent literature. *International Journal of Environmental Research and Public Health, 11*, 4745–4767.

Flores, M., Corral, S., Cano-García, L., Salvador, A., & Belloch, C. (2015). Yeast strains as potential aroma enhancers in dry fermented sausages. *International Journal of Food Microbiology, 212*, 16–24.

Fonseca, S., Cachaldora, A., Gómez, M., Franco, I., & Carballo, J. (2013). Monitoring the bacterial population dynamics during the ripening of Galician chorizo, a traditional dry fermented Spanish sausage. *Food Microbiology, 33*, 77–84.

Fougy, L., Desmonts, M. H., Coeuret, G., Fassel, C., Hamon, E., Hézard, B., Champomier-Vergès, M. C., & Chaillou, S. (2016). Reducing salt in raw pork sausages increases spoilage and correlates with reduced bacterial diversity. *Applied and Environmental Microbiology, 82*, 3928–3939.

Foulquié Moreno, M. R., Sarantinopoulos, P., Tsakalidou, E., & De Vuyst, L. (2006). The role and application of enterococci in food and health. *International Journal of Food Microbiology, 106*, 1–24.

Franz, C. M. A. P., Huch, M., Abriouel, H., Holzapfel, W., & Gálvez, A. (2011). Enterococci as probiotics and their implications in food safety. *International Journal of Food Microbiology, 151*, 125–140.

Geeraerts, W., Pothakos, V., De Vuyst, L., & Leroy, F. (2017). Diversity of the dominant bacterial species on sliced cooked pork products at expiration date in the Belgian retail. *Food Microbiology, 65*, 236–243.

Gerber, P. J., Mottet, A., Opio, C. I., Falcucci, A., & Teillard, F. (2015). Environmental impacts of beef production: Review of challenges and perspectives for durability. *Meat Science, 109*, 2–12.

Greppi, A., Ferrocino, I., La Storia, A., Rantsiou, K., Ercolini, D., & Cocolin, L. (2015). Monitoring of the microbiota of fermented sausages by culture independent rRNA-based approaches. *International Journal of Food Microbiology, 212*, 65–75.

Guran, H. S., Vural, A., & Erkan, M. E. (2014). The prevalence and molecular typing of *Clostridium perfringens* in ground beef and sheep meats. *Journal für Verbraucherschutz und Lebensmittelsicherheit, 9*, 121–128.

Hammes, W. P., & Hertel, C. (1998). New developments in meat starter cultures. *Meat Science, 49*, S125–S138.

Han, Q., Kong, B., Chen, Q., Sun, F., & Zhang, H. (2017). *In vitro* comparison of probiotic properties of lactic acid bacteria isolated from Harbin dry sausages and selected probiotics. *Journal of Functional Foods, 32*, 391–400.

Harada, T., Dang, V. C., Nguyen, D. P., Nguyen, T. A. D., Sakamoto, M., Ohkuma, M., Matooka, D., Nakamura, S., Uchida, K., Jinnai, M., Yonogi, S., Kawahara, R., Kawai, T., Kumeda, Y., & Yamamoto, Y. (2016). *Enterococcus saigonensis* sp. nov., isolated from retail chicken meat and liver. *International Journal of Systematic and Evolutionary Microbiology, 66*, 3779–3785.

Haugaard, P., Hansen, F., Jensen, M., & Grunert, K. G. (2014). Consumer attitudes toward new technique for preserving organic meat using herbs and berries. *Meat Science, 96*, 126–135.

Héquet, A., Laffitte, V., Simon, L., De Sousa-Caetano, D., Thomas, C., Fremaux, C., & Berjeaud, J. M. (2007). Characterization of new bacteriocinogenic lactic acid bacteria isolated using a medium designed to simulate inhibition of *Listeria* by *Lactobacillus sakei* 2512 on meat. *International Journal of Food Microbiology, 113*, 67–74.

Hoffman, L. C., & Dicks, L. M. T. (2011). Preliminary results indicating game meat is more resistant to microbiological spoilage. In P. Paulsen, A. Bauer, M. Vodnansky, R. Winkelmayer, & F. J. M. Smulders (Eds.), *Game meat hygiene in focus: Microbiology, epidemiology, risk analysis and quality assurance* (pp. 137–139). Wageningen, The Netherlands: Wageningen Academic Publishers.

Holko, I., Hrabe, J., Salakova, A., & Rada, V. (2013). The substitution of a traditional starter culture in mutton fermented sausages by *Lactobacillus acidophilus* and *Bifidobacterium animalis*. *Meat Science, 94*, 275–279.

Höll, L., Behr, J., & Vogel, R. F. (2016). Identification and growth dynamics of meat spoilage microorganisms in modified atmosphere packaged poultry meat by MALDI-TOF MS. *Food Microbiology, 60*, 84–91.

Hospital, X. F., Hierro, E., Stringer, S., & Fernández, M. (2016). A study on the toxigenesis by *Clostridium botulinum* in nitrate and nitrite-reduced dry fermented sausages. *International Journal of Food Microbiology, 218*, 66–70.

Huang, H., Brooks, B. W., Lowman, R., & Carrillo, C. D. (2015). *Campylobacter* species in animal, food, and environmental sources, and relevant testing programs in Canada. *Canadian Journal of Microbiology, 61*, 701–721.

Huang, Y., Ye, K., Yu, K., Wang, K., & Zhou, G. (2016). The potential influence of two *Enterococcus faecium* on the growth of *Listeria monocytogenes*. *Food Control, 67*, 18–24.

Hue, O., Allain, V., Laisney, M. J., Le Bouquin, S., Lalande, F., Petetin, I., Rouxel, S., Quesne, S., Gloaguen, P. Y., Picherot, M., Santolini, J., Bougeard, S., Salvat, G., & Chemaly, M. (2011). *Campylobacter* contamination of broiler caeca and carcasses at the slaughterhouse and correlation with *Salmonella* contamination. *Food Microbiology, 28*, 862–868.

Hugas, M., Garriga, M., & Aymerich, M. T. (2003). Functionality of enterococci in meat products. *International Journal of Food Microbiology, 88*, 223–233.

Hultman, J., Rahkila, R., Ali, J., Rousu, J., & Björkroth, K. J. (2015). Meat processing plant microbiome and contamination patterns of cold-tolerant bacteria causing food safety and spoilage risks in the manufacture of vacuum-packaged cooked sausages. *Applied and Environmental Microbiology, 81*, 7088–7097.

Iacumin, L., Chiesa, L., Boscolo, D., Manzano, M., Cantoni, C., Orlic, S., & Comi, G. (2009). Moulds and ochratoxin A on surfaces of artisanal and industrial dry sausages. *Food Microbiology, 26*, 65–70.

Jääskeläinen, E., Johansson, P., Kostiainen, O., Nieminen, T., Schmidt, G., Somervuo, P., Mohsina, M., Vanninen, P., Auvinen, P., & Björkroth, J. (2013). Significance of heme-based respiration in meat spoilage caused by *Leuconostoc gasicomitatum*. *Applied and Environmental Microbiology, 79*, 1078–1085.

Jacome, S. L., Fonseca, S., Pinheiro, R., Todorov, S. D., Noronha, L., Silva, J., Gomes, A., Pintado, M., Morais, A. M. M. B., Teixeira, P., & Vaz-Velho, M. (2014). Effect of lactic acid bacteria on quality and safety of ready-to-eat sliced cured/smoked meat products. *Chemical Engineering Transactions, 38*, 403–408.

Jacquot, A., Neveu, D., Aujoulat, F., Mercier, G., Marchandin, H., Jumas-Bilak, E., & Picaud, J. C. (2011). Dynamics and clinical evolution of bacterial gut microflora in extremely premature patients. *Journal of Pediatrics, 158*, 390–396.

Jafari, M., Mortazavian, A. M., Hosseini, H., Safaei, F., Mousavi Khaneghah, A., & Sant'Ana, A. S. (2017). Probiotic *Bacillus*: Fate during sausage processing and storage and influence of different culturing conditions on recovery of their spores. *Food Research International, 95*, 46–51.

Janssens, M., Myter, N., De Vuyst, L., & Leroy, F. (2012). Species diversity and metabolic impact of the microbiota are low in spontaneously acidified Belgian sausages with an added starter culture of *Staphylococcus carnosus*. *Food Microbiology, 29*, 167–177.

Janssens, M., Myter, N., De Vuyst, L., & Leroy, F. (2013). Community dynamics of coagulase-negative staphylococci during spontaneous artisan-type meat fermentations differ between smoking and moulding treatments. *International Journal of Food Microbiology, 166*, 168–175.

Jayamanne, V. S., & Adams, M. R. (2006). Determination of survival, identity and stress resistance of probiotic bifidobacteria in bio-yoghurts. *Letters in Applied Microbiology, 42*, 189–194.

Jiang, Y., Gao, F., Xu, X. L., Su, Y., Ye, K. P., & Zhou, G. H. (2010). Changes in the bacterial communities of vacuum-packaged pork during chilled storage analyzed by PCR–DGGE. *Meat Science, 86*, 889–895.

Jimenez, E., Delgado, S., Maldonado, A., Arroyo, R., Albújar, M., García, N., Jariod, M., Fernández, L., Gómez, A., & Rodríguez, J. M. (2008). *Staphylococcus epidermidis*: A differential trait of the fecal microbiota of breast-fed infants. *BMC Microbiology, 8*, 143.

Johansson, P., Paulin, L., Säde, E., Salovuori, N., Alatalo, E. R., Björkroth, K. J., & Auvinen, P. (2011). Genome sequence of a food spoilage lactic acid bacterium, *Leuconostoc gasicomitatum* LMG 18811^T, in association with specific spoilage reactions. *Applied and Environmental Microbiology, 77*, 4344–4351.

Jones, R. J. (2004). Observations on the succession dynamics of lactic acid bacteria populations in chill-stored vacuum-packaged beef. *International Journal of Food Microbiology, 90*, 273–282.

Josephs-Spaulding, J., Beeler, E., & Singh, O. V. (2016). Human microbiome versus food-borne pathogens: Friend or foe. *Applied Microbiology and Biotechnology, 100*, 4845–4863.

Kang, D. H., Koohmaraie, M., & Siragus, G. R. (2001). Application of multiple antimicrobial interventions for microbial decontamination of commercial beef trim. *Journal of Food Protection, 64*, 168–171.

Klingberg, T. D., & Budde, B. B. (2006). The survival and persistence in the human gastrointestinal tract of five potential probiotic lactobacilli consumed as freeze-dried cultures or as probiotic sausage. *International Journal of Food Microbiology, 109*, 157–159.

Klingberg, T. D., Axelsson, L., Naterstad, K., Elsser, D., & Budde, B. B. (2005). Identification of potential probiotic starter cultures for Scandinavian-type fermented sausages. *International Journal of Food Microbiology, 105*, 419–431.

Koutsoumanis, K., Stamatiou, A., Skandamis, P., & Nychas, G.-J. E. (2006). Development of a microbial model for the combined effect of temperature and pH on spoilage of ground meat, and validation of the model under dynamic temperature conditions. *Applied and Environmental Microbiology, 72*, 124–134.

Kudirkienė, E., Bunevičienė, J., Brøndsted, L., Ingmer, H., Olsen, J. E., & Malakauskas, M. (2011). Evidence of broiler meat contamination with post-disinfection strains of *Campylobacter jejuni* from slaughterhouse. *International Journal of Food Microbiology, 145*, 5116–5120.

Kumari, A., Catanzaro, R., & Marotta, F. (2011). Clinical importance of lactic acid bacteria: A short review. *Acta Bio-medica, 82*, 177–180.

La Storia, A., Ferrocino, I., Torrieri, E., Di Monaco, R., Mauriello, G., Villani, F., & Ercolini, D. (2012). A combination of modified atmosphere and antimicrobial packaging to extend the shelf-life of beefsteaks stored at chill temperature. *International Journal of Food Microbiology, 158*, 186–194.

Laursen, B. G., Byrne, D. V., Kirkegaard, J. B., & Leisner, J. J. (2009). Lactic acid bacteria associated with a heat-processed pork product and sources of variation affecting chemical indices of spoilage and sensory characteristics. *Journal of Applied Microbiology, 106*, 543–553.

Lebert, I., Leroy, S., Giammarinaro, P., Lebert, A., Chacornac, J. P., Bover-Cid, S., Vidal, M., & Talon, R. (2007). Diversity of micro-organisms in environments and dry fermented sausages of French traditional small units. *Meat Science, 76*, 1112–1122.

Leisner, J. J., Laursen, B., Provost, H., Drider, D., & Dalgaard, P. (2007). *Carnobacterium*: positive and negative effects in the environment and in foods. *FEMS Microbiology Reviews, 31*, 592–613.

Leitzmann, C. (2014). Vegetarian nutrition: Past, present, future. *American Journal of Clinical Nutrition, 100*, 496–502.

Lemay, M. J., Choquette, J., Delaquis, P. J., Gariépy, C., Rodrique, N., & Saucier, L. (2002). Antimicrobial effect of natural preservatives in a cooked and acidified chicken meat model. *International Journal of Food Microbiology, 78*, 217–226.

Leroy, F., & Degreef, F. (2015). Convenient meat and meat products. Societal and technological issues. *Appetite, 94*, 40–46.

Leroy, F., & Praet, I. (2015). Meat traditions. The co-evolution of humans and meat. *Appetite, 90*, 200–211.

Leroy, F., & Praet, I. (2017). Animal killing and postdomestic meat production. *Journal of Agricultural and Environmental Ethics, 30*, 67–86.

Leroy, F., Verluyten, J., & De Vuyst, L. (2006). Functional meat starter cultures for improved sausage fermentation. *International Journal of Food Microbiology, 106*, 270–285.

Leroy, F., Vasilopoulos, C., Van Hemelryck, S., Falony, G., & De Vuyst, L. (2009). Volatile analysis of spoiled, artisan-type, modified-atmosphere-packaged cooked ham stored under different temperatures. *Food Microbiology, 26*, 94–102.

Leroy, S., Giammarinaro, P., Chacornac, J. P., Lebert, I., & Talon, R. (2010). Biodiversity of indigenous staphylococci of naturally fermented dry sausages and manufacturing environments of small-scale processing units. *Food Microbiology, 27*, 294–301.

Leroy, F., Geyzen, A., Janssens, M., De Vuyst, L., & Scholliers, P. (2013). Meat fermentation at the crossroads of innovation and tradition: A historical outlook. *Trends in Food Science and Technology, 31*, 130–137.

Leroy, F., Scholliers, P., & Amilien, V. (2015). Elements of innovation and tradition in meat fermentation: conflicts and synergies. *International Journal of Food Microbiology, 212*, 2–8.

Libera, J., Karwowska, M., Stasiak, D. M., & Dolatowski, Z. J. (2015). Microbiological and physicochemical properties of dry-cured neck inoculated with probiotic of *Bifidobacterium animalis* ssp. *lactis* BB-12. *International Journal of Food Science and Technology, 50*, 1560–1566.

Lingbeck, J. M., Cordero, P., O'Bryan, C. A., Johnson, M. G., Ricke, S. C., & Crandall, P. G. (2014). Functionality of liquid smoke as an all-natural antimicrobial in food preservation. *Meat Science, 97*, 197–206.

Lorenzo, J. M., & Gómez, M. (2012). Shelf life of fresh foal meat under MAP, overwrap and vacuum packaging conditions. *Meat Science, 92*, 610–618.

Lowry, P. D., & Gill, C. O. (1984). Mould growth on meat at freezing temperatures. *International Journal of Refrigeration, 7*, 133–136.

Lucquin, L., Zagorec, M., Champomier-Vergès, M., & Chaillou, S. (2012). Fingerprint of lactic acid bacteria population in beef carpaccio is influenced by storage process and seasonal changes. *Food Microbiology, 29*, 187–196.

Lytou, A. E., Panagou, E. Z., & Nychas, G.-J. E. (2017). Effect of different marinating conditions on the evolution of spoilage microbiota and metabolomic profile of chicken breast fillets. *Food Microbiology, 66*, 141–149.

Magistà, D., Susca, A., Ferrara, M., Logrieco, A. F., & Perrone, G. (2017). *Penicillium* species: Crossroad between quality and safety of cured meat production. *Current Opinion in Food Science, 17*, 36–40.

Marco, M. L., Heeney, D., Binda, S., Cifelli, C. J., Cotter, P. D., Foligné, B., Gänzle, M., Kort, R., Pasin, G., Pihlanto, A., Smid, E. J., & Hutkins, R. (2017). Health benefits of fermented foods: Microbiota and beyond. *Current Opinion in Biotechnology, 44*, 94–102.

Marin, C., Hernandez, A., & Lainez, M. (2009). Biofilm development capacity of *Salmonella* strains isolated in poultry risk factors and their resistance against disinfectants. *Poultry Science, 88*, 424–431.

Martín, B., Garriga, M., Hugas, M., & Aymerich, T. (2005). Genetic diversity and safety aspects of enterococci from slightly fermented sausages. *Journal of Applied Microbiology, 98*, 1177–1190.

Martín, B., Garriga, M., Hugas, M., Bover-Cid, S., Veciana-Nogues, M. T., & Aymerich, T. (2006). Molecular, technological and safety characterization of Gram-positive catalase-positive cocci from slightly fermented sausages. *International Journal of Food Microbiology, 107*, 148–158.

Martínez-Onandi, N., Castioni, A., San Martín, E., Rivas-Cañedo, A., Torriani, S., & Picon, A. (2017). Microbiota of high-pressure-processed Serrano ham investigated by culture-dependent and culture-independent methods. *International Journal of Food Microbiology, 241*, 298–307.

Matamoros, S., André, S., Hue, I., Prévost, H., & Pilet, M. P. (2010). Identification of lactic acid bacteria involved in the spoilage of pasteurized "foie gras" products. *Meat Science, 85*, 467–471.

Mauriello, G., Casaburi, A., Blaiotta, G., & Villani, F. (2004). Isolation and technological properties of coagulase negative staphylococci from fermented sausages of Southern Italy. *Meat Science, 67*, 149–158.

Mc Nulty, K., Soon, J. M., Wallace, C. A., & Nastasijevic, I. (2016). Antimicrobial resistance monitoring and surveillance in the meat chain: a report from five countries in the European Union and European Economic Area. *Trends in Food Science and Technology, 58*, 1–13.

Mendonça, R. C. S., Gouvea, D. M., Hungano, H. M., Sodre, A. F., & Querol-Simon, A. (2013). Dynamics of the yeasts flora in artisanal country style and industrial dry cured sausage. *Food Control, 29*, 143–148.

Metaxopoulos, J., Mataragas, M., & Drosinos, E. H. (2002). Microbial interaction in cooked cured meat products under vacuum or modified atmosphere at 4°C. *Journal of Applied Microbiology, 93*, 363–373.

Milios, K., Drosinos, E. H., & Zoiopoulos, P. E. (2014). Carcass decontamination methods in slaughterhouses: A review. *Journal of the Hellenic Veterinarian Medical Society, 65*, 65–78.

Morales, P. A., Aguirre, J. S., Troncoso, M. R., & Figueroa, G. O. (2016). Phenotypic and genotypic characterization of *Pseudomonas* spp. present in spoiled poultry fillets sold in retail settings. *Food Science and Technology, 73*, 609–614.

Müller, A., Reichhardt, R., Fogarassy, G., Basse, R., Gibis, M., Weiss, J., Schmidt, H., & Weiss, A. (2016). Safety assessment of selected *Staphylococcus carnosus* strains with regard to their application as meat starter culture. *Food Control, 66*, 93–99.

Muthukumarasamy, P., & Holley, R. A. (2007). Survival of *Escherichia coli* O157:H7 in dry fermented sausages containing micro-encapsulated probiotic lactic acid bacteria. *Food Microbiology, 24*, 82–88.

Nieminen, T. T., Vihavainen, E., Paloranta, A., Lehto, J., Paulin, L., Auvinen, P., Solsimaa, M., & Björkroth, K. J. (2011). Characterization of psychrotrophic bacterial communities in modified atmosphere-packed meat with terminal restriction fragment length polymorphism. *International Journal of Food Microbiology, 144*, 360–366.

Nieminen, T. T., Koskinen, K., Laine, P., Hultman, J., Säde, E., Paulin, L., Paloranta, A., Johansson, P., Björkroth, J., & Auvinen, P. (2012). Comparison of microbial communities in marinated and unmarinated broiler meat by metagenomics. *International Journal of Food Microbiology, 157*, 142–149.

Nieminen, T. T., Dalgaard, P., & Björkroth, J. (2016). Volatile organic compounds and *Photobacterium phosphoreum* associated with spoilage of modified-atmosphere-packaged raw pork. *International Journal of Food Microbiology, 218*, 86–95.

Nogueira Ruiz, J., Montes Villanueva, N. D., Pavaro-Trindade, C. S., & Contreras-Castillo, C. J. (2014). Physicochemical, microbiological and sensory assessments of Italian salami sausages with probiotic potential. *Scientia Argricola, 71*, 204–211.

Núñez, F., Lara, M. S., Peromingo, B., Delgado, J., Sánchez-Montero, L., & Andrade, M. J. (2015). Selection and evaluation of *Debaryomyces hansenii* isolates as potential bioprotective agents against toxigenic penicillia in dry-fermented sausages. *Food Microbiology, 46*, 114–120.

Nyquist, O. L., McLeod, A., Brede, D. A., Snipen, L., Aakra, Å., & Nes, I. F. (2011). Comparative genomics of *Lactobacillus sakei* with emphasis on strains from meat. *Molecular Genetics and Genomics, 285*, 297–311.

O'Hara, A. M., & Shanahan, F. (2006). The gut flora as a forgotten organ. *EMBO Reports, 7*, 688–693.

Oakley, B. B., Morales, C. A., Line, J., Berrang, M. E., Meinersmann, R. J., Tillman, G. E., Wise, M. G., Siragusa, G. R., Hiet, K. L., & Seal, B. S. (2013). The poultry-associated microbiome: Network analysis and farm-to-fork characterizations. *PLoS One, 8*, 1–11.

Olivares, M., Díaz-Ropero, M. P., Gómez, N., Sierra, S., Lara-Villoslada, F., Martín, R., Rodríguez, J. M., & Xaus, J. (2006). Dietary deprivation of fermented foods causes a fall in innate immune response. Lactic acid bacteria can counteract the immunological effect of this deprivation. *Journal of Dairy Research, 73*, 492–498.

Ortega, E., Abriouel, H., Lucas, R., & Galvez, A. (2010). Multiple roles of *Staphylococcus aureus* enterotoxins: Pathogenicity, superantigenic activity, and correlation to antibiotic resistance. *Toxins, 2*, 2117–2131.

Ortiz, S., López, V., Garriga, M., & Martínez-Suárez, J. V. (2014). Antilisterial effect of two bioprotective cultures in a model system of Iberian chorizo fermentation. *International Journal of Food Science and Technology, 49*, 753–758.

Papagianni, M., Ambrosiadis, I., & Filiousis, G. (2007). Mould growth on traditional Greek sausages and penicillin production by *Penicillium* isolates. *Meat Science, 76*, 653–657.

Patsias, A., Chouliara, I., Badeka, A., Savvaidis, I. N., & Kontominas, M. G. (2006). Shelf-life of a chilled precooked chicken product stored in air and under modified atmospheres: Microbiological, chemical, sensory attributes. *Food Microbiology, 23*, 423–429.

Patterson, M. P., Mckay, A. M., Connolly, M., & Linton, M. (2010). Effect of high pressure on the microbiological quality of cooked chicken during storage at normal and abuse refrigeration temperatures. *Food Microbiology, 27*, 266–273.

Pavelková, A., Kačániová, M., Horská, E., Rovná, K., Hleba, L., & Petrová, J. (2014). The effect of vacuum packaging, EDTA, oregano and thyme oils on the microbiological quality of chicken's breast. *Anaerobe, 29*, 128–133.

Pennacchia, C., Ercolini, D., & Villani, F. (2011). Spoilage-related microbiota associated with chilled beef stored in air or vacuum pack. *Food Microbiology, 28*, 84–93.

Pidcock, K., Heard, G. M., & Henriksson, A. (2002). Application of nontraditional meat starter cultures in production of Hungarian salami. *International Journal of Food Microbiology, 76*, 75–81.

Piotrowska-Cyplik, A., Myszka, K., Czarny, J., Ratajczak, K., Kowalski, R., Bieanska-Marecik, R., Staninska-Pieta, J., Nowak, J., & Cryplik, P. (2017). Characterization of specific spoilage organisms (SSOs) in vacuum-packed ham by culture-plating techniques and MiSeq next-generation sequencing technologies. *Journal of the Science of Food and Agriculture, 97*, 689–668.

Plé, C., Breton, J., Daniel, C., & Foligné, B. (2015). Maintaining gut ecosystems for health: Are transitory food bugs stowaways or part of the crew? *International Journal of Food Microbiology, 213*, 139–143.

Pothakos, V., Snauwaert, C., De Vos, P., Huys, G., & Devlieghere, F. (2014). Psychrotrophic members of *Leuconostoc gasicomitatum, Leuconostoc gelidum* and *Lactococcus piscium* dominate at the end of shelf-life in packaged and chilled-stored food products in Belgium. *Food Microbiology, 39*, 61–67.

Pothakos, V., Devlieghere, F., Villani, F., Björkroth, J., & Ercolini, D. (2015). Lactic acid bacteria and their controversial role in fresh meat spoilage. *Meat Science, 109*, 66–74.

Quercia, S., Candela, M., Giuliani, C., Turroni, S., Louiselli, D., Rampelli, S., Brigidi, P., Franceschi, C., Bacalini, M. G., Garagnani, P., & Pirazzini, C. (2014). From lifetime to evolution: Timescales of human gut microbiota adaptation. *Frontiers in Microbiology, 5*, 587.

Raeisi, M., Tabaraei, A., Hashemi, M., & Behnampour, N. (2016). Effect of sodium alginate coating incorporated with nisin, *Cinnamomum zeylanicum*, and rosemary essential oils on microbial quality of chicken meat and fate of *Listeria monocytogenes* during refrigeration. *International Journal of Food Microbiology, 238*, 139–145.

Rahkila, R., Johansson, P., Säde, E., & Björkroth, J. (2011). Identification of enterococci from broiler products and a broiler processing plant and description of *Enterococcus viikkiensis* sp. nov. *Applied and Environmental Microbiology, 77*, 1196–1203.

Rantsiou, K., Urso, R., Iacumin, L., Cantoni, C., Cattaneo, P., Comi, G., & Cocolin, L. (2005). Culture-dependent and -independent methods to investigate the microbial ecology of Italian fermented sausages. *Applied and Environmental Microbiology, 84*, 1043–1049.

Ravyts, F., De Vuyst, L., & Leroy, F. (2012). Bacterial diversity and functionalities in food fermentations. *Engineering in Life Sciences, 12*, 356–367.

Remenant, B., Jaffrès, E., Dousset, X., Pilet, M. F., & Zagorec, M. (2015). Bacterial spoilers of food: Behavior, fitness and functional properties. *Food Microbiology, 45*, 45–53.

Rimaux, T., Vrancken, G., Vuylsteke, B., De Vuyst, L., & Leroy, F. (2011). The pentose moiety of adenosine and inosine is an important energy source for the fermented-meat starter culture *Lactobacillus sakei* CTC 494. *Applied and Environmental Microbiology, 77*, 6539–6550.

Rimaux, T., Rivière, A., Illeghems, K., Weckx, S., De Vuyst, L., & Leroy, F. (2012). Expression of the arginine deiminase pathway genes in *Lactobacillus sakei* is strain dependent and is affected by environmental pH. *Applied and Environmental Microbiology, 78*, 4874–4883.

Roseiro, L. C., Gomes, A., & Santos, C. (2011). Influence of processing in the prevalence of polycyclic aromatic hydrocarbons in a Portuguese traditional meat product. *Food and Chemical Toxicology, 49*, 1340–1345.

Rosenstein, R., & Götz, F. (2013). What distinguishes highly pathogenic staphylococci from medium- and non-pathogenic? In U. Dobrindt, J. Hacker, & C. Svanborg (Eds.), *Between pathogenicity and commensalism. Current topics in microbiology and immunology* (Vol. 358, pp. 33–89). Berlin, Germany: Springer.

Rouger, A. (2017). *Déscription et comportement des communautés bactériennes de la viande de poulet conserve sous atmosphère protectrice*. PhD Thesis. Université Bretagne Loire, Rennes, France.

Rouger, A., Remenant, B., Prévost, H., & Zagorec, M. (2017). A method to isolate bacterial communities and characterize ecosystems from food products: Validation and utilization in a reproducible chicken meat model. *International Journal of Food Microbiology, 247*, 38–47.

Rouhi, M., Sohrabvandi, S., & Mortazavian, M. (2013). Probiotic fermented sausage: Viability of probiotic microorganisms and sensory characteristics. *Critical Reviews in Food Science and Nutrition, 53*, 331–348.

Rubio, R., Aymerich, T., Bover-Cid, S., Guàrdia, M. D., Arnau, J., & Garriga, M. (2013a). Probiotic strains *Lactobacillus plantarum* 299V and *Lactobacillus rhamnosus* GG as starter cultures for fermented sausages. *Food Science and Technology, 54*, 51–56.

Rubio, R., Bover-Cid, S., Martin, B., Garriga, M., & Aymerich, T. (2013b). Assessment of safe enterococci as bioprotective cultures in low-acid fermented sausages combined with high hydrostatic pressure. *Food Microbiology, 33*, 158–165.

Rubio, R., Jofré, A., Aymerich, T., Guàrdia, M. D., & Garriga, M. (2014a). Nutritionally enhanced fermented sausages as a vehicle for potential probiotic lactobacilli delivery. *Meat Science, 96*, 937–942.

Rubio, R., Jofré, A., Martín, B., Aymerich, T., & Garriga, M. (2014b). Characterization of lactic acid bacteria isolated from infant faeces as potential probiotic starter cultures for fermented sausages. *Food Microbiology, 38*, 303–311.

Rubio, R., Martín, B., Aymerich, T., & Garriga, M. (2014c). The potential probiotic *Lactobacillus rhamnosus* CTC1679 survives the passage through the gastrointestinal tract and its use as starter cultures results in safe nutritionally enhanced fermented sausages. *International Journal of Food Microbiology, 186*, 55–60.

Ruiz, J. N., Villanueva, N. D. M., Favaro-Trindade, C. S., & Contreras-Castillo, C. J. (2014). Physicochemical, microbiological and sensory assessments of Italian salami sausages with probiotic potential. *Scientia Agricola, 71*, 204–211.

Ruiz-Moyano, S., Martír, A., Benito, M. J., Hernández, A., Casquete, R., Serradilla, M. J., & Córdoba, M. G. (2009). Safety and functional aspects of pre-selected lactobacilli for probiotic use in Iberian dry-fermented sausages. *Meat Science, 83*, 460–467.

Ruiz-Moyano, S., Martír, A., Benito, M. J., Hernández, A., Casquete, R., Serradilla, M. J., & Córdoba, M. G. (2010). Safety and functional aspects of pre-selected pediococci for probiotic use in Iberian dry-fermented sausages. *International Journal of Food Science and Technology, 45*, 1138–1145.

Ruiz-Moyano, S., Martír, A., Benito, M. J., Hernández, A., Casquete, R., & Córdoba, M. G. (2011). Application of *Lactobacillus fermentum* HL57 and *Pediococcus acidilactici* SP979 as potential probiotics in the manufacture of traditional Iberian dry-fermented sausages. *Food Microbiology, 28*, 839–847.

Säde, E., Murros, A., & Björkroth, J. (2013). Predominant enterobacteria on modified-atmosphere packaged meat and poultry. *Food Microbiology, 34*, 252–258.

Säde, E., Lassila, E., & Björkroth, J. (2016). Lactic acid bacteria in dried vegetables and spices. *Food Microbiology, 53*, 110–114.

Salminen, S., Endo, A., Isolauri, E., & Scalabrin, D. (2015). Early gut colonization with lactobacilli and *Staphylococcus* in infants: The hygiene hypothesis extended. *Journal of Pediatric Gastroenterology and Nutrition, 62*, 80–86.

Samelis, J., Kakouri, A., & Rementzis, J. (2000). The spoilage microflora of cured, cooked turkey breasts prepared commercially with or without smoking. *International Journal of Food Microbiology, 56*, 133–143.

Sánchez Mainar, M., Xhaferi, R., Samapundo, S., Devlieghere, F., & Leroy, F. (2016). Opportunities and limitations for the production of safe fermented meats without nitrate and nitrite using an antibacterial *Staphylococcus sciuri* starter culture. *Food Control, 69*, 267–274.

Sánchez Mainar, M., Stavropoulou, D. A., & Leroy, F. (2017). Exploring the metabolic heterogeneity of coagulase-negative staphylococci to improve the quality and safety of fermented meats: A review. *International Journal of Food Microbiology, 247*, 24–37.

Sanz, Y., Sánchez, E., Marzotto, M., Calabuig, M., Torriani, S., & Dellaglio, F. (2007). Differences in faecal bacterial communities in coeliac and healthy children as detected by PCR and denaturing gradient gel electrophoresis. *FEMS Immunology and Medical Microbiology, 51*, 562–568.

Schirmer, B. C., Heir, E., & Langsrud, S. (2009). Characterization of the bacterial spoilage flora in marinated pork products. *Journal of Applied Microbiology, 106*, 1364–5072.

Selgas, M. D., & García, M. L. (2014). Yeasts. In F. Toldrá (Ed.), *Handbook of fermented meat and poultry* (2nd ed., pp. 139–146). Hoboken, NJ: Wiley-Blackwell.

Sharma, S., Thind, S. S., & Kaur, A. (2015). In vitro meat production system: Why and how? *Journal of Food Science and Technology, 52*, 7599–7607.

Shori, A. B. (2015). The potential applications of probiotics on dairy and non-dairy foods focusing on viability during storage. *Biocatalysis and Agricultural Biotechnology, 4*, 423–431.

Šimko, P. (2005). Factors affecting elimination of polycyclic aromatic hydrocarbons from smoked meat foods and liquid smoke flavorings. *Molecular Nutrition and Food Research, 49*, 637–647.

Simoncini, N., Rotelli, D., Virgili, R., & Quintavalla, S. (2007). Dynamics and characterization of yeasts during ripening of typical Italian dry-cured ham. *Food Microbiology, 24*, 577–584.

Sivertsvik, M., Rosnes, J. T., & Jeksrud, W. K. (2004). Solubility and absorption rate of carbon dioxide into non-respiring foods. Part 2. Raw fish fillets. *Journal of Food Engineering, 63*, 451–458.

Škaljac, S., Petrović, L., Tasić, T., Ikonić, P., Jokanović, M., Tomović, V., Džinić, N., Šojić, B., Tjapkin, A., & Škrbić, B. (2014). Influence of smoking in traditional and industrial conditions on polycyclic aromatic hydrocarbons content in dry fermented sausages (*Petrovská klobása*) from Serbia. *Food Control, 40*, 12–18.

Smil, V. (2013). *Should we eat meat? Evolution and consequences of modern carnivory*. Chichester, UK: Wiley-Blackwell.

Smolander, M., Alakomi, H., Ritvanen, T., Vainionpää, J., & Ahvenainen, R. (2004). Monitoring of the quality of modified atmosphere packaged broiler chicken cuts stored in different temperature conditions. A. Time-temperature indicators as quality-indicating tools. *Food Control, 15*, 217–229.

Sparo, M., Nuñez, G. G., Castro, M., Calcagno, M. L., Allende, M. A. G., & Ceci, M. (2008). Characteristics of an environmental strain, *Enterococcus faecalis* CECT7121, and its effects as additive on craft dry-fermented sausages. *Food Microbiology, 25*, 607–615.

Sparo, M. D., Confalonieri, A., Urbizu, L., Ceci, M., & Bruni, S. F. (2013). Bio-preservation of ground beef meat by *Enterococcus faecalis* CECT7121. *Brazilian Journal of Microbiology, 44*, 43–49.

Srey, S., Jahid, I. K., & Ha, S. D. (2013). Biofilm formation in food industries: A food safety concern. *Food Control, 31*, 572–585.

Stella, S., Soncini, G., Ziino, G., Panebianco, A., Pedonese, F., Nuvoloni, R., Giannatale, E. D., Colavita, G., Alberghini, L., & Giaccone, V. (2017). Prevalence and quantification of thermophilic *Campylobacter* spp. in Italian retail poultry meat: Analysis of influencing factors. *Food Microbiology, 62*, 232–238.

Stellato, G., La Storia, A., De Flippis, F., Borriello, G., Villani, F., & Ercolini, D. (2016). Overlap of spoilage-associated microbiota between meat and the meat processing environment in small-scale and large-scale retail distributions. *Applied and Environmental Microbiology, 82*, 4045–4054.

Stoops, J., Ruyters, S., Busschaert, P., Spaepen, R., Verreth, C., Claes, J., Lievens, B., & Van Campenhout, L. (2015). Bacterial community dynamics during cold storage of minced meat packaged under modified atmosphere and supplemented with different preservatives. *Food Microbiology, 48*, 192–199.

Sunesen, L. O., & Stahnke, L. H. (2003). Mould starter cultures for dry sausages-selection, application and effects. *Meat Science, 65*, 935–948.

Toldrá, F. (2014). *Handbook of fermented meat and poultry* (2nd ed.). Hoboken, NJ: Wiley-Blackwell.

Toldrá, F. (2017). *Lawrie's meat science*. Duxford, UK: Woodhead Publishing.

Tremonte, P., Sorrentino, E., Pannella, G., Tipaldi, L., Sturchio, M., Masucci, A., Maiuro, L., Coppola, R., & Succi, M. (2017). Detection of different microenvironments and *Lactobacillus sakei* biotypes in Ventricina, a traditional fermented sausage from Central Italy. *International Journal of Food Microbiology, 242*, 132–140.

Tripathi, M. K., & Giri, S. K. (2014). Probiotic functional foods: Survival of probiotics during processing and storage. *Journal of Functional Foods, 9*, 225–241.

Troy, D. J., & Kery, J. P. (2010). Consumer perception and the role of science in the meat industry. *Meat Science, 86*, 214–226.

Urso, R., Rantsiou, K., Cantoni, C., Comi, G., & Cocolin, L. (2006). Technological characterization of a bacteriocin-producing *Lactobacillus sakei* and its use in fermented sausages production. *International Journal of Food Microbiology, 110*, 232–239.

Vasilopoulos, C., Ravyts, F., De Maere, H., De Mey, E., Paelinck, H., De Vuyst, L., & Leroy, F. (2008). Evaluation of the spoilage lactic acid bacteria in modified-atmosphere-packaged artisan-type cooked ham using culture-dependent and culture-independent approaches. *Journal of Applied Microbiology, 104*, 1341–1353.

Vasilopoulos, C., De Maere, H., De Mey, E., Paelinck, H., De Vuyst, L., & Leroy, F. (2010a). Technology-induced selection towards the spoilage microbiota of artisan-type cooked ham packed under modified atmosphere. *Food Microbiology, 27*, 77–84.

Vasilopoulos, C., De Mey, E., Pevulf, L., Paelinck, H., De Smedt, A., Vandendriessche, F., De Vuyst, L., & Leroy, F. (2010b). Interactions between bacterial isolates from modified-atmosphere-packaged artisan-type cooked ham in view of the development of a bioprotective culture. *Food Microbiology, 27*, 1086–1094.

Vasilopoulos, C., De Vuyst, L., & Leroy, F. (2015). Shelf-life reduction as an emerging problem in cooked hams underlines the need for improved preservation strategies. *Critical Reviews in Food Science and Nutrition, 55*, 1425–1443.

Verbeke, W. (2015). Profiling consumers who are ready to adopt insects as a meat substitute in a Western society. *Food Quality and Preference, 39*, 147–155.

Vermeiren, L., Devlieghere, F., Vandekinderen, I., & Debevere, J. (2006). The interaction of the non-bacteriocinogenic *Lactobacillus sakei* 10A and lactocin S producing *Lactobacillus sakei* 148 towards *Listeria monocytogenes* on a model cooked ham. *Food Microbiology, 23*, 511–518.

Vihavainen, E. J., & Björkroth, K. J. (2007). Spoilage of value-added, high-oxygen modified-atmosphere packaged raw beef steaks by *Leuconostoc gasicomitatum* and *Leuconostoc gelidum*. *International Journal of Food Microbiology, 119*, 340–345.

Villani, F., Casaburi, A., Pennacchia, C., Filosa, L., Russo, F., & Ercolini, D. (2007). Microbial ecology of the Soppressata of Vallo di Diano, a traditional dry fermented sausage from Southern Italy, and in vitro and in situ selection of autochthonous starter cultures. *Applied and Environmental Microbiology, 73*, 5453–5463.

Virgili, R., Simoncini, N., Roscani, T., Leggierei, M. C., Formenti, S., & Battilani, P. (2012). Biocontrol of *Penicillium nordicum* growth and ochratoxin A production by native yeasts of dry cured ham. *Toxins, 4*, 68–82.

Vranken, L., Avermaete, T., Petalios, D., & Mathijs, E. (2014). Curbing global meat consumption. Emerging evidence of a second nutrition transition. *Environmental Science and Policy, 39*, 95–106.

Walter, J., Hertel, C., Tannock, G. W., Lis, C. M., Munro, K., & Hammes, W. P. (2001). Detection of *Lactobacillus, Pediococcus, Leuconostoc*, and *Weissella* species in human feces by using group-specific PCR primers and denaturing gradient gel electrophoresis. *Applied and Environmental Microbiology, 67*, 2578–2585.

Wang, X., Lin, X., Ouyang, Y. Y., Liu, J., Zhao, G., Pan, A., & Hu, F. B. (2015). Red and processed meat consumption and mortality: Dose-response meta-analysis of prospective cohort studies. *Public Health Nutrition, 19*, 893–905.

Wang, G. J., Wang, H. H., Han, Y. W., Xing, T., Ye, K. P., Xu, X. L., & Zhou, G. H. (2017). Evaluation of the spoilage potential of bacteria isolated from chilled chicken *in vitro* and *in situ*. *Food Microbiology, 63*, 139–146.

Wójciak, K. M., Libera, J., Stasiak, D. M., & Kolozyn-krajewska, D. (2016). Technological aspect of *Lactobacillus acidophilus* Bauer, *Bifidobacterium animalis* BB-12 and *Lactobacillus rhamnosus* LOCK900 use in dry-fermented pork neck and sausage. *Journal of Food Processing and Preservation, 41*, 1–9.

Yalcin, S., Nizamlioclu, M., & Gurbuz, U. (2001). Fecal coliform contamination of beef carcasses during the slaughtering process. *Journal of Food Safety, 21*, 225–231.

Zhang, H., Kong, B., Xiong, Y. L., & Sun, X. (2009). Antimicrobial activities of spice extracts against pathogenic and spoilage bacteria in modified atmosphere packaged fresh pork and vacuum packaged ham slices stored at 4°C. *Meat Science, 81*, 686–692.

Zhang, C., Derrien, M., Levenez, F., Brazeilles, R., Ballal, S. A., Kim, J., Degivry, M. C., Quéré, G., Carault, P., van Hylckama Vlieg, J. E. T., Garrett, W. S., Doré, J., & Veiga, P. (2016). Ecological robustness of the gut microbiota in response to ingestion of transient food-borne microbes. *The ISME Journal, 10*, 2235–2245.

Zhong, L., Zhang, X., & Covasa, M. (2014). Emerging roles of lactic acid bacteria in protection against colorectal cancer. *World Journal of Gastroenterology, 20*, 7878–7886.

Fermented Vegetables as Vectors for Relocation of Microbial Diversity from the Environment to the Human Gut

Ilenys M. Pérez-Díaz

Abstract The discovery of yeasts as living cells able to produce ethanol in fermented foods and beverages in the 1920s continues to captivate our imagination with respect to the functionality and role of microbes in food preservation and human health. Mounting evidence confirms the ability of microbes to deliver nutrition, flavor and many bio-functionalities to fermented foods and the gastrointestinal (GI) tract of mammals. The microbial diversity found in fermented foods, particulalrly vegetables, can benefit the human GI tract microbiome. Critical functions for microbes associated with fresh vegetables include the contribution to growth, development and defense of host plants. In parallel, plants have evolved to select and maintain beneficial microbes, including those within their tissue. Fermentation then serves as an instrument to pre-adapt beneficial microbes indigenous to fresh vegetables to the acidic pH and high lactic acid concentration characteristic of the colon and to the metabolism of dietary fiber, particularly pectic substances naturally present in the plant material and the gut. Fermented vegetable products enjoy a long-lasting record of safety upon consumption and are an appropriate vector for the translocation of microbial diversity from plants to the gut. Fermented vegetables can enhance prebiotic fiber and beneficial microbes content and consequently augment the catalog of metabolic functions needed in and available to the gut for building resilience in a healthy individual. It is the indigenous microbiota of fermented vegetables and intrinsic chemical composition of substrates, particularly dietary fibers, which can enable beneficial health claims from the consumption of pickles.

Keywords Vegetables microbiome · Natural fermentation of vegetables · Spoilage · Cucumber fermentation · Sauerkraut · Lactic Acid Bacteria (LAB)

I. M. Pérez-Díaz (✉)
USDA-ARS Food Science Research Unit, Raleigh, NC, USA
e-mail: Ilenys.Perez-Diaz@ARS.USDA.GOV

© Springer Nature Switzerland AG 2019
M. A. Azcarate-Peril et al. (eds.), *How Fermented Foods Feed a Healthy Gut Microbiota*, https://doi.org/10.1007/978-3-030-28737-5_4

Introduction

Studies of the indigenous microbiota in fermented vegetables began in the 1920s, a few centuries after the discoveries of microbes as living cells capable of producing ethanol and fermenting milk by Antonie van Leeuwenhoek, Cagnard-Latour, Louis Pasteur and Joseph Lister occurred (Brock 1961; Nanniga 2010). There was no concept or understanding of microbial diversity in the early 1900s and the tools available for microbiological studies were limited compared to the knowledge base and tools available in the twenty-first century.

Lacking an understanding of the microbiology behind a desirable fermentation, past generations performed what is known as "back slopping" or the use of cover brines or doughs from fermentations with desirable attributes to initiate fresh fermentations in an attempt to perpetuate specific organoleptic attributes in the desired fermented foods (Cogan 1996). The fermentation of vegetables in the twentieth century consisted of dry-salting of shredded cabbage, turnips and lettuce and whole grains of corn, lima beans and green peas to support a vigorous conversion of the sugars to lactic and acetic acids and possibly ethanol (Etchells et al. 1947). Bulky vegetables, some with a low water content, were chopped prior to brining or brined whole with varied sodium chloride concentrations (Etchells et al. 1947). When and exactly how the preservation of vegetables by fermentation began is unknown but it is chronologically situated between the first and third centuries before Christ (B.C.). Records of mixed vegetable fermentations date back to the third century B. C. during the construction of the Great Wall in China (Anderson et al. 1988; Lee 2001). Sauerkraut production was described as early as the first century by Plinius the elder (Buckenhüskes et al. 1990). The diverse preparation forms of table olives were also described by Columela in the book *The Re Rustica* in the I century (Columela 1979, 45). Early written records of cucumber pickles come from surviving fragments of a play (*The Taxiarchs*) by the Greek writer Eupolis (429–412 BC), and pickles are mentioned several times in the Christian bible. Today the consumption of vegetables is widespread in the world and represents an important component of the human diet.

This chapter addresses the advances made in understanding the indigenous microbiota in fermenting and fermented vegetables and the influence of modern industrial production practices on microbial diversity. The consequent role of fermented vegetables as a delivery vehicle for microbes to the human gut is also described. The many metabolic and physiological functionalities of the cultures present in fermented vegetables is beyond the scope of this chapter.

For the purpose of this chapter fermented vegetables are defined as low acid vegetables subjected to the action of acid producing microorganisms that will naturally achieve and maintain a pH of 4.6 or lower, regardless of whether acid is added (Pérez-Díaz et al. 2014). If the fermentation proceeds to completion and good manufacturing practices are applied, spoilage organisms capable of rising the pH above 4.6 are prevented from growing in the product and pathogens of public health significance are destroyed during the process, thus making the final product safe for consumption (Ito et al. 1976; Breidt and Caldwell 2011).

The Microbiota of Fresh Vegetables

Microbial diversity on fresh vegetables primarily derives from the soil. A low incidence of lactic acid bacteria occurs in fresh vegetables.

The cucurbits rhizoplane is known to predominantly host *Rhizobium* and *Cellvibrio* and to a lesser extent *Saccharophagus, Devosia*, and *Pseudomonas* (Ofek et al. 2014). *Cellvibrio*, a Pseudomonadaceae, may reach up to 20% of the cucumber plant rhizoplane microbial population and can degrade plant cell wall components and other complex polysaccharides which enables microbes to penetrate and colonize the plant tissue (DeBoy et al. 2008). Dried seeds used for planting a new crop are known vectors of microbial diversity for plants, flowers and fruits (Lemanceau et al. 2017). Although microbial diversity could also come from the soil and/or bioaerosols, it is documented that seeds richer in oil (50%), protein (35%) and DNA, contribute the most to the selection and evolution of endophytes, which are microbes that reside in the internal plant tissues without adverse effects on their host (Lemanceau et al. 2017).

Microbes that colonize plant tissue contribute to the host growth and development in multiple ways. *Bacillus* species contribute to cucumber plants through nitrogen-fixation and scavenging (converting such gas into a solid and usable form), deaminase activity and protease, pectinase or cellulose activity. *Bacillus* spp., Enterobacteriaceae and LAB assist with phosphate solubilization through the production of organic acids or phosphatases (Khalaf and Raizada 2016). The Enterobacteriaceae family and *Pseudomonas*, are also able to produce auxin, a plant growth hormone, and siderophores used to chelate iron. Interestingly, *Bacillus* species isolated from cucumber seeds cluster apart from Bacilli isolated from other cucurbit seeds (Khalaf and Raizada 2016). Given the specialized functionality of the plant derived endophytes, it is speculated that they may be transmitted by seeds and conserved for future generations to help secure this important symbiotic relationship between plants and their microbiomes.

The relationship between microbes and plants is enabled by the nutritional and anti-nutritional factors (i.e. oxalate, lectins, tannins, phytic acid) intrinsically present in the later (Filannino et al. 2018). The response of lactic acid bacteria (LAB) to various stresses on the vegetation results in nutrient enhancement, stress reduction and consequently plant growth promotion (Filannino et al. 2018). Concomitantly, the plant-produced food is enriched in bioavailable and bioactive compounds (Filannino et al. 2018). Plants select their microbiome in the vicinity of the plant roots or rhizosphere, and seeds are involved in the transmission of microorganisms to future generations (Lemanceau et al. 2017). Microbes associated with vegetation have evolved to benefit from specific plants because of the easy access to nutrients (Khalaf and Raizada 2016). Microbes with a positive impact on plant growth and health are selected and maintained to evolve within the system.

Plant growth and development involves both biotic and abiotic factors. The range of abiotic factors produced by plants including oxygen, organic acids, vitamins and sugars can be used as nutrients and signals by microbes. Conversely, abiotic factors produced by microbes including hormones, volatile compounds and small molecules impact plants immunity and growth. The close association of vegetables with the soil promotes higher microbial density and diversity in fresh produce (Samish and Etinger-Tulczynsky 1962). The availability of oxygen promotes microbial colonization of the blossom end, seed cavity and the outer 6 mm layer of cucumbers, including the exocarp and a portion of the mesocarp, consisting of 5–6 log of CFU/g of total aerobic microbes (Mattos et al. 2005). In tomatoes, microbial colonization is more frequently found near the stem-scar and central core and decreases closer to the exocarp (Lemanceau et al. 2017; Rastogi et al. 2012). Cabbage contains the greatest numbers of bacteria on the outer leaves and lower numbers toward the center of the head (Pederson and Albury 1969). The adhesion of bacteria to cucumber exocarp depends on contact time, cell species and density, and temperature. These factors impact the adhesion of *Salmonella, Staphylococcus, Lactobacillus* and *Listeria* to fresh cucumber surfaces in an aqueous solution (Reina et al. 2002). Bacterial adhesion to the cucumber exocarp is less extensive at lower temperatures and shorter contact times (Reina et al. 2002). While Gram-negative bacteria, which are mostly motile, migrate to the cucumber mesocarp and persist commensally, inoculated gram positive LAB establish on the exocarp (Samish and Etinger-Tulczynsky 1962). Enterobacteriaceae, in particular the motile rods *Erwinia* spp., are known to colonize the internal cucumber tissue and produce carbon dioxide (CO_2) from fermentative metabolism in the presence of oxygen (Samish and Etinger-Tulczynsky 1962).

The core bacteriome of fresh vegetables including fresh cucumbers, corn, cabbage, carrots, spinach and peas is composed of the two taxonomical families, Enterobacteriaceae and Pseudomonadaceae (Lopez-Velasco et al. 2013; Manani et al. 2006; Samish et al. 1963; Samish and Etinger-Tulczynsky 1962; Shi et al. 2009; Weiss et al. 2007) (Figs. 1 and 2). However, fresh produce contains a diverse range of epiphytic microbiota. Average aerobic colony counts for fresh cucumbers, cabbage, and olives are estimated at 5.16 ± 0.76, 4.84 ± 0.26 and 1.90 ± 0.50 CFU/g, respectively (Pérez-Díaz et al. 2014). Colony counts from Violet Red Bile agar plates for lactose fermenting coliforms from cucumber, cabbage and olive have been estimated at 4.58 ± 0.98, 4.36 ± 0.06 and below detectable levels, respectively (Pérez-Díaz et al. 2014). *Pseudomonas, Pantoea, Chryseomonas* and *Enterobacter* colonize tomatoes, lettuce and Chinese cabbage (Lee et al. 2017; Ottesen et al. 2016; Shi et al. 2009). *Bacillus* species are present in lettuce, tomatoes and cucumbers (Ottesen et al. 2016; Shi et al. 2009; Rastogi et al. 2012; Ofek et al. 2014). The microbiome of tomatoes include *Microvirga, Sphingomonas, Brachybacterium, Rhizobiales, Paracocccus, Microbacterium, Cyanobacterium, Hafnia,* and *Erwinia* (Ottesen et al. 2016; Shi et al. 2009). The lettuce core microbiota includes the microorganisms listed above for tomatoes plus *Massilia* and *Arthrobacter* (Rastogi et al. 2012). Furthermore, *Acinetobacter, Burkholderia, Dickeya, Klebsiella, Pectobacterium, Rahnella, Serratia* and *Stenotrophomonas* colonize fresh lettuce and Chinese cabbage (Lee et al. 2017). A study of the fresh cucumber microbiota showed the dominance of the Gram-negative bacteria *Rhizobium, Pseudomonas, Acinetobacter, Sphingomonas, Sphingobacterium, Methylobacterium,*

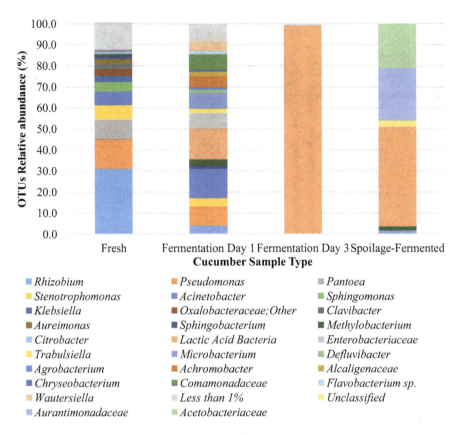

Fig. 1 Description of the relative abundance of OTUs found in samples of fresh and fermented cucumbers as determined by high throughput sequencing. Cucumber fermentation cover brine samples were obtained from batches brined with 6% sodium chloride. Only the bacterial communities are represented. Data was adapted from Medina-Pradas et al. (2016) and Pérez-Díaz et al. (2016, 2018)

Stenotrophomonas, *Citrobacter* and *Klebsiella* spp. (Pérez-Díaz et al. 2018). Despite the variations in the microbiota of specific vegetables in terms of density and diversity induced by type, variety, lot and harvesting location (Samish et al. 1963), more recent investigations revealed the occurrence of a vegetable-specific core microbiota (Rastogi et al. 2012). However, a noticeable difference exist between the microbial load of vegetables cultivated in greenhouses and those grown in the field (Meneley and Stanghellini 1974; Leben 1972; Geldreich and Bordner 1971). The numbers of internally-borne bacteria in healthy cucumbers grown in trellises and covered by a glasshouse was below detection levels (Meneley and Stanghellini 1974). Coliforms are absent in indoor-cultivated-foliage as well (Geldreich and Bordner 1971).

Fig. 2 Summary of microbes found in a cucumber plant and the fresh fruit (left panel), in cucumber fermentations (central panel) and the human gut (right panel), followed by an abbreviated list of functions needed to compete in the various habitats

Of interest is the fact that the LAB that dominate in vegetable fermentations are present in the corresponding fresh substrate in minimal numbers. Colony counts from MRS agar plates, typically used for the enumeration of LAB, particularly lactobacilli, from fresh cucumbers and cabbage are estimated at 3.84 ± 1.21 and 3.18 ± 0.33 Log CFU/g, respectively (Pérez-Díaz et al. 2014). In fresh olives the colony counts from MRS agar plates tend to be below detection levels, presumably due to its natural antimicrobial phenolic content (Pérez-Díaz et al. 2014). A characterization of the population forming colonies on MRS agar plates inoculated with fresh cucumber homogenate revealed lack of selectivity by such medium (Pérez-Díaz et al. 2018). *Enterococcus* (25%), *Exiguobacterium* (15%), *Lactococcus* (15%), *Staphylococcus* (13%), *Lactobacillus* (11%) and *Leuconostoc* (10%) are frequently isolated from MRS plates and three genera are infrequently encountered including *Bacillus* (6%), *Aerococcus* (4%) and *Clostridium* (1%) (Pérez-Díaz et al. 2018). Only 8.6% of the colonies formed in MRS agar plates belonged to the *Lactobacillus plantarum* cluster, a species that prevails in vegetable fermentations and are widely used as starter cultures. However, presumptive LAB have been found in plant material in substantial numbers with seasonal-related variability (Mundt 1970; Mundt and Hammer 1968).

The Microbiota of Natural Fermentations

LAB Consistently Prevail in Vegetable Fermentations

The diversity of the microbial population found associated with fresh vegetables is drastically reduced during the fermentation process, which supports the safety record of such preserved foods. It is generally accepted that the vegetable fermentation microbiota is dominated by three stages: initiation, primary fermentation and secondary fermentation (Garrido Fernández et al. 1997). During the initiation stage, the various Gram-positive and Gram-negative bacteria that colonize the fresh vegetable compete for dominance. The Enterobacteriaceae, *Bacillus* spp., LAB and a few other bacteria and yeasts may be active for several days or weeks depending on the temperature, water content, oxygen availability and salt concentration (Fuccio et al. 2016; Nychas et al. 2002; Panagou et al. 2008; Pérez-Díaz et al. 2016, 2018; Botta and Cocolin 2012; De Angelis et al. 2015). Eventually, the LAB prevail during primary and secondary fermentation due to the low pH from the conversion of sugars to organic acids. Seven species of LAB prevail in vegetable fermentations: *Enterobacter* spp., *Leuconostoc mesenteroides*, *Pediococcus* spp., *Lactobacillus brevis*, *Lb. plantarum*, *Lb. pentosus* and *Weissella* spp. (Bleve et al. 2015; Botta and Cocolin 2012; De Angelis et al. 2015; Etchells et al. 1973; Hong et al. 2016; Kyung et al. 2015; Lee et al. 2016, 2017; Nychas et al. 2002; Park et al. 2012; Plengvidhya et al. 2007; Pérez-Díaz et al. 2016, 2018). *Leuconostoc*, *Lactococcus* and *Weissella* tend to lead during the primary fermentation of vegetables such as cabbage. *Lactobacillus plantarum* and *Lb. pentosus* are typically found in the finished vegetable fermentations due to their resistance to extreme acidic pH (McDonald et al. 1990). A succession of microbes is often needed to complete a vegetable fermentation. The complete fermentation of cabbage requires the activity of *Leuc. mesenteroides*, *Leuc. citreum* and *Weissella* spp. which decrease the pH to approximately 6.5, and are followed by *Lb. plantarum*, *Lb. curvatus* and other lactobacilli, which continue to drop the pH to about 4.5 (Plengvidhya et al. 2007). *L. brevis* ends the production of acids in a cabbage fermentation spearheading the final decrease in pH to ~4.0. The fermentation of kimchi proceeds at 18 °C for a few days followed by a longer incubation period at refrigerated temperatures to promote microbial stability and reduce the development of excess sourness (Pérez-Díaz et al. 2014). This type of temperature control provides advantageous conditions for the proliferation of heterofermentative *Leuconostoc* spp. at the outset, followed by the growth of homofermentative lactobacilli and *Weissella* spp. (Jung et al. 2012). The use of *Leuc. citreum* as a starter culture for kimchi fermentation has proven to prevent over-ripening and growth of yeasts during refrigerated storage (Chang and Chang 2010).

Even though, salting is a critical step in vegetable fermentations, the function of sodium chloride with regards to the fermentation microbes is intrinsically restricted to the modulation of the density of certain species (Pérez-Díaz et al., submitted). The fermentation of cucumbers in water and 0.1% potassium sorbate to inhibit yeasts results in the dominance of the heterofermentative lactic acid bacterium, *Leuconostoc* and the homofermentor, *Lactococcus* (Pérez-Díaz et al., submitted). Supplementation of cover brines with NaCl compromises the proliferation of *Leuconostoc* and *Lactococcus,* opening an opportunity for *Weissella* to prevail. Additionally, in systems brined with NaCl, *Weissella*, a heterofermentative lactic acid bacterium, competes with a number of Gram negative bacteria that are likely to compromise the quality of the finished product. *Lactobacillus* and *Pediococcus* are noted in salt free fermentations by day 3. The addition of a salt in cucumber fermentation cover brines results in the dominance of *Lactobacillus* by day 7. Conversely, a lack of a salt in the fermentation cover brines yields comparatively more microbial diversity with less acid production even after 14 days. *Pediococcus, Leuconostoc, Lactococcus* and *Weissella* are present in salt free cucumber fermentations by day 14. The spoilage associated *Enterobacter* is also present in salt free cucumber fermentations by day 14. The microbes present in cucumber fermentations can produce acid and tolerate some salt and extremely acidic pH.

Bacteria present in fresh cucumbers such as *Rhizobium, Pseudomonas, Acinetobacter, Sphingomonas, Stenotrophomonas, Sphingobacterium, Methylobacterium, Klebsiella, Pantoea* and *Citrobacter* are also present on the first day of cucumber fermentations (Pérez-Díaz et al. 2018) (Figs. 1 and 2). However, the density of such populations start to decrease as a function of time and acid production. The presence of opportunistic pathogens such as *Citrobacter freundii, C. brakii, Enterobacter* spp., *Pseudomonas fluorescens, Stenotrophomonas maltophilia* and *Kluyvera cryocrescens*, and the antibiotic resistant pathogen *Serratia marcescens* during the initial stage of commercial cucumber fermentations is jeopardized by temperature, sodium chloride content of at least 3.5%, a pH below 4.5 and oxygen availability (Rothwell et al., unpublished; Olsen and Pérez-Díaz 2009). The population density corresponding to these organisms reaches undetectable levels in commercial cucumber fermentations by day 10 (Pérez-Díaz et al. 2018). It is presumed that upon initiation of the fermentation, the majority of the microbial population on fresh cucumbers, which localizes on the cucumbers exocarp (Mattos et al. 2005), is exposed to the full strength cover brine containing between 12 and 18% NaCl, a known preservative. A study of the salt tolerance of various LAB isolated from Spanish-style fermented olives and natural black olive fermentations suggests that 66% are inhibited by 6% NaCl and those that are resistant cannot grow in the presence of 9% of the salt (Balatsouras 1985). Salt content is gradually increased to7% in Spanish-style table olive fermentation as a function of equilibration to maintain stability (Garrido Fernández et al. 1997).

Some particularities apply to the fermentation of certain vegetables. The microbiota of watery kimchi is dominated by *Leuconostoc*, Enterobacteriaceae, Lactobacillaceae and *Pseudomonas* regardless of fermentation temperature (Kyung et al. 2015). A study of the microbiota in traditional Korean cabbage kimchi revealed the presence of, in order of prevalence, *Pediococcus pentosaceus, Leuconostoc citreum, Leuconostoc gelidum, Leuconostoc mesenteroides, Tetragenococcus, Pseudomonas* and *Weissella* (Hong et al. 2016). Household and commercial kimchi fermentations were found to harbor the organisms listed above and *Psychrobacter*,

Hafnia, Lactococcus, Rahnella, Enterobacter, and *Pantoea* (Lee et al. 2017). A study of ten representative kinds of kimchi that were refrigerated at 4 °C for 30–35 days found that although some microbial diversity is present in the early stage of the fermentation, *Weissella, Leuconostoc* and *Lactobacillus* are dominant in the later stage (Park et al. 2012). Sauerkraut fermentations are characterized by the presence of *Lc. mesenteroides, Lc. citreum, Lc. argentinum, Lactobacillus plantarum, Lb. paraplantarum, Lb. coryniformis, Pediococcus pentosaceous, Lb. brevis* and *Weissella* (Plengvidhya et al. 2007). Commercial cucumber fermentations are dominated by *Lactobacillus pentosus, Lb. brevis* and *Lb. plantarum* regardless of salt type or content, and are followed, in relative abundance, by *Pediococcus, Leuconostoc, Lactococcus* and *Weissella* (Pérez-Díaz et al. 2018). The fermentation of other vegetables products such as capers consist of lactic acid production by predominantly *Lb. plantarum* and to a lesser extent *Lb. paraplantarum, Lb. pentosus, Lb. brevis, Lb. fermentum, P. pentosaceus, P. acidilactici*, and *Enterococcus faecium* (Pérez Pulido et al. 2005).

Although, many researchers associated yeast with cucumber fermentations as early as the 1910s, given that gas production and bubbling was observed in the commercial process, a systematic classification study was not conducted until 1941. The average yeast population count in commercial cucumber fermentations is estimated to initiate at 3 log of CFU/g and increases to 5–6 log of CFU/g in about 10 days (Etchells 1941). Such population remains somewhat stable until about days 20–30 of the fermentation followed by a gradual decline to undetectable levels (Etchells 1941). Fluctuations in yeast counts are commonly observed in cucumber fermentations with salt concentrations between 5 and 10% (Etchells 1941), with a peak of activity on days 15 and 20, respectively (Etchells 1941). Yeast cells are primarily found on the skin of the cucumber fruits, exposed mesocarps and the fermentation cover brines. The size of yeast cells prevents their penetration into the mesocarp through the fruits exocarp (Daeschel et al. 1985). A study classified 47 surface film-forming yeasts isolated from fermentations containing between 5 and 19% NaCl from various locations in the USA in the following genera: *Debaryomyces membranaefaciens* var. *Hollandicus* (18), *Debaryomyces* sp. (4), *Endomycopsis ohmeri* (12), *Zygosaccharomyces halomembranis* (9) and *Candida krusei* (4) (Etchells and Bell 1950b, Fig. 3). *Pichia* was also isolated by the same group from fermentations containing less than 5% NaCl (Etchells and Bell 1950b). A study carried out to identify 1226 subsurface yeast isolated from 42 commercial cucumber fermentations revealed four predominating genera in increasing order of abundance: the tiny yeast *Torulopsis caroliniana* (718), acid producing *Brettanomyces versatilis* (559), hyperosmophilic *Zygosaccharomyces halomembranis* (59), *Hansenula subpelliculosa* (49), *Torulaspora rosei* (6), *Torulopsis holmii* (4), *Brettanomyces sphaericus* (2), and *Kloeckera magna* (1) (Etchells and Bell 1950a, Fig. 4). In the fermentations studied, *T. caroliniana* dominated until day 30 of the fermentations and was followed by *B. versatilis* and *Zygosaccharomyces* spp. until the end of the study at 100 days (Etchells and Bell 1950a). *T. holmii* and *Torulaspora rosei* were detected towards the end of the *T. caroliniana* fermentation. *Brettanomyces* spp. were found to

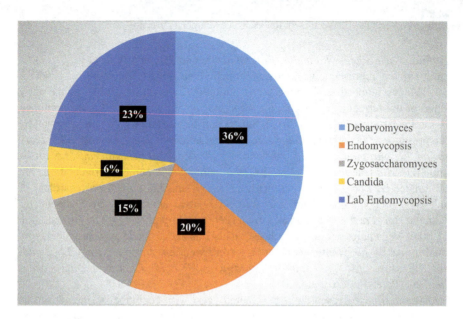

Fig. 3 Populations of surface oxidative yeasts isolated from commercial cucumber fermentations brined with 5–19% NaCl and conducted in 40 outdoor vats packed between 1947 and 1948. Data adapted from Etchells and Bell (1950b)

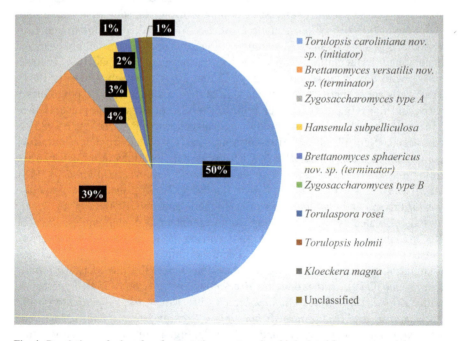

Fig. 4 Populations of subsurface fermentative yeasts and molds isolated from commercial cucumber fermentations brined with 6 ± 2% NaCl and conducted in outdoor vats packed between 1946 and 1947. A total of 1226 cultures were classified. Data adapted from Etchells and Bell (1950a). The yeasts *Torulopsis caroliana*, *Brettanomyces versatilis*, *Brettanomyces sphaericus*, and *Torulaspora rosei* are currently known as *Candida lactiscondensi*, *Candida versatilis* and *Torulopsis versatilis*, *Pichia subpelliculosa*, *Candida etchelsii* and *Torulaspora delbruckii*, respectively

overlap with those species present between days 23 and 77 and suspected to derive energy from compounds other than glucose or fructose, possibly acetic acid, lactic acid or ethanol (Etchells and Bell 1950a). A follow up study confirmed the following genera dominate in cucumber fermentations, in particular those performed in the USA midwest: *Brettanomyces* (*versatilis* and *sphaericus*; 29%), *Torulopsis* (*holmii* and *caroliniana*; 23%), *Torulaspora* (*rosei*; 15%), *Hansenula* (*subpellicu-losa*; 13%), *Zygosaccharomyces* (*halomembranis, pastoris, globiformis*; 10%), *Saccharomyces* (*globosus*; 5%), *Candida* (*krusei*; 2%) and *Debaryomyces* (*mem-branaefaciens*; 2%) (Etchells et al. 1952). Attempts to study the population of yeasts present in commercial cucumber fermentations using DNA sequencing technology have been hampered by the inability to classify yeasts OTUs to the genera and species taxonomical levels. Such studies have been restricted to the identification of *Capnodiales* and *Tremellales* in fresh cucumbers with colony counts from Yeast and Molds agar plates to 2.82 ± 0.95 CFU/g, and *Saccharomycetales*, mostly *Candida* spp. and *Aureobasidium, Hanseniaspora, Torulaspora, Cryptococcus* and *Hannaella* in commercial cucumber fermentation cover brines containing 4.15 ± 0.68 CFU/g (Pérez-Díaz et al., unpublished).

Although colony counts for the population of yeast and molds from olives are below the detection level, their presence in fermentations is desirable for the development of specific flavors (Arroyo-López et al. 2008). Yeasts only propagate and predominate in olive fermentations (5.1 ± 0.86 CFU/g) if the fruits are neither properly lye treated nor heat shocked before brining (Mark et al. 1956). Yeasts present in olive fermentations include species of the genera, in order of prevalence, *Candida, Debaryomyces, Pichia, Saccharomyces, Rhodotorula* and *Kluyveromyces* (Mark et al. 1956; Marquina et al. 1992). Yeasts naturally present in olives may form mixed biofilms with the LAB prevailing in fermentations, which is evaluated for the delivery of probiotics to the human digestive tract upon the ingestion of fermented olives (De Bellis et al. 2010).

Significantly less information is available with regards to the yeast and molds population in cabbage and sauerkraut given that most of the research pertaining to such fermentations focuses on the LAB and quality issues. Yeast and mold colony counts from fresh cabbage is 2.87 ± 0.79 CFU/g (Pérez-Díaz et al. 2014). The biosurfactant producing yeast, *Pseudozyma*, was isolated from fruits and vegetables of the *Brassica* family and is been studied for the production of mannosylerythritol lipids (Konishi et al. 2014).

The community of bacteriophages, viruses that infect bacteria, in fermented foods is significantly less diverse than the counterparts found in many environmental habitats such as seawater, marine sediment, the human gut and soil (Park et al. 2010). Although, many bacteriophages have been isolated from fermented vegetables, particularly sauerkraut and cucumber fermentations, three viral families have been implicated including *Myoviridae, Siphoviridae* and *Podoviridae* (Barrangou et al. 2002; Lu et al. 2003a, b, 2005, 2010, 2012; Yoon et al. 2002, 2007). Bacteriophages are known to have a relevant function in the modulation of the microbiota during commercial fermentations, particularly in the manufacture of dairy products. The presence of bacteriophages in vegetable fermentations is concomitant with the abundance of the hosting LAB species. In cucumber fermentations, bacteriophages infect primarily *Lb. plantarum/Lb. pentosus* and *Lb. brevis*, the most abundant LAB (Lu et al. 2012; Pérez-Díaz et al. 2016). The genomic organization of bacteriophages from fermented vegetables and dairy products is similar (Lu et al. 2005, 2010). It is estimated that about 10% of the LAB population is sensitive to bacteriophages (Lu et al. 2012). The infection ability of specific bacteriophages found in a commercial cucumber fermentation may be limited to a species or strain, however, few of the viruses can also infect multiple species (Lu et al. 2012). Bacteriophages able to infect multiple genera are rare (Lu et al. 2012). Bacteriophages able to infect *Enterobacter* spp. naturally present in commercial cucumber fermentations have also been found (Lu et al., unpublished).

Microbially Induced Spoilage of Fermenting and Fermented Vegetables

Increased microbial diversity exists in spoiled fermented vegetables at the cost of acceptable sensorial attributes.

Industrial vegetable fermentations mostly rely on the indigenous microbiota, which occasionally results in the growth of undesired microbes or microbial spoilage during long term storage. The presence of residual sugars and viable microbes in finished products can alter appearance. Undesirable microbes in vegetable fermentations include *Enterobacter, Pantoea, Eschericia, Acetobacter, Clostridium, Propionibacterium, Desulfovibrio, Pichia, Zygosaccharomyces, Saccharomyces, Wicherhamomyces, Rhodotorula, Alternaria, Mucor, Fusarium, Aerobacter, Aeromonas, Achromobacter, Paracolobactrum* and *Lactobacillus buchneri* (Garrido Fernández et al. 1997; Nychas et al. 2002; Panagou et al. 2008; Gililland and Vaughn 1943; Levin and Vaughn 1966; Moon et al. 2014; Duran-Quintana et al. 1979; Vaughn et al. 1969, 1972; Ruiz-Cruz and Gonzalez-Cancho 1969; Franco and Pérez-Díaz 2012; Hernandez et al. 2007; Moon et al. 2014; Costilow et al. 1980; Arroyo-López et al. 2008; Golomb et al. 2013; King and Vaughn 1961; Franco and Pérez-Díaz 2012; Johanningsmeier and McFeeters 2013; Medina-Pradas et al. 2016; Fred and Peterson 1922).

The initiation of a vegetable fermentation is prone to produce butyric acid and/or sulphrydic compounds under anaerobiosis. A putrid olive fermentation, characterized by a manure-like and decomposing aroma, and a rancid butter odor develops as the consequence of the proliferation of *Clostridium* and *Desulfovibrio* species (Gililland and Vaughn 1943). Prevention of anaerobiosis at the bottom of fermentation tanks by air purging is used to impede the growth of the strict anaerobe and culprit, *Clostridium*. The emergence of sulphur-like aroma generated by *Desulfovibrio* spp. is commonly prevented by the implementation of good sanitation practices for fermentation vessels and potable water (Levin and Vaughn 1966).

A delayed decrease in pH during the initiation of a vegetable fermentation results in the proliferation of *Enterobacteriaceae* (Garrido Fernández et al. 1997; Nychas et al. 2002; Panagou et al. 2008; West et al. 1941) regardless of oxygen content.

Certain *Enterobacteriaceae*, such as *Enterobacter* spp., can metabolize the sugars naturally present in the vegetables and produce lactic acid, acetic acid and CO_2 (Garrido Fernández et al. 1997; Etchells et al. 1945; Samish and Tulczynsky 1962). Production of CO_2 in olive and cucumber fermentations leads to bloater defect caused by the accumulation of the gas just below the epidermis or in the meso- and endocarps forming hollow cavities that mimic an internal bubble (Garrido Fernández et al. 1997; Fleming 1979). Cucumber and olive bloating represents the most costly spoilage in the production of the fermented products. Uncontrolled growth of *Saccharomyces cerevisiae* and *Wicherhamomyces anomalus* can also cause bloater defect in olive fermentations given the ability to produce CO_2 from sugars (Duran-Quintana et al. 1979; Garrido Fernández et al. 1997; Vaughn et al. 1972).

Softening of vegetables during fermentation is a defect associated mainly with the activity of yeast and is difficult to prevent. *Rhodotorula minuta, W. anomalus, Debaryomyces hansenii, P. kudriavzevii, Alternaria, Fusarium* and *Mucor* are known to produce proteases, xylanases or pectinases in vegetable fermentations causing softening of the plant tissues by degrading their pectin, cellulose, hemicellulose and polysaccharide content (Hernandez et al. 2007; Moon et al. 2014; Costilow et al. 1980; Arroyo-López et al. 2008; Golomb et al. 2013). Pectinolytic bacteria, able to cause tissue softening have also been isolated from vegetable fermentations. Gram negative bacteria such as *Aerobacter, Aeromonas, Achromobacter, Escherichia* and *Paracolobactrum* produce tissue softening of black olives during the oxidation step, if applied at high temperature (Garrido Fernández et al. 1997; King and Vaughn 1961; Vaughn et al. 1969).

Pink sauerkraut is one of the most common defects in the production of such commodity. Although pink sauerkraut has been reported to be edible and is often sold at a lower price, it has been related to undesirable changes in texture, flavour and odour (Fred and Peterson 1922). Sodium chloride concentrations above 3%, high acidity, and extrinsic factors such as temperature and the supply of oxygen can be manipulated to control the pink sauerkraut defect (Fred and Peterson 1922).

As described above oxidative yeast are capable of proliferating on the surface of cucumber fermentation cover brines and cause a rise in pH, tissue softening and/or off odors and taste. *Pichia manshurica* and *Issatchenkia occidentalis* consume lactic acid in aerobic or air purged cucumber and olive fermentations, which induces an increase in pH (Ruiz-Cruz and Gonzalez-Cancho 1969; Franco et al. 2012). *Pichia kudriavzevii* causes kimchi spoilage characterized by undesirable changes in organoleptic properties (Moon et al. 2014).

The extreme acidic pH, high organic acid concentrations and lack of monomeric fermentable sugars ensure the long-term stability of fermented vegetables. Even though conditions are unfavourable for microbial growth at the end of a complete fermentation, some especially unique microbes can initiate spoilage. *Propionibacterium* spp. metabolize sugars, or the lactic acid formed during primary fermentation, to produce propionic acid, acetic acid and CO_2, inducing an increase in pH and volatile acidity (Gonzalez-Cancho et al. 1970). Rising pH spoilage of fermented cucumbers results in the development of cheese and manure-like aromas (Franco et al. 2012). *Acetobacter* spp. and *Lb. buchneri* are present in fermented cucumber spoilage at pH 3.3 (Franco and Pérez-Díaz 2012; Johanningsmeier and

McFeeters 2013; Medina-Pradas et al. 2016). *Acetobacter* spp. are commonly used in the production of vinegar and are known for converting lactic acid to water and carbon dioxide or acetic acid, concomitantly with the conversion of ethanol to acetic acid (Raj et al. 2001; Lefeber et al. 2010). *Lb. buchneri* is able to produce acetic acid and 1,2-propanediol from lactic acid during the first stage of the undesired secondary fermentation (Johanningsmeier and McFeeters 2015). Should the undesired secondary fermentation enable the increase in pH above 4.2, *Propionibacterium* and *Pectinatus* species, and *Clostridium bifermentans* and *Enterobacter cloacae* are able to convert lactic acid to propionic acid and butyric acid imparting the characteristic putrid aromas (Breidt et al. 2013b; Franco et al. 2012). The activity of *Propionibacterium* spp. in olive fermentation spoilage and of *L. buchneri* in fermented cucumber spoilage can be prevented by controlling the end of fermentation pH and salt concentrations (Garrido Fernández et al. 1997; Johanningsmeier and McFeeters 2015).

Spoilage associated LAB include *Lb. plantarum*, *Lb. brevis*, *Lb. casei*, *Lb. paracasei* and as mentioned above, *Lb. buchneri*. *Lb. brevis* is associated with the formation of a water-soluble red pigment in sterile cabbage juice at a pH of 5.7 ± 0.5 and it is suppressed by anaerobic conditions (Stamer et al. 1973). *Lb. casei* and *Lb. paracasei* have been implicated in sporadic cases of red colored fermented cucumber spoilage in fermented cucumber products (Díaz-Muñiz et al. 2007). *Lb. casei* and *Lb. paracasei* are able to degrade the azo dye tartrazine (FD&C yellow no. 5) used as a yellow colouring in cover brines. This type of spoilage is prevented by the addition of 0.1% sodium benzoate (Díaz-Muñiz et al. 2007). *Lb. plantarum* is the cause of the so-called yeast spots in fermented olives (Vaughn et al. 1953, 1969). A strains of *Lb. plantarum* (3.2.8) capable of producing exopolysaccharides and dominant over *Lb. pentosus* in cucumber fermentations was also found responsible for the production of yeast spots in such fruit (Pérez-Díaz et al., unpublished).

Mass Production Parameters for Vegetable Fermentations That Consistently Yield Finished Products with Acceptable Attributes for Consumers

A rapid decrease in pH and stability during long term storage are key to controlling the fermentation of vegetables.

There are three parameters that must be controlled in vegetable fermentations to obtain an acceptable product including a rapid initiation and acid production, the complete conversion of freely available sugars to organic acids and/or ethanol and a stable post-fermentation pH. The quick production of acids from sugars ensures a drop in pH below 4.6, which is critical for preventing growth of the deadly toxin producer, *Clostridium botulinum* (Ito et al. 1976). This is particularly relevant in vegetable fermentations conducted in closed vessels that support the development of anaerobiosis, a strict requirement for the proliferation of the spore former, *Cl.*

botulinum. At least 0.9% acetic acid in cover brines is needed to achieve the inhibition of the clostridial species in a cucumber fermentation (Ito et al. 1976). Artisanal preparations of table olives have been associated with botulinum outbreaks (Medina-Pradas and Arroyo-López 2015). As mentioned above a delayed decrease in pH creates an ideal scenario for the Enterobacteriaceae, naturally present in vegetables, to proliferate and produce excess CO_2 conducive to bloater defect. A drop in pH to levels that are inhibitory for growth of *Lb. plantarum/ Lb. pentosus* before a complete sugar conversion occurs, results in the availability of energy sources for spoilage microbes as described above. Unstable vegetable fermentations result in the establishment of conditions that allow growth of undesired microbes capable of utilizing lactic acid and thus, of rising the pH above 4.6, generating a public health concern from the potential production of the botulinum toxin. Thus, control of pH before, during and after a vegetable fermentation is a critical parameter and the most important one in the production of a safe preserved food.

The use of sodium chloride in vegetable fermentation cover brines has five main functions including the densitometric modulation of specific members of the microbiota, the speed of the fermentation, prevention of softening caused by salt sensitive microbes, imparting a salty flavour in the finished fermented product and assisting with the equilibration of the vegetable content with the cover brine by weakening the tissue membranes (Bell and Etchells 1961; Bell et al. 1950). For instance, more than 8% sodium chloride is needed in table olive fermentation for long term bulk storage to inhibit spoilage by *Propionibacterium* spp. (Garrido Fernández et al. 1997). A combination of pH 3.3 and 4% salt is needed in cucumber fermentations to prevent spoilage by *Lb. buchneri* (Johanningsmeier and McFeeters 2013). Although, sodium chloride has become synonymous with vegetable fermentations for centuries, the fermentation of certain vegetables, such as cucumber, was demonstrated in closed containers with cover brines supplemented with calcium chloride as the only salt (McFeeters and Pérez-Díaz 2010). Sodium chloride-free cucumber fermentations in commercial open top tanks requires the addition of a preservative to inhibit yeasts and the growth of undesired microbes (Pérez-Díaz et al. 2015). Cucumber fermentations in closed containers without salt and with potassium sorbate results in a complete conversion of the sugars to lactic acid and a more diverse community of LAB, sustaining the growth not only of *Lb. plantarum* and *Lb. pentosus*, but also that of *Pediococcus*, *Weissella*, *Lactococcus* and *Leuconostoc* after 14 days (Pérez-Díaz et al., *submitted*). Although the absence of sodium chloride in a fermentation can support a greater microbial diversity, other factors must be applied to control the growth of microbes that can impart less appealing organoleptic characteristics in the finished product and compromise safety. Sauerkraut fermentations with reduced sodium chloride, from 2 to 0.5%, resulted in an undesirable flavour profile (Johanningsmeier et al. 2005). Consuming vegetables that have been fermented without sodium chloride or any additive other than water is like playing the lottery where the winning price is a delicious and freshly putrefied fermented vegetable and the loosing ticket is food poisoning or diarrhea.

A starter culture is a critical factor in low salt vegetable fermentations and in achieving finished product consistency for mass production (Etchells et al. 1964;

Etchells et al. 1973; Vega Leal-Sánchez et al. 2003). It is also a main factor in the reduction of the microbial diversity in the industrial production of fermented vegetable products. Pure starter cultures were introduced commercially in New Zealand in 1934 (Cogan and Hill 1993) beginning the era of "controlled" fermentations. Utilization of *Lc. mesenteroides* as a starter culture for low salt sauerkraut fermentations prevents off-flavour and odors (Johanningsmeier et al. 2005). The use of a *Lb. plantarum* starter culture for caper berry fermentation induces a consistent process and faster sugar catabolism (Palomino et al. 2015). Fermented vegetable products manufactured with a starter culture enjoy higher acceptability by consumers and improved nutritional characteristics (Martínez-Villaluenga et al. 2012; Di Cagno et al. 2012). Starter cultures of *Lb. delbrueckii* and *Lb. paracasei* reduce the nitrite content in Chinese cabbage (Han et al. 2014). *Lb. pentosus* and *Lb. plantarum* are among the LAB species with major applications as starter cultures in fermented vegetables such as cucumber, capers and tables olives, albeit *Lc. mesenteroides* is also occasionally used in low salt fermentations such as sauerkraut (Corsetti et al. 2012; Pérez-Díaz et al. 2015). However, the dominance of a starter culture in a vegetable fermentation intrinsically eliminates or reduces biodiversity selecting for those microbes that can tolerate a number of stresses associated with the specific habitat.

The exclusion of air from cucumber fermentations with nitrogen purging results in a higher quality pickle (Costilow et al. 1980). Air purging is commonly applied in cucumber fermentations to displace the carbon dioxide produced during the fermentation and prevent bloater defect characterize by the formation of hollow cavities within the flesh (Fleming 1979). The incorporation of air in cucumber fermentations results in the proliferation of yeasts and molds and tissue softening (Costilow et al. 1980). Although, air purging could potentially increase the diversity of yeasts and molds in a cucumber fermentation, the resulting sensory attributes are undesirable. The need to reduce bloater defect and maintain product quality demands the incorporation of preservatives such as sorbic acid in cucumber fermentation cover brines to inhibit yeasts (Borg et al. 1972). Incorporation of sorbic acid additionally aids in the elimination of film-forming yeasts on the superficial cover brines that occasionally leads into a decreased acidity. Additionally, natural cucumber fermentations without preservatives can sustain yeast growth after the primary fermentation stage by LAB is initiated (Etchells 1941).

Processing Parameters for Vegetable Fermentations That Could Increase the Delivery of Microbial Diversity to the Gut

Lower salt content and more diversified starter cultures are key parameters in augmenting the diversity of desirable microbes in commercial scale fermentations.

Obviously, the processing parameters currently in place for the mass production of fermented vegetables have evolved to accommodate for consistency, acceptable sensorial characteristics and food safety. The incorporation of a more diverse microbiota in a fermented vegetable product has not been prioritized. A commonality of vegetable fermentations, as it is done to date, is the reduction of biodiversity to accomplish a controlled fermentation that consistently delivers a healthy product of quality.

There is an emerging sector of the population with a preference for the consumption of raw spoiled vegetables, coagulated soups and putrid fermented foods hoping to gain a beneficial health effect from the consumption of a broader microbial diversity. People willing to rescue vegetables from the dumpsters, irrigate their root crops in their gardens with the sink effluent and pickle leftovers to enable the consumption of live foods expected to inoculate the gut, at the risk of sporadic diarrhea and food poisoning, to boost the immunological system (Holes 2010). Fermentation is considered by this sector of the population as one of the initial conditions of civilization that facilitated a sedentary lifestyle, before refrigeration developed (Holes 2010).

While there may be some truth in Sandor Katz saying that "we are killing ourselves with cleanliness by killing every microbe that could enter our stainless-steel kitchens"; an intermediate point that preserves the safety hurdles in a fermentation process must be found. A middle point between the industrial perspective and the interest to boost the microbial biodiversity in the gut is achievable with the application of science and a more flexible consumer base. Compromises are needed from both sectors to achieve healthy products with long term stability. But, how would the fermented vegetable product with a diverse microbiota be prepared? What would this elixir contain?

The main objective of food fermentation is to achieve long term preservation and extend the shelf-life of a fresh or about to spoil food. Thus, it is an intrinsic function of the fermentation to reduce or minimize microbial diversity and activity. A primary objective of fermentation is to transform the sugars naturally present in the vegetables and render them unavailable as an energy source for the proliferation of a number of microbes. Additionally the conversion of sugars to acids and alcohol causes a decrease in the pH of a vegetable matrix which ends up suppressing a number of microbes resulting in microbial stability. So, can fermented vegetables truly be carriers of a diverse microbiota?

Only the resistant microbes survive in a fermentation, which happen to belong to the same taxonomical genera present in the gut, as the colonic habitat is also characterized by an acidic pH, limiting oxygen and the availability of complex undigestible sugar polymers and amino acids as primary energy source for microbes. If only fermented vegetables are considered, the list of survivors tends to be short, leading with *Lb. plantarum* and *Lb. pentosus* and followed by *Leuconostoc, Weissella* and *Pediococcus*. A number of other lactobacilli are also present in fermented vegetables in lower abundance. Thus, if the microbial diversity in a fermented vegetable is to be expanded the obvious choices are those microbes that survive a fermentation, which often times end up causing spoilage. Suffice it to remind the reader that spoilage is defined as undesirable changes in organoleptic properties as define by consumers. As described above, spoilage microbes in fermented vegetables tend to convert the lactic acid, acetic acid and ethanol commonly produced in the bioconversion to propionic and butyric acids imparting putrid and manure-like aromas to an otherwise perfectly edible food (Gonzalez-Cancho et al. 1970; Raj et al. 2001; Lefeber et al. 2010; Johanningsmeier and McFeeters 2015). The fermented vegetable spoilage bacteria tend to be those that can use lactic and acetic acids and ethanol including among others *Clostridium*, and *Propionibacterium*, which are also able to colonize the human gut. Such microbes assist in the conversion of undigestible fibers in the gut to short chain fatty acids like propionic and butyric acids that are utilized as energy sources by the epithelium. So, do we need to adapt to eating stinky fermented vegetables?

Preservation by fermentation must continue to remove the sugars naturally present in the produce and yield a finished product with an extended shelf-life. Factors that science and consumers can modify, at least in theory, is the desirable metabolic product in a vegetable fermentation and the definition of an acceptable finished product, respectively.

A vegetable fermentation processing parameter that cannot be changed is the initiation of the fermentation in a closed container at a pH below 4.6. This parameter not only represents the exclusion of the deadly botulinum toxin in fermented vegetables (Ito et al. 1976) but also assures the eventual die off of microbes of public health significance, such as *Listeria monocytogenes*, *Salmonella* spp. and acid resistant strains of *Escherichia coli* (Breidt and Caldwell 2011). *Listeria monocytogenes*, a food-borne pathogen, has become a major concern to the food industry over the past 30 years, mainly for refrigerated and ready-to-eat products. The bacterium is commonly found in the environment and has been isolated from various plant materials, including silage (Fenlon 1985), soybeans, corn (Welshimer 1968; Welshimer and Donker-Voet 1971) and cabbage (Seelinger and Jones 1986; Beuchat et al. 1986). Outbreak strains of *L. monocytogenes* are also able to grow on raw cabbage and in cabbage juice (Beuchat et al. 1986), but not if the pH is adjusted to 4.6 or below (Conner et al. 1986). *Listeria monocytogenes* can additionally survive and grow in green table olives after 2 months of storage (Caggia et al. 2004). The die off of *Escherichia coli* O157:H7 in cucumber fermentation cover brines takes 3 and 24 days at a pH of 3.2 and 4.6, respectively (Breidt and Caldwell 2011). The development of fermentation in the same cover brine reduces the die off of this pathogen to 1 and 16 days. The survival of *Salmonella* spp. at a pH below 4.6 is also assured, given that it is less sensitive to acidic conditions than *E. coli* (Breidt et al. 2013a, b). Fermentation should continue to function as a sanitizer for fresh vegetables that could be potentially contaminated with agricultural run offs or the fecal matter from handlers. It is not necessary to push its limits by inoculating vegetables with microbes that are outside of the plants habitat. Compromises with regards to food safety parameters jeopardizes consumer's health and are thus not negotiable.

There is flexibility within the existing processing parameters for vegetable fermentations to host a diverse microbial population. A fully or partially fermented vegetable could be safely consumed by healthy adults if the equilibrated pH is between 4.5 and 3.3. The closer the pH is to 4.5 the greater the likelihood of consuming the most diverse microbiota a fermented vegetable can provide, which would be likely restricted to some *Enterobacteriaceae, Pseudomonadaceae* and

LAB. *Enterobacteriaceae* such as *Enterobacter* and presumptive *Klebsiella* and *Pseudomonas* were isolated from cucumber fermentations with a pH of 4.04 ± 0.15 (Pérez-Díaz et al. 2018). These microbes are also present at the initiation of numerous vegetable fermentations as described above. The reduction of sodium chloride in the fermentation of some vegetables may enable a more diverse fermentative microbiota. However, such an approach may not consistently deliver an edible and safe product. Additionally, with the elimination of salt other more permissive factors would have to be incorporated to achieve a complete and stable fermentation that produces an acceptable product.

The use of mixed starter cultures also offers an opportunity to enhance biodiversity in a fermented product. Yeasts have robust enzymatic diversity, are considered bioprotectants in vegetable fermentations, enhance the growth of LAB and improve the organoleptic properties of certain pickles (Arroyo-López et al. 2008). The application of starter cultures composed of yeasts and LAB, such as *Lb. plantarum* and *Saccharomyces oleaginosus*, leads to more complete sugar consumption in the fermentation of carrots, cabbages, beets and onions, with a consequently higher acidity as compared to spontaneous fermentation (Gardner et al. 2001; Montaño et al. 1997). The development of green and black olives containing probiotics has been achieved by the selection of compatible yeasts and LAB that can form biofilms on the fruits (Rodríguez-Gómez et al. 2014; Bleve et al. 2015). Mixed starter cultures of yeasts and LAB in green table olives also modify the concentration of free amino acids, phenols and volatile compounds and generates finished products with increased consumer's acceptability (De Angelis et al. 2015). Additionally, the successful fermentation of certain vegetables such as green beans necessitates mixed starter cultures to remove different sugars such as glucose, fructose, mannitol and cellobiose (Chen et al. 1983a, b). The use of *Lb. plantarum* LPCO10 as a starter culture and of *Enterococcus casseliflavus* cc45 and *Lb. pentosus* 5138A in sequential inoculations has proven effective in accelerating acid production and the die off of pathogenic microbes as compared to spontaneous fermentation of green table olives (De Castro et al. 2002; Leal-Sánchez et al. 2002; Vega Leal-Sánchez et al. 2003). Further advancement in the understanding of the contributions of individual microbes to specific vegetable fermentations will offer the opportunity to develop fully functional and safe products that taste like fresh produce.

Can Fermented Vegetables Aid in Augmenting Biodiversity in the Gut or Repopulating It?

Lactobacillus plantarum will continue to be central to the ability of fermented vegetables to deliver beneficial health effects.

The human body is estimated to host 10^{14} bacteria, with the stomach and lower small intestine contributing the lowest amount (10^7 each) and the colon contributing the highest (10^{11}) (Sender et al. 2016). Intermediate values of bacterial counts are contributed by the skin, saliva, dental plaque and the upper small intestine to the human microbiome (Sender et al. 2016). It is also estimated that the number of eukaryotic cells are equal to the number of bacteria at 10^{14}, with woman and newborns carrying twice as many bacteria as eukaryotic cells (Sender et al. 2009). The number of LAB rarely reach 10^9 CFU/g in a fermenting vegetable, which translates into 0.001% of the gut microbiota (Fig. 1). Fully fermented cucumbers host a microbial load of lactobacilli and yeasts at 5.01 ± 3.75 (MRS agar plates) and 5.28 ± 4.81 (Yeast and Mold agar plates) log of CFU/g, respectively, during long term storage (Pérez-Díaz et al. 2014). If the most active microbial population in a still fermenting vegetable survives passage through the digestive tract and reaches the colon, it will encounter a microbial jungle. A particular niche for the transiting microbes would have to exist, so that colonization can take place. The newly formed colony would have to metabolize the fibers, proteins, fat and polyphenols that are not digested by the host and are thus available in the gut to establish itself in the new habitat, sense its surrounding and efficiently compete with the indigenously diverse population. Alternatively, the transiting microbes may have the ability to simply attach to the existing microbial mass or epithelial gut, conquer a niche and establish itself in the gut. A more complex model for establishment in the gut would be through the association of certain less fitted microbes to a robust colonizer. It is likely that the result of this type of establishment in the gut would result in the production of compounds that could be beneficial or detrimental to the host's health. However, there is evidence suggesting that the human gut microbiome is resistant to colonization by foreign species (Salonen and deVos 2014; a review).

The human microbiome project directed by the US National Institute of Health reported that the metabolic activity of the microbes in the gut produce beneficial compounds such as vitamins and anti-inflammatories that the human genome cannot produce (Lloyd-Price et al. 2016). In achieving the metabolism of compounds that are undigestible by humans, it becomes more relevant for the gut microbiome to contain a complete set of metabolic enzymes rather than specific microbes. Consequently, a variety of microbes would fulfill the need to metabolize fat or polyphenols and become a part of a stable healthy gut microbiome.

The human gut microbiome changes among healthy people with age, medical interventions and diet. Taking antibiotics causes an imbalance in the microbiome resulting in lower microbial diversity (McDonald et al. 2018). The necessary functionalities are restored as a function of time even if the new microbiome composition is different. Despite the many advancements that have been made in understanding the human gut microbiome, its holistic microbial diversity is still unknown (McDonald et al. 2018). The influence of lifestyle, health state and diet on the composition of the human microbiome is still unclear. However, recent evidence has unexpectedly emerged on the influence of the consumption of 30 types of plants versus 10 on the human gut microbiome. This is specifically related to short-chain fatty acid fermenters (McDonald et al. 2018). The microbial fermentation of undigested plant derived components suggests that diversity in the microbiome is related to the availability of a variety of dietary fibers and resistant starches.

The microbial diversity in the gut has been recognized as limited when compared to the environmental counterpart, but enormous if compared to the indigenous fermented foods microbiota (McDonald 2018). With the variability rate in the human gut metagenome the definition of a healthy human gut will continue to be elusive (Lloyd-Price et al. 2016), giving birth to the need for personalized nutrition. Contrary to the microbial diversity in the gut microbiome, with only a 30% conserved metagenome (Lloyd-Price et al. 2016), the fermentome is composed of less than 50 genera in the initial stage of the process and a handful of genera during the active fermentation period. However, expansion of the diversity in the fermentome promises to better position fermented foods to contribute to the human gut metagenome.

Consumption of various vegetables in significant amounts is a main component of the food consumption guidelines around the world. But, should this translate into a recommendation for the consumption of a higher diversity of fermented vegetables regularly? A definitive answer to this question may not exist. As discussed above the raw vegetables microbiome is more diverse than that found in a fully fermented vegetable with acceptable sensorial characteristics by consumers. However, fermented vegetables have the potential to host more of those microbes commonly found in the human gut such as *Clostridium*, *Lactobacillus*, *Enterococcus*, *Dialister*, *Veillonella*, *Prevotella*, *Bacteroides*, *Escherichia* and *Shigella* at comparatively higher abundance (Barko et al. 2018). About to spoil of spoiling fermented cucumbers undergoing secondary fermentation at a pH of 3.7 contain all the genera listed above, except for *Escherichia* and *Shigella* (Medina-Pradas et al. 2016). However, even if fermented cucumbers undergoing secondary fermentation with exotic aromas were to be consumed, the effectiveness of such an elixir in the microbial diversity of the gut would depend on the individual microbiome composition and need to fulfil a metabolic niche.

Data generated from human feces suggests that a stable community of lactobacilli is found in the human gut (Rossi et al. 2016). *Lb. rhamnosus*, *Lb. ruminis*, *Lb. delbrueckii*, *Lb. plantarum*, *Lb. casei* and *Lb. acidophilus* are among the 58 lactobacilli species that have been found in human feces at densities fluctuating between 6 and 8 log CFU/g (Rossi et al. 2016). Other lactobacilli species have been found at concentrations between 4 and 5 log of CFU/g of feces. Among the most prevalent lactobacilli found in the gut, only *Lb. plantarum* is consistently present in fermented vegetables in high concentrations (8 log of CFU/g).

The delivery of *Lb. plantarum* as probiotic to the human gut by consuming fermented or partially fermented vegetables is considered a low-calorie, lactose-free alternative for obtaining beneficial health effects (Cauley 2016). Challenges exist with regard to the delivery of an effective dose of *L. plantarum* per serving size of a fermented vegetable to the gut including the subsequent establishment of the specific culture in the gut and positioning in a way that expresses the necessary genes associated with probiotic properties. Many studies have been conducted to elucidate the mechanism by which *Lb. plantarum* could impact human health (Hemert et al. 2010; McDonald et al. 2018; Kishino et al. 2013; Marco et al. 2006). *Lb. plantarum* is one of the most competitive LAB with the ability to resist extremely acidic pH, high salt concentrations (>8% sodium chloride), colonize a variety of habitats, possesses a comparatively large genome among LAB and acquires genes by horizontal transfer (McDonald et al. 1990; Siezen and van Hylckama Vlieg 2011). Several strains of *Lb. plantarum* contain genes coding for N-acetyl-glucosamine/galactosamine phosphotransferase system, LamBDCA quorum sensing system, and components of the plantaricin biosynthesis and transport system potentially responsible for the stimulation of anti- or pro-inflammatory immune response in the gut (Hemert et al. 2010). *Lb. plantarum* is also known to convert linoleic acid to conjugated linoleic acid, a metabolic reaction identified as a marker for microbes impacting gut health in individuals consuming more than 30 plants as part of a regular diet (McDonald et al. 2018; Kishino et al. 2013). Strains of *Lb. plantarum* isolated from various fermented vegetables are able to survive in simulated gastric and intestinal conditions, adhere to intestinal Caco-2 and HT29 MTX cell tissues, catabolize fructoligosaccharides as the only carbon source and cholesterol, and inhibit pathogens from human sources (De Angelis et al. 2017; a review). *Lb. plantarum* is known to transit the mouse gastrointestinal tract in about 4 h. This probiotic maintains a presence in the stomach and small intestine and the cecum and colon for 4 and 8 h, respectively, in addition to displaying specific and differential responses at various sites along such mammalian gastrointestinal tract (Marco et al. 2006). Although, a strong body of evidence has been generated with regard to the potential of *Lb. plantarum* to deliver a beneficial health effect in the human gut, its ability to colonize a healthy gut and modulate specific responses/needs is still somewhat elusive. This task is further complicated by the fact that the healthy microbiome composition varies among individuals, the lack of in-depth knowledge of the metabolic potential needed in the gut to effectively process a plethora of food-derived undigestible compounds and access to a developing wealth of knowledge on how the gut microbiome impacts body functions at large.

Potential Impacts of Fermented Vegetables in the Human Gut Microbiome

Fermented vegetables can deliver prebiotics and pre-adapted probiotics to the Western gut.

The concept of nutrition has changed from the consumption of foods that satisfy our biological needs to personalized probiotic and prebiotic containing diets that boost our gut microbiome and health. Support of the gastrointestinal tract microbiome diversity imparts a resilience that buffers against dysbiosis, a transient change in permeability, inflammation, pre-disposition to illness and infection and psychological imbalance (Karl et al. 2018). As described above, vegetable fermentations sustain a diverse bacterial, bacteriophage and yeast ecosystem that can serve to expand the catalog of reactions available to the gut microbiome during a perturbation of health. The health promoting lactobacilli naturally prevailing in vegetable fermentations offer basic functionalities related to simple and complex carbohydrate catabolism and short chain fatty acid production to the gut (Gänzle 2015; a review). These functions are associated with the reduction of the gut pH to inhibit pathogens and the production of energy for the epithelium, respectively. Similarly, as a vegetable fermentation proceeds a number of undesirable acid sensitive microbes in the indigenous microbiota are suppressed reducing the probability of the fermented finished product to deliver pathogenic microbes to the gut and consequently functions associated with protein fermentation, production of sulfate and sulfites and the induction of inflammation (Pérez-Díaz et al. 2018; Gililland and Vaughn 1943; Karl et al. 2018, a review).

Microbes in the gut derive energy from dietary components that are not digested (degraded nor absorbed) by the host and are secreted in the intestine or carbohydrates produced by the gut microbiome itself (Tingirikari 2018; a review). A substantial proportion of carbohydrates available to the microbiome in the human gut derives from plant material, particularly dietary fiber which is composed of cellulose, hemicellulose, pectic substances, and lignin (Rincón-León 2003). Cellulose, hemicellulose and lignin can trigger and regulate bowel movement (Viuda-Martos et al. 2010). Pectic substances are water-soluble and abundant in the soft tissue of vegetables and fruits (15–20%) (Grigelmo-Miguel et al. 1999; García et al. 1995). Pectic substances influence the gel-forming and water holding capacity of the gut and serve as energy sources for the microbiome to induce an acidic pH in the colon and the production of short chain fatty acids and gases (Roberfroid 1993). Together the delivery of natural dietary fibers and of a diversity of microbes by a fermented vegetable represents a theoretical elixir for the gut. Dietary fibers that remain whole after transiting the digestive tract can be digested by the indigenous vegetables microbiome. Co-existence of dietary fibers and the vegetable microbiome in a fermentation process prior to consumption is an opportunity to pre-adapt the microbes to the degradation of such complex carbohydrates and thus enable them to make a difference in and acidic habitat such as the gut upon colonization or transient passage. Consumption of un-pasteurized fermented vegetables is thus a natural vehicle for the re-introduction of energy sources for the microbiome and microbial diversity not commonly present in the Western-like individuals with low intake of plant-derived-foods, such as *Prevotella* (Sonnenburg et al. 2016). While the enrichment of cucumber fermentations with *Prevotella* indicates the development of spoilage, in the gut it can serve as a biomarker for dietary interventions (Medina-Pradas et al. 2016; Salonen and deVos 2014; Gorvitovskaia et al. 2016; Kovatcheva-Datchary et al. 2015; Verbeke et al. 2015). The dominance of *Prevotella* in the human gut is associated with the exposure to complex plant-derived carbohydrates (Salonen and deVos 2014).

The delivery of lactic acid producing microbes and possibly of lactic acid itself by fermented vegetables to the gut can also be advantageous. Production of lactic and acetic acids by LAB in vegetable fermentations consequently generates a need to resist the negative effect of the acids on the cells. LAB are notorious for their ability to produce mM concentrations of such acids and tolerate pH as low as 3.3 (McDonald et al. 1990). Consequently, fermented vegetables can deliver significant concentrations of L- and D-lactic acid to the GI tract. Some lactobacilli incorporate D-lactic acid on the cell wall (Delcour et al. 1999). Nanomolar concentrations of D-lactic acid are produced by the human body from methylglyoxal metabolism

(Ewaschuk et al. 2005; Spencer et al. 2009). To date D-lactate acidosis is a rare condition in human and has not been associated with the consumption of fermented foods, but with surgical intervention (Uribarri et al. 1998). The millimolar concentration of L-lactic acid produced in mammals can be increased by excess microbial activity in the gut (Ewaschuk et al. 2005). L-lactic acid is currently recognize as an energy source for the human skeletal muscles (Lund et al. 2018). Lactic acid is microbially converted to propionic and butyric acids in the gut which are energy sources for the gut epithelium and other organs (Fitch and Fleming 1999). Thus the availability of lactic acid in the gut, should it not be absorbed in the upper digestive tract, could serve as an energy source for the microbiome and the epithelium.

Conclusion: Can Fermented Vegetables Seed the Gut-Associated Microbiota?

The ability of fermented vegetables to deliver bacterial consortia to the human gut is still undefined. Logically, one would think that the higher microbial diversity a fermented vegetable can sustain the higher the probability of such product to deliver diversity to the gut. It can also be deduced that a freshly fermented vegetable containing viable cells of *Lb. plantarum* could serve as a vehicle for inoculation of the gut. Once in the gut, *Lb. plantarum* could establish itself, should a niche exist for its many genome encoded functionalities or leave a footprint in the gut as it transits. While it seems to be premature to hypothesize whether other LAB found in vegetable fermentations such as *Leuconostoc*, *Pediococcus*, *Lactococcus*, *Weisella* would colonize the gut, there is circumstantial evidence implicating a niche for fermented vegetable spoilage associated microbes in the gut including *Prevotella*, *Veillonella*, *Dialister*, and *Clostridium* among others. Regardless of the specific microorganisms delivered to the gut by fermented vegetables, such microbes would be advantageously pre-adapted to the utilization of dietary fibers, particularly pectic substances, an acidic pH and to substantial concentrations of lactic acid, acetic acid and ethanol. Such pre-adaptation could represent a competitive advantage for their establishment in the gut.

References

Anderson, R. E., Daeschel, M. A., & Ericksson, C. E. (1988). Controlled lactic acid fermentation of vegetables. In G. Durand, L. Babichon, & J. Florent (Eds.), *Proceedings of the 8th International Symposium* (pp. 855–868). Paris, France: Societé Fraçaise de Microbiologie.

Arroyo-López, F. N., Querol, A., Bautista-Gallego, J., & Garrido-Fernández, A. (2008). Role of yeasts in table olive production. *International Journal of Food Microbiology, 128*, 189–196.

Balatsouras, G. (1985). Taxonomic and physiological characteristics of the facultative rod type lactic acid bacteria isolated from fermenting green and black olives. *Grasas y Aceites, 36*, 239–249.

Barko, P. C., McMichael, M. A., Swanson, K. S., & Williams, D. A. (2018). The gastrointestinal microbiome: A review. *Journal of Veterinary Internal Medicine, 32*, 9–25.

Barrangou, R., Yoon, S. S., Breidt, F., Fleming, H. P., & Klaenhammer, T. R. (2002). Characterization of six *Leuconostoc fallax* bacteriophages isolated from an industrial sauerkraut fermentation. *Applied and Environmental Microbiology, 68*, 5452–5458.

Bell, T. A., & Etchells, J. L. (1961). Influence of salt (NaCl) on pectinolytic softening of cucumbers. *Journal of Food Science, 26*, 84–90.

Bell, T. A., Etchells, J. L., & Jones, I. D. (1950). Softening of commercial cucumber salt-stock in relation to polygalacturonase activity. *Food Technology, 4*, 157–163.

Beuchat, L. R., Brackett, R. E., Hao, D. Y., & Conner, D. E. (1986). Growth and thermal inactivation of *Listeria monocytogenes* in cabbage and cabbage juice. *Canadian Journal of Microbiology, 32*, 791–795.

Bleve, G., Tufariello, M., Durante, M., Grieco, F., Ramires, F. A., Mita, G., Tasioula-Margari, M., & Logrieco, A. F. (2015). Physico-chemical characterization of natural fermententation process of Conservolea and Kalamata table olives and development of a protocol for the preselection of fermentation starters. *Food Microbiology, 46*, 368–382.

Borg, A. F., Etchells, J. L., Bell, T. A. (1972). Microbial examination of solar, rock, and granulated salts and the effect of these salts on the growth of certain species of lactic acid bacteria. *Pickle Pak Sci 2*(1), 11–17.

Botta, C., & Cocolin, L. (2012). Microbial dynamics and biodiversity in table olive fermentation: Culture-dependent and independent approaches. *Frontiers in Microbiology, 3*, 245. https://doi.org/10.3389/fmicb.2012.00245.

Breidt, F., & Caldwell, J. M. (2011). Survival of *Escherichia coli* O157:H7 in cucumber fermentation brines. *Journal of Food Science, 76*(3), M198–M203.

Breidt, F., Kay, K., Cook, J., Osborne, J., Ingham, B., & Arritt, F. (2013a). Determination of 5-log reduction times for *Escherichia coli* O157:H7, *Salmonella enterica*, or *Listeria monocytogenes* in acidified foods with pH 3.5 or 3.8. *Journal of Food Protection, 76*(7), 1245–1249.

Breidt, F., Medina-Pradas, E., Wafa, D., Pérez-Díaz, I. M., Franco, W., Huang, H., Johanningsmeier, S. D., & Kim, J. (2013b). Characterization of cucumber fermentation spoilage bacteria by enrichment culture and 16S rDNA cloning. *Journal of Food Science, 78*(3), M470–M476.

Brock, T. D. (1961). *Milestones in microbiology*. Englewood Cliffs, NJ: Prentice Hall.

Buckenhüskes, H., Jensen, H. A., Anderson, R., Garrido Fernández, A., & Rodrigo, M. (1990). Fermented vegetables. In P. Zeuthen, J. C. Cheftel, C. Eriksson, T. R. Gormley, P. Linko, & K. Paulus (Eds.), *Processing and quality of foods. Food biotechnology: Avenue to healthy and nutritious products* (Vol. 2, pp. 2162–2188). London: Elsevier Appl. Science.

Caggia, C., Randazzo, C. L., Di Salvo, M., Romeo, F., & Giudici, P. (2004). Occurrence of *Listeria monocytogenes* in green table olives. *Journal of Food Protection, 67*, 2189–2194.

Cauley, S. M. 2016. Survival of commercially available lyophiized *Lactobacillus plantarum* and *Pediococcus acidilactici* probiotic cultures in acidified, refrigerated cucumbers. Thesis Dissertation at North Carolina State University.

Chang, J. Y., & Chang, H. C. (2010). Improvements in the quality and shelf-life of kimchi by fermentation with induced bacteriocin-producing strain, *Leuconostoc citreum* GJ7 as a starter. *Journal of Food Science, 75*, M103–M110.

Chen, K. H., McFeeters, R. F., & Fleming, H. P. (1983a). Stability of mannitol to *Lactobacillus plantarum* degradation in green beans fermented with *Lactobacillus cellobiosus*. *Journal of Food Science, 48*(3), 972–974.

Chen, K. H., McFeeters, R. F., & Fleming, H. P. (1983b). Complete heterolactic acid fermentation of green beans by *Lactobacillus cellobiosus*. *Journal of Food Science, 48*(3), 967–971.

Cogan, T. M. (1996). History and taxonomy of starter cultures. In T. M. Cogan & J. P. Accolas (Eds.), *Dairy starter cultures* (pp. 1–23). New York: VCH Publishers.

Cogan, T. M., & Hill, C. (1993). Cheese starter cultures. In P. F. Fox (Ed.), *Cheese: Chemistry, physics, and microbiology* (Vol. 1, 2nd ed., pp. 193–194). London: Chapman and Hall.

Columela, L. J. M. (1979). *45. De Re Rustica* (Vol. II). Spain: Nestlé, A.E.P.A. Santander.

Conner, D. E., Brackett, R. E., & Beuchat, L. R. (1986). Effect of temperature, sodium chloride and pH on growth of *Listeria monocytogenes* in cabbage juice. *Applied and Environmental Microbiology, 52*, 59–63.

Corsetti, A., Perpetuini, G., Schirone, M., Tofalo, R., Suzzi, G. (2012). Application of starter cultures to table olive fermentation: an overview on the experimental studies. *Front. Microbiol., 19*(3), 248.

Costilow, R. N., Gates, K., & Lacy, M. L. (1980). Molds in brine cucumbers. Cause of softening during air-purging of fermentations. *Applied and Environmental Microbiology, 40*, 417–422.

Daeschel, M. A., Fleming, H. P., & Potts, E. A. (1985). Compartmentalization of LAB and yeasts in the fermentation of brined cucumbers. *Food Microbiology, 2*(1), 77–84.

De Angelis, M., Campanella, D., Cosmai, L., Summo, C., Rizzello, C. G., & Caponio, F. (2015). Microbiota and metabolome of un-started and started Greek-type fermentation of Bella di Cerignola table olives. *Food Microbiology, 52*, 8–30.

De Angelis, M., Garruti, G., Minervini, F., Bonfrate, L., Portincasa, P., & Gobbetti, M. (2017). The food-gut human axis: The effects of diet on gut microbiota and metabolome. *Current Medicinal Chemistry*. https://doi.org/10.2174/0929867324666170428103848.

De Bellis, P., Valerio, F., Sisto, A., Lonigro, S. L., & Lavermicocca, P. (2010). Probiotic table olives: Microbial populations adhering on olive surface in fermentation sets inoculated with the probiotic strain *Lactobacillus paracasei* IMPC2.1 in an industrial plant. *International Journal of Food Microbiology, 140*, 6–13.

De Castro, A., Montaño, A., Casado, F. J., Sánchez, A. H., & Rejano, L. (2002). Utilization of *Enterococcus casseliflavus* and *Lactobacillus pentosus* as starter culture for Spanish-style green olive fermentation. *Food Microbiology, 19*, 637–644.

DeBoy, R. T., Mongodin, E. F., Fouts, D. E., Tailford, L. E., Khouri, H., Emerson, J. B., Nohanoud, Y., Watkins, K., Henrissat, B., Gilbert, H. J., & Nelson, K. E. (2008). Insights into plant cell wall degradation from the genome sequence of the soil bacterium *Cellvibrio japonicus*. *Journal of Bacteriology, 190*, 5545–5463.

Delcour, J., Ferain, T., Deghorain, M., Palumbo, E., & Hols, P. (1999). The biosynthesis and functionality of the cell-wall of lactic acid bacteria. *Antonie Van Leeuwenhoek, 76*, 159–184.

Di Cagno, R., Coda, R., De Angelis, M., & Gobbetti, M. (2012). Exploitation of vegetables and fruits through lactic acid fermentation. *Food Microbiology, 33*(1), 1–10. https://doi.org/10.1016/j.fm.2012.09.003. Epub 2012 Sep 17. Review.

Díaz-Muñiz, I., Kelling, R., Hale, S., Breidt, F., & McFeeters, R. F. (2007). Lactobacilli and tartrazine as causative agents of a red colored spoilage in cucumber pickle products. *Journal of Food Science, 72*, M240–M245.

Duran-Quintana, M. C., Gonzalez-Cancho, F., & Garrido-Fernandez, A. (1979). Natural black olives in brine. IX. Production of alambrado by some microorganisms isolated from fermentation brines. *Grasas y Aceites, 30*, 361–367.

Etchells, J. L. (1941). Incidence of yeasts in cucumber fermentations. *Food Research, 6*(1), 95–104.

Etchells, J. L., & Bell, T. A. (1950a). Classification of yeasts from the fermentation of commercially brined cucumbers. *Farlowia, 4*(1), 87–112.

Etchells, J. L., & Bell, T. A. (1950b). Film yeasts on commercial cucumber brines. *Food Technology, 4*(3), 77–83.

Etchells, J. L., Fabian, F. W., & Jones, I. D. (1945). *The Aerobacter fermentation of cucumbers during salting*. Mich Agric. Expt Sta Tech Bull No. 200. 56 p.

Etchells, J. L., Jones, I. D., & Lewis, W. M. (1947). *Bacteriological changes during the fermentation of certain brined and salted vegetables*. USDA Tech. Bull. No. 947: 64.

Etchells, J. L., Costilow, R. N., & Bell, T. A. (1952). Identification of yeasts from commercial cucumber fermentations in northern brining areas. *Farlowia, 4*(2), 249–264.

Etchells, J. L., Costilow, R. N., Anderson, T. E., & Bell, T. A. (1964). Pure culture fermentation of brined cucumbers. *Applied Microbiology, 12*(6), 523–535.

Etchells, J. L., Bell, T. A., Fleming, H. P., Kelling, R. E., & Thompson, R. L. (1973). Suggested procedure for the controlled fermentation of commercially brined pickling cucumbers—The use of starter cultures and reduction of carbon dioxide accumulation. *Pickle Pak Science, 3*(1), 4–14.

Ewaschuk, J., Naylor, J., & Zello, G. (2005). D-lactate in human and ruminant metabolism. *Journal of Nutrition, 135*, 1619–1625.

Fenlon, D. R. (1985). Wild birds and silage as reservoirs of *Listeria* in the agricultural environment. *Journal of Applied Microbiology, 59*, 537–543.

Filannino, P., Di Cagno, R., & Gobbetti, M. (2018). Metabolic and functional paths of LAB in plant foods: Get out of the labyrinth. *Current Opinion in Biotechnology, 49*, 64–72.

Fitch, M. D., & Fleming, S. E. (1999). Metabolism of short-chain fatty acids by rat colonic mucosa in vivo. *American Journal of Physiology, 277*, G31–G40.

Fleming, H. P. (1979). Purging carbon dioxide from cucumber brines to prevent bloater damage—A review. *Pickle Pak Science, 6*(1), 8–22.

Franco, W., & Pérez-Díaz, I. M. (2012). Role of selected oxidative yeasts and bacteria in cucumber secondary fermentation associated with spoilage of the fermented fruit. *Food Microbiology, 32*, 338–344.

Franco, W., Pérez-Díaz, I. M., Johanningmeier, S. D., & McFeeters, R. F. (2012). Characteristics of spoilage-associated secondary cucumber fermentation. Applied and Environmental Microbiology, 78 (4), 1273–1284.

Fred, E. B., & Peterson, W. H. (1922). The production of pink sauerkraut by yeasts. *Journal of Bacteriology, 7*, 257–269.

Fuccio, F., Bevilacqua, A., Sinigaglia, M., & Corbo, M. R. (2016). Using a polynomial model for fungi from table olives. *International Journal of Food Science & Technology, 51*, 1276–1283.

Gänzle, M. G. (2015). Lactic metabolism revisited: Metabolism of lactic acid bacteria in food fermentations and food spoilage. *Current Opinion in Food Science, 2*, 106–117.

García, M., Serra, N., Pujola, M., & García, J. (1995). Analisis de la fibra alimentaría y sus fracciones por el método de Englyst. *Alimentaria, 95*, 45–50.

Gardner, N. C., Savard, T., Obermeier, P., Caldwell, G., & Champagne, C. P. (2001). Selection and characterization of mixed starter cultures for lactic acid fermentation of carrot, cabbage, beet, and onion vegetable mixtures. *International Journal of Food Microbiology, 64*, 261–275.

Garrido Fernández, A., Fernández Díez, M. J., & Adams, R. M. (1997). *Table olives: Production and processing*. London: Chapman & Hall.

Geldreich, E. E., & Bordner, R. H. (1971). Fecal contamination of fruits and vegetables during cultivation and processing for market. A review. *Journal of Milk and Food Technology, 34*, 184–195.

Gililland, J. R., & Vaughn, R. H. (1943). Characteristics of butyric acid bacteria from olives. *Journal of Bacteriology, 46*, 315–322.

Golomb, B. L., Morales, V., Jung, A., Yau, B., Boundy-Mills, K. L., & Marco, M. L. (2013). Effects of pectinolytic yeast on the microbial composition and spoilage of olive fermentations. *Food Microbiology, 33*, 97–106.

Gonzalez-Cancho, F., Nosti-Vega, M., Fernandez-Diez, M. J., & Buzcu, N. (1970). *Propionibacterium* spp. associated with olive spoilage. Factors influencing their growth. *Microbiología Española, 23*, 233–252.

Gorvitovskaia, A., Holmes, S. P., & Huse, S. M. (2016). Interpreting *Prevotella* and *Bacteroides* as biomarkers of diet and lifestyle. *Microbiome, 4*, 15. https://doi.org/10.1186/s40168-016-0160-7.

Grigelmo-Miguel, N., Gorinstein, S., & Martín-Belloso, O. (1999). Characterization of peach dietary fiber concentrate as a food ingredient. *Food Chemistry, 65*, 175–181.

Han, X., Yi, H., Zhang, L., Huang, W., Zhang, Y., Zhang, L., & Du, M. (2014). Improvement of fermented Chinese cabbage characteristics by selected starter cultures. *Journal of Food Science, 79*, M1387–M1392.

Hemert, S. V., Meijerink, M., Molenaar, D., Bron, P. A., deVos, P., Kleerebezem, M., Wells, J. M., & Marco, M. L. (2010). Identification of *Lactobacillus plantarum* genes modulating the cytokine response of human peripheral blood mononuclear cells. *BMC Microbiology, 10*, 293. https://doi.org/10.1186/1471-2180-10-293.

Hernandez, A., Martin, A., Aranda, E., Perez-Nevado, F., & Cordoba, M. G. (2007). Identification and characterization of yeast isolated from the elaboration of seasoned green table olives. *Food Microbiology, 24*, 346–351.

Holes, R. (2010). Nature's spoils. *The New Yorker*. Retrieved November 22, 2010, from https://www.newyorker.com/magazine/2010/11/22/natures-spoils

Hong, S. W., Choi, Y. J., Lee, H. W., Yang, J. H., & Lee, M. A. (2016). Microbial community structure of Korean cabbage kimchi and ingredients with denaturing gradient gel electrophoresis. *Journal of Microbiology and Biotechnology, 26*(6), 1057–1062. https://doi.org/10.4014/jmb.1512.12035.

Ito, K. A., Chen, J. K., Lerke, P. A., Seeger, M. L., & Unverferth, J. A. (1976). Effect of acid and salt concentration in fresh-pack pickles on the growth of *Clostridium botulinum* spores. *Applied and Environmental Microbiology, 32*(1), 121–124.

Johanningsmeier, S. D., & McFeeters, R. F. (2013). Metabolism of lactic acid in fermented cucumbers by *Lactobacillus buchneri* and related species, potential spoilage organisms in reduced salt fermentations. *Food Microbiology, 35*(2), 129–135.

Johanningsmeier, S. D., & McFeeters, R. F. (2015). Metabolic footprinting of *Lactobacillus buchneri* strain LA1147 during anaerobic spoilage of fermented cucumbers. *International Journal of Food Microbiology, 215*, 40–48. https://doi.org/10.1016/j.ijfoodmicro.2015.08.004.

Johanningsmeier, S. D., Fleming, H. P., Thompson, R. L., & McFeeters, R. F. (2005). Chemical and sensory properties of sauerkraut produced with *Leuconostoc mesenteroides* starter cultures of differing malolactic phenotypes. *Journal of Food Science, 70*(5), S343–S349.

Jung, J. Y., Lee, S. H., Lee, H. J., Seo, H. Y., Park, W. S., & Jeon, C. O. (2012). Effects of *Leuconostoc mesenteroides* starter cultures on microbial communities and metabolites during kimchi fermentation. *International Journal of Food Microbiology, 153*, 378–387.

Karl, J. P., Hatch, A. M., Arcidiacono, S. M., Pearce, S. C., Pantoja-Feliciano, I. G., Doherty, L. A., & Soares, J. W. (2018). Effects of psychological, environmental and physical stressors on the gut microbiota. *Frontiers in Microbiology, 9*, article 2013. https://doi.org/10.3389/fmicb.2018.02013.

Khalaf, E. M., & Raizada, M. N. (2016). Taxonomic and functional diversity of cultured seed associated microbes of the cucurbit family. *BMC Microbiology, 16*, 131. https://doi.org/10.1186/s12866-016-0743-2.

King, A. D., & Vaughn, R. H. (1961). Media for detecting pectolytic Gram-negative bacteria associated with the softening of cucumbers, olives and other plant tissues. *Journal of Food Science, 26*, 635–643.

Kishino, S., Takeuchi, M., Park, S. B., Hirata, A., Kitamura, N., Kunisawa, J., Kiyono, H., Iwamoto, R., Isobe, Y., Arita, M., Arai, H., Ueda, K., Shima, J., Takahashi, S., Yokozeki, K., Shimizu, S., & Ogawa, J. (2013). Polyunsaturated fatty acid saturation by gut LAB affecting host lipid composition. *Proceedings of the National Academy of Sciences of the United States of America, 110*, 17808–17813. https://doi.org/10.1073/pnas.1312937110.

Konishi, M., Maruoka, N., Furuta, Y., Morita, T., Fukuoka, T., Imura, T., & Kitamoto, D. (2014). Biosurfactant-producing yeasts widely inhabit various vegetables and fruits. *Bioscience, Biotechnology, and Biochemistry, 78*(3), 516–523. https://doi.org/10.1080/09168451.2014.882754. Epub 2014 Apr 16.

Kovatcheva-Datchary, P., Nilsson, A., Akrami, R., Lee, Y. S., De Vadder, F., Arora, T., Hallen, A., Martens, E., Björck, I., Bäckhed, F. (2015). Dietaryfiber-induced improvement in glucose metabolism is associated with increased abundance of *Prevotella*. *Cell Metab. 22*(6), 971–982. https://doi.org/10.1016/j.cmet.2015.10.001. Epub2015Nov6.

Kyung, K. H., Medina Pradas, E., Kim, S. G., Lee, Y. J., Kim, K. H., Choi, J. J., Cho, J. H., Chung, C. H., Barrangou, R., & Breidt, F. (2015). Microbial ecology of watery kimchi. *Journal of Food Science, 80*(5), M1031–M1038. https://doi.org/10.1111/1750-3841.12848.

Leal-Sánchez, M. V., Jiménez-Díaz, R., Maldonado-Barragán, A., Garrido-Fernández, A., & Ruiz-Barba, J. L. (2002). Optimization of bacteriocin production by batch fermentation of *Lactobacillus plantarum* LPCO10. *Applied and Environmental Microbiology, 68*, 4465–4471.

Leben, C. (1972). Micro-organisms associated with plant buds. *Journal of General Virology, 71*, 327–331.

Lee, C. H. (2001). *Fermentation technology in Korea*. Seoul, South Korea: Korea University Press.

Lee, S. A., Park, J., Chu, B., Kim, J. M., Joa, J. H., Sang, M. K., Song, J., & Weon, H. Y. (2016). Comparative analysis of bacterial diversity in the rhizosphere of tomato by culture-dependent and -independent approaches. *Journal of Microbiology, 54*(12), 823–831.

Lee, M., Song, J. H., Jung, M. Y., Lee, S. H., & Chang, J. Y. (2017). Large-scale targeted metagenomics analysis of bacterial ecological changes in 88 kimchi samples during fermentation. *Food Microbiology, 66*, 173–183. https://doi.org/10.1016/j.fm.2017.05.002.

Lefeber, T., Janssens, M., Camu, N., & De Vuyst, L. (2010). Kinetic analysis of strains of lactic acid bacteria and acetic acid bacteria in cocoa pulp simulation media toward development of a starter culture for cocoa bean fermentation. *Applied and Environmental Microbiology, 76*(23), 7708–7716. https://doi.org/10.1128/AEM.01206-10.

Lemanceau, P., Barret, M., Mazurier, S., Mondy, S., Pivato, B., Fort, T., & Vacher, C. (2017). Plant communication with associated microbiota in the spermosphere, rhizosphere and phyllosphere. *Advances in Botanical Research, 82*, 101–133. https://doi.org/10.1016/bs.abr.2016.10.007.

Levin, R. E., & Vaughn, R. H. (1966). *Desulfovibrio aestuarii*, the causative agent of hydrogen sulfide spoilage of fermenting olive brines. *Journal of Food Science, 31*, 768–772.

Lloyd-Price, J., Abu-Ali, G., & Huttenhower, C. (2016). The healthy human microbiome. *Genome Medicine Review, 8*(1), 51. https://doi.org/10.1186/s13073-016-0307-y.

Lopez-Velasco, G., Carder, P. A., Welbaum, G. E., & Ponder, M. A. (2013). Diversity of the spinach (Spinacia oleracea) spermosphere and phyllosphere bacterial communities. *FEMS Microbiology Letters, 346*(2), 146–54.

Lu, Z., Breidt, F., Fleming, H. P., Altermann, E., & Klaenhammer, T. R. (2003a). Isolation and characterization of a *Lactobacillus plantarum* bacteriophage, FJL-1, from a cucumber fermentation. *International Journal of Food Microbiology, 84*, 225–235.

Lu, Z., Breidt, F., Plengvidhya, V., & Fleming, H. P. (2003b). Bacteriophage ecology in commercial sauerkraut fermentations. *Applied and Environmental Microbiology, 69*, 3192–3202.

Lu, Z., Altermann, E., Breidt, F., Predki, P., Fleming, H. P., & Klaenhammer, T. R. (2005). Sequence analysis of the *Lactobacillus plantarum* bacteriophage FJL-1. *Gene, 348*, 45–54.

Lu, Z., Altermann, E., Breidt, F., & Kozyavkin, S. (2010). Sequence analysis of *Leuconostoc mesenteroides* bacteriophage (phi)1-A4 isolated from industrial vegetable fermentation. *Applied and Environmental Microbiology, 76*, 1955–1966.

Lu, Z., Pérez-Díaz, I. M., Hayes, J. S., & Breidt, F. (2012). Bacteriophage ecology in a commercial cucumber fermentation. *Applied and Environmental Microbiology, 78*(24), 8571–8578.

Lund, J., Aas, V., Tingstad, R. H., Van Hees, A., & Nikolić, N. (2018). Utilization of lactic acid in human myotubes and interplay with glucose and fatty acid metabolism. *Nature, 8*, 9814. https://doi.org/10.1038/s41598-018-28249-5.

Manani, T. A., Collison, E. K., & Mpuchane, S. (2006). Microflora of minimally processed frozen vegetables sold in Gaborone, Botswana. *Journal of Food Protection, 69*(11), 2581–2586.

Marco, M. L., Bongers, R. S., deVos, W. M., & Kleerebezem, M. (2006). Spatial and temporal expression of *Lactobacillus plantarum* genes in the gastrointestinal tracts of mice. *Food Microbiology, 73*(1), 124–132. https://doi.org/10.1128/AEM.01475-06.

Mark, E. M., Vaughn, R. H., Miller, M. W., & Phaff, H. I. (1956). Yeasts occurring in brines during the fermentation and storage of green olives. *Food Technology, 10*, 416.

Marquina, D., Peres, C., Caldas, F. V., Marqjes, J. F., Peipjado, J. M., & Spencer-Martins, I. (1992). Characterization of the yeast population in olive brines. *Letters in Applied Microbiology, 14*, 279–283.

Martínez-Villaluenga, C., Peñas, E., Sidro, B., Ullate, M., Frias, J., & Vidal-Valverde, C. (2012). White cabbage fermentation improves ascorbic content, antioxidant and nitric oxide production inhibitory activity in LPS-induced macrophages. *LWT- Food Science and Technology, 46*, 77–83.

Mattos, F. R., Fasina, O. O., Reina, L. D., Fleming, H. P., Breidt, F., Damasceno, G. S., & Passos, F. V. (2005). Heat transfer and microbial kinetics modeling to determine the location of microorganisms within cucumber fruit. *Journal of Food Science, 70*(5), E324–E330.

McDonald, L. C., Fleming, H. P., & Hassan, H. M. (1990). Acid tolerance of *Leuconostoc mesenteroides* and *Lactobacillus plantarum*. *Applied and Environmental Microbiology, 56*(7), 2120–2124.

McDonald, D., Hyde, E., Debelius, J. W., Morton, J. T., Gonzalez, A., Ackermann, G., Aksenov, A. A., Behsaz, B., Brennan, C., Chen, Y., DeRight Goldasich, L., Dorrestein, P. C., Dunn, R. R., Fahimipour, A. K., Gaffney, J., Gilbert, J. A., Gogul, G., Green, J. L., Hugenholtz, P., Humphrey, G., Huttenhower, C., Jackson, M. A., Janssen, S., Jeste, D. V., Jiang, L., Kelley, S. T., Knights, D., Kosciolek, T., Ladau, J., Leach, J., Marotz, C., Meleshko, D., Melnik, A. V., Metcalf, J. L., Mohimani, H., Montassier, E., Navas-Molina, J., Nguyen, T. T., Peddada, S., Pevzner, P., Pollard, K. S., Rahnavard, G., Robbins-Pianka, A., Sangwan, N., Shorenstein, J., Smarr, L., Song, S. J., Spector, T., Swafford, A. D., Thackray, V. G., Thompson, L. R., Tripathi, A., Vázquez-Baeza, Y., Vrbanac, A., Wischmeyer, P., Wolfe, E., Zhu, Q., American Gut Consortium, & Knight, R. (2018). American gut: An open platform for citizen. *mSystems, 3*(3), pii: e00031-18. https://doi.org/10.1128/mSystems.00031-18.

McFeeters, R. F., & Pérez-Díaz, I. M. (2010). Fermentation of cucumbers brined with calcium chloride instead of sodium chloride. *Journal of Food Science, 75*(3), C291–C296.

Medina-Pradas, E., & Arroyo-López, F. N. (2015). Presence of toxic microbial metabolites in table olives. *Frontiers in Microbiology, 6*, 873. https://doi.org/10.3389/fmicb.2015.00873.

Medina-Pradas, E., Pérez-Díaz, I. M., Breidt, F., Hayes, J. S., Franco, W., Butz, N., & Azcarate-Peril, A. (2016). Bacterial ecology of fermented cucumber rising pH spoilage as determined by non-culture based methods. *Journal of Food Science, 80*(1), M121–M129. https://doi.org/10.1111/1750-3841.13158.

Meneley, J. C., & Stanghellini, M. E. (1974). Detection of enteric bacteria within locular tissue of healthy cucumbers. *Journal of Food Science, 39*, 1267.

Montaño, A., Sánchez, A. H., Rejano, L., & de Castro, A. (1997). Processing and storage of lye-treated carrots fermented by a mixed starter culture. *International Journal of Food Microbiology, 35*, 83–90.

Moon, S. H., Chang, M., Kim, H. Y., & Chang, H. C. (2014). *Pichia kudriavzevii* is the major yeast involved in film-formation, off-odor production, and texture-softening in over-ripened Kimchi. *Food Science and Biotechnology, 23*, 489–497.

Mundt, O. (1970). LAB associated with raw plant food material. *Journal of Milk and Food Technology, 33*, 550–553.

Mundt, J. O., & Hammer, J. L. (1968). Lactobacilli on plants. *Journal of Applied Microbiology, 16*(9), 1326–1330.

Nanniga, N. (2010). Did van Leeuwenhoek observe yeast cells in 1680? In *Small things considered*. American Society for Microbiology. Retrieved from http://schaechter.asmblog.org/schaechter/2010/04/did-van-leeuwenhoek-observe-yeast-cells-in-1680.html

Nychas, G. J. E., Panagou, E. Z., Parker, M. L., Waldron, K. W., & Tassou, C. C. (2002). Microbial colonization of naturally black olives during fermentation and associated biochemical activities in the cover brine. *Letters in Applied Microbiology, 34*, 173–177.

Ofek, M., Voronov-Goldman, M., Hadar, Y., & Minz, D. (2014). Host signature effect on plant root-associated microbiomes revealed through analyses of resident vs. active communities. *Environmental Microbiology, 16*(7), 2157–2167. https://doi.org/10.1111/1462-2920.12228.

Olsen, M., & Pérez-Díaz, I. M. (2009). Influence of microbial growth in the redox potential of fermented cucumbers. *Journal of Food Science, 74*(4), M149–M153.

Ottesen, A., R., Gorham, S., reed, E., Newell, M. J., Ramachandran, P., Canida T., Allard, M., Evans, P., Brown, E., White, J. R. (2016). Using a control to better understand phyllosphere micorbiota. *PloS ONE 11*(9), e0163482. https://doi.org/10.1371/journal.pone.0163482.

Palomino, J. M., Toledo del Árbol, J., Benomar, N., Abriouel, H., Martínez Cañamero, M., Gálvez, A., & Pérez Pulido, R. (2015). Application of *Lactobacillus plantarum* Lb9 as starter culture in caper berry fermentation. *LWT-Food Science and Technology, 60*, 788–794.

Panagou, E. Z., Schillinger, U., Franz, C. M. A. P., & Nychas, G. J. E. (2008). Microbiological and biochemical profile of cv. Conservolea naturally black olives during controlled fermentation with selected strains of LAB. *Food Microbiology, 25*, 348–358.

Park, E. J., Kim, K. H., Abell, G. C. J., Kim, M. S., Roh, S. W., & Bae, J. W. (2010). Metagenomic analysis of the viral communities in fermented foods. *Applied and Environmental Microbiology, 77*(4), 1284–1291. https://doi.org/10.1128/AEM.01859-10.

Park, E. J., Chun, J., Cha, C. J., Park, W. S., Jeon, C. O., & Bae, J. W. (2012). Bacterial community analysis during fermentation of ten representative kinds of kimchi with barcoded pyrosequencing. *Food Microbiology, 30*(1), 197–204. https://doi.org/10.1016/j.fm.2011.10.011.

Pederson, C. S., & Albury, M. N. (1969). *The sauerkraut fermentation.* N.Y. Agric. Expt. Sta. Bull. 824.

Pérez Pulido, R., Omar, N. B., Abriouel, H., Lucas López, R., Martínez Cañamero, M., & Gálvez, A. (2005). Microbiological study of lactic acid fermentation of caper berries by molecular and culture-dependent methods. *Applied and Environmental Microbiology, 71*, 7872–7879.

Pérez-Díaz, I. M., Breidt, F., Buescher, R. W., Arroyo-López, F. N., Jimenez-Diaz, R., Bautista-Gallego, J., Garrido-Fernandez, A., Yoon, S., & Johanningsmeier, S. D. (2014). Fermented and acidified vegetables (Chapter 51). In F. Pouch Downes & K. A. Ito (Eds.), *Compendium of methods for the microbiological examination of foods* (5th ed.). American Public Health Association.

Pérez-Díaz, I. M., McFeeters, R. F., Moeller, L., Johanningsmeier, S. D., Hayes, J. S., Fornea, D., Gilbert, C., Custis, N., Beene, K., & Bass, D. (2015). Commercial scale cucumber fermentations brined with calcium chloride instead of sodium chloride. *Journal of Food Science, 80*(12), M2827–M2836. https://doi.org/10.1111/1750-3841.13107.

Pérez-Díaz, I. M., Hayes, J. S., Medina-Pradas, E., Anekella, K., Daughtry, K. V., Dieck, S., Levi, M., Price, R., Butz, N., Lu, Z., & Azcarate-Peril, M. (2016). Reassessment of the succession of lactic acid bacteria in commercial cucumber fermentations and physiological and genomic features associated with their dominance. *Food Microbiology, 63*, 217–227. https://doi.org/10.1016/j.fm.2016.11.025.

Pérez-Díaz, I. M., Hayes, J. S., Medina, E., Webber, A. M., Butz, N., Dickey, A. N., Lu, Z., & Azcarate-Peril, M. A. (2018). Assessment of the non-LAB microbiota in fresh cucumbers and commercially fermented cucumber pickles brined with 6% NaCl. *Food Microbiology, 77*, 10–20. https://doi.org/10.1016/j.fm.2018.08.003.

Plengvidhya, V., Breidt, F., Lu, Z., & Fleming, H. P. (2007). DNA fingerprinting of LAB in sauerkraut fermentations. *Applied and Environmental Microbiology, 73*(23), 7697–7702.

Raj, K. C., Ingram, L. O., & Mauphin-Furlow, J. A. (2001). Pyruvate decarboxylase: A key enzyme for the oxidative metabolism of lactic acid by *Acetobacter pasteurianus. Archives of Microbiology, 176*, 443–451.

Rastogi, G., Sbodio, A., Tech, J. J., Suslow, T. V., Coaker, G. L., & Leveau, J. H. (2012). Leaf microbiota in an agroecosystem: spatiotemporal variation in bacterial community composition on field-grown lettuce. *The ISME Journal, 6*(10), 1812–1822. https://doi.org/10.1038/ismej.2012.32.

Reina, L. D., Fleming, H. P., & Breidt, F. Jr. (2002). Bacterial contamination of cucumber fruit through adhesion. *Journal of Food Protection, 65*(12), 1881–1887.

Rincón-León, F. (2003). Functional foods. In B. Caballero (Ed.), *Encyclopedia of food science and nutrition* (2nd ed., pp. 2827–2832). New York: Academic Press. https://doi.org/10.1016/B0-12-227055-X/01328-6.

Roberfroid, M. (1993). Dietary fiber, inulin, and oligofructose: A review comparing their physiological effects. *Critical Reviews in Food Science and Nutrition, 33*(2), 103–148. Review. Erratum in: Critical Reviews in Food Science and Nutrition 1993; 33(6):553.

Rodríguez-Gómez, F., Romero Gil, V., Bautista Gallego, J., García García, P., Garrido Fernández, A., & Arroyo López, F. N. (2014). Production of potential probiotic Spanish-style green table olives at pilot plant scale using multifunctional starters. *Food Microbiology, 44*, 278–287.

Rossi, M., Martinez-Martinez, D., Amaretti, A., Ulrici, A., Raimondi, S., & Moya, A. (2016). Mining metagenomics whole genome sequences revealed subdominant but constant *Lactobacillus* population in the human gut microbiota. *Environmental Microbiology Reports, 8*, 399–406.

Ruiz-Cruz, J., & Gonzalez-Cancho, F. (1969). The metabolism of yeasts isolated from the brine of pickled Spanish-type green olives. I. The assimilation of lactic, acetic and citric acids. *Grasas y Aceites, 20*, 6–11.

Salonen, A., & deVos, W. M. (2014). Impact of diet on human intestinal microbiota and health. *Annual Review of Food Science and Technology, 5*, 239–262. https://doi.org/10.1146/annurev-food-030212-182554.

Samish Z., Etinger-Tulczynsky R. (1962). Bacteria within fermenting tomatoes and cucumbers. *In*: Leitch J. M. Proc. 1st Int. Conm. Food Scie. Technol. Gordon & Breach Science Publications, New York. 2: 373.

Samish, Z., Etinger-Tulczynska, R., & Bick, M. (1963). The microflora within the tissue of fruits and vegetables. *Journal of Food Science, 28*(3), 259–266.

Seelinger, H. P. R., & Jones, D. (1986). Genus *Listeria*, p. In P. H. A. Sneath, N. S. Mair, M. E. Sharpe, & J. G. Holt (Eds.), *Bergey's manual of systematic bacteriology* (Vol. 2, p. 1235). Baltimore, MD: Williams and Wilkins.

Sender, R., Fuchs, S., & Milo, R. (2016). Revised estimates for the number of human and bacterial cells in the body. *PLoS Biology, 14*(8), e1002533. https://doi.org/10.1371/journal.pbio.1002533.

Shi, X., Wu, Z., Namvar, A., Kostrzynska, M., Dunfield, K., & Warriner, K. (2009). Microbial population profiles of the microflora associated with pre- and postharvest tomatoes contaminated with *Salmonella typhimurium* or *Salmonella montevideo*. *Journal of Applied Microbiology, 107*(1), 329–338.

Siezen, R. J., & van Hylckama Vlieg, J. E. (2011). Genomic diversity and versatility of *Lactobacillus plantarum*, a natural metabolic engineer. *Microbial Cell Factories, 10* (Suppl 1), S3. PMC3271238. https://doi.org/10.1186/1475-2859-10-S1-S3.

Sonnenburg, E. D., Smits, S. A., Tikhonov, M., Higginbottom, S. K., Wingreen, N. S., & Sonnenburg, J. L. (2016). Diet-induced extinctions in the gut microbiota compound over generations. *Nature, 529*, 212–215. https://doi.org/10.1038/nature16504.

Spencer, C., Randic, L., Butler, J. (2009). Survival following profound lactic acidosis and cardiac arrest: does metformin really induce lactic acidosis?J Int. Care Soc. 10(2):115–117.

Stamer, J. R., Hrazdina, G., & Stoyla, B. O. (1973). Induction of red color formation in cabbage juice by *Lactobacillus brevis* and its relationship to pink sauerkraut. *Applied Microbiology, 26*, 161–166.

Tingirikari, J. M. R. (2018). Microbiota-accessible pectic poly- and oligosaccharides in gut health. Food & Function. https://doi.org/10.1039/c8fo01296b.

Uribarri, J., Oh, M., & Carroll, H. (1998). D-lactic acidosis. *Medicine, 77*, 73–82.

Vaughn, R. H., Won, W. D., Spencer, F. B., Pappagranis, D., Foda, I. O., & Krumperman, P. M. (1953). *Lactobacilllus plantarum*, the cause of yeast spots on olives. *Applied Microbiology, 1*, 82–85.

Vaughn, R. H., King, A. D., Nagel, C. W., Ng, H., Levin, R. E., Macmilla, J. D., & York, G. K. (1969). Gram-negative bacteria associated with sloughing, a softening of Californian ripe olives. *Journal of Food Science, 34*, 224–227.

Vaughn, R. H., Stevenson, K. E., Dave, B. A., & Park, H. C. (1972). Fermenting yeast associated with softening and gas-pocket formation in olives. *Applied Microbiology, 23*, 316–320.

Vega Leal-Sánchez, M., Ruiz Barba, J. L., Sánchez, A. H., Rejano, L., Jiménez Díaz, R., & Garrido-Fernandez, A. (2003). Fermentation profile and optimization of green olive fermentation using *Lactobacillus plantarum* LPCO10 as a starter culture. *Food Microbiology, 20*, 421–430.

Verbeke, K. A., Boobis, A. R., Chiodini, A., Edwards, C. A., Franck, A., Kleerebezem, M., Nauta, A., Raes, J., van Tol, E. A., & Tuohy, K. M. (2015). Towards microbial fermentation metabolites as markers for health benefits of prebiotics. Nutrition Research Reviews, 28, 42–66. https://doi.org/10.1017/S095442415000037.

Viuda-Martos, M., López-Marcos, M. C., Fernández-López, J., Sendra, E., López-Vargas, J. H., & Pérez-Alvarez, J. A. (2010). Role of fiber in cardiovascular diseases: A review. *Comprehensive Reviews in Food Science and Food Safety, 9*(2), 240–258.

Weiss, A., Hertel, C., Grothe, S., Ha, D., Hammes, W. P. (2007). Characterization of the cultivable micorbiota of sprouts and their potential for application as protective cultures. *Syst. Appl. Micorbiol. 30*(6), 483–493.

Welshimer, H. J. (1968). Isolation of *Listeria monocytogenes* from vegetation. *Journal of Bacteriology, 95*, 300–303.

Welshimer, H. J., & Donker-Voet, I. (1971). *Listeria monocytogenes* in nature. *Applied Microbiology, 21*, 516–519.

West, N. S., Gililland, J. R., & Vaughn, R. H. (1941). Characteristics of coliform bacteria from olives. *Journal of Bacteriology, 41*, 341–353.

Yoon, S. S., Barrangou-Poueys, R., Breidt, F., Klaenhammer, T. R., & Fleming, H. P. (2002). Isolation and characterization of bacteriophages from fermenting sauerkraut. *Applied and Environmental Microbiology, 68*, 973–976.

Yoon, S. S., Barrangou-Poueys, R., Breidt, F., & Fleming, H. P. (2007). Detection and characterization of a lytic *Pediococcus* bacteriophage from the fermenting cucumber brine. *Journal of Microbiology and Biotechnology, 17*, 262–270.

Production and Conservation of Starter Cultures: From "Backslopping" to Controlled Fermentations

Hunter D. Whittington, Suzanne F. Dagher, and José M. Bruno-Bárcena

Abstract As human society has evolved from small, nomadic groups of hunter-gatherers to large, stationary civilizations, there has been an increased reliance on the preservation of foods to sustain populations through periods of reduced agricultural productivity. Microbial fermentations have been used for millennia to preserve high water activity foods such as fruits, vegetables, and meats. Originally, a process called "backslopping", in which a small portion of a previously successful fermentation is used to inoculate fresh substrate was used to generate starter cultures for future fermentations. However, these processes fell from favor in the nineteenth century concurrently with the rise in public interest and governmental regulations concerning food safety. Starter cultures for mass-produced fermented foods were subsequently required to be produced from defined GRAS microorganisms, triggering a systematic reduction of microbial diversity seeding the digestive tract. Recently, several landmark studies have highlighted the importance of a healthy gut microbiome leading to a renewed interest in more traditional (artisanal) methods of food fermentations. New methods of mixed-strain starter culture production, particularly immobilized cell reactors, present attractive alternatives to the more traditional batch reactors due to their ability to produce a more robust and diverse starter all in one step. Additionally, advances in culture preservation technology, like freeze- and spray-drying, have increased the long-term viability and reduced the cost of starter cultures.

Keywords Backslopping · Food preservation · Food biotransformation · Fermentation · Traditional fermentation · Industrial fermentation · Culture production

---

H. D. Whittington · S. F. Dagher · J. M. Bruno-Bárcena (✉)
Department of Plant and Microbial Biology, North Carolina State University, Raleigh, NC, USA
e-mail: jbbarcen@ncsu.edu

© Springer Nature Switzerland AG 2019
M. A. Azcarate-Peril et al. (eds.), *How Fermented Foods Feed a Healthy Gut Microbiota*, https://doi.org/10.1007/978-3-030-28737-5_5

Introduction

Food preservation, ethnic traditions, and microbial coevolution are intimately correlated to the shaping of human populations. At the beginning of human history, migration provided our hunter-gatherer ancestors the opportunity to consume from diverse food sources and their associated microbes (Cordain et al. 2005). At some point, the communities reached a critical size making this type of feeding practice unsustainable for overall population growth (Buchanan 2018). Thus, human growth stagnated and, to achieve an extended population density, a pause for technological innovations was required (Boserup 1976). Agricultural development then became a vital practice, appearing when the human population was estimated at a global maximum of ten million (Holliday et al. 2014). Direct consequences of agricultural practices were the reduction of population mobility, the reduced consumption of freshly harvested foods, and a need to store food. At that time, feeding from stored foods became not only essential but also a physiologically rewarding process that led to an adequate and balanced diet.

Foods of every origin ripen, rot, or spoil; processes that accelerate once the food is harvested. This natural attribute leads to microbial pre-digestion, which after consumption has an unquestionably large impact on human health and wellbeing (Marco et al. 2017). During the unfavorable hunting and cultivating periods, food maintenance and preservation required methods to transiently extend food stability and availability. To overcome this food limitation, natural processes to challenge accelerated rotting due to microbial spoilage and contamination were required (Hammond et al. 2015). Accordingly, to conquer a new critical maximum in population size, every human group surviving from the sustainable production of indigenous foods (from limited local diversity) was again forced to innovate. They implemented procedures that reduced and delayed spoilage specific for each food and climate, and it is at this time when cyclic methods became an integral part of community identity and expertise. Humans that inhabited extreme cold climates froze whale meat, those in hot climates learned to sundry foods, those at high altitudes seized upon the extreme daily variations of temperature to freeze-dry meats or vegetables, and individuals in temperate climates learned the magic of fermentation (*fervere*). All of them developed methods to extend and stabilize foods by biotransforming them into new, enriched products (Amit et al. 2017; Food Preservation 2018). Sadly, at present, the fermentation practices of "boiling without heat" are being neglected, forgotten, and in some cases banned. Methods belonging to our treasured ethnic heritage, communicated through generations, are at war with modern practices to control microbial spoilage and contamination, which often lead to food waste (Wu et al. 2018).

Initially, food storage practices of low-water activity seeds, tubers, and root crops dominated agricultural methods of production and harvest (Black and Pritchard 2002; Blomstedt et al. 2018). However, since high water activity is ideal for microbial proliferation, the preservation technologies concerning fruit, vegetables, and meats demanded new innovations (Doulgeraki et al. 2012; Leff and Fierer

2013). Accordingly, the observation of reproducible food decomposition between food harvest periods linked to microbial growth played a central role in preservation and storage (Prokopov and Tanchev 2007). Empirical trial-error processes led to "backslopping" (culture) methods which were assimilated into the human cultural practices but are restrictively exercised today. Nevertheless, these practices are regaining favour confirming the heightened interest in ancient food processing traditions (Sieuwerts et al. 2008).

Thus, for communities to settle down, artisanal fermentation practices needed to be established leading to novel foods and new food safety challenges. In an environment prone to microbial contamination, microbial-human coevolution warranted relative safety against food-borne diseases. Moreover, we could theorize that the continuous replenishment of non-pathogenic microbes through diet limited the impact of pathogens. Additionally, the selected microbes, individually or as a consortium, offered dedicated functions and generate essential nutrients critical for a balanced diet and human survival (Nair et al. 2016).

In the 1800s, the French microbiologist Louis Pasteur provided the basis for understanding food-born disease causation by developing *pasteurization* methods (Dubos 1960), which resulted in a radical extension of human lifespan. Implementation of pasteurization delayed food spoilage, limited community exposure to food-borne diseases and reduced death rates. It is important to state here that pasteurization methods, as they are applied today, are not sterilization techniques; however, the original sterilization intent conceived by Pasteur is still correctly interpreted today. Consequently, a direct connection between heat treatment, *pasteurization*, and avoidance of food spoilage will remain forever assimilated into our vocabulary as synonyms of safe preservation. This health-associated correlation also shaped a social fascination for food sterilization that survives today. Food preservation practices employing heat enhanced human lifespan and over time have been positively associated to our common germophobic interpretation of the microbial universe. Not surprising is our society's persistent perception of the close relationship between non-pathogenic and pathogenic microbes with the outcome of a permanent war on microbes, and a cultural, pervasive and progressive elimination of microbes and their metabolic functions from our diet.

While pasteurization was justified a century ago, in an environment prone to microbial contamination and food-borne diseases, we are now entering a period of renovated interest in reincorporating key microbes to our diet. Moreover, society is demanding the implementation of knowledge-based hygiene practices and the return to low-heat food treatments. Although it is extremely problematic from the safety perspective to restore fully traditional artisanal preservation practices, controlled fermentation can save ethnic preservation traditions and add back essential ingredients to our food, including live microbes, with positive impacts to our health (van Hylckama Vlieg et al. 2011; Marco et al. 2017). Furthermore, different types of microbial cultures and their specific metabolic functions are intricately linked to their food and biotransformation process.

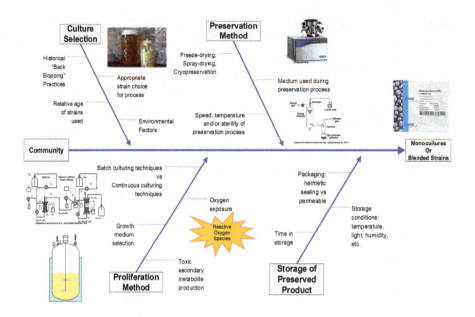

"Backslopping" (Culture) Practices

Liquid remnants from previous spontaneous processes or undefined mixtures of live microorganisms are the essential ingredients for some food biotransformations. Backslopping is defined as the transference of a portion of a successful fermentation with desired traits to fresh raw material as inoculant for initiation of a new fermentation. Since aging of a microbial culture is cyclical and transient in nature, large amounts of cells (viable and non-viable) will accumulate in sugar-containing water solutions until the carbon source is depleted. Time will determine final cell viability of the culture since the population will be exposed to its own toxic metabolites over time. This cycle will restart as soon as sufficient numbers of viable cells are added (or land and settle) in a new sugar-containing solution and the abundance of viable microbes comprising the initial mix will play a fundamental role in the technological characteristics of the final product. Obviously, this method adds uncertainty to the final product, which plays an important role in its final unique flavor (Irlinger and Mounier 2009; Gilbert et al. 2014; Schornsteiner et al. 2014; González-Córdova et al. 2016; Zheng et al. 2018). Specifically, it is this inconsistency and capacity for adaptation to ever changing environmental conditions (raw materials, season, temperature) that create the different and fascinating "individuality" of artisanal products.

The historically consistent success of food fermentations and their associated backslopping practices generated significant interest within the scientific community starting in the mid-twentieth century, which subsequently led to efforts to isolate the microorganisms responsible for food fermentations. This happened for a variety of reasons, including (1) to allow for convenient and consistent food fermentations permitting larger production, (2) to investigate strains for properties relevant to human health and industry, and (3) strains shown to be safe enabled the food industry to comply to emerging food safety regulations (Altieri et al. 2017).

Until recently, identification and classification of microorganisms in fermented foods, and indeed in all ecosystems, relied on culturing techniques and biochemical testing. This led to the development of various growth media specifically formulated for the cultivation of microbes associated with food (De Man et al. 1960; Kline and Sugihara 1971; Vandenberg et al. 1993). The isolation of strains associated with desirable fermented foods provided a unique opportunity in the food industry and allowed companies to use single strains or blends of strains to achieve specific characteristics in their foods. In addition to streamlining the production of fermented foods, these isolations led to the discovery of antimicrobial compounds such as Reuterin (Talarico et al. 1988) and small antimicrobial peptides called bacteriocins (Klaenhammer 1988; Axelsson et al. 1989).

With the vastly increased availability of affordable DNA sequencing options in the early- to mid-2000s came a renewed interest in the characterization of know strains and search for novel strains connected to food fermentations. The applications of next-generation sequencing as applied to fermented foods vary from 16S rRNA amplicon-based community surveys to the whole genome sequencing (Klaenhammer et al. 2005; van Hijum et al. 2013). The information gained from these technologies, along with culturing techniques, can be used to formulate defined mixed-strain starter cultures that simulate the diversity found in artisanal fermented foods, thereby conferring much, if not all, of the health benefits conveyed by these foods.

Traditional Methods to Generate Monocultures

Microbiological research has provided an understanding of the microbial ecology singularities associated with each food system (Cocolin and Ercolini 2015). The microbes responsible for the transformation of specific foods have been screened for decades and are currently further investigated using molecular and next generation sequencing methods (Dolci et al. 2013; Yang et al. 2014; Bokulich et al. 2016; Chombo-Morales et al. 2016; Chen et al. 2017; Murugesan et al. 2018). Traditionally, the identification of niche-specific strains capable of reinitiating fermentations was a required step before today's mandated application of heat treatment to foods. Industrial simplicity and economy led to the generation of monocultures or blended strains using large scale pure culture techniques and were followed by the development of methods of conservation to allow cell survival and activity retention during long-distance transportation while offering flexibility and convenience during application. Generating the desirable uniformity among fermented foods triggered a systematic reduction of the microbial diversity entering our gastrointestinal tract.

Thus, the traditional and most of the contemporary methods of cell proliferation are usually performed using selected mono- or mixed cultures blended after production, on the premise that we can consistently culture and harvest planktonic homogeneous viable and non-viable cell suspensions. The challenge is then to maintain survival, efficacy, and stability of the strains over time. As described earlier, during production strains are exposed to cyclical challenges. These cyclic periods of cell proliferation, inferred from nature and reproduced in our laboratories, are methods presently denoted as a batch mode of operation. In some cases, this conventional production method has been combined with challenging cells to sub-lethal stresses to enhance stability, followed by the addition of protectants and/or microencapsulation to increase cell protection and stability. Unavoidably, batch cultures generate populations containing mixtures of cells in different physiological states that frequently lead to high experimental variability. This heterogeneity is exacerbated when cell harvest time is prolonged due to the link between cell viability and stability, and the mode of operation process (Mikelsaar et al. 2011). In summary, traditional batch methods are prone to the generation of variable cultures due to compromised cell viability and stability during bacterial growth, preparation, and harvest, hence probably leading to ineffective cell preparations.

Current Methods of Culture Production

There is a growing demand for technologies that generate high cell counts while also ensuring cell stability, consistency, and process scalability (Di Cagno et al. 2013). In nature, microorganisms often live as highly organized communities of sessile cells enclosed in a self-produced polymeric matrix (or biofilm) that adheres to an inert or living surface (Costerton et al. 1999; Junter and Jouenne 2004). The formation of biofilms is a major strategy for bacterial survival. As an example, natural biofilm structures are found attached to food particles and to the gut mucosa (Macfarlane et al. 1997; Cinquin et al. 2004). Once biofilms reach certain density, sessile cells undergo complex cell-to-cell interactions and quorum-sensing signals accumulate to threshold concentrations triggering the expression of genes in different parts of the films and at different stages of their development (Costerton et al. 1999). The triggering of multiple resistance mechanisms leads to more resistant cells comprising the biofilm (Junter and Jouenne 2004; Rangel 2011).

A parallel to natural biofilms is found in immobilized cell reactors where the biological catalyst is kept fixed in a natural or artificial matrix while substrates, products, and a small fraction of cells (planktonic cells) continuously flow in the mobile phase (Champagne et al. 1994). Cell culture methods that simulate biofilm behavior have been employed to perform high cell density fermentations for both cell and metabolite production using reactor configurations that include batch reactors, continuous stirred tank rectors, fluidized bed reactors, and packed bed reactors. Previous publications have described biofilm operation in a packed bed reactors filled with porous solid supports (Bruno-Bárcena et al. 1999, 2001) and reviewed the biofilm as a technology applied to production (Dagher et al. 2010). Several studies have also shown that immobilized (sessile) and released (planktonic) bacteria exhibit changes in growth, morphology, and physiology compared with cells produced in conventional cell-free cultures (Doleyres et al. 2002, 2004).

What Parameters Should Be Controlled During Culture Production?

Despite the importance of viability and stability of bacteria in food, studies have shown large fluctuations and poor viability of these microbes in food matrices (Schillinger 1999; Masco et al. 2005) while there seems to be an overall lack of studies that have assessed cell viability and stability during storage. The factors that impact viability of starters ultimately decrease their efficacy during food transformations and their beneficial impact after consumption (Macori and Cotter 2018).

One of the most critical factors to consider for improving viability and stability early in the production process is the growth phase at which cultures should be harvested. Obviously, sufficient cell concentration and a cost-effective efficiency of the process restrict the harvest to times not earlier than the end of the logarithmic growth phase during batch culture or cyclic processes. Specifically, idiophase or early stationary phases are the preferred harvesting times. However, stability needs to be evaluated for most strains since there is a clear relationship between time of harvest and shelf-life. Therefore, cultures should be harvested and cooled when they reach homogeneous physiological states to ensure the efficacy of the food transformation as well as technological reproducibility (Bruno-Bárcena et al. 1998).

Another impactful component on viability and stability of fermented foods is the growth medium used to generate starter cultures. Typical media routinely used at laboratory scale are not cost-effective for large-scale production and, since different strains differ in their auxotrophic requirements, the traditional approach for providing growth factors cost-effectively has been to utilize complex nutrients such as peptones, whey permeates, yeast extracts, and others. Additions of Tween80, different combinations of salts, and reducing agents traditionally complete the formulations (Bruno-Bárcena et al. 1998; Salminen et al. 2004). It is important to consider that a cost-effective growth medium should not be the only criterion when selecting an appropriate platform. Instead, suitable microbial functional capabilities can only be assured by the optimal medium (Tavan et al. 2002; Lahtinen et al. 2012).

The physical and chemical variables historically recognized as the most critical for controlling specific growth rate and maximizing viability are strain-dependent and include oxygen, pH, and temperature (Pont and Holloway 1968; Cogan et al. 1971). However, optimal physical and chemical variables, which can guarantee higher specific growth rates, do not always deliver process-resistant viable cells that are also stable during storage.

Complex microbial communities involved in food fermentation usually proliferate under conditions of limited oxygen indicating that these microorganisms have not evolved or acquired complete sets of enzymatic mechanisms for protection against oxygen radicals (Condon 1987). Consequently, controlling oxygen concentrations during production have consequences on the final viability numbers after proliferation and during future stability. Additionally, viability of the generated monocultures or blend of strains will be compromised if exposed to aerobic environments during storage.

The characteristic microaerophilic or anaerobic conditions required for culturing create a shortage of electron acceptors, a limitation leading to the metabolic biotransformation of sugars to incomplete oxidized products (organic acids or alcohols). Thus, oxygen limitation has determined an evolutive adaptation of the strains dominating traditional food fermentations, and these types of organisms are only able to carry out fermentative metabolic processes generating organic acids as end products (Septembre-Malaterre et al. 2018; Ai et al. 2019). The generated organic acids, which also have antimicrobial activity, are ultimately responsible for a pH reduction in the culture, which acts as a growth-limiting factor as well. While the acidophilic features of lactic acid bacteria are well known in comparison with other taxonomic groups, this tolerance does not prevent a decrease in cell viability leading to poor stability. Additionally, this reduction is directly proportional to the time of exposure to the undissociated form of the acid at low pH values. Thus, control of pH during growth at values from 5.4 to 6.5 depending on the specific microbe usually increases total biomass and the number of viable cells (Salminen et al. 2004). Moreover, oxygen and acid tolerance are more interconnected than initially predicted since evidence has showed that antioxidant enzymes like superoxide dismutases and peroxidases actually provide protection against acid stress (Bruno-Bárcena et al. 2010; Leite et al. 2014) and it is becoming clear that industrially friendly strains possessing more antioxidant defense capabilities retain viability and exhibit greater stability during storage.

Finally, the optimal culture temperature will allow cells to proliferate at their maximum specific growth rate, maximizing process productivity. According to their origin microbial strains can be classified anywhere from psychrophiles or "cold-loving" to thermophiles, which thrive at higher than normal temperatures. Interestingly, a study has shown that a decrease in culture temperature can favor biomass accumulation due to a reduction in wasted energy for cell maintenance (Cogan et al. 1971; Bruno-Bárcena et al. 1998).

Preservation of Starter Cultures

Traditionally, starter cultures used in households and in the early food industry consisted of liquid cultures or fermented foods that needed to be continually transferred to preserve viability. It was not until the 1960s that the food industry began to show an increased interest in the quality (i.e. cell viability) and preservation of starter cultures (Mäyrä-Mäkinen 2004) as the growing market required better methods for preservation of microbial cultures before the culture or the fermented food reached the end user.

Current methods of culture preservation include freezing, freeze-drying, and spray-drying, each with its own benefits and drawbacks. It is important to note that the successful preservation of starter cultures is completely dependent on the quality of the input culture, which in turn is dependent on a robust production process. The earliest form of preserved and concentrated cultures used in the food industry were frozen concentrated cultures. The introduction of this technology allowed fermented food manufacturers to forgo the process of in-plant culture amplification, which is not only expensive, but is prone to microbial contamination and bacteriophage (bacterial viruses) infection (Mäyrä-Mäkinen 2004; Santivarangkna et al. 2007). However, frozen, concentrated starter cultures must be maintained at sub-zero temperatures from −20 to −40 °C during storage and transport, which increases costs. Freeze-drying is an attractive alternative to freezing, particularly because storage temperatures are not a major factor in the stability of the culture in the short term. However, freeze-drying requires the use of sophisticated equipment and relatively long process times and it is hence expensive to perform (Santivarangkna et al. 2007). An alternative temperature-stable preservation process that has been investigated for use with concentrated starter cultures is spray drying. This process involves atomizing a concentrated liquid culture and passing it though a heated stream of air to remove the water from the tiny droplets. Spray drying is considerably cheaper (approximately 80% reduction in costs) than freeze-drying, but the process generally leads to reduced cellular viability compared to freeze-drying (Özilgen 1996; Santivarangkna et al. 2007).

The maintenance of cellular viability is critical when considering preservation methods and is perhaps the most important factor in the preservation process as diminished activity of cultures is deleterious to downstream processes. Also, a high concentration of the starter culture appears to have a protective effect since concentrated cultures are more active than non-concentrated cultures when exposed to the same preservation method (Özilgen 1996). Of particular interest in the freezing and freeze-drying methods, are cryoprotectant compounds. Cryoprotectants are water-sequestering compounds ranging from sugars and sugar alcohols (i.e. lactose and sorbitol), to amino acid salts (i.e. monosodium glutamate). There is substantial evidence that cryoprotectants are able to protect cellular viability by preventing the formation of intracellular ice crystals that form solute gradients and damage the structure of the cell (Carvalho et al. 2004; Mäyrä-Mäkinen 2004; Santivarangkna et al. 2007).

A Summary and Conclusions

The evolution of human society from nomadic hunter-gatherer groups to the civilization we see today has been directly influenced by the microbially-mediated preservation of foods. The development of agriculture practices reduced the tendency of early human populations to relocate, and as these stationary populations grew, long-term storage of produced foodstuff was mandatory so that communities could survive the winter and the occasional drought. High initial water activity in foods like vegetables and meats mean that these foods are particularly susceptible to spoilage immediately after harvest and it is these foods that benefitted most from microbial fermentation and subsequent preservation. Traditionally, remnants from previous successful fermentations were used to perpetuate the preservation of different types of foods, often for generations, in a process called "backslopping". These processes fell out of favor in the nineteenth century alongside growing concerns about food safety and with the development of more modern food preservation techniques such as pasteurization. However, in more recent history, there has been a resurgence of interest in the beneficial impact of fermented foods, leading to an entire food industry surrounding these products. Governmental safety regulations require defined ingredients, and as such, most industrially fermented foods require the use of starter cultures composed of GRAS strains.

Currently, and for the past several decades, starter cultures are produced using large industrial batch fermenters. Even when a fermented food requires multiple strains, the strains are produced separately and mixed after expansion to produce the starter. However, batch fermentations have an inherent variability that can lead to compromised cellular viability between batches. Thus, there has been an increasing demand for new starter culture production processes in which cell viability and stability is improved. One such alternative process is the use of immobilized cell reactors. These systems tend to produce cells that are more robust and can potentially be used to produce a mixed culture starter in a single step. To produce vigorous and stable starter cultures, several additional variables must be controlled, including growth medium choice, oxygen levels, pH, and temperature during fermentation and storage, and method of preservation. Due to the large-scale nature of modern industrial fermentations and the physical separation between culture production plants and food production plants, concentration and preservation of starter cultures is a necessity. Freezing, freeze-drying, and spray drying are preservation methods used in the industry, each having their own benefits and drawbacks.

Acknowledgements This work was supported by the College of Agriculture and Life Science and the Department of Plant and Microbial Biology at North Carolina State University. HW was supported by a fellowship generously provided by the NC State Graduate Student Support Plan.

References

Ai, M., Qiu, X., Huang, J., Wu, C., Jin, Y., & Zhou, R. (2019). Characterizing the microbial diversity and major metabolites of Sichuan bran vinegar augmented by *Monascus purpureus*. *International Journal of Food Microbiology, 292*, 83–90.

Altieri, C., Ciuffreda, E., Di Maggio, B., & Sinigaglia, M. (2017). Lactic acid bacteria as starter cultures. In B. Speranza, A. Bevilacqua, M. R. Corbo, & M. Sinigaglia (Eds.), *Starter cultures in food production*. Chichester: Wiley.

Amit, S. K., Uddin, M. M., Rahman, R., Islam, S. M. R., & Khan, M. S. (2017). A review on mechanisms and commercial aspects of food preservation and processing. *Agriculture & Food Security, 6*(1), 51.

Axelsson, L. T., Chung, T. C., Dobrogosz, W. J., & Lindgren, S. E. (1989). Production of a broad spectrum antimicrobial substance by *Lactobacillus reuteri*. *Microbial Ecology in Health and Disease, 2*(2), 131–136.

Black, M., & Pritchard, H. W. (2002). *Desiccation and survival in plants: Drying without dying.* Wallingford: CABI Publishing.

Blomstedt, K. C., Griffiths, A. C., Gaff, F. D., Hamill, D. J., & Neale, D. A. (2018). Plant desiccation tolerance and its regulation in the foliage of resurrection "flowering-plant" species. *Agronomy, 8*(8).

Bokulich, N. A., Lewis, Z. T., Boundy-Mills, K., & Mills, D. A. (2016). A new perspective on microbial landscapes within food production. *Current Opinion in Biotechnology, 37*, 182–189.

Boserup, E. (1976). Environment, population, and technology in primitive societies. *Population and Development Review, 2*(1), 21–36.

Bruno-Bárcena, J. M., Ragout, A. L., & Sineriz, F. (1998). Microbial physiology applied to process optimisation: Lactic acid bacteria. In E. Galindo & O. T. Ramirez (Eds.), *Advances in bioprocess engineering II* (Vol. II, pp. 97–110). Netherlands: Kluwer Academic Publishers.

Bruno-Bárcena, J. M., Ragout, A. L., Cordoba, P. R., & Sineriz, F. (1999). Continuous production of L(+)-lactic acid by *Lactobacillus casei* in two-stage systems. *Applied Microbiology and Biotechnology, 51*(3), 316–324.

Bruno-Bárcena, J. M., Ragout, A. L., Cordoba, P. R., & Sineriz, F. (2001). Reactor configuration for fermentation in immobilized continuous system. In J. F. T. Spencer & A. L. R. de Spencer (Eds.), *Food microbiology protocols*. Totowa, NJ: Humana Press.

Bruno-Bárcena, J. M., Azcarate-Peril, M. A., & Hassan, H. M. (2010). Role of antioxidant enzymes in bacterial resistance to organic acids. *Applied and Environmental Microbiology, 76*(9), 2747–2753.

Buchanan, R. A. (2018). *History of technology. Encyclopaedia Britannica.* Encyclopaedia Britannica, Inc.

Carvalho, A. S., Silva, J., Ho, P., Teixeira, P., Malcata, F. X., & Gibbs, P. (2004). Relevant factors for the preparation of freeze-dried lactic acid bacteria. *International Dairy Journal, 14*(10), 835–847.

Champagne, C. P., Lacroix, C., & Sodini-Gallot, I. (1994). Immobilized cell technologies for the dairy industry. *Critical Reviews in Biotechnology, 14*(2), 109–134.

Chen, G., Chen, C., & Lei, Z. (2017). Meta-omics insights in the microbial community profiling and functional characterization of fermented foods. *Trends in Food Science & Technology, 65*, 23–31.

Chombo-Morales, P., Kirchmayr, M., Gschaedler, A., Lugo-Cervantes, E., & Villanueva-Rodríguez, S. (2016). Effects of controlling ripening conditions on the dynamics of the native microbial population of Mexican artisanal Cotija cheese assessed by PCR-DGGE. *LWT-Food Science and Technology, 65*, 1153–1161.

Cinquin, C., Le Blay, G., Fliss, I., & Lacroix, C. (2004). Immobilization of infant fecal microbiota and utilization in an in vitro colonic fermentation model. *Microbial Ecology, 48*(1), 128–138.

Cocolin, L., & Ercolini, D. (2015). Zooming into food-associated microbial consortia: A 'cultural' evolution. *Current Opinion in Food Science, 2*, 43–50.

Cogan, T. M., Buckley, D. J., & Condon, S. (1971). Optimum growth parameters of lactic streptococci used for the production of concentrated cheese starter cultures. *The Journal of Applied Bacteriology, 34*(2), 403–409.

Condon, S. (1987). Responses of lactic acid bacteria to oxygen. *FEMS Microbiology Letters, 46*(3), 269–280.

Cordain, L., Eaton, S. B., Sebastian, A., Mann, N., Lindeberg, S., Watkins, B. A., O'Keefe, J. H., & Brand-Miller, J. (2005). Origins and evolution of the Western diet: Health implications for the 21st century. *The American Journal of Clinical Nutrition, 81*(2), 341–354.

Costerton, J. W., Stewart, P. S., & Greenberg, E. P. (1999). Bacterial biofilms: A common cause of persistent infections. *Science, 284*(5418), 1318–1322.

Dagher, S. F., Ragout, A. L., Sineriz, F., & Bruno-Bárcena, J. M. (2010). Cell immobilization for production of lactic acid biofilms do it naturally. *Advances in Applied Microbiology, 71*, 113–148.

De Man, J. C., Rogosa, M., & Sharpe, M. E. (1960). A medium for the cultivation of lactobacilli. *The Journal of Applied Bacteriology, 23*(1), 130–135.

Di Cagno, R., Coda, R., De Angelis, M., & Gobbetti, M. (2013). Exploitation of vegetables and fruits through lactic acid fermentation. *Food Microbiology, 33*(1), 1–10.

Dolci, P., Zenato, S., Pramotton, R., Barmaz, A., Alessandria, V., Rantsiou, K., & Cocolin, L. (2013). Cheese surface microbiota complexity: RT-PCR-DGGE, a tool for a detailed picture? *International Journal of Food Microbiology, 162*(1), 8–12.

Doleyres, Y., Paquin, C., LeRoy, M., & Lacroix, C. (2002). *Bifidobacterium longum* ATCC 15707 cell production during free- and immobilized-cell cultures in MRS-whey permeate medium. *Applied Microbiology and Biotechnology, 60*(1–2), 168–173.

Doleyres, Y., Fliss, I., & Lacroix, C. (2004). Increased stress tolerance of *Bifidobacterium longum* and *Lactococcus lactis* produced during continuous mixed-strain immobilized-cell fermentation. *Journal of Applied Microbiology, 97*(3), 527–539.

Doulgeraki, A. I., Ercolini, D., Villani, F., & Nychas, G.-J. E. (2012). Spoilage microbiota associated to the storage of raw meat in different conditions. *International Journal of Food Microbiology, 157*(2), 130–141.

Dubos, R. (1960). *Pasteur and modern science*. New York: Anchor Books Doubleday & Company, Inc.

Food Preservation. (2018). *Food preservation*. Dictionary of American History. Encyclopedia. com.

Gilbert, J. A., van der Lelie, D., & Zarraonaindia, I. (2014). Microbial terroir for wine grapes. *Proceedings of the National Academy of Sciences of the United States of America, 111*(1), 5–6.

González-Córdova, A. F., Yescas, C., Ortiz-Estrada, Á. M., De la Rosa-Alcaraz, M. Á., Hernández-Mendoza, A., & Vallejo-Cordoba, B. (2016). Invited review: Artisanal Mexican cheeses. *Journal of Dairy Science, 99*(5), 3250–3262.

Hammond, S. T., Brown, J. H., Burger, J. R., Flanagan, T. P., Fristoe, T. S., Mercado-Silva, N., Nekola, J. C., & Okie, J. G. (2015). Food spoilage, storage, and transport: Implications for a sustainable future. *Bioscience, 65*(8), 758–768.

Holliday, T. W., Gautney, J. R., & Friedl, L. (2014). Right for the wrong reasons reflections on modern human origins in the post-Neanderthal genome era. *Current Anthropology, 55*(6), 696–724.

Irlinger, F., & Mounier, J. (2009). Microbial interactions in cheese: Implications for cheese quality and safety. *Current Opinion in Biotechnology, 20*(2), 142–148.

Junter, G. A., & Jouenne, T. (2004). Immobilized viable microbial cells: From the process to the proteome em leader or the cart before the horse. *Biotechnology Advances, 22*(8), 633–658.

Klaenhammer, T. R. (1988). Bacteriocins of lactic acid bacteria. *Biochimie, 70*(3), 337–349.

Klaenhammer, T. R., Barrangou, R., Buck, B. L., Azcarate-Peril, M. A., & Altermann, E. (2005). Genomic features of lactic acid bacteria effecting bioprocessing and health. *FEMS Microbiology Reviews, 29*(3), 393–409.

Kline, L., & Sugihara, T. F. (1971). Microorganisms of the San Francisco sour dough bread process. II. Isolation and characterization of undescribed bacterial species responsible for the souring activity. *Applied Microbiology, 21*(3), 459–465.

Lahtinen, S., Salminen, S., von Wright, A., & Ouwehand, A. (2012). *Lactic acid bacteria: Microbiological and functional aspects* (4th ed.). New York: Taylor & Francis Group/CRC Press.

Leff, J. W., & Fierer, N. (2013). Bacterial communities associated with the surfaces of fresh fruits and vegetables. *PLoS One, 8*(3), e59310.

Leite, M. C. T., Troxell, B., Bruno-Bárcena, J. M., & Hassan, H. M. (2014). Biology of reactive oxygen species, oxidative stress, and antioxidants in lactic acid bacteria. In K. Venema & A. P.

de Carmo (Eds.), *Probiotics and prebiotics: Current research and future trends*. Norfolk, VA: Caister Academic Press.

Macfarlane, S., McBain, A. J., & Macfarlane, G. T. (1997). Consequences of biofilm and sessile growth in the large intestine. *Advances in Dental Research, 11*(1), 59–68.

Macori, G., & Cotter, P. D. (2018). Novel insights into the microbiology of fermented dairy foods. *Current Opinion in Biotechnology, 49*, 172–178.

Marco, M. L., Heeney, D., Binda, S., Cifelli, C. J., Cotter, P. D., Foligné, B., Gänzle, M., Kort, R., Pasin, G., Pihlanto, A., Smid, E. J., & Hutkins, R. (2017). Health benefits of fermented foods: Microbiota and beyond. *Current Opinion in Biotechnology, 44*, 94–102.

Masco, L., Huys, G., De Brandt, E., Temmerman, R., & Swings, J. (2005). Culture-dependent and culture-independent qualitative analysis of probiotic products claimed to contain bifidobacteria. *International Journal of Food Microbiology, 102*(2), 221–230.

Mäyrä-Mäkinen, A. B. M. (2004). Industrial use and production of lactic acid bacteria. In S. Salminen, A. von Wright, & A. Ouwehand (Eds.), *Lactic acid bacteria: Microbiological and functional aspects* (pp. 175–193). New York: Marcel Dekker, Inc.

Mikelsaar, M., Lazar, V., Onderdonk, A. B., & Donelli, G. (2011). Do probiotic preparations for humans really have efficacy? *Microbial Ecology in Health and Disease, 22*(1), 10128.

Murugesan, S., Reyes-Mata, M. P., Nirmalkar, K., Chavez-Carbajal, A., Juárez-Hernández, J. I., Torres-Gómez, R. E., Piña-Escobedo, A., Maya, O., Hoyo-Vadillo, C., Ramos-Ramírez, E. G., Salazar-Montoya, J. A., & García-Mena, J. (2018). Profiling of bacterial and fungal communities of Mexican cheeses by high throughput DNA sequencing. *Food Research International, 113*, 371–381.

Nair, M. R. B., Chouhan, D., Sen Gupta, S., & Chattopadhyay, S. (2016). Fermented foods: Are they tasty medicines for *Helicobacter pylori* associated peptic ulcer and gastric cancer? *Frontiers in Microbiology, 7*(1148).

Özilgen, M. (1996). Kinetics of food processes involving pure or mixed cultures of lactic acid bacteria. In T. F. Bozoglu & B. Ray (Eds.), *Lactic acid bacteria: Current advances in metabolism, genetics and applications* (pp. 367–378). Berlin: Springer.

Pont, E. G., & Holloway, G. L. (1968). A new approach to production of cheese starter—Some preliminary investigations. *Australian Journal of Dairy Technology, 23*(1), 22.

Prokopov, T., & Tanchev, S. (2007). *Methods of food preservation. Food safety*. Boston, MA: Springer.

Rangel, D. E. (2011). Stress induced cross-protection against environmental challenges on prokaryotic and eukaryotic microbes. *World Journal of Microbiology and Biotechnology, 27*(6), 1281–1296.

Salminen, S., von Wright, A., & Ouwehand, A. (2004). *Lactic acid bacteria: Microbiological and functional aspects* (3rd ed.). New York: Marcel Dekker, Inc./CRC Press.

Santivarangkna, C., Kulozik, U., & Foerst, P. (2007). Alternative drying processes for the industrial preservation of lactic acid starter cultures. *Biotechnology Progress, 23*(2), 302–315.

Schillinger, U. (1999). Isolation and identification of lactobacilli from novel-type probiotic and mild yoghurts and their stability during refrigerated storage. *International Journal of Food Microbiology, 47*(1–2), 79–87.

Schornsteiner, E., Mann, E., Bereuter, O., Wagner, M., & Schmitz-Esser, S. (2014). Cultivation-independent analysis of microbial communities on Austrian raw milk hard cheese rinds. *International Journal of Food Microbiology, 180*, 88–97.

Septembre-Malaterre, A., Remize, F., & Poucheret, P. (2018). Fruits and vegetables, as a source of nutritional compounds and phytochemicals: Changes in bioactive compounds during lactic fermentation. *Food Research International, 104*, 86–99.

Sieuwerts, S., de Bok, F. A. M., Hugenholtz, J., & van Hylckama Vlieg, J. E. T. (2008). Unraveling microbial interactions in food fermentations: From classical to genomics approaches. *Applied and Environmental Microbiology, 74*(16), 4997–5007.

Talarico, T. L., Casas, I. A., Chung, T. C., & Dobrogosz, W. J. (1988). Production and isolation of reuterin, a growth inhibitor produced by *Lactobacillus reuteri*. *Antimicrobial Agents and Chemotherapy, 32*, 1854–1858.

Tavan, E., Cayuela, C., Antoine, J. M., & Cassand, P. (2002). Antimutagenic activities of various lactic acid bacteria against food mutagens: Heterocyclic amines. *The Journal of Dairy Research, 69*(2), 335–341.

van Hijum, S. A. F. T., Vaughan, E. E., & Vogel, R. F. (2013). Application of state-of-art sequencing technologies to indigenous food fermentations. *Current Opinion in Biotechnology, 24*(2), 178–186.

van Hylckama Vlieg, J. E., Veiga, P., Zhang, C., Derrien, M., & Zhao, L. (2011). Impact of microbial transformation of food on health—From fermented foods to fermentation in the gastrointestinal tract. *Current Opinion in Biotechnology, 22*(2), 211–219.

Vandenberg, D. J. C., Smits, A., Pot, B., Ledeboer, A. M., Kersters, K., Verbakel, J. M. A., & Verrips, C. T. (1993). Isolation, screening and identification of lactic-acid bacteria from traditional food fermentation processes and culture collections. *Food Biotechnology, 7*(3), 189–205.

Wu, S., Xu, S., Chen, X., Sun, H., Hu, M., Bai, Z., Zhuang, G., & Zhuang, X. (2018). Bacterial communities changes during food waste spoilage. *Scientific Reports, 8*(1), 8220.

Yang, Y., Shevchenko, A., Knaust, A., Abuduresule, I., Li, W., Hu, X., Wang, C., & Shevchenko, A. (2014). Proteomics evidence for kefir dairy in Early Bronze Age China. *Journal of Archaeological Science, 45*, 178–186.

Zheng, X., Liu, F., Shi, X., Wang, B., Li, K., Li, B., & Zhuge, B. (2018). Dynamic correlations between microbiota succession and flavor development involved in the ripening of Kazak artisanal cheese. *Food Research International, 105*, 733–742.

Part II
The Oral Microbiota: The Seeder and Gatekeeper

Introduction to the Oral Cavity

Roland R. Arnold and Apoena A. Ribeiro

Abstract The oral cavity plays a critical role as the seeder and gatekeeper of the microbiome that populates the continuum of mucosal surfaces of the gastrointestinal tract, as well as that of the respiratory tract. The mouth has a variety of discrete niches and environmental conditions (microhabitats) that select for and discriminate against a vast array of microorganisms that ultimately determine the microbiome. The oral microbiome is an important contributor to host health and refers specifically to the microorganisms that reside on or in the human oral cavity and its contiguous mucosal surfaces to the distal esophagus. The oral microbiome is composed of approximately 700 species of bacteria, and also includes viruses, fungi, protozoa and archaea associated with the varied microhabitats that define the oral microbial ecosystem. The normal microbiota of the mouth is responsible for maintaining homeostasis of the oral cavity, but is also responsible for two of the most common diseases of bacterial etiology in humans—dental caries and periodontal diseases. Oral diseases have also been linked to systemic chronic diseases including: cardiovascular disease, stroke, abnormal pregnancy outcomes, diabetes, aspiration pneumonia, cancers and Alzheimer's disease. This Chapter aims to highlight the unique features of the main niches that compose the oral cavity and have influence on its microbiome composition.

Keywords Oral microbiome · Oral environment · Salivary microbiome · Teeth surfaces microbiome · Oral tissue surfaces microbiome

In its position at the beginning of the digestive tract, the oral cavity plays a critical role as the seeder and gatekeeper of the microbiome that populates the continuum of mucosal surfaces of the gastrointestinal tract, as well as that of the respiratory tract. The oral environment includes several unique features that create a variety of ecological niches that select for microorganisms that establish in consortia

---

R. R. Arnold (✉) · A. A. Ribeiro
Division of Diagnostic Sciences, Adams School of Dentistry, University of North Carolina at Chapel Hill, Chapel Hill, NC, USA
e-mail: Roland_Arnold@unc.edu

© Springer Nature Switzerland AG 2019
M. A. Azcarate-Peril et al. (eds.), *How Fermented Foods Feed a Healthy Gut Microbiota*, https://doi.org/10.1007/978-3-030-28737-5_6

communities. Other orally-introduced microorganisms fail to establish in this highly selective environment and after initial processing in the oral cavity transit through the mechanical and chemical gauntlet of washing sheer forces, shedding surfaces, mucins, antimicrobials, reactive oxygen and nitrogen species, lytic enzymes, acids, bile salts etc. from the esophagus (Kongara and Soffer 1999) through the stomach and to the small intestines and lower GI tract. There is a variety of discrete niches and environmental conditions (microhabitats) encountered along the way that select for and discriminate against a vast array of microorganisms that ultimately determine the gut microbiome.

The oral microbiome is an important contributor to host health and refers specifically to the microorganisms that reside on or in the human oral cavity and its contiguous mucosal surfaces to the distal esophagus (Dewhirst et al. 2010; Peterson et al. 2013). The oral microbiome is second only to that of the colon in microbial density, richness and diversity with around 700 species of bacteria, and also includes a variety of viruses, fungi, protozoa and archaea associated with the varied microhabitats that define the oral microbial ecosystem (The Human Microbiome Consortium 2012; Aas et al. 2005; Paster et al. 2006; Chen et al. 2010; Dewhirst et al. 2010). The normal microbiota of the mouth is responsible for maintaining homeostasis of the oral cavity, but is also responsible for two of the most common diseases of bacterial etiology in humans—dental caries and periodontal diseases (read chapter "Dysbiosis of the Oral Microbiome" of this section for more details about these diseases). In addition, there is mounting evidence that links oral diseases to systemic chronic diseases including: cardiovascular disease, stroke, abnormal pregnancy outcomes, diabetes, aspiration pneumonia, cancers and Alzheimer's disease (Scannapieco and Binkley 2012; Seymour et al. 2007; Whitmore and Lamont 2014; Atanasova and Yilmaz 2014; Dominy et al. 2019). Thus, an understanding of the determinants that shape and define the oral microbiome is crucial to address both oral and broader systemic health.

Likewise, the human diet, which begins its processing in the oral cavity, is a central determinant in shaping the structures and activities of the gut microbiome (reviewed in Singh et al. 2017). However, despite the global roles of diet in oral microbiome evolution and the awareness that diet plays a vital and dynamic role in shaping the gut microbiome (Singh et al. 2017), relatively less is known regarding direct or indirect influences of diet on the oral microbiome (Kato et al. 2017). Throughout human evolution, our environment and societal norms have had demonstrable influences on the composition of our microbiome, increasingly so during the Neolithic, industrial revolution and modern eras (Gillings et al. 2015). Microbial DNA sequencing of ancient calcified dental plaque suggests that when humans turned to agriculture, and later to modern starch- and sugar-rich diets, the microbes colonizing their teeth changed drastically (Adler et al. 2013). Sequencing of microbial DNA from the calcified dental plaque of mesolithic to medieval human skeletons indicated that the oral microbial populations of individuals who lived in early farming communities were much less diverse than those of hunter-gatherers, and harbored more bacteria linked to diseases such as periodontal diseases (Costalonga and Herzberg 2014). Contemporary microbial populations are, in turn, less diverse than those of earlier farming communities and are dominated by bacteria linked to diseases such as those that cause dental caries. With the introduction of refined sugar to our diet, certain bacteria such as the mutans streptococci genetically adapted their metabolism to successfully compete against other bacteria through increased acid tolerance and acquisition of sucrose-specific enzymes (glucosyl transferases) that make water-insoluble extracellular carbohydrate polymers (glucans) that provide a selective (glucan-specific receptors and acid tolerance) matrix-defined environment for specific microbiome development (plaque accumulation) on teeth (Cornejo et al. 2013). The introduction of antibiotics and the promotion of the oral hygiene practices of brushing and flossing also likely had profound effects on the composition of the oral microbiome (Marsh 2010). Modern day excesses in consumption of acidic drinks and refined sugars have impacted the oral ecosystem contributing to both caries and periodontal diseases (Adler et al. 2013).

Clues as to the influences of the environment and the unique features of the oral cavity on the acquisition and shaping of the oral microbiota can be gained by following the consequences of initial environmental exposure of an infant. The microbiome of the mother is naturally the first source for the establishment of a normal oral microbiome and thus the delivery mode (vaginal *vs* caesarean) is an important determinant for initial microbial exposure (Dominguez-Bello et al. 2010). Furthermore, this initial exposure also shapes the diversity of the oral microbiome later in the infant's life as vaginally-born children show a proportionally significant difference in oral taxa at 3 months that discriminates them from children born by caesarean section (Lif Holgerson et al. 2011). The nature of feeding also has demonstrable effects on oral microbiome composition with 3-month-old breast-fed infants having significantly higher proportions of lactobacilli than formula-fed infants (Holgerson et al. 2013). The eruption of teeth provides a variety of new surfaces to select for microbial colonization that reveal strains that are obligate hard (non-shedding) surface colonizers as well as creating new retention sites at the epithelia-tooth interface (gingival crevice) (Sampaio-Maia and Monteiro-Silva 2014). By the age of three, the oral microbiome is already complex and becomes increasingly diverse with age (Crielaard et al. 2011). The loss of primary teeth and the eruption of permanent teeth represent other major dynamic alterations in oral microhabitats that are associated with shifting compositions of the oral microbial communities (Xu et al. 2015).

Once established, the microbiome should be sustained. However, the oral microbiome is in constant fight for survival, against host-protection mechanisms. For the resident microbiota, the oral cavity is not homogenous, presenting a variety of microhabitats that challenge microbial colonization and persistence (Aas et al. 2005; Dewhirst et al. 2010) including constant bathing of all accessible surfaces with the exocrine secretions of salivary glands and the highly disruptive and potentially hazardous activities involved in the processing of food and the mechanics of speech. The teeth and their specialized epithelial surroundings, the tongue, lips, cheeks, tonsils and hard and soft palates are all adapted to facilitate a variety of oral functions and each presenting its own highly heterogenous and dynamic ecological landscape that is reflected in the significantly different microbial communities that have successfully occupied all of these sites (Xu et al. 2015; Lamont et al. 2018). The oral cavity presents a warm and moist environment that by its nature is indiscriminately opened to exposure to external challenges. Ironically, while the food that provides nourishment to the body is initially introduced through the oral cavity, the availability of diet-derived nutrients to the populating microbiome is limited by design. Successful colonizers must adapt to acquiring sustenance from host-derived sources such as saliva and gingival crevicular fluid (a serum-like tissue exudate from where the gingiva (gums) attach to the teeth) and/or from the intermittent availability of food debris and dietary carbohydrates (Xu et al. 2015). Another challenge is the availability of stable sites for attachment and biofilm development. Despite the fact that the oral mucosal epithelial cells are rapidly and totally turning over, the newly exposed cells are continually repopulated with microbial colonizers. In contrast, the teeth provide the only natural non-shedding surface in the human body and provide unique microhabitats that permit persistent and extensive biofilm development (Marsh and Devine 2011).

Due to the high complexity of the oral cavity, represented by an ecosystem with different inner-ecosystems, and in view of the aforementioned, this Chapter aims to highlight the unique features of the main niches that compose the oral cavity and have influence on its microbiome composition.

Characterization of the Distinct Ecological Niches That Determine the Oral Environment

The oral ecosystem is broadly defined by its diverse structures and tissues including saliva, the non-shedding surfaces of teeth and their surrounding soft tissues, the dorsal and lateral surfaces of the tongue, the mucosal epithelial surfaces of cheeks, lips, palate and teeth, that each provide distinct microhabitats, growth conditions and nutrient availability that select for discrete and highly specialized populations of microorganisms (Paster et al. 2006; Zaura et al. 2009; Lamont et al. 2018). Therefore, microbiomes from the same sites of different individuals had greater similarities than different sites within the same individual. In addition, microflora attach to surfaces continuously shed into the saliva, making salivary microbiota the "fingerprint" of the oral microbiome inhabiting on the oral mucosal surfaces (Fábián et al. 2008). The profiles of 40 cultivable bacterial species were able to distinguish saliva, oral soft tissue surfaces, and supragingival and subgingival plaque samples in healthy subjects (Mager et al. 2003).

Saliva

The oral cavity is continuously bathed in saliva that is critical to the preservation of oral tissues and maintenance of oral health and makes major contributions as the initial interface to our environment and diet. It is a clear, slightly acidic mucoserous exocrine secretion that is a complex mix from multiple glands strategically located throughout the oral and pharyngeal cavities. The major salivary glands include paired parotid glands that deliver serous secretion through a duct in the buccal mucosa of the cheeks opposite the maxillary first molars, and the submandibular and sublingual glands found in the floor of the mouth that deliver mixed serous and mucous secretions. The numerous minor glands are mucous secreting and distributed throughout the oral cavity including the lower lip, tongue, hard and soft palate, cheeks and pharynx (Humphrey and Williamson 2001). These glands together secrete an average daily flow of whole saliva varying, in health, between 1 and 1.5 L.

Salivary glands are innervated by both sympathetic and parasympathetic nerve fibers and secretion is controlled by a salivary center in the medulla. There is an unstimulated salivary flow that follows a circadian rhythm, peaking in late afternoon with almost no flow during sleep and a normal day time flow that ranges from 0.3 to 0.4 mL/min, with any flow <0.1 mL/min considered to be abnormal (reviewed in Dawes et al. 2015). The submandibular glands contribute 65% of this unstimulated flow with 20% coming from parotid, 8% from sublingual and <10% from minor glands. Salivary flow is stimulated by at least three types of stimuli associated with food ingestion: mastication (the act of chewing), gustatory (with acid the most stimulatory and sweet the least) and olfactory (a relatively poor stimulator). Under stimulation, salivary flow rates can peak at 7 mL/min and there is a shift in the glandular contributions with the serous (watery) parotid glands now contributing >50% of the total volume. Stimulated saliva can contribute as much as 80–90% of the daily cumulative flow and the composition of stimulated saliva changes dramatically from that of unstimulated flow.

Although saliva is a very dilute fluid (approximately 99% water) and hypotonic relative to plasma, it contains a number of critical constituents that contribute to its many functions (reviewed in Dawes et al. 2015). There are a variety of electrolytes including sodium, potassium, calcium, magnesium, bicarbonate, thiocyanate, fluoride and phosphates. In addition, there are proteins shared with other exocrine secretions, immunoglobulins, enzymes, mucins and nitrogenous products including urea and ammonia. While the washing and diluting effects of the water in saliva are important contributors to the oral environment and the processing of food, the salivary mucins, principally MUC5B (high molecular weight) and MUC7 (low molecular weight), coat all of the surfaces of the oral cavity including the forming food bolus with a slimy, viscoelastic lubricant that reduces the friction between opposing surfaces during such processes as mastication, swallowing and speaking. Furthermore, these mucins form an important part of the acquired enamel pellicle that rapidly coats the exposed tooth surface and the mucosal film that forms a thin surface on mucosal epithelial cells. In addition to protecting these surfaces from environmental insults such as the demineralizing actions of acids, they present a selective challenge to colonizing microorganisms. The rapid (within seconds) adsorption of these and other protein and peptide constituents of saliva to these surfaces limits direct accessibility for microbial attachment and would favor those microorganisms with receptors for these pellicle coatings. The selective binding of these molecules to these surfaces would also serve to concentrate and retain these molecules from their highly dilute concentrations in saliva. The hypotonic nature of saliva would also facilitate essential ionic attractions to the favor of pellicle formation. In addition to creating potential binding sites for colonization, the mucins potentially in concert with other surface-active molecules in saliva rapidly bind to microbial surfaces clumping them together, limiting their surface attachment and mechanically favoring clearance by the washing action of saliva flow during stimulation. These saliva and subsequent esophageal coatings might also serve to protect orally-introduced microorganisms in their transit through the digestive tract to microhabitats more favorable to their colonization (Sarosiek and McCallum 2000).

The bicarbonate, phosphate, urea and ammonia serve to modulate the pH and buffering capacity of saliva. Because of low concentrations (~5 mmol/L) during unstimulated salivary flow, bicarbonate does not provide a strong acid-buffering capacity. In contrast, when salivary flow is stimulated, especially with acid, the bicarbonate levels dramatically increase in proportion to the flow rate to levels that exceed that found in plasma (reviewed in Dawes et al. 2015). Bicarbonate as the primary buffer in saliva is ideal as its reaction with hydrogen ions yields carbonic acid that is converted to water and the volatile gas carbon dioxide by salivary carbonic anhydrase VI (Kivela et al. 1999). This arrangement results in an overall slightly acidic pH range of 6–7 that fluctuates from a low of 5.3 (unstimulated flow) to 7.8 (peak flow). Demineralization of the enamel surface occurs when acids diffuse through the plaque and pellicle into the liquid phase between the enamel crystal structures of the tooth. This crystalline structure begins dissolution at a pH below 5.5, the critical pH for caries development (Edgar 1990). Saliva greatly influences the pH of the plaque environment orchestrated by the metabolism of the acidogenic bacteria that constitute this biofilm population, thereby limiting caries progression (Stephan 1944). Therefore, determining the buffering capacity of saliva rather than its direct pH is considered a useful tool as a component of the diagnostic assessment of a patient's caries activity/risk (Larmas 1992).

Maintaining tooth integrity is also a function that depends heavily on the composition and distribution of saliva. In addition to buffering the demineralizing actions of bacterially generated and diet introduced acids, the high salivary concentrations (supersaturated with respect to hydroxyapatite) of calcium and phosphate under the regulation of salivary proteins such as statherin, histatins, cystatins and proline-rich proteins are engaged in an ongoing process of replacement of minerals lost during demineralization of the enamel surfaces of the tooth (reviewed in Dawes et al. 2015). These proteins form a protective barrier as part of the pellicle on the hydroxyapatite surfaces of teeth that regulates the demineralization-remineralization process including the incorporation of fluoride into the maturing enamel surfaces. This allows remodeling of the apatite crystal including the displacement of magnesium and carbonate in the enamel structure with the stronger and more acid-resistant fluoride-apatite crystals (reviewed in Edgar 1990).

Saliva is central to the oral processing and the sensory aspects of food including not only the pleasant aspects of taste, but also screening and reflex responses to potentially noxious substances (reviewed in Dawes et al. 2015). Saliva provides the solvent for tastants to be delivered to their respective receptors on taste buds distributed throughout the mouth and surrounding oral apparatus (Matsuo 2000). There are five recognized basic taste receptors that in addition to taste sensation serve as salivary stimulants including in order of their stimulant potency: sour, salt, bitter, sweet and umami (savory) (reviewed in Dawes et al. 2015). Saliva provides a first line of defense against potential noxious substances by dilution through reflexive increase in salivary flow and either elimination of a noxious taste before swallowing by spitting or secondarily by coughing or vomiting before further progression through the digestive tract. These taste stimulants as well as the mastication process make significant contributions to the initial processing of food. The more food is chewed the greater the mixing of saliva with the food bolus and the more gustatory, as well as olfactory stimulation occurs. One of the main proteins in saliva is alpha-amylase that splits starches into maltotriose, maltotetrose and some higher oligosaccharides (Kaczmarek and Rosenmund 1977). As its pH optimum is near neutral, it is not thought to play much of a digestive role once the food bolus reaches the gastric juices. Its exact role in the oral cavity is not clear, although it might facilitate more rapid clearance of retentive food starches, thus minimizing the availability of this potential metabolite for the oral microbiome. There are also lingual lipases in the secretions from the glands of von Ebner on the human tongue (Spielman et al. 1993), but their low levels do not support a significant role for these lipases in human digestion of fats. In summary the major contribution of saliva to the initial digestion of food is likely in the formation of a cohesive bolus covered by mucin that facilitates swallowing.

In terms of microbiome development, one of the most important functions is the facilitation of the removal of food, drink and food debris from the mouth. The clearance of fermentable carbohydrates including sucrose, fructose and glucose and acidic drinks and foods as soon as possible is critical to the maintenance of homeostasis of the oral microbiome by limiting microbial metabolism. This is facilitated by relatively high salivary flow rates during food processing versus low residual volumes by the incomplete syphon operation of swallowing (Dawes 2012).

Despite the fact that saliva has no indigenous microbiota, whole saliva contains up to 10^8 colony forming units per mL of cultivable bacteria. Compared to other body fluids, saliva provides a nutrient poor milieu for bacterial growth and its transient presence in the oral cavity is more consistent with the role of washing and removal of shedding surfaces. All epithelial surfaces in the oral cavity desquamate, releasing cell associated bacteria into the bathing saliva. It has been estimated that the entire surface layer of the oral mucosa is replaced every 3 h (Dawes 2003). Most of these cells are associated with ~100 bacteria, suggesting that newly exposed epithelial surfaces must be continuously and rapidly recolonized. This rapid turnover limits the opportunity for the development of diversity and bacterial density through biofilm stability and secondary colonization on most oral epithelial surfaces. The retentive and protected surfaces of the tongue are the exception permitting dense and diverse microbial growth. These properties are indeed consistent with the fact that the microbial profile of saliva is most similar to that of the shedding soft tissues and disproportionately reflective of the more retentive and thus densely populated papillate surfaces of the tongue (Mager et al. 2003; Keijser et al. 2008).

As discussed earlier saliva constituents contribute in many ways to the composition of dental biofilms on both soft and hard surfaces. Selected proteins from saliva populate the bathed enamel surface as acquired enamel pellicle. These do not bind uniformly, but associate in discrete patches with one protein influencing the binding of another through protein-protein interactions (Siqueira et al. 2012; Schweigel et al. 2016). This in turn creates discrete sites for interaction with the microbial adhesins of pioneering colonizers in the initiation of biofilm formation on the tooth surface (Gong and Herzberg 1997; Gong et al. 2000) and for interspecies interactions to facilitate plaque maturation (Cavalcanti et al. 2016). Likewise, salivary constituents also rapidly coat the mucosal epithelial surfaces of the oral cavity and likely also contribute to the surface mucosal layers of the esophagus that also contains 600–700 mucous glands that secrete bicarbonate and mucous (Sarosiek and McCallum 2000). In common with other exocrine secretions that bathe mucosal surfaces, saliva delivers a variety of antimicrobial molecules any of which could affect oral biofilm formation. These include the acquired secretory IgA antibodies (Brandtzaeg 2013) and the innate or constitutive defense proteins (van't Hof et al. 2014) including lactoferrin, lysozyme, salivary peroxidase, cationic peptides, proline-rich proteins, defensins and the previously mentioned mucins (Offner and Troxler 2000) and the salivary agglutinin GP340 (Leito et al. 2008). In addition to influencing the composition of the oral microbiome, these salivary constituents could initiate antimicrobial influences on potential colonizers of both the digestive and respiratory tracts. As the tooth surface approaches, the gingival margin, there is less salivary access and more access to gingival crevicular fluid shifting nutrient and binding site availabilities to favor a more fastidious and less acid- and oxygen-tolerant microbiota.

Teeth

Unlike the shedding surfaces of the oral epithelia, the tooth surfaces are the only non-shedding surfaces naturally available in the oral cavity. This property allows for stable anchoring and long-term biofilm development and maturation. As discussed above, the hydroxyapatite mineral of the enamel of the crowns and any exposure of the root and subgingival regions of the tooth are rapidly coated with salivary proteins or in the case of subgingival surfaces with admixes of salivary and serum proteins. These protein-rich pellicles become the actual sites of initial adherence by the more mechanically-restricted pioneering microbial colonizers (Siqueira et al. 2012; Kolenbrander et al. 2010). These pioneer streptococci have evolved specialized adhesins that specifically bind to ligands presented by the acquired pellicle that require sheer-induced conformational changes of the adhesins (Ding et al. 2010). These initial colonizers provide new opportunities for the introduction of secondary species initially through specific interspecies co-aggregation. As the biofilm matures, the changing architecture (e.g. extracellular matrices), metabolic products (lactic acid) and altered atmospheric conditions become prime determinants of the developing plaque composition (Valm et al. 2011). The dental plaque is built in a continual, order process characterized by succession of different bacterial species, each one with relevant roles in every step of biofilm construction that imaging studies have revealed as an oral microbial biogeography (Welch et al. 2016). Species participating in the process of biofilm formation are traditionally characterized as "early" and "late" colonizers. Among early colonizers the viridans streptococci group is considered a cornerstone of the oral biofilm puzzle given its ability to bind saliva proteins through Antigens I and II. In this manner, streptococcal species become the first colonizers able to bind tooth surfaces and promote arrival of secondary colonizers by intergeneric co-aggregation (Kolenbrander et al. 2002). *Actinomyces naeslundii* is one of the secondary colonizers and a well-known co-aggregation partner of streptococci (Palmer Jr et al. 2003). *Fusobacterium nucleatum* is considered a key player given its capability to co-aggregate both with early and late colonizers of the oral biofilm, the latter group characterized by species belonging to Bacteroidetes and Spirochaetes (Kolenbrander et al. 1989). Recently, the microbiome composition of dental biofilm assigned the amplicon reads in to 10 bacterial phyla, 25 classes, 29 orders, 58 families, 107 genera and 723 species (Ribeiro et al. 2017).

As suggested earlier, the plaque that forms at the gum margin (supragingival) is more influenced by saliva-derived acquired enamel pellicle than its more saliva sequestered subgingival counterpart. The growth of the microbial community in the subgingival sulcus, if undisturbed can extend its growth along the root and away from the salivary environment facilitated by the host inflammatory response provoked by bacterial growth and metabolism resulting in loss of tissue integrity. This environment becomes more protein-rich, more anaerobic with a more stable pH and temperature and reduced salivary and mechanical sheer forces. Consequently, the subgingival environment matures to favor more anaerobic, asaccharolytic and proteolytic Gram-negative species rather than the facultative, saccharolytic Gram-positive species that characterize the supragingival biofilms.

The tooth crown can be further divided into five distinct ecological niches: occlusal or chewing surfaces; the approximal surfaces or contact points between teeth; the supragingival surfaces; the buccal or cheek-contacting surfaces, and the lingual surfaces impacted by the tongue. The occlusal and approximal surfaces provide sequestered sites that are mechanically less challenging to microbial colonization and favor retention of growth promoting food debris not cleared by the washing action of saliva. The approximal and supragingival surfaces near the gum line are less accessible to tooth brushing permitting stagnation and promotion of a serum-like exudate, gingival crevicular fluid that further alters the proportional composition of the microbiota. The occlusal surfaces are not influenced by crevicular fluid access, but the enamel surfaces do form pits and fissures that provide retention areas for sustained plaque growth. These surfaces are thus the most susceptible to tooth decay. The approximal and lingual surfaces of the molar teeth harbor microbial communities of greater diversity than that of the flat surfaces of buccal and anterior teeth. The composition of the tooth microbiota is determined not only by its location in the mouth and its proximity to salivary flow from nearby ducts, but also by the anatomy of the tooth surface and proximity to the surrounding gingiva. Those niches that present opportunities for undisturbed, sustained establishment of biofilm on the tooth result in more diverse, but potentially more pathogenic microbial communities (Zaura et al. 2009; Simón-Soro et al. 2013). This speaks to the efficacy of repetitious oral hygiene procedures to provide timely disruption of the maturation process of plaque as both a caries and periodontal disease intervention.

The Surfaces of Soft Tissues

Despite continual turnover of the superficial epithelial surfaces, the oral mucosa is constantly populated with microorganisms (Costalonga and Herzberg 2014). The populating of the newly exposed surfaces demands selective proficiencies of these species that likely translate in to their ability to efficiently colonize other surfaces such as those found in the esophagus and the trachea. The transient nature of the surfaces of the cheeks and palate and the efficiency of the washing action of saliva compared to other more stable ecological niches results in limited microbial colonization of minimal diversity with bacterial monolayers on shedding epithelial cells (He et al. 2015). In contrast, the surface topography of the tongue offers more protected microhabitats that permit the buildup of multilayered microbial biofilms of much greater density comprised of considerable diversity including highly fastidious, strict anaerobes. These anaerobes can flourish in the protected and nutritionally rich environments provided by the crypts of the tongue dorsa, and their metabolic end-products are an established source of halitosis (Scully and Greenman 2008).

References

Aas, J. A., Paster, B. J., Stokes, L. N., Olsen, I., & Dewhirst, F. E. (2005). Defining the normal bacterial flora of the oral cavity. *Journal of Clinical Microbiology, 43*, 5721–5732.

Adler, C. J., Dobney, K., Weyrich, L. S., Kaidonis, J., Walker, A. W., Haak, W., Bradshaw, C. J., Townsend, G., Sołtysiak, A., Alt, K. W., Parkhill, J., & Cooper, A. (2013). Sequencing ancient calcified dental plaque shows changes in oral microbiota with dietary shifts of the Neolithic and Industrial revolutions. *Nature Genetics, 45*, 450–455.e1. https://doi.org/10.1038/ng.2536.

Atanasova, K. R., & Yilmaz, O. (2014). Looking in the *Porphyromonas gingivalis* cabinet of curiosities: The microbium, the host and cancer association. *Molecular Oral Microbiology, 29*, 55–66. https://doi.org/10.1111/omi.12047. Epub 2014.

Brandtzaeg, P. (2013). Secretory IgA: Designed for anti-microbial defense. *Frontiers in Immunology, 4*, 222. https://doi.org/10.3389/fimmu.2013.00222.

Cavalcanti, I. M., Nobbs, A. H., Ricomini-Filho, A. P., Jenkinson, H. F., & Del Bel Cury, A. A. (2016). Interkingdom cooperation between *Candida albicans, Streptococcus oralis* and *Actinomyces oris* modulates early biofilm development on denture material. *Pathogens and Disease, 74*. https://doi.org/10.1093/femspd/ftw002.

Chen, T., Yu, W. H., Izard, J., Baranova, O. V., Lakshmanan, A., & Dewhirst, F. E. (2010). The human oral microbiome database: A web accessible resource for investigating oral microbe taxonomic and genomic information. *Database, 2010*, baq013. https://doi.org/10.1093/database/baq013.

Cornejo, O. E., Lefébure, T., Bitar, P. D., Lang, P., Richards, V. P., Eilertson, K., Do, T., Beighton, D., Zeng, L., Ahn, S. J., Burne, R. A., Siepel, A., Bustamante, C. D., & Stanhope, M. J. (2013). Evolutionary and population genomics of the cavity causing bacteria Streptococcus mutans. *Molecular Biology and Evolution, 30*, 881–893. https://doi.org/10.1093/molbev/mss278. Epub 2012 Dec 10.

Costalonga, M., & Herzberg, M. C. (2014). The oral microbiome and the immunobiology of periodontal disease and caries. *Immunology Letters, 162*, 22–38. https://doi.org/10.1016/j.imlet.2014.08.017.

Crielaard, W., Zaura, E., Schuller, A. A., Huse, S. M., Montijn, R. C., & Keijser, B. J. (2011). Exploring the oral microbiota of children at various developmental stages of their dentition in the relation to their oral health. *BMC Medical Genomics, 4*, 22. https://doi.org/10.1186/1755-8794-4-22. PMID: 21371338.

Dawes C. (2003). Estimates, from salivary analyses, of the turnover time of the oral mucosal epithelium in humans and the number of bacteria in an edentulous mouth. *ArchOral Biol, 48*(5), 329–336.

Dawes C. (2012). Salivary clearance and its effects on oral health. In: Edgar M, Dawes C. O'Mullane D, editors. *Saliva and Oral Health.* 4th ed. London: Stephen Jancocks Ltd. p. 81–96. [Chapter 5].

Dawes, C., Pedersen, A. M., Villa, A., Ekström, J., Proctor, G. B., Vissink, A., Aframian, D., McGowan, R., Aliko, A., Narayana, N., Sia, Y. W., Joshi, R. K., Jensen, S. B., Kerr, A. R., & Wolff, A. (2015). The functions of human saliva: A review sponsored by the World Workshop on Oral Medicine VI. *Archives of Oral Biology, 60*, 863–874. https://doi.org/10.1016/j.archoralbio.2015.03.004.

Dewhirst, F. E., Chen, T., Izard, J., Paster, B. J., Tanner, A. C. R., Yu, W., Lakshmanan, A., & Wade, W. G. (2010). The human oral microbiome. *Journal of Bacteriology, 192*, 5002–5017.

Ding, A. M., Palmer, R. J., Jr., Cisar, J. O., & Kolenbrander, P. E. (2010). Shear-enhanced oral microbial adhesion. *Applied and Environmental Microbiology, 76*, 1294–1297. https://doi.org/10.1128/AEM.02083-09.

Dominguez-Bello, M. G., Costello, E. K., Contreras, M., Magris, M., Hidalgo, G., Fierer, N., & Knight, R. (2010). Delivery mode shapes the acquisition and structure of the initial microbiota across multiple body habitats in newborns. *Proceedings of the National Academy of Sciences of the United States of America, 107*, 11971–11975. https://doi.org/10.1073/pnas.1002601107.

Dominy, S. S., Lynch, C., Ermini, F., et al. (2019). Porphyromonas gingivalis in Alzheimer's disease brains: Evidence for disease causation and treatment with small-molecule inhibitors. *Science Advances, 5*, eaau3333. https://doi.org/10.1126/sciadv.aau3333.

Edgar, W. M. (1990). Saliva and dental health. Clinical implications of saliva: Report of a consensus meeting. *British Dental Journal, 169*(3–4), 96–98.

Fábián, T. K., Fejérdy, P., & Csermely, P. (2008). Salivary genomics, transcriptomics and proteomics: The emerging concept of the oral ecosystem and their use in the early diagnosis of cancer and other diseases. *Current Genomics, 9*, 11–21.

Gillings, M. R., Paulsen, I. T., & Tetu, S. G. (2015). Ecology and evolution of the human microbiota: Fire, farming and antibiotics. *Genes, 6*, 841–857. https://doi.org/10.3390/genes6030841.

Gong, K., & Herzberg, M. C. (1997). Streptococcus sanguis expresses a 150-kilodalton two-domain adhesin: Characterization of several independent adhesin epitopes. *Infection and Immunity, 65*, 3815–3821.

Gong, K., Mailloux, L., & Herzberg, M. C. (2000). Salivary film expresses a complex, macromolecular binding site for Streptococcus sanguis. *The Journal of Biological Chemistry, 275*, 8970–8974.

He, J., Li, Y., Cao, Y., Xue, J., & Zhou, X. (2015). The oral microbiome diversity and its relation to human diseases. *Folia Microbiologica, 60*, 69–80. https://doi.org/10.1007/s12223-014-0342-2.

Holgerson, P. L., Vestman, N. R., Claesson, R., Ohman, C., Domellof, M., Tanner, A. C., et al. (2013). Oral microbial profile discriminates breast-fed from formula-fed infants.

Journal of Pediatric Gastroenterology and Nutrition, 56, 127–136. https://doi.org/10.1097/MPG.0b013e31826f2bc6. PMID: 22955450.

Humphrey, S. P., & Williamson, R. T. (2001). A review of saliva: Normal composition, flow, and function. *The Journal of Prosthetic Dentistry, 85,* 162–169.

Kaczmarek, M. J., & Rosenmund, H. (1977). The action of human pancreatic and salivary isoamylases on starch and glycogen. *Clinica Chimica Acta, 79,* 69–73.

Kato, I., Vasquez, A., Moyerbrailean, G., Land, S., Djuric, Z., Sun, J., Lin, H. S., & Ram, J. L. (2017). Nutritional correlates of human oral microbiome. *Journal of the American College of Nutrition, 36,* 88–98. https://doi.org/10.1080/07315724.2016.1185386.

Keijser, B. J. F., Zaura, E., Huse, S. M., van der Vossen, J. M. B. M., Schuren, F. H. J., Montijn, R. C., ten Cate, J. M., & Crielaard, W. (2008). Pyrosequencing analysis of the oral microflora of healthy adults. *Journal of Dental Research, 87,* 1016–1020.

Kivela, J., Parkkila, P., Parkkila, A. K., Leinonen, J., & Rajaniemi, H. (1999). Salivary carbonic anhydrase isoenzyme VI. *The Journal of Physiology, 520,* 315–320.

Kolenbrander, P. E., Andersen, R. N., & Moore, L. V. (1989). Coaggregation of Fusobacterium nucleatum, Selenomonas flueggei, Selenomonas infelix, Selenomonas noxia, and Selenomonas sputigena with strains from 11 genera of oral bacteria. *Infection and Immunity, 57,* 3194–3203.

Kolenbrander, P. E., Andersen, R. N., Blehert, D. S., Egland, P. G., Foster, J. S., & Palmer, R. J., Jr. (2002). Communication among oral bacteria. *Microbiology and Molecular Biology Reviews, 66,* 486–505.

Kolenbrander, P. E., Palmer, R. J., Jr., Periasamy, S., & Jakubovics, N. S. (2010). Oral multispecies biofilm development and the key role of cell-cell distance. *Nature Reviews. Microbiology, 8,* 471–480. https://doi.org/10.1038/nrmicro2381.

Kongara, K. R., & Soffer, E. E. (1999). Saliva and esophageal protection. *The American Journal of Gastroenterology, 94,* 1446–1452.

Lamont, R. J., et al. (2018). The oral microbiota: Dynamic communities and host interactions. *Nature Reviews, 16,* 745–759.

Larmas, M. (1992). Saliva and dental caries: Diagnostic tests for normal dental practice. *International Dental Journal, 42,* 199–208.

Leito, J. T., Ligtenberg, A. J., Nazmi, K., de Blieck-Hogervorst, J. M., Veerman, E. C., & Nieuw Amerongen, A. V. (2008). A common binding motif for various bacteria of the bacteria-binding peptide SRCRP2 of DMBT1/gp-340/salivary agglutinin. *Biological Chemistry, 389,* 1193–1200. https://doi.org/10.1515/BC.2008.135.

Lif Holgerson, P., Harnevik, L., Hernell, O., Tanner, A. C., & Johansson, I. (2011). Mode of birth delivery affects oral microbiota in infants. *Journal of Dental Research, 90,* 1183–1188. https://doi.org/10.1177/0022034511418973. PMID: 21828355.

Mager, D. L., Ximenez-Fyvie, L. A., Haffajee, A. D., & Socransky, S. S. (2003). Distribution of selected bacterial species on intraoral surfaces. *Journal of Clinical Periodontology, 30,* 644–654.

Marsh, P. D. (2010). Controlling the oral biofilm with antimicrobials. *Journal of Dentistry, 38*(Suppl 1), S11–S15. https://doi.org/10.1016/S0300-5712(10)70005-1.

Marsh, P. D., & Devine, D. A. (2011). How is the development of dental biofilms influenced by the host? *Journal of Clinical Periodontology, 38*(Suppl 11), 28–35. https://doi.org/10.1111/j.1600-051X.2010.01673.x.

Matsuo, R. (2000). Role of saliva in the maintenance of taste sensitivity. *Critical Reviews in Oral Biology and Medicine, 11,* 216–229.

Offner, G. D., & Troxler, R. F. (2000). Heterogeneity of high-molecular-weight human salivary mucins. *Advances in Dental Research, 14,* 69–75.

Palmer, R. J., Jr., Gordon, S. M., Cisar, J. O., & Kolenbrander, P. E. (2003). Coaggregation-mediated interactions of streptococci and actinomyces detected in initial human dental plaque. *Journal of Bacteriology, 185,* 3400–3409.

Paster, B. J., Olsen, I., Aas, J. A., & Dewhirst, F. E. (2006). The breadth of bacterial diversity in the human periodontal pocket and other oral sites. *Periodontology 2000, 42,* 80–87.

Peterson, S. N., Snesrud, E., Liu, J., Ong, A. C., Kilian, M., Schork, N. J., et al. (2013). The dental plaque microbiome in health and disease. *PLoS One, 8*, e58487.

Ribeiro, A. A., Azcarate-Peril, M. A., Cadenas, M. B., Butz, N., Paster, B. J., Chen, T., & Arnold, R. A. (2017). The oral bacterial microbiome of occlusal surfaces in children and its association with diet and caries. *PLoS One, 12*, e0180621.

Sampaio-Maia, B., & Monteiro-Silva, F. (2014). Acquisition and maturation of oral microbiome throughout childhood: An update. *Journal of Dental Research, 11*, 291–301.

Sarosiek, J., & McCallum, R. W. (2000). Mechanisms of oesophageal mucosal defence. *Baillière's Best Practice & Research. Clinical Gastroenterology, 14*, 701–717.

Scannapieco, F. A., & Binkley, C. J. (2012). Modest reduction in risk for ventilator-associated pneumonia in critically ill patients receiving mechanical ventilation following topical oral chlorhexidine. *The Journal of Evidence-Based Dental Practice, 12*, 103–106. https://doi.org/10.1016/j.jebdp.2012.03.010.

Schweigel, H., Wicht, M., & Schwendicke, F. (2016). Salivary and pellicle proteome: A datamining analysis. *Scientific Reports, 6*, 38882. https://doi.org/10.1038/srep38882.

Scully, C., & Greenman, J. (2008). Halitosis (breath odor). *Periodontology 2000, 48*, 66–75. https://doi.org/10.1111/j.1600-0757.2008.00266.x.

Seymour, G. J., Ford, P. J., Cullinan, M. P., Leishman, S., & Yamazaki, K. (2007). Relationship between periodontal infections and systemic disease. *Clinical Microbiology and Infection, 13*(Suppl 4), 3–10.

Simón-Soro, A., Tomás, I., Cabrera-Rubio, R., Catalan, M. D., Nyvad, B., & Mira, A. (2013). Microbial geography of the oral cavity. *Journal of Dental Research, 92*, 616–621. https://doi.org/10.1177/0022034513488119.

Singh, R. K., Chang, H. W., Yan, D., Lee, K. M., Ucmak, D., Wong, K., Abrouk, M., Farahnik, B., Nakamura, M., Zhu, T. H., Bhutani, T., & Liao, W. (2017). Influence of diet on the gut microbiome and implications for human health. *Journal of Translational Medicine, 15*, 73–90.

Siqueira, W. L., Custodio, W., & McDonald, E. E. (2012). New insights into the composition and functions of the acquired enamel pellicle. *Journal of Dental Research, 91*, 1110–1118. https://doi.org/10.1177/0022034512462578.

Spielman, A. I., D'Abundo, S., Field, R. B., & Schmale, H. (1993). Protein analysis of human von Ebner saliva and a method for its collection from the foliate papillae. *Journal of Dental Research, 72*, 1331–1335.

Stephan, R. M. (1944). Intra-oral hydrogen ion concentrations associated with dental caries activity. *Journal of Dental Research, 23*, 257–266.

The Human Microbiome Consortium. (2012). Structure, function and diversity of the healthy human microbiome. *Nature, 486*, 207–214.

van't Hof, W., Veerman, E..C. I., Nieuw Amerongen, A. V., Ligtenberg, A. J. M. (2014). Antimicrobial defense systems in saliva. In Ligtenberg AJM, Veerman ECI (eds): Saliva: Secretion and functions. Monogr Oral Sci. Basel, Karger, 2014, vol 24, pp 40–51. https://doi.org/10.1159/000358783

Valm, A. M., Mark Welch, J. L., Rieken, C. W., Hasegawa, Y., Sogin, M. L., Oldenbourg, R., Dewhirst, F. E., & Borisy, G. G. (2011). Systems-level analysis of microbial community organization through combinatorial labeling and spectral imaging. *Proceedings of the National Academy of Sciences of the United States of America, 108*(10), 4152–4157. https://doi.org/10.1073/pnas.1101134108.

Welch, M., Rosetti, B. J., Rieken, C. W., Dewhirst, F. W., & Borisy, G. G. (2016). Biogeography of a human oral microbiome at the micron scale. *Proceedings of the National Academy of Sciences of the United States of America, 113*, E791–E800.

Whitmore, S. E., & Lamont, R. J. (2014). Oral bacteria and cancer. *PLoS Pathogens, 10*, e1003933. https://doi.org/10.1371/journal.ppat.1003933. eCollection 2014.

Xu, X., He, J., Xue, J., Wang, Y., Li, K., Zhang, K., et al. (2015). Oral cavity contains distinct niches with dynamic microbial communities. *Environmental Microbiology, 17*, 699–710. https://doi.org/10.1111/1462-2920.12502.

Zaura, E., Keijser, B. J., Huse, S. M., et al. (2009). Defining the healthy "core microbiome" of oral microbial communities. *BMC Microbiology, 9*, 259–271.

Defining the Healthy Oral Microbiome

G. M. S. Soares and M. Faveri

Abstract The human mouth harbors one of the most diverse microbiomes in the human body and includes bacteria, viruses, fungi, protozoa, and archaea. This chapter revises the traditional methods utilized to study the mouth microbiota as well as novel technologies that allowed to have a complete picture of the oral microbial repertoire. The differences in microbiota composition between the different oral microenvironments saliva and mucosa and a broad stroke of the non-bacterial members of the oral microbiome are also presented.

Keywords Oral microbiome · Methods to study the oral microbiome · Oral biofilms · Open-ended molecular diagnostic · Oral health · Non-bacterial oral microbiome

Introduction

Microorganisms (animalcules) of the oral cavity have been studied with great interest since Anton van Leeuwenhoek first examined the plaque between his teeth with his early version of the microscope in 1683 (van Leeuwenhoek 1683). A variety of conventional methods have since been used to analyze the composition of the oral microbiome, including increasingly more sophisticated microscopy, cultural analyses, enzymatic assays and immunoassay (Dzink et al. 1989; Moore and Moore 1994). From these studies we have learned that the human mouth harbors one of the

G. M. S. Soares
Department of Stomatology, Dental Research Division, Federal University of Parana, Curitiba, Parana, Brazil
e-mail: geisla.soares@ufpr.br

M. Faveri (✉)
Department of Periodontology, Dental Research Division, Guarulhos University, Guarulhos, São Paulo, Brazil
e-mail: mfaveri@prof.ung.br

© Springer Nature Switzerland AG 2019
M. A. Azcarate-Peril et al. (eds.), *How Fermented Foods Feed a Healthy Gut Microbiota*, https://doi.org/10.1007/978-3-030-28737-5_7

most diverse microbiomes in the human body, including viruses, fungi, protozoa, archaea as well as bacteria. Using cultivable methods, approximately 300 species from the human oral cavity have been isolated, characterized and formally named. We also recognized early on that there were many more morphotypes present than could be cultivated. The data produced during these studies were extremely informative and laid the foundation for the next phase, which started in the 1990s, when new molecular targeted techniques were introduced and enabled major advances in the study of the microbiota associated with oral health and disease. During the 2000s, the introduction of new open-ended sequencing techniques rekindled characterization of the oral microbiome (Sakamoto et al. 2000; Paster et al. 2001; Kumar et al. 2006; Keijser et al. 2008; Faveri et al. 2008; Petrosino et al. 2009). Using these new approaches, several bacterial species that have yet to be cultivated, have been associated with different states of periodontal health or disease (Sakamoto et al. 2000; Kumar et al. 2006; Keijser et al. 2008; Faveri et al. 2008).

During at least three decades different methods of culture independent methods of identification have been used to characterize the oral microbiome, such as PCR (Ashimoto et al. 1996; Li et al. 2005; Tanner et al. 2006), real time-PCR (Sakamoto et al. 2001, 2004; Gomes et al. 2008) and *in-situ* hybridization (Thurnheer et al. 2004; Al-Ahmad et al. 2007; Zijnge et al. 2012; Mark Welch et al. 2016). According to the relevance of bacterial species to periodontal health and disease, a panel of 40 strains was prepared to be used with the Checkerboard DNA-DNA hybridization technique (Socransky et al. 1998). This microbiological tool has the advantage of allowing analysis of thousands of oral samples in a relatively short period of time, but the Checkerboard technique is a bacterial culture-driven tool, and important data of sets of species that had yet to be cultivated were still lacking in the literature. Therefore, open-ended techniques were welcomed and became useful, and some research groups have begun to use cloning and sequencing to evaluate the oral microbiome of healthy and diseased subjects (Kroes et al. 1999; Sakamoto et al. 2000; Paster et al. 2001). The earlier data provided by these microbiological tools estimated that 500 taxa could colonize the oral cavity, and found that 347 species/ phylotypes could be found in the subgingival environment. Of these, 215 were novel phylotypes and 140 were detected only once. Moreover, the presence of members of phyla never previously detected in oral samples, such as OP11 (SR1) and TM7, as well as *Deferribacteres* (*Synergistetes*) were detected (Paster et al. 2001). It is important to note that these pioneering studies, which had used complete or partial sequencing of the 16S ribosomal RNA (16S rRNA) gene by Sanger sequencing (Paster et al. 2001; Aas et al. 2005; Faveri et al. 2008; Shchipkova et al. 2010); or by pyrosequencing (Keijser et al. 2008; Zaura et al. 2009; Griffen et al. 2012); as well as the emergence of new genomic technologies, including next-generation sequencing (NGS) and bioinformatics tools, have provided powerful approaches toward understanding the oral microbiome.

Indeed, culture-independent methods have provided great insight into the diversity of the microbiome, but to investigate the properties and potential of an organism, it needs to be grown in culture and therefore culture methods, once again, became essential for the characterization of the oral microbiota. This new era of culturing oral microorganisms has incorporated new techniques such as *in vitro* biofilm models, and more recently, *ex vivo* biofilm models, primarily with the purpose of understanding the function of each microorganism within the biofilm community (Soares et al. 2015; Klug et al. 2016). The understanding of biofilms as multispecies communities has shifted the focus from the role of individual species within the biofilm to a view of how they behave and how they change from a condition of heath to one of disease. Another important point in trying to understand the oral microbiome lies in observing the unique characteristic of the oral cavity composed of both hard and soft tissues, and both covered by saliva (see chapter "Introduction to the Oral Cavity").

Therefore, this chapter presents a current overview of the composition of the healthy oral microbiome, with focus on current knowledge about the diversity of these biofilms, based on the results of the studies that have used cutting-edge open-ended approaches. In addition, a brief discussion regarding the strengths and weaknesses of these new diagnostic techniques is also presented. This body of information might help to understand the shifts occurring in the composition of healthy biofilm structures that may lead to the development of an oral imbalance, and consequently, a state of dysbiosis that may lead to some types of oral diseases or even contribute to some changes in human non-oral microbiota.

The Role of Open-Ended Molecular Diagnostic Methods in the Study of Oral Biofilm Diversity

During the last few years, great progress has been made toward the application of novel molecular microbiological tools in studies of the human microbiota, including the oral microbiome. The cutting-edge open-ended molecular techniques allow for genome mapping of the entire microbial spectrum in a sample, and provide comprehensive characterization of both the cultivable and not-yet-cultivable microbiota associated with health and disease. These techniques permit an overview of the microbial communities as a whole, which represents an important advantage over culture and even over other molecular targeted tools, such as polymerase chain reaction (PCR), DNA probes and microarrays (Hiyari and Bennett 2011; Wade 2011). The large body of information derived from these sequencing techniques has revealed new species that could act as pathogens in several oral infections (Paster et al. 2001; Faveri et al. 2008; Kumar et al. 2012), including different conditions of the oral cavity such as endodontic infections (Siqueira Jr and Rôças 2005; Li et al. 2010), peri-implantitis (Koyanagi et al. 2010, 2013; Kumar et al. 2012), and periodontitis (Faveri et al. 2008; Griffen et al. 2012).

From 2001 to 2010, Sanger sequencing was the most widely used DNA sequencing method for studying the microbial diversity of the oral biofilm (Paster et al. 2001; Kumar et al. 2006; Faveri et al. 2008; Shchipkova et al. 2010). Several studies published in the 1990s indicated that sequencing of the small ribosomal subunit gene (16S rDNA) could be useful for microbial identification (Weisburg et al. 1991; Green and Giannelli 1994; Cilia et al. 1996). The 16S rDNA gene is present in all prokaryotes and contains variable regions that are unique between microorganisms providing a means of identification. The 16S rRNA genes can be extracted from heterogeneous samples, amplified and sequenced, and then compared with databases such as the Human Oral Microbiome Database (Chen et al. 2010). Therefore, the construction and analyses of ribosomal gene libraries are very important undertakings that will provide tools for studying microbial ecology. Although a large body of phylogenetic data for microbial identification has been generated via Sanger sequencing, new sequencing technologies that offer a series of additional benefits have recently emerged. One of these new sequencing technologies, pyrosequencing and Illumina MiSeq, is faster and more cost-effective than Sanger sequencing (Rastogi et al. 2013; Harrington et al. 2013) and allows thousands to hundreds of thousands of sequence reads, with up to 27 million sequences being generated in a single run (compared with a few hundred by means of the traditional method) (Harrington et al. 2013).

Sanger sequencing and NGS are powerful methods for evaluating oral biodiversity; however, DNA extraction and PCR amplification have been reported to be potential sources of bias associated with these techniques (Diaz et al. 2012; Abusleme et al. 2014). The understanding of possible limitations, intrinsic bias and inherent variability of the different diagnostic methods is crucial for the proper evaluation and interpretation of the results of the various studies. Diaz et al. (2012) evaluated the possible bias of DNA isolation and PCR amplification of 454-sequencing of 16S rRNA gene. The authors used three different laboratory-created samples (mocks) of seven bacterial species (*Streptococcus oralis, Streptococcus mutans, Lactobacillus casei, Actinomyces oris, Fusobacterium nucleatum, Porphyromonas gingivalis* and *Veillonella* sp.). Mock 1 contained equal numbers of *16S rDNA* molecules, mock 2 equal numbers of cells and mock 3 unequal numbers of cells of these seven bacterial oral species. In theory, no difference in the number of readouts of these species would be expected in mock 1, since they comprised equal amounts of genomic DNA for each species. On the other hand, mocks 2 and 3 could potentially be affected either by some bias of the PCR or sequencing processes or the cell lysis procedures. However, mock 1 did not show the estimated results, as *F. nucleatum* produced a higher number of reads and *A. oris* and *L. casei* a lower number of reads than expected. In addition to being under-represented in mock 1, *A. oris* and *L. casei* were also under-represented in mocks 2 and 3, which could be due to some PCR bias. Both *S. mutans* and *P. gingivalis* were shown in lower abundance than expected only in mocks 2 and 3, suggesting that these species were less effectively lysed. Other research groups have also observed some of these biases associated with the Sanger or 454-sequencing techniques (de Lillo et al. 2004; Abusleme et al. 2014).

The results of the above-mentioned studies suggest that although 454-pyrosequencing is a powerful technique for investigating the oral microbial diversity, the abundance of species is subject to empirical bias introduced through the methods used for DNA isolation and amplification. Investigators should be aware of these limitations to minimize technical errors by accounting for them while designing the studies and evaluating their data.

Microbiome of Oral Health

Using a target microbial method, the checkerboard DNA-DNA hybridization technique, Socransky et al. (1998) described the definition of microbial complexes that distinguished between health and disease. Thus, the species associated with periodontal health were grouped in the yellow, purple and green complexes as well as the group of *Actinomyces* ssp. All these bacterial species were named initial colonizers of the tooth surface and were considered of greater importance during biofilm initiation (Socransky et al. 1998). The yellow complex is composed of a group of *streptococci*: *Streptococcus mitis, Streptococcus sanguinis, S. oralis, Streptococcus gordonii and Streptococcus intermedius*. The purple complex includes *Actinomyces odontolyticus* and *Veillonella parvulla*. *Finally,* the green complex comprises *Capnocytophaga sputigena, Capnocytophaga gingivalis, Capnocytophaga ochracea, Eikenella corrodens* and *Aggregatibacter actinomycetemcomitans* serotype a. Following these four groups, the authors described the orange complex as being a bridge for the colonization of bacterial species associated with disease. The orange complex is composed of *Fusobacterium nucleatum nucleatum, Fusobacterium nucleatum polymorphum, Fusobacterium nucleatum vincentii, Fusobacterium periodonticum, Prevotella intermedia, Prevotella nigrescens, Parvimonas micra, Eubacterium nodatum, Campylobacter rectus, Campylobacter showae, Campylobacter gracilis* and *Streptococcus constellatus*. The red complex harbors the bacterial species considered as overt periodontal pathogens and include *Tannerella forsythia, P. gingivalis* and *Treponema denticola*. The last group of bacterial species were regularly present, but were not assignable to a specific health status. This group was named "Others" and included *Eubacterium saburreum, Gemella morbillorum, Leptotrichia buccalis, Peptostreptococcus acnes, Prevotella melaninogenica, Neisseria mucosa, Streptococcus anginosus, Selenomonas noxia* and *Treponema socranskii*. Although the red and especially the orange complex species could also be found in healthy dental biofilm samples, their frequency and proportions were lower in comparison with those in biofilm samples of periodontal disease (Socransky and Haffjee 2005). Therefore, a microbial profile compatible with health would be formed by similar bacterial species as those in disease, but in different levels and proportions. Following these concepts, several studies such as Ximénez-Fyvie et al. (2000), Ramberg et al. (2003), Haffajee et al. (2005) performed in-depth analyses of oral samples, and the knowledge about oral microbiota took a great leap forward in a few years, however with the limitation in use of culture-driven techniques.

Considering the need for open-ended analyses of the oral microbiota several studies (Kroes et al. 1999; Sakamoto et al. 2000; Paster et al. 2001; Aas et al. 2005) started to describe microorganisms that had not yet been cultivated by the sequencing method. The data derived from such studies, in association with the original panel described by Socransky et al. (1998), estimated that up to 500 species may colonize the oral cavity. Therefore, with the purpose of contributing to this topic, Kumar et al. (2006), using cloning and the Sanger sequencing method compared the subgingival microbiota of periodontal health and disease and found significant differences in the microbial profiles that discriminated these conditions. They reported higher levels of *S. mitis* and *S. sanguinis* in healthy subgingival biofilms and suggested that these species were the dominant taxa in periodontal health. These results corroborated the findings of earlier studies of culture-dependent techniques (Socransky et al. 1998). In addition, Zaura et al. (2009) using 454 pyrosequencing, studied the biofilm of the healthy oral microbiome, and characterized some bacterial phylotypes as the core microbiome. This study selected three healthy subjects and collected samples of supragingival dental biofilm, saliva and mucosal swabs from the cheek, hard palate and tongue surfaces. Samples from tooth surfaces showed higher diversity and abundance of OTUs. The authors demonstrated that core microbiome harbored mainly the known bacterial species from the genera *Streptococcus*, *Corynebacterium*, *Neisseria*, *Rothia*, *Veillonella*, *Actinomyces*, *Granulicatella*, *Porphyromonas* and *Firmicutes*. Finally, the authors showed that on an average, each individual sample harbored 266 "species-level" phylotypes.

In order to organize the enormous bank of data from the sequencing studies, a database was created and named "The human oral microbiome database" (http://www.ehomd.org) (Chen et al. 2010). This database provided a naming scheme whereby each human oral taxon (HOT) was given a unique number. This HOT number was linked to their source, sequence information, synonyms, taxonomic hierarchy, bibliographic information and status. About 35,000 clones of different samples of oral conditions, as well as 1000 clinical oral isolates from culture collections (Dewhirst et al. 2010) were analyzed and sequenced. The expanded human oral microbiome database contains approximately 645 taxa, of those: 50% are named species, about 16% are unnamed cultured taxa and, 34% are uncultured phylotypes. After all this progress, an individual's healthy oral microbiome was identified as harboring between 100 and 200+ bacterial species (Dewhirst et al. 2010; Griffen et al. 2012).

Bik et al. (2010), on an average, described 236 different OTUs sequenced from dental biofilm and saliva samples from 10 healthy individuals, of which the majority belonged to 9 bacterial phyla already identified, and 24 were new OTUs. The authors noticed different patterns of genus dominance when each individual was compared; five subjects presented dominance of *Streptococcus sp.*; two, *Prevotella*; and of the other three individuals one each presented either more *Neisseria*, *Haemophilus* or *Veillonella*. The diversity of microbiota from different subjects has frequently been described, however, as explained in the section *The role of open-ended molecular diagnostic methods in the study of oral biofilm diversity*, each method has some limitations that might cause such diversity, in addition to other factors that might define each individual microbiome, such as human genetics or lifestyles.

The data generated from the oral microbiome were assembled in a database with the aim of creating a core human oral microbiome (Griffen et al. 2011). A group of seven phyla was described as commonly found and another seven, were named rare phyla. Of these, 14 phyla, five, have only uncultivated species. The same group of authors (Griffen et al. 2012) used 454 sequencing of 16S rRNA genes to compare the subgingival microbiota of 29 patients with periodontitis with those of 29 periodontally healthy individuals. Once more, lower community diversity was

Defining the Healthy Oral Microbiome

identified in health than in disease. Abusleme et al. (2014) sequenced two subgingival samples from ten periodontally healthy patients, to try once more to identify the most prevalent bacterial species. The results of the study showed the genera *Rothia* and *Actinomyces* as the core microbiome in periodontal health, and other health-associated OTUs included five *Actinomyces spp.*, a *Streptococcus spp.* closely related to *S. sanguinis*, two *Proteobacteria* and a *Porphyromonas spp.* closely related to *Porphyromonas catoniae*, and another group of genera observed both in heath and in disease, mainly represented by *F. nucleatum* ss. These results corroborated the findings of a previous analysis (Zaura et al. 2009; Griffen et al. 2012) and are presented in detail in Table 1.

With the aim of observing the changes between each step from oral health to disease, Kistler et al. (2013) used 454 pyrosequencing of 16S rRNA genes to characterize the composition of plaque during the transition from periodontal health to gingivitis of 20 subjects. The authors observed an increase in the number and diversity of OTUs according to the establishment of gingivitis, and an average of up to 299 OTUs per sample were detected, from which a cluster analysis described a microbiome compatible with previous results (Griffen et al. 2012; Abusleme et al. 2014) and added a new bacterial species. The higher relative abundance observed was from taxa *S. sanguinis*, *Rothia dentocariosa*, *V. parvula*, *F. nuc. subsp. polymorphum*, *S. mitis*/HOT064/HOT423/HOTA95/HOTE14, *Streptococcus cristatus*/HOT071, *F. nuc. subsp. vincentii*, *Lautropia mirabilis*, *P. gingivalis* and *L. buccalis*.

Recently, Belstrøm et al. (2016) used the molecular technique named HOMINGS to describe the composition of saliva samples. This methodology uses Human Oral Microbe Identification with next generation sequencing. A total of 30 saliva samples were analyzed and the results showed a range of 120–260 bacterial species in orally

Table 1 The most abundant OTUs in periodontal health

Rothia sp. (*Rothia dentocariosa* OT 587)	*Fusobacterium sp* (*F. nucleatum ss. nucleatum* OT 698)
Actinomyces sp. (*Actinomyces sp.* OT 170)	*Fusobacterium nucleatum ss. vincentii* OT 200
Actinomyces gerenceriae OT 618	*Fusobacterium nucleatum ss. animalis* OT 420
Actinomyces sp. (*A. naeslundii OT 176*)	*Veillonella parvulla OT 161*
Actinomyces sp. OT 177	*Eikenella corrodens OT 577*
Actinomyces sp. (*A. odontolyticus OT 701*)	Unclassified *Pasteurellaceae* (*Haemophilus sp.* OT 8266)
Streptococcus sp. (*S. sanguinis OT 758*)	*Pseudomonas sp.* (*P. pseudoalcaligenes OT 740*)
Porphyromonas sp. (*P. catoniae OT 283*)	*Corynebacterium matruchotii OT 666*
Unclassified *Xanthomonadaceae*	*Campylobacter gracilis OT 623*
Burkholderia cepacia OT 571	*Lautropia mirabilis OT 022*
	Streptococcus sp. (*S. mitis OT 677*)
	Granulicatella adjacens OT 534

Name in parentheses depicts the oral taxon (OT) from the HOMD with the highest hit (>97%) to the OTU representative sequence, in OTUs with no consensus taxonomy. Data from Abusleme et al. (2014)

Table 2 Species detected in subgingival biofilm samples

Streptococcus mitis pneumoniae	*Treponema vincentii medium*
Streptococcus sanguinis	*Anaeroglobus geminatus*
Moraxella osloensis	*Prevotella denticola*
Acinetobacter junii	*Prevotella intermedia*
Granulicatella adiacens	*Desulfobulbus R004*
Acinetobacter sp RUH1139	*Treponema socranskii subsp*
Streptococcus intermedius	*Selenomonas sputigena*
Arthrobacter woluwensis	*Eubacterium brachy*
Actinomyces viscosus naeslundii	*Prevotella tannerae*
Brachybacterium rhamnosus	*Tannerella forsythia*
Lautropia AP009	*Bacteroidales oral taxon 274*
Gemella morbillorum	*TM7 oral taxon 437*
Rothia dentocariosa	*Porphyromonas endodontalis*
Rothia aeria	*TM7oral taxon 349*
Comamonadaceae nbu379c11c1	*Peptostreptococcus stomatis*
Haemophilus parahaemolyticus	*Campylobacter rectus*
Lautropia mirabilis	*Aggegatibacter AY349380*
Actinomyces massiliensis	*Eubacterium yurii subsp*
Actinomyces oral taxon 171	*Treponema maltophilum*
Haemophilus P3D1 620	*TM7 401H12*
Streptococcus oral taxon B66	*Selenomonas EY047*
Comamonadaceae 98 63833	
Comamonadaceae VE3A04	

The species listed were detected $\geq$0.2% different in health and periodontitis samples
P $\leq$ 0.05 after FDR correction for all taxa shown. Data from Griffen et al. (2012)

healthy individuals. The authors observed statistically significant different mean numbers of bacterial taxa from subjects with periodontitis (n = 220) and dental caries (n = 221) compared with the numbers in orally healthy individuals (n = 174). However, the ten most predominant genera and the 20 most predominant bacterial species agreed with the findings of previous studies (Griffen et al. 2011; Abusleme et al. 2014) (Table 2).

In summary, the overall results of the studies indicated an enormous variability of microbial species, however a CORE of the most predominant bacterial species, for now, was defined. For the future, efforts continue to be made by distinct microbiological study groups to identify, name and characterize all these microorganisms.

Microbiome of Different Habitats of the Oral Cavity

It is well known that the oral cavity comprises different niches of microorganisms, and studies have observed a selected microbiota related to different oral surfaces (See Part II chapter "Introduction to the Oral Cavity"; Mager et al. 2003; Zaura et al. 2009; Faveri et al. 2008). Zaura et al. (2009) compared biofilm samples collected from tooth surfaces with different niches in the oral cavity, including the mucosa of the tongue, cheek and palate. The principal component analysis was used and distinct microbial profiles were found when samples originating from shedding surfaces were compared with samples obtained from solid surfaces (teeth). Fifteen taxa were found in common in all patients and in all sites analyzed, including: the genera *Streptococcus, Neisseria, Corynebacterium, Rothia, Actinomyces, Haemophilus, Prevotella, Fusobacterium, Granulicatella, Capnocytophaga*, representatives of the *Veillonellaceae, Neisseriaceae* and *Pasteurellaceae* families, the *Bacteroidales* order and unclassified *Firmicutes*, and an additional four taxa were found in all but one sample: genus *Porphyromonas, Leptotrichia*, TM7 genera *incertae sedis* and *Campylobacter.*

The biodiversity of oral bacteria in saliva and on oral mucosa was again analyzed in 2012 (Diaz et al. 2012). The authors collected samples from five individuals and used a 454-pyrosequencing method to describe the microbial profiles of these samples. The authors identified 455 OTUs and the genera, *Streptococcus* and *Gemella*, were the most common in mucosa, whereas 26 genera were highlighted in saliva, including *Eubacterium, Mogibacterium, Catonella, Oribacterium, Peptostreptococcus, Megasphaera, Selenomonas* and *Solobacterium. The species levels analysis showed V. parvulla, P. melaninogenica, F. periodonticum and S. mitis* as predominant strains in the saliva of healthy subjects as well as *Neisseria sicca* and *Neisseria flavescens,* which were observed in saliva and had low affinity for mucosa.

More recently, Belstrøm et al. (2016), used the Human Oral Microbe Identification with Next Generation Sequencing (HOMINGS), to analyze the salivary microbiota of five healthy individuals. Samples were collected at intervals of 4 h over a period of 24 h; and after 7 days the same protocol was repeated. The authors wanted to study the variability within and between subjects during plaque formation. They observed almost no variation within individuals, and a range of 153–307 bacterial species between them. The predominant genera identified were *Streptococcus, Haemophilus, Prevotella, Rothia* and *Neisseria.*

In summary, the oral health microbiomes of different sites other than the tooth surface have been well characterized, and the affinity of specific species for different sites has been identified.

The Definition of the Non-bacterial Species in the Healthy Oral Microbiome

The majority of efforts to define the healthy oral microbiome have focused on identifying the bacterial species; however, other microorganisms are clearly living in these same habitats. Over the last few decades, viruses, protozoa, fungi and archaea have been investigated relative to their presence, co-aggregation and function in the oral environment. Wade (2013) described the human mouth as one of the most diverse microbiomes in the human body.

The presence of viruses has been identified in different oral health conditions; however, viruses are primarily associated with disease. Although viruses may colonize and replicate without causing any symptoms, viruses were mostly identified in periodontal disease, which did not define the virus as a pathogen, but exposed the difficulty of routinely identifying viruses. Future metatranscriptomic analysis of biofilm or tissue samples might help to elucidate a potential association between active viral and oral conditions. Among the viruses most frequently identified are: Human papilloma virus (HPV), Epstein–Barr virus and human cytomegalovirus (Teles et al. 2013).

Protozoan species were found as part of the normal microbiome, with most of the cases identified being *Entamoeba gingivalis* and *Trichomonas tenax* found in subjects with poor oral hygiene and gingival disease, and therefore considered potential opportunistic pathogens. Fungal species are represented by *Candida* species; they present without symptoms in approximately half of the individuals tested, and their prevalence increases with age. The fungal oral microbiome characterization in healthy individuals found 85 fungal genera (Ghannoum et al. 2010). Twenty subjects were evaluated and the predominant genera identified were *Candida*, *Cladosporium*, *Aureobasidium*, *Saccharomycetales*, *Aspergillus*, *Fusarium*, and *Cryptococcus*.

Archaea members are restricted to a small number of species/phylotypes, all of which are methanogens. The most frequently found *archaea* are *Methanobrevibacter oralis* and un-named *Methanobrevibacter phylotypes*, *Methanobacterium curvum/ congolense* and *Methanosarcina mazeii*. Matarazzo et al. (2011) investigated the presence of *Archaea* in subgingival oral biofilm samples from periodontally healthy and diseased subjects. The microbial diversity of samples was defined by sequencing archaeal 16S rRNA, and the presence of *Archaea* was detected in 26 of 30 subjects and in the majority of sites. A similar result was observed in periodontal disease biofilm samples, but with a higher level of presence. The most frequent species identified was *M. oralis*. Considering the results, the authors highlighted the possible relationship of increased numbers of *Archaea* species in diseased sites.

Overall, the oral microbial species other than bacteria found in the oral cavity are still in the process of being evaluated. This does not reduce their potential importance in the oral cavity with further studies using promising new techniques are justified and necessary to better understand the complex ecosystems that define the unique niches in the oral cavity.

A Current Overview of Understanding Microbial Diversity

Incomplete coverage of the richness of microbial diversity, even at greater sequencing depths, was demonstrated as partly being the cause of the high variability of species observed with different sequencing methods (Diaz et al. 2012). The same study suggested that comparing communities by their structure was more effective than comparisons based solely on membership. The inter-subject variability found was lower in community structures than in sites, and so were the differences between salivary and mucosal communities within subjects. Moreover, different requirements from different sites might clearly be the cause of microbial variability, which includes concentration of oxygen and pH. Rosier et al. (2018) described mechanisms of stability and variances in oral microbiota, and added to previous aspects, the variations in age, lifestyle, genetics and even environment of individuals (Kilian et al. 2016). However, the composition of oral microbiome changes as the biofilm matures, for instance, compared with other biofilm communities of the body, the oral microbiome in health is normally considered the most stable community over time (Zhou et al. 2013).

The oral microbiome is so complex that it might be considered almost impossible to assign a function to each organism within the community. Most important, however, is the functional role of each microorganism and it is probable that the range of such functions is more limited than the phylogenetic range of species present. Within each habitat, and depending on the health or disease status, a restricted range of functions may be needed and could be provided by a variety of organisms. The most important functions are presumed to be nutritional. For instance, the methanogens (Archaea) use short chain fatty acids in the production of methane (Lepp et al. 2004), and different species of sulphate-reducing bacteria, such as *Desulfobulbus* and *Desulfomicrobium* and *Desulfovibrio* can be found (Langendijk et al. 2001).

Oxygen sensitivity is another aspect that defines microbiome variability. There are relatively few species of obligate aerobes in the mouth. The principal aerobic genera are *Neisseria* and *Rothia* species, which are the earliest colonizers of tooth surfaces (Diaz et al. 2006). The most numerous genera found in the oral cavity are facultative anaerobes including *Streptococcus* and *Actinomyces*. As oral biofilms grow, and become mature communities, they rapidly become anaerobic, which explains the high proportions and large number of species of obligate anaerobes that are found in the mouth. One important aspect has been demonstrated: that is, members of an oral bacterial community can cooperate to protect each other from atmospheric stresses (Socransky and Haffjee 2005). The presence of the obligate aerobe *Neisseria* has been shown to enable obligate anaerobes to grow in a mixed-culture biofilm under aerobic conditions (Bradshaw et al. 1996).

This diversity is likely to lend versatility to the community providing for the ability to respond to environmental stresses in the most appropriate ways. Xie et al. (2010) used a combination of 454 and Illumina sequencing platforms with the aim of assembling a gene catalog of the dental plaque microbiota with the final objective not only of identifying, but also of defining the ecological roles of most of the species/phylotypes in mediating plaque homeostasis. The development of genetic methods has allowed the functional potential of the oral microbiome to be assessed. For instance, metagenomic sequencing using next generation sequencing methods has established significant differences between oral health and disease by coding sequences of a variety of virulence factors (Alcaraz et al. 2012; Liu et al. 2012).

Concluding Remarks

The oral health microbiome may be considered the commensal microbiota, which plays an important role in maintaining oral and systemic health. The presence of the oral microbiota itself in the oral cavity inhibits the colonization of potential pathogens and opportunistic microorganisms. However, to define the precise composition of the oral microbiome is difficult considering that the mouth is an open system, and is frequently exposed to exogenous bacteria from food, water, air and the normal microbiota of other persons via proximal contact. Studies on the composition of the oral microbiome started in the 1990s and from the beginning, the main focus of these studies has been the search for presence, counts, proportion and function of each of the microorganisms. A sequence of evolutions in methodologies has allowed the evaluation of an enormous number of samples, by numbers of groups of researchers around the world. A considerable amount of data from studies using culture and molecular targeted techniques supported the notion that most oral strains detected in oral health were also found in disease. Thus, it has been widely accepted that there was a great similarity between the compositions of the oral microbiome in different heath conditions, and more than the profile, the counts and proportion of species are the most important differences distinguishing health from disease. In the last few decades, the use of cutting-edge open-ended diagnostic techniques to study the diversity of oral microbiota has brought new insights on this subject. The overall results of these studies have revealed that the structure of the microbiome has a higher diversity of microorganisms than initially appreciated. Hundreds of new bacterial species have been identified, and association of other domains of microorganisms demonstrated, giving us an overview of the oral heath microbiome. However, studies continue with the most ambitious aims to fully define the exquisite complexity of the oral ecosystem and its central role as a determinant of both local and systemic health.

References

Aas, J. A., Paster, B. J., Stokes, L. N., Olsen, I., & Dewhirst, F. E. (2005). Defining the normal bacterial flora of the oral cavity. *Journal of Clinical Microbiology, 43*(11), 5721–5732.

Abusleme, L., Hong, B. Y., Dupuy, A. K., Strausbaugh, L. D., & Diaz, P. I. (2014). Influence of DNA extraction on oral microbial profiles obtained via 16S rRNA gene sequencing. *Journal of Oral Microbiology, 23*, 1–7.

Al-Ahmad, A., Wunder, A., Auschill, T. M., Follo, M., Braun, G., Hellwig, E., & Arweiler, N. B. (2007). The in vivo dynamics of *Streptococcus spp.*, *Actinomyces naeslundii*, *Fusobacterium nucleatum* and *Veillonella spp.* in dental plaque biofilm as analysed by five-colour multiplex fluorescence in situ hybridization. *Journal of Medical Microbiology, 56*, 681–687.

Alcaraz, L. D., Belda-Ferre, P., Cabrera-Rubio, R., Romero, H., Simon-Soro, A., Pignatelli, M., et al. (2012). Identifying a healthy oral microbiome through metagenomics. *Clinical Microbiology and Infection, 18*(Suppl. 4), 54–57.

Ashimoto, A., Chen, C., Bakker, I., & Slots, J. (1996). Polymerase chain reaction detection of 8 putative periodontal pathogens in subgingival plaque of gingivitis and advanced periodontitis lesions. *Oral Microbiology and Immunology, 11*, 266–273.

Belstrøm, D., Paster, B. J., Fiehn, N. E., Bardow, A., & Holmstrup, P. (2016). Salivary bacterial fingerprints of established oral disease revealed by the Human Oral Microbe Identification using Next Generation Sequencing (HOMINGS) technique. *Journal of Oral Microbiology, 14*(8), 30170. https://doi.org/10.3402/jom.v8.30170.

Bik, E. M., Long, C. D., Armitage, G. C., Loomer, P., Emerson, J., Mongodin, E. F., Nelson, K. E., Gill, S. R., Fraser-Liggett, C. M., & Relman, D. A. (2010). Bacterial diversity in the oral cavity of 10 healthy individuals. *The ISME Journal, 4*(8), 962–974. https://doi.org/10.1038/ismej.2010.30.

Bradshaw, D. J., Marsh, P. D., Allison, C., & Schilling, K. M. (1996). Effect of oxygen, inoculum composition and flow rate on development of mixed-culture oral biofilms. *Microbiology, 142*(Pt 3), 623–629.

Chen, T., Yu, W. H., Izard, J., Baranova, O. V., Lakshmanan, A., & Dewhirst, F. E. (2010). The human oral microbiome database: A web accessible resource for investigating oral microbe taxonomic and genomic information. *Database: The Journal of Biological Databases and Curation*, baq013.

Cilia, V., Lafay, B., & Christen, R. (1996). 148. Sequence heterogeneities among 16S ribosomal RNA sequences, and their effect on phylogenetic analyses at the species level. *Molecular Biology and Evolution, 13*(3), 451–461.

de Lillo, A., Booth, V., Kyriacou, L., Weightman, A. J., & Wade, W. G. (2004). Culture-independent identification of periodontitis-associated Porphyromonas and Tannerella populations by targeted molecular analysis. *Journal of Clinical Microbiology, 42*(12), 5523–5527.

Dewhirst, F. E., Chen, T., Izard, J., Paster, B. J., Tanner, A. C., Yu, W. H., Lakshmanan, A., & Wade, W. G. (2010). The human oral microbiome. *Journal of Bacteriology, 192*, 5002–5017.

Diaz, P. I., Chalmers, N. I., Rickard, A. H., Kong, C., Milburn, C. L., Palmer, R. J., Jr., & Kolenbrander, P. E. (2006). Molecular characterization of subject-specific oral microflora during initial colonization of enamel. *Applied and Environmental Microbiology, 72*(4), 2837–2848.

Diaz, P. I., Dupuy, A. K., Abusleme, L., Reese, B., Obergfell, C., Choquette, L., Dongari-Bagtzoglou, A., Peterson, D. E., Terzi, E., & Strausbaugh, L. D. (2012). Using high throughput sequencing to explore the biodiversity in oral bacterial communities. *Molecular Oral Microbiology, 27*, 182–201.

Dzink, J. L., Gibbons, R. J., Childs, W. C., III, & Socransky, S. S. (1989). The predominant cultivable microbiota of crevicular epithelial cells. *Oral Microbiology and Immunology, 4*(1), 1–5.

Faveri, M., Mayer, M. P., Feres, M., de Figueiredo, L. C., Dewhirst, F. E., & Paster, B. J. (2008). Microbiological diversity of generalized aggressive periodontitis by 16S rRNA clonal analysis. *Oral Microbiology and Immunology, 23*, 112–118.

Ghannoum, M. A., Jurevic, R. J., Mukherjee, P. K., Cui, F., Sikaroodi, M., Naqvi, A., et al. (2010). Characterization of the oral fungal microbiome (mycobiome) in healthy individuals. *PLoS Pathogens, 6*, e1000713.

Gomes, S. C., Nonnenmacher, C., Susin, C., Oppermann, R. V., Mutters, R., & Marcantonio, R. A. (2008). The effect of a supragingival plaque-control regimen on the subgingival microbiota in smokers and never-smokers: Evaluation by real-time polymerase chain reaction. *Journal of Periodontology, 79*(12), 2297–2304. https://doi.org/10.1902/jop.2008.070558.

Green, P. M., & Giannelli, F. (1994). Direct sequencing of PCR-amplified DNA. *Molecular Biotechnology, 1*(2), 117–124.

Griffen, A. L., Beall, C. J., Firestone, N. D., Gross, E. L., Difranco, J. M., Hardman, J. H., Vriesendorp, B., Faust, R. A., Janies, D. A., & Leys, E. J. (2011). CORE: A phylogenetically-curated 16S rDNA database of the core oral microbiome. *PLoS One, 6*(4), e19051. https://doi.org/10.1371/journal.pone.0019051.

Griffen, A. L., Beall, C. J., Campbell, J. H., Firestone, N. D., Kumar, P. S., Yang, Z. K., Podar, M., & Leys, E. J. (2012). Distinct and complex bacterial profiles in human periodontitis and health revealed by 16S pyrosequencing. *The ISME Journal, 6*, 1176–1185.

Haffajee, A. D., Japlit, M., Bogren, A., Kent, R. L., Jr., Goodson, J. M., & Socransky, S. S. (2005). Differences in the subgingival microbiota of Swedish and USA subjects who were periodontally healthy or exhibited minimal periodontal disease. *Journal of Clinical Periodontology, 32*(1), 33–39.

Harrington, C. T., Lin, E. I., Olson, M. T., & Eshleman, J. R. (2013). Fundamentals of pyro-sequencing. *Archives of Pathology & Laboratory Medicine, 137*(9), 1296–1303. https://doi.org/10.5858/arpa.2012-0463-RA.

Hiyari, S., & Bennett, K. M. (2011). Dental diagnostics: Molecular analysis of oral biofilms. *Journal of Dental Hygiene, 85*(4), 256–263.

Keijser, B. J., Zaura, E., Huse, S. M., van der Vossen, J. M., Schuren, F. H., Montijn, R. C., ten Cate, J. M., & Crielaard, W. (2008). Pyrosequencing analysis of the oral microflora of healthy adults. *Journal of Dental Research, 87*, 1016–1020.

Kilian, M., Chapple, I. L., Hannig, M., Marsh, P. D., Meuric, V., Pedersen, A. M., Tonetti, M. S., Wade, W. G., & Zaura, E. (2016). The oral microbiome—An update for oral healthcare professionals. *British Dental Journal, 221*(10), 657–666. https://doi.org/10.1038/sj.bdj.2016.865.

Kistler, J. O., Booth, V., Bradshaw, D. J., & Wade, W. G. (2013). Bacterial community development in experimental gingivitis. *PLoS One, 8*(8), e71227. https://doi.org/10.1371/journal.pone.0071227.

Klug, B., Santigli, E., Westendorf, C., Tangl, S., Wimmer, G., & Grube, M. (2016). From mouth to model: combining in vivo and in vitro oral biofilm growth. *Frontiers in Microbiology, 7*, 1448. eCollection.

Koyanagi, T., Sakamoto, M., Takeuchi, Y., Ohkuma, M., & Izumi, Y. (2010). Analysis of microbiota associated with peri-implantitis using 16S rRNA gene clone library. *Journal of Oral Microbiology, 24*, 2. https://doi.org/10.3402/jom.v2i0.5104.

Koyanagi, T., Sakamoto, M., Takeuchi, Y., Maruyama, N., Ohkuma, M., & Izumi, Y. (2013). Comprehensive microbiological findings in peri-implantitis and periodontitis. *Journal of Clinical Periodontology, 40*(3), 218–226. https://doi.org/10.1111/jcpe.12047.

Kroes, I., Lepp, P. W., & Relman, D. A. (1999). Bacterial diversity within the human subgingival crevice. *Proceedings of the National Academy of Sciences of the United States of America, 96*, 14547–14552.

Kumar, P. S., Leys, E. J., Bryk, J. M., Martinez, F. J., Moeschberger, M. L., & Griffen, A. L. (2006). Changes in periodontal health status are associated with bacterial community shifts as assessed by quantitative 16S cloning and sequencing. *Journal of Clinical Microbiology, 44*, 3665–3673.

Kumar, P. S., Mason, M. R., Brooker, M. R., & O'Brien, K. (2012). Pyrosequencing reveals unique microbial signatures associated with healthy and failing dental implants. *Journal of Clinical Periodontology, 39*, 425–433.

Langendijk, P. S., Kulik, E. M., Sandmeier, H., Meyer, J., & van der Hoeven, J. S. (2001). Isolationof *Desulfomicrobium orale sp. Nov.* and *Desulfovibrio strain NY682*, oral sulfate-reducing bacteria involved in human periodontal disease. *International Journal of Systematic and Evolutionary Microbiology, 51*, 1035–1044.

Lepp, P. W., Brinig, M. M., Ouverney, C. C., Palm, K., Armitage, G. C., & Relman, D. A. (2004). Methanogenic archaea and human periodontal disease. *Proceedings of the National Academy of Sciences of the United States of America, 101*, 6176–6181.

Li, Y., Ku, C. Y., Xu, J., Saxena, D., & Caufield, P. W. (2005). Survey of oral microbial diversity using PCR-based denaturing gradient gel electrophoresis. *Journal of Dental Research, 84*(6), 559–564.

Li, L., Hsiao, W. W., Nandakumar, R., Barbuto, S. M., Mongodin, E. F., Paster, B. J., Fraser-Liggett, C. M., & Fouad, A. F. (2010). Analyzing endodontic infections by deep coverage pyrosequencing. *Journal of Dental Research, 89*(9), 980–984. https://doi.org/10.1177/0022034510370026.

Liu, B., Faller, L. L., Klitgord, N., Mazumdar, V., Ghodsi, M., Sommer, D. D., et al. (2012). Deep sequencing of the oral microbiome reveals signatures of periodontal disease. *PLoS One, 7*, e37919.

Mager, D. L., Ximenez-Fyvie, L. A., Haffajee, A. D., & Socransky, S. S. (2003). Distribution of selected bacterial species on intraoral surfaces. *Journal of Clinical Periodontology, 30*(7), 644–654.

Mark Welch, J. L., Rossetti, B. J., Rieken, C. W., Dewhirst, F. E., & Borisy, G. G. (2016). Biogeography of a human oral microbiome at the micron scale. *Proceedings of the National*

Academy of Sciences of the United States of America, 113(6), E791–E800. https://doi.org/10.1073/pnas.1522149113.

Matarazzo, F., Ribeiro, A. C., Feres, M., Faveri, M., & Mayer, M. P. (2011). Diversity and quantitative analysis of Archaea in aggressive periodontitis and periodontally healthy subjects. *Journal of Clinical Periodontology, 38*, 621–627.

Moore, W. E., & Moore, L. V. (1994). The bacteria of periodontal diseases. *Periodontology 2000, 5*, 66–77.

Paster, B. J., Boches, S. K., Galvin, J. L., Ericson, R. E., Lau, C. N., Levanos, V. A., Sahasrabudhe, A., & Dewhirst, F. E. (2001). Bacterial diversity in human subgingival plaque. *Journal of Bacteriology, 183*, 3770–3783.

Petrosino, J. F., Highlander, S., Luna, R. A., Gibbs, R. A., & Versalovic, J. (2009). Metagenomic pyrosequencing and microbial identification. *Clinical Chemistry, 55*, 856–866.

Ramberg, P., Sekino, S., Uzel, N. G., Socransky, S., & Lindhe, J. (2003). Bacterial colonization during de novo plaque formation. *Journal of Clinical Periodontology, 30*(11), 990–995.

Rastogi, G., Coaker, G. L., & Leveau, J. H. (2013). New insights into the structure and function of phyllosphere microbiota through high-throughput molecular approaches. *FEMS Microbiology Letters, 348*(1), 1–10.

Rosier, B. T., Marsh, P. D., & Mira, A. (2018). Resilience of the oral microbiota in health: Mechanisms that prevent dysbiosis. *Journal of Dental Research, 97*(4), 371–380. https://doi.org/10.1177/0022034517742139.

Sakamoto, M., Umeda, M., Ishikawa, I., & Benno, Y. (2000). Comparison of the oral bacterial flora in saliva from a healthy subject and two periodontitis patients by sequence analysis of 16S rDNA libraries. *Microbiology and Immunology, 44*, 643–652.

Sakamoto, M., Takeuchi, Y., Umeda, M., Ishikawa, I., & Benno, Y. (2001). Rapid detection and quantification of five periodonto-pathic bacteria by real-time PCR. *Microbiology and Immunology, 45*, 39–44.

Sakamoto, M1., Huang, Y., Ohnishi, M., Umeda, M., Ishikawa, I., & Benno, Y. (2004). Changes in oral microbial profiles after periodontal treatment as determined by molecular analysis of 16S rRNA genes. *Journal of Medical Microbiology, 53*(Pt 6), 563–571.

Shchipkova, A. Y., Nagaraja, H. N., & Kumar, P. S. (2010). Subgingival microbial profiles of smokers with periodontitis. *Journal of Dental Research, 89*, 1247–1253.

Siqueira, J. F., Jr., & Rôças, I. N. (2005). Uncultivated phylotypes and newly named species associated with primary and persistent endodontic infections. *Journal of Clinical Microbiology, 43*(7), 3314–3319. https://doi.org/10.1128/JCM.43.7.3314-3319.2005. PMCID: PMC1169097.

Soares, G. M., Teles, F., Starr, J. R., Feres, M., Patel, M., Martin, L., & Teles, R. (2015). Effects of azithromycin, metronidazole, amoxicillin, and metronidazole plus amoxicillin on an in vitro polymicrobial subgingival biofilm model. *Antimicrobial Agents and Chemotherapy, 59*(5), 2791–2798. https://doi.org/10.1128/AAC.04974-14.

Socransky, S. S., & Haffjee, A. D. (2005). Periodontal microbial ecology. *Periodontology 2000, 38*, 135–187.

Socransky, S. S., Haffajee, A. D., Cugini, M. A., Smith, C., & Kent, R. L., Jr. (1998). Microbial complexes in subgingival plaque. *Journal of Clinical Periodontology, 25*, 134–144.

Tanner, A. C., Paster, B. J., Lu, S. C., Kanasi, E., Kent, R., Jr., Van Dyke, T., & Sonis, S. T. (2006). Subgingival and tongue microbiota during early periodontitis. *Journal of Dental Research, 85*(4), 318–323.

Teles, R., Teles, F., Frias-Lopez, J., Paster, B., & Haffajee, A. (2013). Lessons learned and unlearned in periodontal microbiology. *Periodontology 2000, 62*, 95–162.

Thurnheer, T., Gmür, R., & Guggenheim, B. (2004). Multiplex FISH analysis of a six-species bacterial biofilm. *Journal of Microbiological Methods, 56*, 37–47. https://doi.org/10.1016/j.mimet.2003.09.003.

van Leeuwenhoek, A. (1683). *Letter of 17 September 1683 to the Royal Society, London.* Royal Society (MS L 1. 69)

Wade, W. G. (2011). Has the use of molecular methods for the characterization of the human oral microbiome changed our understanding of the role of bacteria in the pathogenesis of periodontal disease? *Journal of Clinical Periodontology, 38*(Suppl 11), 7–16. https://doi.org/10.1111/j.1600-051X.2010.01679.x. Review.

Wade, W. G. (2013). The oral microbiome in health and disease. *Pharmacological Research, 69*(1), 137–143. https://doi.org/10.1016/j.phrs.2012.11.006. Review.

Weisburg, W. G., Barns, S. M., Pelletier, D. A., & Lane, D. J. (1991). 16S ribosomal DNA amplification for phylogenetic study. *Journal of Bacteriology, 173*, 697–703.

Xie, G., Chain, P. S., Lo, C. C., Liu, K. L., Gans, J., Merritt, J., & Qi, F. (2010). Community and gene composition of a human dental plaque microbiota obtained by metagenomic sequencing. *Molecular Oral Microbiology, 25*(6), 391–405. https://doi.org/10.1111/j.2041-1014.2010.00587.x.

Ximénez-Fyvie, L. A., Haffajee, A. D., & Socransky, S. S. (2000). Microbial composition of supra- and subgingival plaque in subjects with adult periodontitis. *Journal of Clinical Periodontology, 27*(10), 722–732.

Zaura, E., Keijser, B. J., Huse, S. M., & Crielaard, W. (2009). Defining the healthy "core microbiome" of oral microbial communities. *BMC Microbiology, 9*, 259. https://doi.org/10.1186/1471-2180-9-259.

Zhou, Y., Gao, H., Mihindukulasuriya, K. A., La Rosa, P. S., Wylie, K. M., Vishnivetskaya, T., Podar, M., Warner, B., Tarr, P. I., Nelson, D. E., et al. (2013). Biogeography of the ecosystems of the healthy human body. *Genome Biology, 14*(1), R1.

Zijnge, V., Ammann, T., Thurnheer, T., & Gmür, R. (2012). Subgingival biofilm structure. *Frontiers of Oral Biology, 15*, 1–16. https://doi.org/10.1159/000329667.

Dysbiosis of the Oral Microbiome

Apoena A. Ribeiro and Roland R. Arnold

Abstract The oral cavity is influenced by the dietary characteristics of each individual. It is in the oral cavity that food will cause the first impact within the human body and its microbiome, due to its composition and consistency. On the other hand, the oral microbiome will affect food processing and impact the human gut microbiome, since bacterial biofilm that is processed within saliva forms the food bolus, which will then be swallowed. The mouth is one of the most heavily colonized parts of our bodies and its microbiome consists of microorganisms that live in symbiosis with healthy individuals who have adequate dietary and oral hygiene habits. Nevertheless, perturbations in the microbiome due to certain stress factors, such as high carbohydrate intake and biofilm accumulation, can lead to dysbiosis and the development of oral diseases. The most prevalent diseases in the oral cavity are dental caries and periodontal diseases including gingivitis and periodontitis, but endodontic (pulp) and soft tissue infections are also prevalent. Thus, this chapter will describe the influence of dietary habits on the oral microbiome, the development of prevalent oral diseases, and their relation to the gut microbiome.

Keywords Oral microbiome dysbiosis · Dental caries · *Streptococcus* ·
Endodontic infection · Periodontal disease

Introduction

The famous quote "we are what we eat" suggests the direct impact of dietary habits and lifestyle on our systemic health. As portal of entry for the digestive system, the oral cavity is greatly influenced by the dietary characteristics of each individual: it is in the oral cavity that food will cause the first impact within the human body,

---

A. A. Ribeiro (✉) · R. R. Arnold
Division of Diagnostic Sciences, Adams School of Dentistry, University of North Carolina at
Chapel Hill, Chapel Hill, NC, USA
e-mail: apoena@email.unc.edu

© Springer Nature Switzerland AG 2019
M. A. Azcarate-Peril et al. (eds.), *How Fermented Foods Feed a Healthy Gut Microbiota*, https://doi.org/10.1007/978-3-030-28737-5_8

since it influences the mouth environment by its composition (leading to pH fluctuation) and consistency (sticky, liquid or hard food). Food is first introduced to the oral cavity to initiate the processing pathway and then swallowed to 'travel' though the gastrointestinal tract. For this reason, a two-way relationship exists: the dietary composition can have direct impact on the oral microbiome composition, activity, and local disease development; on the other hand, the oral microbiome, the transient microorganisms and the functional components of the oral cavity will affect food processing and impact the human gut microbiome, since oral biofilm is the major microbial constituent of the saliva that facilitates the formation of the food bolus that is subsequently swallowed (see chapter "Baby's First Microbes: The Microbiome of Human Milk").

The oral cavity provides different oral structures and tissues for bacterial colonization and community development, including saliva, gingival fluid, and keratinized/non-keratinized epithelial or mineralized tooth surfaces, such as the tongue, gingiva and teeth (Kolenbrander 2000; Aas et al. 2005; Simón-Soro et al. 2013a, b). The mouth is one of the most heavily colonized parts of our bodies and, as explained in the first chapter of this section, the microbiome of the oral structures consists of microorganisms that live in symbiosis with healthy individuals who have favorable dietary and oral hygiene habits. This balance is possible due to the diverse microbial communities that prevent the colonization of foreign pathogens and contribute to a healthy host physiology (Hezel and Weitzberg 2015). Nevertheless, perturbations in the microbiome due to certain stress factors, such as high carbohydrate intake, undisturbed biofilm development, and/or saliva alterations in volume or composition, can lead to imbalances in the symbiotic composition of the commensal populations and the development of oral diseases (Marsh 1994, 2016). There are both shifts in species and their functional expressions associated with dysbiosis characteristic of both caries and periodontal diseases (Belda-Ferre et al. 2012; Griffen et al. 2012; Wang et al. 2013; Jorth et al. 2014).

The oral cavity must be recognized as a complex macro ecosystem, composed of different microhabitats with distinct characteristics that either favor or prevent different species from establishing. It has been shown that different microbiome profiles can be verified among different tooth surfaces in the same individual, but very similar microbiome profiles are observed on the same tooth surfaces within different individuals. For example, *Streptococcus* were found at high abundance on the buccal surfaces of teeth and sulci, but were found at lower levels on the lingual surfaces of the same tooth (Simón-Soro et al. 2013a, b). In terms of structure, mucosal surfaces (constantly shedding) will favor microorganisms that can express unique receptors with high affinity to rapidly re-adhere to the newly exposed mucosal cell; whereas, other microorganisms accumulate in dental plaque mainly in protected areas of the tooth, such as the occlusal and the interproximal surfaces. Teeth are the only natural non-shedding surfaces in the human body and provide unique opportunities for undisturbed biofilm formation and sustained fermentation of dietary carbohydrates sufficient to permit accumulation of metabolic end-products such as lactic acid to alter the environmental pH (Marsh and Devine 2011). Teeth can also be recognized as harboring different micro-ecosystems, since they present differentiated habitats with singularities such as differences in saliva access and flow, oxygen availability, different temperatures, pH and food retention. For example, smooth free surfaces of the tooth (buccal and lingual) are constantly being "washed" by salivary flow, are highly aerated and are subjected to sheer forces from the lips, tong and other teeth during mastication. It leads to a development of biofilm in an ordered fashion, located closely to the gingival margin, markedly with bacteria that have strong adhesins to the pellicle that coats the enamel. On the other hand, protected surfaces, such as that of the pit and fissures on occlusal surfaces, and the interproximal spaces (in between contacting teeth) are characterized by a compacted biofilm formed mostly by short rods and often *Actinomyces* spp. in a condensed inner layer, and a looser biofilm layer is seen with a random arrangement of bacteria, including *S. mitis*, *Veillonella* spp. and *Fusobacterium* spp. (Dige et al. 2014). The most prevalent diseases in the oral cavity are dental caries and periodontal diseases including gingivitis and periodontitis, but endodontic (pulp) and soft tissue infections are also prevalent. These diseases are mainly caused by the oral microbiome imbalance, which may also play an important role in altering the homeostasis of systemic conditions, including gut diseases and its microbiome. Thus, this chapter will describe the influence of dietary habits on the oral microbiome, the development of these diseases, and their relation to the gut microbiome.

Dental Caries as a Dysbiosis

Despite all the knowledge and years of research, dental caries remains the most common chronic disease in the United States (NCHS—National Center for Health Statistics 2017), as well as in the rest of the world (Bagramian et al. 2009; Bourgeois and Llodra 2014). Recent data from the CDC showed that, in the US, the prevalence of untreated cavities among children remains high, affecting 19.5% of children between the ages of 2 and 5 years and 22.9% of children and adolescents aged 6–19 years. Dental caries is four times more common than asthma among adolescents aged 14–17 years, and it also affects 9 out of 10 adults older than 20 years (NCHS—National Center for Health Statistics 2017). Although the percentage of untreated dental caries declined steadily from 39.0% in 1988–1994 to 24.7% in 2011–2014 for children and adolescents aged 5–19 living below the federal poverty level, this percentage was similar in 1988–1994 and 2011–2014 for adults of all income levels (NCHS—National Center for Health Statistics 2017).

Dental caries can be defined as a biofilm-mediated dysbiosis (Fig. 1a). It is characterized by the dissolution of tooth tissues (enamel and dentin) by acid produced by oral bacteria as a result of the fermentation of dietary carbohydrates. When the fermentation process is enhanced by the excessive and/or frequent ingestion of fermentable sugars, the buffering capacity of saliva overwhelmed and the sustained local reduction in pH leads to the demineralization of enamel, cementum, and dentin. Due to the highly dynamic nature of the disease, resulting from continuous physical-chemical interactions between the tooth surface and biofilm that covers the

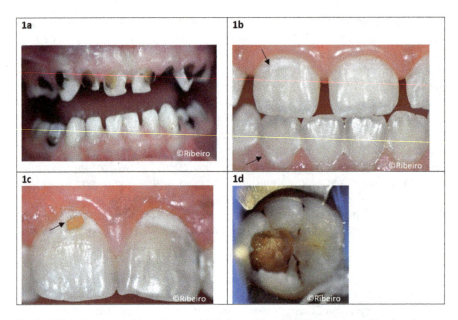

Fig. 1 Clinical pictures of dental caries. (**a**) Patient with multiple lesions caused by dental caries disease; (**b**) Incipient caries lesion, known as active white spot lesion (arrows), is the first clinical sign of caries disease; (**c**) Dentin cavity, a clinical sign of lesion development, from white spot lesion, to cavitation (arrow); (**d**) dentin tissue destruction caused by proteolytic enzymes produced by bacteria from the biofilm

surface, multiple pH fluctuations in the biofilm lead to episodes of mineral loss (demineralization) and mineral gain (remineralization) of the teeth. If equilibrium of these episodes is not achieved over time, demineralization will reach the level when an incipient lesion, known as active white spot lesion (Fig. 1b), can be visually detected by a trained professional (Xu et al. 2014).

Chemically, the lesion is characterized by the dissolution of the calcium and phosphate constituents of enamel. The acid production depends on the carbohydrate intake from dietary sources. If the demineralization episodes exceed the remineralization process, the development of the disease will not be controlled, and the lesion will not be arrested. Clinically, the destruction of the enamel tissue progresses with a breakdown of the superficial layer, leading to a cavity that is more prone to accumulate biofilm (Fig. 1c) and the lesion is more likely to rapidly progress, affecting the underlying tissue, called dentin. At this point, a change in the bacterial composition and metabolic profiles can be observed, with species that produce proteolytic enzymes capable of affecting the collagen fibers that compose dentin (Fig. 1d).

The bacterial microbiome from dental biofilms can harbor more than 720 unique species. 800 to 1000 different oral bacterial taxa (as sharing >98.5% 16S ribosomal RNA (rRNA) sequence identity) can be identified with more modern techniques with differences in abundance and diversity patterns across age, sample quality and origin, and health status (Dewhirst et al. 2010). Since diversity on oral health was the theme of the previous chapter, the focus now will be towards the core group of phylotypes found under diseased conditions (Becker et al. 2002; Ling et al. 2010; Ribeiro et al. 2017), that are different from those observed in healthy conditions (Bik et al. 2010; Dewhirst et al. 2010; Zaura and Mira 2015). However, it is important to recognize that it is still necessary to find consistent bacterial markers across studies and cohorts.

A key aspect to consider in relation to caries dysbiosis is that the development of carious lesions does not occur in all teeth nor on all surfaces of a tooth at the same time nor with the same intensity. Although many studies compared the microbiome associated with caries activity by using pooled samples (for example, saliva and pooled biofilm from multiple tooth surfaces) (Li et al. 2007; Aas et al. 2008; Bik et al. 2010; Peterson et al. 2013), it is important to recognize the highly local nature of the microhabitats responsible for disease progression and that pooled sampling will compromise the resolution of the microbial composition at the involved sites. A study conducted by Dige et al. (2014) showed the spatial distribution of bacterial taxa *in vivo* at various stages of occlusal caries, applying a molecular methodology involving fluorescence in situ hybridization (FISH) and confocal microscopy. *S. mutans* could be observed on sites with both active and inactive caries, but not on clinically sound enamel; whereas, *Bifidobacterium* spp. were only detected in sites with active caries. *Lactobacillus* spp. was not detected on clinically sound and non-cavitated sites.

By using 16S rRNA amplicon sequencing combined with the BLASTN-based search algorithm for species identification, a recent study compared healthy and caries active occlusal tooth surfaces from 12-year-old children. It was found that the sites varied not only among individuals, but also among caries samples from the same individual. Interestingly, the same levels of members of the genera *Streptococcus, Pseudomonas, Granulicatella, Actinomyces, Prevotella and Veillonella*, traditionally associated with caries active patients, were found on both sound surfaces and active white spot lesions, but the percentage of *Actinobaculum* and *Porphyromonas* were higher in active white spot lesions. On the other hand, the numbers of *Klebsiella and Acinetobacter* species were higher on sound surfaces. The presence of eight bacterial taxa were observed in active carious sites (*Abiotrophia defectiva, Actinomyces sp._Oral_Taxon_448, Propionibacterium acidifaciens, Actinobaculum sp._Oral_Taxon_183, Streptococcus gordonii, Streptococcus sp._Oral_Taxon_064, Streptococcus oralis, Streptococcus pneumoniae* and *Rothia dentocariosa*). Five bacterial taxa (*Lactobacillus johnsonii, Actinomyces gerencseriae, Actinomyces naeslundii, Cardiobacterium hominis* and *Streptococcus sp._Oral_Taxon_B66*) were present at significantly higher proportions in the biofilm from healthy occlusal surfaces (Ribeiro et al. 2017).

Another oral microbiome investigation, conducted on saliva samples from adults with dental caries, reported that higher levels of two bacterial taxa (*Streptococcus salivarius* and *Solobacterium moorei*) and three bacterial clusters (*Streptococcus parasanguinis I* and *II* and sp. clone BE024_ot057/411/721, *Streptococcus parasanguinis I* and *II* and sinensis_ot411/721/767, *S. salivarius* and sp. clone FO042_ot067/755) were found compared to individuals without caries activity (Belstrøm et al. 2016).

As a recognized complex disease, dental caries has been identified as a two-step process, namely initiation/demineralization of enamel followed by progression through dentin, characterized by a succession of microorganisms (Simón-Soro et al. 2013a, b). Healthy biofilms are characterized by high numbers of species, while mature biofilms express lower bacterial diversity because it requires special abilities from the microorganism to survive and to overcome a hostile environment. For example, *S. mutans*, one of the most studied bacterial specie involved in caries initiation and progression, can harbor different virulence factors such as the ability to form biofilm by the synthesis of adhesive glucans from sucrose by the action of three glucosyltransferases (GtfB, GtfC, GtfD; encoded by *gtfB*, *gtfC* and *gtfD*, respectively) (Merritt and Qi 2012) and glucan-binding protein (Banas and Vickerman 2003). This biofilm-forming capacity varies widely among strains (Merritt and Qi 2012; Banas and Vickerman 2003). Other virulence factors include the cell surface protein antigen c (Pac), responsible for bacterial adherence to the salivary pellicle (Palmer et al. 2013); production and excretion of organic acids, such as lactic acid; and the production of antibacterial bacteriocins, such as mutacins I and IV (Paes Leme et al. 2006). Species other than *S. mutans,* such as *S. sobrinus, Rothia dentocariosa, Actinomyces* species and *S. salivarius* are also related to the early stages of dental caries due to the genetic virulence repertoire that allows these species to set up the environment for more acid-tolerant and acidogenic species, including *Scardovia wiggsiae* and *Actinomyces sp.* HOT 448 (Kressirer et al. 2018).

The cariogenic biofilm is characterized, then, by bacterial species with the ability to: (1) adhere to the tooth surface, (2) produce water-insoluble exopolysaccharides (EPS)-rich matrix, which will limit the diffusion of the carbohydrate fermentation end products (acids) and, (3) survive in this environment with organic acid accumulation due to the presence of a diffusion-limiting EPS-rich matrix. This causes an acid dissolution of the enamel mineral due to the localized acidic pH microenvironments across the biofilm structure and at the tooth-biofilm interface (Ilie et al. 2012; Xiao et al. 2017).

Knowledge obtained through NGS technology has enabled research of traditional bacterial species, which have been investigated for years and are considered the most cariogenic species (Loesche 1986; Lang et al. 1987; Alaluusua et al. 1996; Burt et al. 1998; Harris et al. 2004; Palmer et al. 2010; Kanasi et al. 2010). A recent study showed that both healthy and diseased sites show high relative abundance of *Streptococcus mutans* and low abundance of *Streptococcus sobrinus* and a relationship between these species and the presence of active white spot lesions could not be observed (Ribeiro et al. 2017). Bacterial species other than *S. mutans* and *S. sobrinus*, e.g., species of the genera *Lactobacillus, Prevotella, Propionibacterium*, non-*mutans* streptococci and *Actinomyces* spp., may also play important roles in caries initiation and biofilm community interactions (Aas et al. 2008; Simón-Soro et al. 2014; Ribeiro et al. 2017).

Since initiation and progression of dental caries are carbohydrate-dependent, some studies also investigated the influence of dietary habits on the oral microbiome related to dental caries. The first dietary stimuli have a strong influence on the etiology of caries at later developmental stages. For example, meta-analyses have shown that early childhood caries (ECC) is more frequent in bottle-fed children. The disease is characterized by a high caries activity and rapid tooth destruction in 3-month-old babies up to the age of 3 years (Avila et al. 2015) due to frequent sucrose intake at an early age, influencing increased colonization with acidogenic (acid producing) and aciduric (acid tolerant) cariogenic bacteria. Many studies also found that *S. mutans* and *S. sobrinus* are the predictive factors for dental caries (Alaluusua et al. 1996; Burt et al. 1998; Kanasi et al. 2010; Harris et al. 2004; Loesche 1986; Palmer et al. 2010). However, bacterial species from genera *Lactobacillus, Prevotella, Propionibacterium* and *Actinomyces* are also related to caries initiation and progression (Aas et al. 2008; Simón-Soro et al. 2014; Ribeiro et al. 2017). Additionally, sucrose is a substrate for the production of extracellular and intracellular polysaccharides, two components that determine biofilm formation and structure (Paes Leme et al. 2006). Thus, the constant intake of fermentable sugars in daily diet results in increased carbohydrate fermentation by acidogenic bacteria, which results in lactic acid production, followed by longer periods of low pH and the selection of aciduric bacteria, such as *S. mutans*, Lactobacilli and Bifidobacteria, that survive under these conditions (Marsh 1994, 2016).

Among the highly abundant species observed in biofilm from adolescent patients with high and frequent carbohydrate intake, *Lactobacillus* spp. showed higher counts in dental biofilms *in situ* in the presence of glucose + fructose and sucrose, and correlations were also found between intake of confectionery-eating events and lactobacillus levels among 12-year-old schoolchildren (Beighton et al. 1996). In addition, the association between relative abundance of bacterial species and frequency of carbohydrate intake (high vs low consumption) was shown by Ribeiro et al. (2017): among 12-year-old patients with high frequency of carbohydrate intake (more than two times between meals), statistically significant differences in the increased relative abundance were observed among *Actinomyces gerencseriae, Actinomyces naeslundii, Lactobacillus crispatus* and *Streptococcus vestibulares*.

The relationship between fermentable carbohydrate intake and oral microbiome in adult populations showed that mutans and non-mutans streptococci of several types, including *S. sanguinis* and *S. salivarius*, are known to be extremely abundant in the mouth and present acidogenic and acid-tolerant properties (Guggenheim 1968; Nyvad and Kilian 1990). However, concerning their relation the development of caries, some data suggest an inverse relationship of *S. sanguinis* and abundance of mutans streptococci (Loesche and Straffon 1979). On the other hand, lactobacilli are known as highly acidogenic from carbohydrates as well as being extremely acid tolerant.

As for the influence of dietary habits on the oral bacterial metabolism, a meta-transcriptomic approach was used to investigate the active oral microbiota before and after a carbohydrate meal (Benítez-Páez et al. 2014). It was found that the metabolism of the microbiota changed, irrespective of the quality of the diet of the individual. Interestingly, no changes were observed in one individual who had never had dental caries, indicating a strong resilience (that is, a high capacity to overcome stress factors and recover from perturbations). Thus, it is an important factor for oral homeostasis since inadequate resilience can lead to oral diseases when disease drivers are strong or sufficiently persistent.

Fermentable carbohydrates will lead to decreased biofilm pH due to bacterial metabolism. Microbial communities located in acidic pH strata biofilms show low diversity of microbial populations, with *Lactobacillus* species being prominent. In comparison, the distinctive species of a more diverse flora are associated with more neutral pH regions of carious lesions, including *Alloprevotella tanerrae*, *Leptothrix* sp., *Sphingomonas* sp. and *Streptococcus anginosus* (Kianoush et al. 2014). These findings were also observed by a more recent study that showed that the high consumption of fermentable carbohydrates was associated with a reduction in bacterial diversity.

Altogether, these observations highlight the non-specific source, polymicrobial nature, and complex metabolic and community dynamics of dental caries and provide a deeper understanding of the differences in bacterial composition associated with health and initial development of caries, and the influence of the diet in the microbiome composition and metabolism.

Endodontic Infections

If dental caries is left untreated, the lesion can progress through the dentine into the root canal that contains the vasculature and innervation (pulp) that maintains the vitality of the tooth. The pulp becomes infected associated with inflammation and pain and ultimately dies. Infectious agents can also reach the pulp through dentinal tubules when the distance between the approaching border of the carious lesion and the pulp is sufficiently small. There can also be direct pulp exposure resulting from fractures or failing restorations and salivary contamination.

Due to the characteristics of the vasculature of the pulp, infections could potentially result from bacteremia. Infections of the dental pulp are generally polymicrobial in nature with anaerobic proteolytic bacteria dominating (Munson et al. 2002) presumably due to the necrotic nature of the environment. Persistent infection can progress through the foramen of the root tip resulting in an abscess in the alveolar bone presenting as a periapical lesion (periapical periodontitis).

Interestingly, in refractory (persistence following treatment) endodontic infections, the most commonly identified species cultured are the Gram positive enterococci. These are considered commensal inhabitants of the gastrointestinal tract and are not normally found in the mouth in health, resulting in some controversy as to the source of this infection. It is possible that they are introduced through transient bacterium from the gut (Goh et al. 2017). In regard to the fermented food topic of this volume, it has been demonstrated that enterococci are found in both unpasteurized and pasteurized cheeses and can persist in the mouth for some time after their consumption (Razavi et al. 2007). Further studies have demonstrated that enterococci can gain access to the root canal by microleakage through the temporary filling materials used between endodontic treatment visits (Kampfer et al. 2007). It therefore has been suggested that cheese and perhaps other fermented dairy and meat products might be a source of enterococci infection challenge during the endodontic treatment course and that patients should avoid eating foods known to be colonized by enterococci (Goh et al. 2017; Wade 2013).

Periodontal Diseases: Gingivitis and Periodontitis

As discussed previously, the biofilm that accumulates on the tooth surface exposed to the oral cavity (supragingival plaque) is saliva-bathed, composed mainly of saccharolytic, acidogenic and aciduric populations of bacteria selected by dietary sugar on nutrient poor enamel surfaces of the teeth. In contrast, there are unique ecological niches created by the architecture and dynamics of the supporting structures of the teeth that select for asaccharolytic, nutritionally fastidious, acid-intolerant, proteolytic anaerobes.

The attachment of epithelium to teeth in a healthy dentition occurs at the transition from the enamel surfaces of the crown of the tooth to the cementum of the root (cemento-enamel junction) forming a thin barrier (junctional epithelium) that protects the underlying supporting structures. The teeth are suspended in sockets in the alveolar bone of the jaws by periodontal ligament. The gingiva (gums) create a sulcus (gingival crevice) surrounding teeth that is composed of unique specialized gingival epithelial cells and keratinocytes. This arrangement creates a close association between the non-sloughing hard surfaces of the teeth and the renewable soft tissue of the gingiva that limits accessibility of saliva and provides microbial attachment sites on both mineral and cell surfaces bathed in a protein rich, tissue-derived gingival crevicular fluid (GCF) of this subgingival space. This relatively sequestered site if undisturbed permits a hierarchical development of complex biofilm communities driven by environmental alterations in nutrient availability, oxygen limitations, specific interspecies co-aggregations, synergisms and antagonisms.

As with the supragingival plaque, the early subgingival colonizers are predominated by facultative anaerobes including the saccharolytic streptococci and actinomycetes. If the biofilm is permitted to develop without mechanical disruption (oral hygiene), robust bacterial species such as *Fusobacteria* and *Prevotella* (Ramberg et al. 2003) neutralize the pH of this subgingival environment by nitrogenous metabolism and stimulate increased efflux of GCF further promoting proteolytic activity allowing a shift in the microbial communities toward the establishment of more acid-intolerant, oxygen-sensitive, more diverse, inflammation-promoting and potentially periodontopathic species.

Gingivitis

Gingivitis is arguably the most common bacterial disease of humans with a prevalence greater than 90% in adults (Coventry et al. 2000). Following meticulous cleaning of the teeth, the gingival margins proximal to the gingival crevice are rapidly repopulated (within hours) with pioneer colonizers predominated by Gram positive, aerotolerant anaerobes including streptococci and actinomycetes (Nyvad and Kilian 1990; Ramberg et al. 2003; Li et al. 2004). This initial adherence is favored by the selective affinity of these bacteria for epitopes of the salivary proteins that specifically adsorb to tooth surfaces (pellicle) and coat the epithelium (Murray et al. 1992). If left undisturbed, the accumulating biofilm of these primary colonizers provides new attachment sites for selected other species through specific co-aggregation interactions (Kolenbrander et al. 2006) and by metabolic reduction in oxygen tension favoring more anaerobic Gram negative species including *Fusobacterium, Treponema* and members of the phylum *Synergistetes* (Zijnge et al. 2010). This increase in the proportions of Gram negative, asaccharolytic and anaerobic bacteria results in the accumulation of endotoxins, metabolic end-products and lytic enzymes that irritate the gingivae activating pro-inflammatory pathways resulting in the clinical signs of gingivitis, including red, swollen and inflamed gums that bleed either spontaneously or on gentle probing. These clinical presentations are entirely reversible with restoration of effective oral hygiene (Loe et al. 1965). It is generally considered that there are no specific pathogens associated with gingivitis, but rather plaque load and especially its level of maturity (transition to Gram negative anaerobes) correlate with disease severity (Socransky 1977). Because established gingivitis is frequently not painful, it can remain undiagnosed in the absence of routine dental care, and thus go untreated for many years without progressing to irreversible periodontitis.

Periodontitis

Periodontitis is a bacterially-induced chronic inflammatory disease of the periodontium that includes not only inflammation of the gingiva, but also destruction of the tissues that surround and support the teeth including the periodontal ligament and the alveolar bone. In susceptible individuals, inflammation in the gingival tissues results in the destruction of the epithelial and connective tissue attachments to the tooth through the activities of neutral proteases, elastases, collagenases and metalloproteinases (Smith et al. 1995; Golub et al. 1997; Hernández et al. 2010). In an attempt to repair, the junctional epithelium responds to the damage by migrating toward the apex of the tooth possibly due to the proteolytic activity of degranulating neutrophils within the gingival environment (Bosshardt and Lang 2005; Eskan et al. 2012). This is measured clinically as attachment loss by calibrated dental probing as a metric of periodontal disease severity. This retreating attachment of the connective tissue results in a deepening of the sulcus forming a periodontal pocket providing an anaerobic environment and neutral pH favoring asaccharolytic, proteolytic anaerobes (Eggert et al. 1991) creating a vicious cycle. The resulting biofilm ultimately provokes a chronic inflammatory response in the surrounding connective tissue that drives the destruction of the alveolar bone that supports the tooth (Armitage 2004).

While accumulation of biofilm triggers gingivitis, the presence of biofilm alone is not sufficient to progress to periodontitis as evidenced by the clinical course observed with untreated chronic gingivitis mentioned above. It is now evident that complex interactions between immune response elements of the host with the biofilm are required for progression to periodontitis. In this scenario, it is proposed that most of the tissue damage is due to a dysbiotic microbial community's subversion of the host response leading to an inappropriate, exaggerated inflammatory response (Darveau 2010; Kilian et al. 2016). The resulting local inflammation provokes an increased flow of the nutrient-rich gingival crevicular fluid possibly associated with bleeding and a reduction in oxygen favoring a shift from a symbiotic microbial population to the more nutritionally fastidious, protein-dependent obligate anaerobic dysbiosis (Marsh et al. 2015). The resulting inflammation damages the sulcular epithelium providing red blood cells for bacterial hemolysis and release of hemoglobin for processing by heme-dependent bacteria such as *Porphyromonas gingivalis*.

Since the 1950s, investigators have sought to identify the microbial species critical for the initiation and progression of periodontitis. From these studies, it was clear that there were profound shifts in the microbial community structures that were associated with the transition from a healthy gingiva to disease and specific organisms were proposed as potential periodontopathogens based on culture biases and virulence properties identified in animal models. In 1994, Socransky and colleagues employed checkerboard DNA-DNA hybridization techniques that permitted enumeration of then relatively large numbers of species in very large numbers of samples. Using 40 species-specific DNA-DNA hybridization probes to quantitate oral bacteria in the subgingival plaque samples from healthy and periodontally diseased sites (Socransky et al. 1998), they first advanced the idea of discrete microorganisms working together to cause disease. They defined five different "complexes" based on their level of association with disease severity. These complexes were color coded with the most highly associated with chronic severe periodontitis identified as the "red complex" that included three species: *P. gingivalis, Tannerella forsythia* and *Treponema denticola*. Although sometimes present in low numbers in healthy subjects (Kumar et al. 2003), the red complex was considered to be responsible for initiation and progression of disease. The disappearance (or significant reduction) of red complex was associated with successful periodontal treatment and again became prominent when inflammation and deep pockets reappeared thus fulfilling a modification of Koch's postulates (Socransky 1977). The "orange complex" demonstrated a less stringent association with disease, but were considered foundational for the subsequent colonization by the "red complex" and included among others *Prevotella* spp., *Fusobacterium* spp. and *Parvimonas micra*. On the other end of the spectrum, members of the "yellow complex" (*Streptococcus gordonii, Streptococcus intermedius, Streptococcus mitis, Streptococcus oralis* and *Streptococcus sanguinis*) and of the "purple complex" (*Actinomyces odontolyticus* and *Veillonella parvula*) were mainly associated with healthy sites. The selection of the probes used in these studies was by necessity based on culture data and were therefore restricted by the same biases that confounded other culture-dependent studies.

Using more contemporary sequencing approaches, the power to study bacterial community compositions has grown exponentially and has facilitated identification of novel associations between periodontitis and previously uncultivable or previously underappreciated species including the Gram positive *Filifactor alocis* (Griffen et al. 2012) and *Peptostreptococcus stomatis* and species from the genera *Prevotella, Synergistes* (Vartoukian et al. 2009), *Megaspaera, Selenomonas* and *Desulfobulbus* (Kumar et al. 2003; Dewhirst et al. 2010). Many of these species correlate as strongly with disease severity as do the classic "red complex". It is now clear that periodontitis is a polymicrobial infection that arises from the expansion of so-called pathobionts within the microbial community that leads to dysbiosis-associated pathologies (Hajishengallis 2014). This shift in dominance from symbionts to pathobionts appears to be driven by low prominence microorganisms (keystone pathogens) that are capable of modulating the host response and possibly the pathobionts directly (Frias-Lopez and Duran-Pinedo 2012) leading to an alteration in the nutrient foundation of the community through a subverted inflammatory response. *P. gingivalis* has long been associated with human periodontitis and is capable of orchestrating disease in a variety of animal models. Recent studies suggest that its role is more consistent with that of a keystone pathogen in that it is not a potent inducer of inflammation, but rather can impair host innate and adaptive defenses in ways that alter the growth, composition and development of the entire microbial community resulting in homeostatic disruption driving commensals toward pathobionts that deregulate inflammation causing bone loss (reviewed in Hajishengallis 2014; Costalonga and Herzberg 2014). The inflammatory destruction of host tissues provides a nutrient-rich inflammatory exudate (e.g. degraded host proteins and hemin) favoring the growth of asaccharolytic and proteolytic bacteria resulting in a dysbiotic shift in the microbiota further altering the environment to create new niches for sustaining and expanding the periodontopathic communities at the expense of the homeostatic symbionts associated with periodontal health.

Dietary Influences on Periodontal Diseases

There is evidence that periodontal diseases are influenced by diet. For example, vitamin C depletion can lead to profuse gingival bleeding, lower serum magnesium/calcium levels, lower antioxidant micronutrient levels, and lower docohexanoic acid intake have also been shown to significantly correlate with higher levels of periodontal diseases (reviewed in Chapple et al. 2017). Vitamin B12 deficiency was also associated with periodontal disease progression and bone and periodontal ligament destruction (Zong et al. 2016).

In relation to carbohydrates, subjects on a high-carbohydrate diet develop gingivitis because it increases the risk of inflammation and thus gingival bleeding (Hujoel 2009; Sidi and Ashley 1984), whereas a switch to a "Stone Age" diet, based on whole grains of barley, wheat, herbs, honey, milk, and meat from domestic animals (goats and hens), resulted in a decrease in gingival bleeding (Baumgartner et al. 2009). Thus, fermentable carbohydrates (sugars and starches) are recognized as the most relevant common dietary risk factor for periodontal diseases, because glycemia drives oxidative stress and advanced glycation end-products may also trigger a hyper inflammatory state (reviewed in Chapple et al. 2017).

Thus, there is evidence that together with sugar restriction, functional foods may improve clinical treatment outcomes following the adjunctive ingestion of fruit and vegetable extracts (Chapple et al. 2012) and probiotics (Martin-Cabezas et al. 2016), although evidence is limited and biological mechanisms not fully elucidated (Chapple et al. 2017). There is also an intriguing study that suggests that a diet that includes frequent ingestion of fermented foods might positively influence periodontal health (Takeshita et al. 2014). This study compared the salivary microbiomes of orally healthy adult participants from a representative community in Japan with that of a cohort from South Korea. This selection was based on national surveys that suggested that South Koreans had better periodontal health than that of Japanese, despite their similar inherent backgrounds. The microbiota of the Japanese individuals comprised a more diverse community, with greater proportions of 17 bacterial genera, including *Veillonella*, *Prevotella* and *Fusobacterium*, compared to the higher proportions of *Neisseria* and *Hemophilus* species found in Korean saliva samples. A previous study by this group found that salivary microbiomes with larger proportions of *Prevotella* and *Veillonella* were associated with periodontitis; whereas, larger proportions of *Neisseria*, *Hemophilus* and *Porphyromonas* were associated with periodontal health (Takeshita et al. 2009). Therefore, the salivary microbiome composition of the Korean cohort could be considered healthier than that of the Japanese subjects, even though all of these individuals were orally healthy. The authors noted that there are major differences in the diets of these two cohorts and suggested that the Korean preferences for spicier foods and especially for the fermented vegetables, kimchi, might contribute to these differences in their microbiomes (Takeshita et al. 2014).

The Influence of Oral Health on the Gut Microbiome

To the best of our knowledge, there are no studies that show that caries dysbiosis has a direct impact on the gut microbiome. Thus, any correlation between presence of active dental caries and gut microbiome remains unclear and further investigations are highly recommended.

However, it is well known that the gut microbiome is highly impacted by the quality of food intake. The microorganisms that reside in the human colon fulfill their energy requirements mainly from diet- and host-derived complex carbohydrates. Individual bacterial species exhibit different preferences for the same set of glycans and this maintains a competitive environment, which promotes stable coexistence and shows that predictable changes in the gut microbiota can improve health through diet (Tuncil et al. 2017).

The use of probiotics can contribute to bacterial resilience (bacterial capacity to recover from perturbations caused by disease drivers) and, thus, represents beneficial functions (e.g., preventing biofilm acidification, biofilm accumulation, or harmful inflammation) (See chapter "Microbial Manipulation of Dysbiosis: Prebiotics and Probiotics for the Treatment of Oral Diseases"). A recent systematic review of 50 studies (3247 participants) concluded that current evidence is insufficient for recommending probiotics for managing dental caries (Gruner et al. 2016). However, instead of using general dairy products or gut-associated bacteria, the identification of new probiotic species without acidogenic characteristics and the development of individualized treatments could improve these results in the future (López-López et al. 2017). The idea for the oral cavity would be to obtain indigenous probiotic species or communities with certain beneficial functions (e.g., arginolytic pathways produce ammonia or denitrification pathways produce nitric oxide) to compete for the bacterial sites and food consumption, and, instead of decreasing the biofilm pH, raising it or maintaining it to a neutral level (Rosier et al. 2017).

Another interesting observation was published recently by Yasuda et al. (2017) while investigating the effect of the use of systemic and topical fluoride in the oral and gut microbiome. Topical and systemic fluoride are regularly used products for dental caries prevention and treatment due to their capacity for increasing mineral remineralization of the dental tissues and inhibiting energy harvest in oral cariogenic bacteria (such as *S. mutans* and *S. sanguinis*). Fluoride also inhibits bacterial growth by inhibiting the enzyme enolase, which catalyzes the conversion of 2-phosphoglycerate to phosphoenolpyruvate (the last step of anaerobic glycolysis), thus leading to bacterial depletion (Marquis 1995; Qin et al. 2006). By treating mice with low or high levels of fluoride over a 12-week period (fluoride exposures at levels commonly found in municipal water and dental products), followed by 16S rRNA gene amplicon and shotgun metagenomic sequencing, they found changes in oral microbiome in both the low- and high-fluoride groups. Several operational taxonomic units (OTUs) belonging to acidogenic bacterial genera (such as *Parabacteroides*, *Bacteroides*, and *Bilophila*) were depleted in the oral community. In addition, fluoride-associated changes in oral community composition resulted in depletion of gene families involved in central carbon metabolism and energy harvest (2-oxoglutarate ferredoxin oxidoreductase, succinate dehydrogenase, and the glyoxylate cycle). However, fluoride treatment, exposure at physiological levels, did not induce a significant shift in the overall composition of the oral microbiome, and even on the established gut microbiome or function, possibly due to absorption in the upper gastrointestinal tract. Fluoride-associated perturbations thus appeared to have a selective effect on the composition of the oral, but not on the gut microbial community.

Systemic Consequences of Oral Dysbiosis

Oral bacteria have been proposed to play a role in a number of human systemic diseases, including atherosclerotic cardiovascular disease (Dietrich et al. 2013), stroke, abnormal pregnancy outcomes, rheumatoid arthritis, respiratory tract infections including pneumonia, meningitis or brain abscesses, inflammatory bowel disease and colorectal cancer (Scannapieco and Binkley 2012; Dewhirst et al. 2010; Han and Wang 2013; Chapple and Genco 2013; de Pablo et al. 2009) and even links to Alzheimer's disease (Shoemark and Allen 2015). Dysbiosis in periodontal disease likely triggers bacteremia facilitating systemic dissemination of oral bacteria (Forner et al. 2006). Dissemination has been demonstrated for select strains (but not all) of *P. gingivalis* through modifications in vascular permeability and septicemia from sequestered sites in animal models (Genco et al. 1991). Oral administration of *P. gingivalis* in a mouse model had direct effect on the gut microbiome and provokes inflammatory changes in various tissues and organs (Arimatsu et al. 2014). Is it possible that *P. gingivalis* can play a role as a keystone pathogen in the gut? It is well known that severe periodontitis negatively impacts glycemic control, not only in diabetes, but in subjects without diabetes. Severe periodontitis is an established risk factor for the onset of type 2 diabetes, and periodontal disease severity correlates with diabetic complications (Chapple and Genco 2013). It therefore follows that good oral hygiene is important not only to dental health maintenance, but should also be considered for controlling total microbial load that disseminates to or influences extra-oral infections and inflammation (Han and Wang 2013).

Summary

In dental caries, biofilm microbiome stability is disturbed by the high and frequent consumption of fermentable carbohydrates, thus this dietary habit is considered a disease driver. The indigenous microbiota ferment these carbohydrates into organic acids and the local pH will drop from 7.0 to below 5.5 when the acid surpasses the buffering capacity of the biofilm and saliva. Acidogenic and aciduric species that are adapted to the acidic conditions in the biofilm environment will gain a selective advantage. Over time, dysbiosis is characterized by a shift in the microbiota, leading to a less diverse and more cariogenic community that is more efficient at fermenting carbohydrates (i.e., saccharolytic) and more adapted to growth and metabolism in low pH (i.e., aciduric). These include, but are not limited to, aciduric representatives of *Lactobacillus*, *Streptococcus*, *Veillonella*, *Bifidobacterium* and *Actinomyces*. During the low pH period, enamel demineralization (mineral loss) exceeds remineralization (mineral gain). If the acidic conditions persist or are repeated frequently without sufficient time for remineralization, then a caries lesion may develop. Thus, a dietary habit defined by frequent carbohydrate intake can lead to a positive feedback loop, causing a shift to a saccharolytic, acidogenic and aciduric microbiota that can cause irreversible dental carious lesions over time.

In periodontal diseases, a microbiome composed of asaccharolytic, nutritionally fastidious, acid-intolerant, proteolytic anaerobes is observed, and is related to a chronic inflammatory response in the surrounding connective tissue that drives the destruction of the alveolar bone that supports the tooth. The microorganisms highly associated with chronic severe periodontitis are identified as the "red complex", which includes three species: *P. gingivalis, T. forsythia* and *T. denticola*. These species and a growing list of others are considered responsible for driving the host compatible commensal microbiome toward dysbiosis leading to the rise of pathobionts to dominance and the initiation and progression of disease. The possibility of specific species orchestrating the shift to dysbiosis resulting in pathology has identified the role of such species as a keystone pathogen. *P. gingivalis* seems especially equipped to play such a role. The disappearance (or significant reduction) of red complex and probably more specifically keystone pathogens was associated with successful periodontal treatment.

References

Aas, J. A., Paster, B. J., Stokes, L. N., Olsen, I., & Dewhirst, F. E. (2005). Defining the normal bacterial flora of the oral cavity. *Journal of Clinical Microbiology, 43*, 5721–5732.

Aas, J. A., Griffen, A. L., Dardis, S. R., Lee, A. M., Olsen, I., Dewhirst, F. E., Leys, E. J., & Paster, B. J. (2008). Bacteria of dental caries in primary and permanent teeth in children and young adults. *Journal of Clinical Microbiology, 46*, 1407–1417. https://doi.org/10.1128/JCM.01410-07.

Alaluusua, S., Mättö, J., Grönroos, L., et al. (1996). Oral colonization by more than one clonal type of mutans streptococcus in children with nursing-bottle dental caries. *Archives of Oral Biology, 41*, 167–173.

Arimatsu, K., Yamada, H., Miyazawa, H., et al. (2014). Oral pathobiont induces systemic inflammation and metabolic changes associated with alteration of gut microbiota. *Scientific Reports, 4*, 4828.

Armitage, G. C. (2004). Periodontal diagnoses and classification of periodontal diseases. *Periodontology 2000, 34*, 9–21.

Avila, W. M., Pordeus, I. A., Paiva, S. M., & Martins, C. C. (2015). Breast and bottle feeding as risk factors for dental caries: A systematic review and meta-analysis. *PLoS One, 10*, e0142922.

Bagramian, R. A., Garcia-Godoy, F., & Volpe, A. R. (2009). The global increase in dental caries. A pending public health crisis. *American Journal of Dentistry, 22*, 3–8.

Banas, J. A., & Vickerman, M. M. (2003). Glucan-binding proteins of the oral streptococci. *Critical Reviews in Oral Biology and Medicine, 14*, 89–99. PMID: 12764072.

Baumgartner, S., Imfeld, T., Schicht, O., Rath, C., Persson, R. E., & Persson, G. R. (2009). The impact of the stone age diet on gingival conditions in the absence of oral hygiene. *Journal of Periodontology, 80*, 759–768.

Becker, M. R., Paster, B. J., Leys, E. J., et al. (2002). Molecular analysis of bacterial species associated with childhood caries. *Journal of Clinical Microbiology, 40*, 1001–1009.

Beighton, D., Adamson, A., & Rugg-Gunn, A. (1996). Associations between dietary intake, dental caries experience and salivary bacterial levels in 12-year-oldEnglish schoolchildren. *Archives of Oral Biology, 41*(3), 271–280.

Belda-Ferre, P., Alcaraz, L. D., Cabrera-Rubio, R., Romero, H., Simón-Soro, A., Pignatelli, M., & Mira, A. (2012). The oral metagenome in health and disease. *ISME Journal, 6*(1), 46–56.

Belstrøm, D., Paster, B. J., Fiehn, N.-E., Bardow, A., & Holmstrup, P. (2016). Salivary bacterial fingerprints of established oral disease revealed by the Human Oral Microbe Identification using Next Generation Sequencing (HOMINGS) technique. *Journal of Oral Microbiology, 8*, 30170. https://doi.org/10.3402/jom.v8.30170.

Benítez-Páez, A., Belda-Ferre, P., Simón-Soro, A., & Mira, A. (2014). Microbiota diversity and gene expression dynamics in human oral biofilms. *BMC Genomics, 15*, 311.

Bik, E. M., Long, C. D., Armitage, G. C., et al. (2010). Bacterial diversity in the oral cavity of 10 healthy individuals. *ISME Journal, 4*, 962–974.

Bosshardt, D. D., & Lang, N. P. (2005). The junctional epithelium: From health to disease. *Journal of Dental Research, 84*, 9–20.

Bourgeois, D. M., & Llodra, J. C. (2014). Global burden of dental condition among children in nine countries participating in an international oral health promotion programme, 2012-2013. *International Dental Journal, 64*(Suppl 2), 27–34. https://doi.org/10.1111/idj.12129.

Burt, B. A., Eklund, S. A., Morgan, K. J., et al. (1998). The effects of sugars intake and frequency of ingestion on dental caries increment in a three-year longitudinal study. *Journal of Dental Research, 67*, 1422–1429.

Chapple, I. L., & Genco, R. (2013). Diabetes and periodontal diseases: Consensus report of the Joint EFP/AAP Workshop on Periodontitis and Systemic Diseases. *Journal of Clinical Periodontology, 40*(Suppl 14), S106–S112.

Chapple, I. L., Milward, M. R., Ling-Mount-ford, N., Weston, P., Carter, K., Askey, K., Dallal, G. E., De Spirt, S., Sies, H., Patel, D., & Matthews, J. B. (2012). Adjunctive daily supplementation with encapsulated fruit, vegetable and berry juice powder concentrates and clinical periodontal outcomes: A double-blind RCT. *Journal of Clinical Periodontology, 39*, 62–72.

Chapple, I. L., Bouchard, P., Cagetti, M. G., Campus, G., Carra, M. C., Cocco, F., et al. (2017). Interaction of lifestyle, behaviour or systemic diseases with dental caries and periodontal diseases: Consensus report of group 2 of the joint EFP/ORCA workshop on the boundaries between caries and periodontal diseases. *Journal of Clinical Periodontology, 44*(Suppl 18), S39–S51. https://doi.org/10.1111/jcpe.12685.

Costalonga M., Herzberg M.C. (2014). The oral microbiome and the immunobiology of periodontal disease and caries. Immunol Lett. 62(2 Pt A):22–38. https://doi.org/10.1016/j.imlet.2014.08.017.

Coventry, J., Griffiths, G., Scully, C., & Tonetti, M. (2000). ABC of oral health: Periodontal disease. *British Medical Journal, 321*(7252), 36–39.

Darveau, R. P. (2010). Periodontitis: A polymicrobial disruption of host homeostasis. *Nature Reviews. Microbiology, 8*(7), 481–490. https://doi.org/10.1038/nrmicro2337.

de Pablo, P., Chapple, I. L., Buckley, C. D., & Dietrich, T. (2009). Periodontitis in systemic rheumatic diseases. *Nature Reviews Rheumatology, 5*, 218–224.

Dewhirst, F. E., Chen, T., Izard, J., et al. (2010). The human oral microbiome. *Journal of Bacteriology, 192*, 5002–5017.

Dietrich, T., Sharma, P., Walter, C., Weston, P., & Beck, J. (2013). The epidemiological evidence behind the association between periodontitis and incident atherosclerotic cardiovascular disease. *Journal of Periodontology, 84*, S70–S84.

Dige, I., Grønkjær, L., & Nyvad, B. (2014). Molecular studies of the structural ecology of natural occlusal caries. *Caries Research, 48*, 451–460. https://doi.org/10.1159/000357920.

Eggert, F. M., Drewell, L., Bigelow, J. A., Speck, J. E., & Goldner, M. (1991). The pH of gingival crevices and periodontal pockets in children, teenagers and adults. *Archives of Oral Biology, 36*, 233–238.

Eskan, M. A., Jorwani, R., Abe, T., Chmelar, J., Lim, J. H., Lian, S., et al. (2012). The leukocyte integrin antagonist Del-1 inhibits IL-17-mediated inflammatory bone loss. *Nature Immunology, 13*, 465–473.

Forner, L., Larsen, T., Kilian, M., & Holmstrup, P. (2006). Incidence of bacteremia after chewing, tooth brushing and scaling in individuals with periodontal inflammation. *Journal of Clinical Periodontology, 33*, 401–407.

Frias-Lopez, J., & Duran-Pinedo, A. (2012). Effect of periodontal pathogens on the metatranscriptome of a healthy multispecies biofilm model. *Journal of Bacteriology, 194*, 2082–2095. https://doi.org/10.1128/JB.06328-11.

Genco, C. A., Cutler, C. W., Kapczynski, D., Maloney, K., & Arnold, R. R. (1991). A novel mouse model to study the virulence of and host response to *Porphyromonas* (*Bacteroides*) *gingivalis*. *Infection and Immunity, 59*, 1255–1263.

Goh, H. M. S., Yong, M. H. A., Chong, K. K. L., & Kline, K. A. (2017). Model systems for the study of Enterococcal colonization and infection. *Virulence, 8*, 1525–1562.

Golub, L. M., Lee, H. M., Greenwald, R. A., et al. (1997). A matrix metalloproteinase inhibitor reduces bone-type collagen degradation fragments and specific collagenases in gingival crevicular fluid during adult periodontitis. *Inflammation Research, 46*, 310–319.

Griffen, A. L., Beall, C. J., Campbell, J. H., Firestone, N. D., Kumar, P. S., Yang, Z. K., Podar, M., & Leys, E. J. (2012). Distinct and complex bacterial profiles in human periodontitis and health revealed by 16S pyrosequencing. *The ISME Journal, 6*, 1176–1185.

Gruner, D., Paris, S., & Schwendicke, F. (2016). Probiotics for managing caries and periodontitis: Systematic review and meta-analysis. *Journal of Dentistry, 48*, 16–25.

Guggenheim, B. (1968). Streptococci of dental plaques. *Caries Research, 2*(2), 147–163.

Hajishengallis, G. (2014). The inflammophilic character of the periodontitis-associated microbiota. *Molecular Oral Microbiology, 29*, 248–257. https://doi.org/10.1111/omi.12065.

Han, Y. W., & Wang, X. (2013). Mobile microbiome: Oral bacteria in extra-oral infections and inflammation. *Journal of Dental Research, 92*, 485–491.

Harris, R., Nicoll, A. D., Adair, P. M., & Pine, C. M. (2004). Risk factors for dental caries in young children: A systematic review of the literature. *Community Dental Health, 21*, 71–85.

Hernández, M., Gamonal, J., Tervahartiala, T., Mäntylä, P., Rivera, O., Dezerega, A., Dutzan, N., & Sorsa, T. (2010). Associations between matrix metalloproteinase-8 and -14 and myeloperoxidase in gingival crevicular fluid from subjects with progressive chronic periodontitis: A longitudinal study. *Journal of Periodontology, 81*, 1644–1652. https://doi.org/10.1902/jop.2010.100196.

Hezel, M. P., & Weitzberg, E. (2015). The oral microbiome and nitric oxide homoeostasis. *Oral Diseases, 21*(1), 7–16.

Hujoel, P. (2009). Dietary carbohydrates and dental-systemic diseases. *Journal of Dental Research, 88*, 490–502.

Ilie, O., van Loosdrecht, M. C., & Picioreanu, C. (2012). Mathematical modelling of tooth demineralisation and pH profiles in dental plaque. *Journal of Theoretical Biology, 309*, 159–175.

Jorth, P., Turner, K. H., Gumus, P., Nizam, N., Buduneli, N., & Whiteley, M. (2014). Metatranscriptomics of the human oral microbiome during health and disease. *MBio, 5*(2), e01012–e01014.

Kampfer, J., Göhring, T. N., Attin, T., & Zehnder, M. (2007). Leakage of food-borne Enterococcus faecalis through temporary fillings in a simulated oral environment. *International Endodontic Journal, 40*, 471–477.

Kanasi, E., Dewhirst, F. E., Chalmers, N. I., et al. (2010). Clonal analysis of the microbiota of severe early childhood caries. *Caries Research, 44*, 485–497.

Kianoush, N., Adler, C. J., Nguyen, K.-A. T., Browne, G. V., Simonian, M., & Hunter, N. (2014). Bacterial profile of dentine caries and the impact of pH on bacterial population diversity. *PLoS One, 9*, e92940. https://doi.org/10.1371/journal.pone.0092940.

Kilian, M., Chapple, I. L., Hannig, M., Marsh, P. D., Meuric, V., Pedersen, A. M., Tonetti, M. S., Wade, W. G., & Zaura, E. (2016). The oral microbiome—An update for oral healthcare professionals. *British Dental Journal, 221*(10), 657–666. https://doi.org/10.1038/sj.bdj.2016.865.

Kolenbrander, P. E. (2000). Oral microbial communities: Biofilms, interactions, and genetic systems. *Annual Review of Microbiology, 54*, 413–437. PMID: 11018133.

Kolenbrander, P. E., Palmer, R. J., Jr., Rickard, A. H., Jakubovics, N. S., Chalmers, N. I., & Diaz, P. I. (2006). Bacterial interactions and successions during plaque development. *Periodontology 2000, 42*, 47–79.

Kressirer, C. A., Chen, T., Harriman, K. L., Frias-Lopez, J., Dewhirst, F. E., Tavares, M. A., & Tanner, A. C. R. (2018). Functional profiles of coronal and dentin caries in children. *Journal of Oral Microbiology, 10*, 1495976. PMCID: PMC6052428.

Kumar, P. S., Griffen, A. L., Barton, J. A., Paster, B. J., Moeschberger, M. L., & Leys, E. J. (2003). New bacterial species associated with chronic periodontitis. *Journal of Dental Research, 82*, 338–344.

Lang, N. P., Hotz, P. R., Gusberti, F. A., & Joss, A. (1987). Longitudinal clinical and microbiological study on the relationship between infection with Streptococcus mutans and the development of caries in humans. *Oral Microbiology and Immunology, 2*, 39–47.

Li, J., Helmerhorst, E. J., Leone, C. W., Troxler, R. F., Yaskell, T., Haffajee, A. D., Socransky, S. S., & Oppenheim, F. G. (2004). Identification of early microbial colonizers in human dental biofilm. *Journal of Applied Microbiology, 97*(6), 1311–1318.

Li, Y., Ge, Y., Saxena, D., & Caufield, P. W. (2007). Genetic profiling of the oral microbiota associated with severe early-childhood caries. *Journal of Clinical Microbiology, 45*(1), 81–87.

Ling, Z., Kong, J., Jia, P., et al. (2010). Analysis of oral microbiota in children with dental caries by PCR-DGGE and barcoded pyrosequencing. *Microbial Ecology, 60*, 677–690.

Loe, H., Theilade, E., & Jensen, S. B. (1965). Experimental gingivitis in man. *Journal of Periodontology, 36*, 177–187.

Loesche, W. J. (1986). Role of Streptococcus mutans in human dental decay. *Microbiological Reviews, 50*(4), 353–380.

Loesche, W. J., & Straffon, L. H. (1979). Longitudinal investigation of the role of Streptococcus mutans in human fissure decay. *Infection and Immunity, 26*(2), 498–507.

López-López, A., Camelo-Castillo, A. J., Ferrer, M. D., Simón-Soro, A., & Mira, A. (2017). Health-associated niche inhabitants as oral probiotics: The case of Streptococcus dentisani. *Frontiers in Microbiology, 8*, 379.

Marquis, R. E. (1995). Antimicrobial actions of fluoride for oral bacteria. *Canadian Journal of Microbiology, 41*, 955–964. https://doi.org/10.1139/m95-133.

Marsh, P. D. (1994). Microbial ecology of dental plaque and its significance in health and disease. *Advances in Dental Research, 8*, 263–271.

Marsh, P. D. (2016). Dental biofilms in health and disease. In M. Goldberg (Ed.), *Understanding dental caries* (pp. 41–52). Berlin: Springer.

Marsh, P. D., & Devine, D. A. (2011). How is the development of dental biofilms influenced by the host? *Journal of Clinical Periodontology, 38*(Suppl 11), 28–35.

Marsh, P. D., Head, D. A., & Devine, D. A. (2015). Ecological approaches to oral biofilms: Control without killing. *Caries Research, 49*(Suppl 1), 46–54.

Martin-Cabezas, R., Davideau, J. L., Tenenbaum, H., & Huck, O. (2016). Clinical efficacy of probiotics as an adjunctive therapy to non-surgical periodontal treatment of chronic periodontitis: A systematic review and meta-analysis. *Journal of Clinical Periodontology, 43*, 520–530.

Merritt, J., & Qi, F. (2012). The mutacins of Streptococcus mutans: regulation and ecology. *Molecular Oral Microbiology, 27*, 57–69. PMCID: PMC3296966.

Munson, M. A., Pitt-Ford, T., Chong, B., Weightman, A., & Wade, W. G. (2002). Molecular and cultural analysis of the microflora associated with endodontic infections. *Journal of Dental Research, 81*, 761–766. Erratum in: Journal of Dental Research. 2003;82:247. Journal of Dental Research. 2003;82:69.

Murray, P. A., Prakobphol, A., Lee, T., Hoover, C. I., & Fisher, S. J. (1992). Adherence of oral streptococci to salivary glycoproteins. *Infection and Immunity, 60*, 31–38.

NCHS—National Center for Health Statistics. (2017). *Health, United States, 2016: With chartbook on long-term trends in health*. Hyattsville, MD.

Nyvad, B., & Kilian, M. (1990). Comparison of the initial streptococcal microflora on dental enamel in caries-active and in caries-inactive individuals. *Caries Research, 24*(4), 267–272.

Paes Leme, A. F., Koo, H., Bellato, C. M., Bedi, G., & Cury, J. A. (2006). The role of sucrose in cariogenic dental biofilm formation—New insight. *Journal of Dental Research, 85*, 878–887. PMCID: PMC2257872.

Palmer, C. A., Kent, R., Jr., Loo, C. Y., Hughes, C. V., Stutius, E., Pradhan, N., Dahlan, M., Kanasi, E., Arevalo Vasquez, S. S., & Tanner, A. C. (2010). Diet and caries-associated bacteria in severe early childhood caries. *Journal of Dental Research, 89*(11), 1224–1229. https://doi.org/10.1177/0022034510376543.

Palmer, S. R., Miller, J. H., Abranches, J., Zeng, L., Lefebure, T., Richards, V. P., Lemos, J. A., Stanhope, M. J., & Burne, R. A. (2013). Phenotypic heterogeneity of genomically-diverse isolates of Streptococcus mutans. *PLoS One, 8*, e61358. PMCID: PMC3628994.

Peterson, S. N., Snesrud, E., Liu, J., Ong, A. C., Kilian, M., Schork, N. J., & Bretz, W. (2013). The dental plaque microbiome in health and disease. *PLoS One, 8*(3), e58487. https://doi.org/10.1371/journal.pone.0058487.

Qin, J., Chai, G., Brewer, J. M., Lovelace, L. L., & Lebioda, L. (2006). Fluoride inhibition of enolase: crystal structure and thermodynamics. *Biochemistry, 45*, 793–800. https://doi.org/10.1021/bi051558s.

Ramberg, P., Sekino, S., Uzel, N. G., Socransky, S., & Lindhe, J. (2003). Bacterial colonization during de novo plaque formation. *Journal of Clinical Periodontology, 30*, 990–995.

Razavi, A., Gmür, R., Imfeld, T., & Zehnder, M. (2007). Recovery of Enterococcus faecalis from cheese in the oral cavity of healthy subjects. *Oral Microbiology and Immunology, 22*(4), 248–251.

Ribeiro, A. A., Azcarate-Peril, M. A., Cadenas, M. B., Butz, N., Paster, B. J., Chen, T., et al. (2017). The oral bacterial microbiome of occlusal surfaces in children and its association with diet and caries. *PLoS One, 12*(7), e0180621.

Rosier, B. T., Marsh, P. D., & Mira, A. (2017). Resilience of the oral microbiota in health: Mechanisms that prevent dysbiosis. *Journal of Dental Research, 97*(4), 371–380. https://doi.org/10.1177/0022034517742139.

Scannapieco, F. A., & Binkley, C. J. (2012). Modest reduction in risk for ventilator-associated pneumonia in critically ill patients receiving mechanical ventilation following topical oral chlorhexidine. *The Journal of Evidence-Based Dental Practice, 12*, 103–106. https://doi.org/10.1016/j.jebdp.2012.03.010.

Shoemark, D. K., & Allen, S. J. (2015). The microbiome and disease: Reviewing the links between the oral microbiome, aging and Alzheimer's disease. *Journal of Alzheimer's Disease, 43*, 725–738.

Sidi, A. D., & Ashley, F. P. (1984). Influence of frequent sugar intakes on experimental gingivitis. *Journal of Periodontology, 55*, 419–423.

Simón-Soro, A., Belda-Ferre, P., Cabrera-Rubio, R., Alcaraz, L. D., & Mira, A. (2013a). A tissue-dependent hypothesis of dental caries. *Caries Research, 47*, 591–600.

Simón-Soro, A., Tomás, I., Cabrera-Rubio, R., Catalan, M. D., Nyvad, B., & Mira, A. (2013b). Microbial geography of the oral cavity. *Journal of Dental Research, 92*, 616–621. https://doi.org/10.1177/0022034513488119.

Simón-Soro, A., Guillen-Navarro, M., & Mira, A. (2014). Metatranscriptomics reveals overall active bacterial composition in caries lesions. *Journal of Oral Microbiology, 6*, 25443. https://doi.org/10.3402/jom.v6.25443.

Smith, G. L. F., Cross, D. L., & Wray, D. (1995). Comparison of periodontal disease in HIV seropositive subjects and controls (1). Clinical features. *Journal of Clinical Periodontology, 22*, 558–568.

Socransky, S. S. (1977). Microbiology of periodontal disease—Present status and future considerations. *Journal of Periodontology, 48*(9), 497–504.

Socransky, S. S., Haffajee, A. D., Cugini, M. A., Smith, C., & Kent, R. L., Jr. (1998). Microbial complexes in subgingival plaque. *Journal of Clinical Periodontology, 25*(2), 134–144.

Takeshita, T., Nakano, Y., Kumagai, T., Yasui, M., Kamio, N., Shibata, Y., Shiota, S., & Yamashita, Y. (2009). The ecological proportion of indigenous bacterial populations in saliva is correlated with oral health status. *The ISME Journal, 3*, 65–78.

Takeshita, T., Matsuo, K., Furuta, M., Shibata, Y., Fukami, K., Shimazaki, Y., Akifusa, S., Han, D.-H., Kim, H.-D., Yokoyama, T., Ninomiya, T., Kiyohara, Y., & Yamashita, Y. (2014). Distinct composition of the oral indigenous microbiota in South Korean and Japanese adults. *Scientific Reports, 4*, 6990. https://doi.org/10.1038/srep06990.

Tuncil, Y. E., Xiao, Y., Porter, N. T., Reuhs, B. L., Martens, E. C., & Hamaker, B. R. (2017). Reciprocal prioritization to dietary glycans by gut bacteria in a competitive environment promotes stable coexistence. *MBio, 8*(5), pii: e01068-17. https://doi.org/10.1128/mBio.01068-17.

Vartoukian, S. R., Palmer, R. M., & Wade, W. G. (2009). Diversity and morphology of members of the phylum "synergistetes" in periodontal health and disease. *Applied and Environmental Microbiology, 75*, 3777–3786. https://doi.org/10.1128/AEM.02763-08.

Wade, W. G. (2013). The oral microbiome in health and disease. *Pharmacological Research, 69*, 137–143. https://doi.org/10.1016/j.phrs.2012.11.006.

Wang, J., Qi, J., Zhao, H., He, S., Zhang, Y., Wei, S., & Zhao, F. (2013). Metagenomic sequencing reveals microbiota and its functional potential associated with periodontal disease. *Scientific Reports, 3*, 1843.

Xiao, J., Hara, A. T., Kim, D., Zero, D. T., Koo, H., & Hwang, G. (2017). Biofilm three-dimensional architecture influences in situ pH distribution pattern on the human enamel surface. *International Journal of Oral Science, 9*(2), 74–79.

Xu, H., Hao, W., Zhou, Q., Wang, W., Xia, Z., Liu, C., Chen, X., Qin, M., & Chen, F. (2014). Plaque bacterial microbiome diversity in children younger than 30 months with or without caries prior to eruption of second primary molars. *PLoS One, 9*(2), e89269. https://doi.org/10.1371/journal. pone.0089269. eCollection 2014.

Yasuda, K., Hsu, T., Gallini, C. A., et al. (2017). Fluoride depletes acidogenic taxa in oral but not gut microbial communities in mice. *mSystems, 2*(4), pii: e00047-17.

Zaura, E., & Mira, A. (2015). Editorial: The oral microbiome in an ecological perspective. *Frontiers in Cellular and Infection Microbiology, 5*, 39. https://doi.org/10.3389/fcimb.2015.00039.

Zijnge, V., van Leeuwen, M. B., Degener, J. E., Abbas, F., Thurnheer, T., Gmür, R., & Harmsen, H. J. (2010 Feb 24). Oral biofilm architecture on natural teeth. *PLoS One, 5*(2), e9321. https:// doi.org/10.1371/journal.pone.0009321.

Zong, G., Holtfreter, B., Scott, A. E., Volzke, H., Petersmann, A., Dietrich, T., Newson, R. S., & Kocher, T. (2016). Serum vitamin B12 is inversely associated with periodontal progression and risk of tooth loss: A prospective cohort study. *Journal of Clinical Periodontology, 43*, 2–9.

Microbial Manipulation of Dysbiosis: Prebiotics and Probiotics for the Treatment of Oral Diseases

Eduardo Montero, Margarita Iniesta, Silvia Roldán, Mariano Sanz, and David Herrera

Abstract Prebiotics and probiotics may have a role in the prevention and treatment of relevant oral diseases and conditions, including dental caries, periodontal and peri-implant diseases and halitosis. Prebiotics and probiotics may be associated with higher numbers or activity of "beneficial" bacterial species, thus controlling microbiological deleterious shifts, although the precise mechanisms of action have not been completely elucidated. For dental caries, most of the studies have not reported effects on caries incidence as a true endpoint. A beneficial effect has been observed in children with high caries-risk, while in low caries-risk children this positive effect is doubtful. In dental biofilm-induced gingivitis, most studies failed to find relevant clinical effects. In periodontitis therapy, the adjunctive use of probiotics has demonstrated relevant benefits in some studies, for with limited consistency. In peri-implant diseases, the available information is very limited, thus definitive conclusion cannot be made. In halitosis treatment, the limitations of the currently available therapies may lead to an important role of prebiotics/probiotics in the control of halitosis. Research is showing some promising results, but studies are needed to confirm this hypothesis.

Keywords Oral probiotics · Oral prebiotics · Halitosis · Caries prevention · Probiotics for gingivitis and periodontitis · Prebiotics for gingivitis and periodontitis · Probiotics for peri-implant diseases · Prebiotics for peri-implant diseases

---

E. Montero · M. Iniesta · S. Roldán · M. Sanz · D. Herrera (✉)
ETEP (Etiology and Therapy of Periodontal Diseases) Research Group, University Complutense, Madrid, Spain
e-mail: davidher@ucm.es

© Springer Nature Switzerland AG 2019
M. A. Azcarate-Peril et al. (eds.), *How Fermented Foods Feed a Healthy Gut Microbiota*, https://doi.org/10.1007/978-3-030-28737-5_9

Introduction

Prebiotics

Concept

Prebiotics are non-digestible food ingredients that improve the health of the host through their physiologic effects of increasing the numbers and/or the activity of beneficial microorganisms (Gibson and Roberfroid 1995). Dietary fibers such as inulin-type, fructans and galacto-oligosaccharides have been considered prebiotics since 1995, as they fulfilled the three criteria used to define a compound as a prebiotic: "(1) resistance to gastric acidity, hydrolysis by mammalian enzymes and gastrointestinal absorption, (2) fermentation by intestinal microbiota and (3) selective stimulation of the growth and/or activity of intestinal bacteria associated with health and wellbeing" (Gibson et al. 2004). Other substances that are not absorbed from the small intestine, such as polyols (or sugar alcohols), or even certain amino acids (such as arginine) are also commonly described as prebiotics.

However, since a cause and effect relationship has not been demonstrated between the consumption of the food and a beneficial physiological effect through the increase in the numbers of gastro-intestinal microbiota, the European Food Safety Authority (EFSA) does not consider individual ingredients as prebiotics, but only as dietary fiber without mentioning the health benefits (Delcour et al. 2016).

Mechanisms of Action

Non-digestible carbohydrates modulate immune functions in different ways. Inulin (IN) and oligofructose (OF), as well as their intestinal fermentation products have been evaluated in clinical trials in subjects with ulcerative colitis and Crohn's disease. Prebiotic (IN and OF) treatment, combined or not with probiotic supplementation (*Bifidobacterium longum*) resulted in a clinical improvement of the inflammatory condition, as well as a significant reduction in the level of pro-inflammatory cytokines such as interleukin (IL)-1β and tumor necrosis factor-α (Furrie et al. 2005). Other randomized controlled trials have suggested that these prebiotics were able to improve postnatal immune development since the incidence of atopic dermatitis in high-risk infants was reduced (Moro et al. 2006).

By reaching the large bowel without being digested, prebiotics increase the number and activity of lactic acid bacteria and produce short-chain fatty acids (SCFA). This shift in the intestinal microbiota towards bifidobacteria and other SCFA-producing bacteria may change the pathogen-associated molecular patterns in the intestinal lumen, including endotoxins or lipopolysaccharides (Seifert and Watzl 2007). SCFA are capable of modifying the gastrointestinal pH, favoring mineral absorption and transport (including calcium) with a reported positive effect in adolescents and older women on the development and maintenance of body mass, preventing osteoporosis in the later (Coxam 2007).

Another possible mechanism of action derives from the interaction of prebiotic carbohydrates and carbohydrate receptors on immune cells (B and T lymphocytes, as well as Natural-Killers cells), which are involved in the recognition of a wide range of pathogens, including fungi and bacteria, and therefore boosting the immune system.

Probiotics

Concept

The term "probiotic" was defined by The Food and Agriculture Organization of the United Nations (FAO) as "Live microorganisms which when administered in adequate amounts confer a health benefit on the host" (FAO/WHO 2001). This definition highlights the fact that probiotic microorganisms must be alive to exert positive health effects. However, viability or survivability may not be indispensable qualities of health-promoting microorganisms, since dead cells and bacterial cell components may also exert some physiological effects (Rachmilewitz et al. 2004).

The rationale of using probiotics in the treatment of oral diseases is similar to their use in gastrointestinal diseases. Administration of oral probiotic bacteria should enhance their adherence and colonization in the oral tissues including hard surfaces, thus preventing the establishment or overgrowth of pathogenic species. However, the available evidence suggests that probiotics may also exert their beneficial effect without colonizing, or only with a temporary colonization of the host (Iniesta et al. 2012), as probiotic bacteria tend to disappear from the oral cavity once their intake is stopped (Caglar et al. 2009).

The mechanisms of action and efficacy of an oral probiotic depend on their interactions with the oral microflora (directly by competitive mechanisms or indirectly by production of antimicrobial substances) and/or immunocompetent host cells.

Mechanisms of Action

Competitive Mechanisms

Probiotic bacteria can compete or antagonize directly or indirectly with pathogenic bacteria for adhesion (for niche colonization) or for nutrients within the biofilm.

Commensal *streptococcal* species, such as *Streptococcus sanguinis, Streptococcus salivarius, Streptococcus mitis* and *Streptococcus cristatus* can inhibit the periodontopathogen *Aggregatibacter actinomycetemcomitans* colonization of soft tissues (Sliepen et al. 2009; Teughels et al. 2007). Also *Lactobacillus brevis* can inhibit *Prevotella melaninogenica* biofilm formation (Vuotto et al. 2014), and similarly, *Weissella cibaria, Enterococcus faecium, Lactobacillus reuteri, Lactobacillus acidophilus, Lactobacillus casei* and *Lactobacillus salivarius* can inhibit the cariogenic species *Streptococcus mutans* biofilm formation (Kang et al.

2006a, 2011; Kumada et al. 2009; Wasfi et al. 2018). Moreover, *Lactobacillus gasseri* and *Lactobacillus rhamnosus* were able to disrupt mature biofilm formation (Tan et al. 2018).

Studies evaluating the action of bifidobacteria in oral health are scarce. But it has been shown that, *Bifidobacterium animalis*, *Bifidobacterium dentium* and *Bifidobacterium longum* can induce a significant reduction in the numbers of *Porphyromonas gingivalis* in *in vitro* biofilms (Jasberg et al. 2016). However, *B. animalis* and *B. longum* can also be potentially cariogenic, having shown to induce enamel demineralization when combined with *S. mutans* and *Streptococcus sobrinus* (Valdez et al. 2016).

The indirect competitive mechanisms can be explained by the alteration in the expression of different adhesion-genes, as well as by the interference with the composition of salivary proteins. Interestingly, *L. acidophilus*, *L. reuteri* and *L. casei* have been shown to down-regulate the expression of genes responsible for the synthesis of the extracellular glucose polymers (glucans) in *S. mutans*, which play an important role in the growth of dental biofilms (Salehi et al. 2014; Savabi et al. 2014; Tahmourespour et al. 2011; Wasfi et al. 2018). In fact, several *Lactobacillus* strains can down-regulate certain *S. mutans* genes related to biofilm formation, such as exopolysaccharides-producing genes and quorum sensing genes (Wasfi et al. 2018). In a similar way, *Lactobacillus paracasei*, *L. rhamnosus* and *L. fermentum* demonstrated the ability to down-regulate expression of *Candida albicans* biofilm-specific genes (Rossoni et al. 2018; Mailander-Sanchez et al. 2017; Ribeiro et al. 2017).

Some probiotic strains can even change the salivary pellicle protein composition by inhibiting an important adhesion protein, salivary agglutinin gp340, which also plays an important role for the *S. mutans* adhesion (Haukioja et al. 2008).

Production of Antimicrobial Substances

Probiotic bacteria can produce a wide range of compounds with antimicrobial activity such as lactic acid, hydrogen peroxide and bacteriocins (Table 1).

Lactic acid, a final product of carbohydrate metabolism, is produced by all *Lactobacillus* species. *In vitro* studies have shown that *L. paracasei*, *L. casei*, *L.*

Table 1 Different strains of oral probiotics and their bacteriocins

Oral strain	Bacteriocin name	References
Lactobacillus salivarius	Bacteriocin LS1	Busarcevic et al. (2008)
	Bacteriocin LS2	Busarcevic and Dalgalarrondo (2012)
Lactobacillus paracasei	Paracasin SD1	Wannun et al. (2014)
	56 kDa bacteriocin	Pangsomboon et al. (2009)
Streptococcus salivarius	Salivaricin A	Tagg (2004)
	Salivaricin A1	
	Salivaricin A2	
	Salivaricin B	
	Salivaricin E	Walker et al. (2016)

salivarius, Lactobacillus plantarum, L. rhamnosus, L. fermentum, L. brevis and *L. reuteri* elicit a inhibitory activities against *S. mutans, S. sobrinus, Streptococcus gordonii, Actinomyces viscosus, Fusobacterium nucleatum, Tannerella forsythia, P. gingivalis* and *A. actinomycetemcomitans* (Baca-Castanon et al. 2015; Samot and Badet 2013; Sookkhee et al. 2001; Teanpaisan et al. 2011; Wasfi et al. 2018).

Some lactic-acid producing bacteria such as *L. gasseri, L. paracasei, L. brevis, L. plantarum, L. acidophilus, L. salivarius, L. reuteri* and *W. cibaria* can also produce hydrogen peroxide, which has direct or indirect microbicidal activities on oral biofilms and may also elicit direct toxicity against a variety of microorganisms (Samot and Badet 2013; Wasfi et al. 2018; Jang et al. 2016).

Bacteriocins are ribosomally synthesized small peptides with antimicrobial properties, mainly directed against *Lactobacillus* species, although antagonisms against other microorganisms have also been reported. The activity of some of these bacteriocins is due to pore formation in the cytoplasmic membrane of target bacteria. Several bacteriocins have been isolated from probiotic bacteria from the gastrointestinal tract, such as the production of reuterin by *L. reuteri* (Kang et al. 2011; Talarico et al. 1988).

Modulation of Host Defenses

Probiotic bacteria can potentially play a significant role in the maintenance of oral health by modulating the immune system through anti-inflammatory mechanisms.

In vitro studies have shown that certain *Streptococcus* species such as *S. cristatus, S. salivarius, S. mitis* and *S. sanguinis* can attenuate the interleukin (IL)-8 response induced by periodontal pathogens such as *A. actinomycetemcomitans* and *F. nucleatum* in epithelial cells (Cosseau et al. 2008; Sliepen et al. 2009; Zhang et al. 2008). *S. cristatus* can also attenuate the production of the proinflammatory cytokines IL-1α, IL-6 and tumor necrosis factor (TNF)-α induced by *F. nucleatum* (Zhang and Rudney 2011). Precise regulatory systems are not clear yet, although there is evidence indicating that these bacteria can inhibit the nuclear factor κβ pathway (Cosseau et al. 2008; Zhang and Rudney 2011). Also, it has been shown that *L. acidophilus* was able to reduce the secretion of IL-1β, IL-6 and IL-8 induced by *P. gingivalis* (Zhao et al. 2012). In addition, *L. rhamnosus* can prevent *P. gingivalis*-induced inflammation, preventing the inhibition of CXCL8 expression by *P. gingivalis* (Mendi et al. 2016).

Interestingly, it was demonstrated that *L. salivarius* and *L. gasseri* were able to attenuate the expression of *CdtB* and *LxtA* genes in *A. actinomycetemcomitans* strains. Since these genes synthesize powerful virulence factors against target cells of the immune system or periodontal tissues, this may be a potential mechanism to reduce the release of bacterial toxins and hence the immune-inflammatory response against this antigenic load (Nissen et al. 2014).

There are several *in vivo* studies designed to assess the immunomodulatory effects of probiotic bacteria. The first group of studies focused on the protective roles of immunoglobulin A (IgA) and human neutrophil peptides 1-3 (HNP1-3) in dental caries. HNP1-3 are antimicrobial peptides that provide the first line of host defense against a broad spectrum of microorganisms. Their preventive role against dental caries has been suggested by the significantly higher salivary HNP1-3 levels in caries-free children in comparison to those experiencing caries (Tao et al. 2005). *E. faecium* can increase the total salivary IgA levels in pre-school children (Surono et al. 2011) and *L. paracasei* can raise the salivary HNP1-3 levels in children (Wattanarat et al. 2015).

The second group of studies have been carried out in adult subjects: *L. brevis* reduced matrix metalloproteinases (MMP), prostaglandin E2 and interferon-γ levels in saliva (Riccia et al. 2007); *L. casei* affected elastase activities and MMP-3 levels in gingival crevicular fluid (GCF) (Staab et al. 2009); and *L. reuteri* decreased TNF-α and IL-8 levels in GCF (Twetman et al. 2009). *L. brevis* did not show effects on salivary IgA levels (Riccia et al. 2007), while *L. reuteri* increased total salivary IgA levels (Ericson et al. 2013).

Treatment and Prevention of Dental Caries

Dental Caries

In 2010, the most prevalent condition in the world was untreated caries in permanent teeth, affecting 35% of the population, or 2.4 billion people. Untreated caries in deciduous/primary teeth was the tenth-most prevalent condition, affecting 9% of the population, or 621 million children (Kassebaum et al. 2015). There is, however, evidence of a significant decline in caries affecting children due to promotion of oral health in school children, increased use of fluoride, reduced sugar intake and improved oral hygiene habits and regular check-ups. This fact seems to indicate that the burden of untreated caries is shifting from children to adults. In fact, the total burden of untreated caries has not declined due to the increase in population growth and longevity and the significant decrease in the prevalence of total tooth loss throughout the world (Frencken et al. 2017; Jepsen et al. 2017; Global Burden of Disease Study Collaborators 2015).

Dental caries is a complex disease characterized by a multifactorial etiology where the interaction of cariogenic bacteria, fermentable carbohydrates and host-related factors such as saliva secretion rates/buffering capacity results in demineralization of the tooth structure. The evidence suggests that in contrast with classical infectious diseases in which a specific pathogen is required for the development of the disease, caries is associated with a dysbiotic shift towards an increase in the number or proportions of acidogenic species (See chapter "Meat and Meat Products"). This shift occurs when sugared food/drinks are ingested frequently, leading to an imbalance between demineralization/remineralization towards net mineral loss (Fejerskov et al. 2015; Marsh 2018).

The lesion can present different extent and severity (initial, moderate or extensive), affecting the crown (coronal caries) or the root (root caries), and may affect the primary and/or the permanent dentition. It can affect enamel, the outer covering of the crown; cementum, the outermost layer of the root; and dentine, the tissue beneath both enamel and cementum.

Pathogenesis, Prevention and Treatment

A characteristic feature of the oral cavity is that teeth offer non-shedding surfaces for the formation of biofilms. These multi-species microbial communities are highly structured not only through their physical co-aggregation, but also through the production of chemical compounds by cell-to-cell metabolic interactions that promote microbial growth and provides resistance from antimicrobial agents and host defenses (Kolenbrander 2011).

Molecular methods, mostly sequence analysis of 16S ribosomal RNA genes, have provided new knowledge on the composition of caries associated biofilms and their role in the etio-pathogenesis of the disease (Gross et al. 2012). In fact, dental biofilms are more complex than previously appreciated, with the identification of several hundred distinct new species (Human Microbiome Project Consortium 2012). *S. mutans*, *S. sobrinus*, and to a lesser extent, *Lactobacillus* spp., have been classically considered the main pathogenic agents of caries (Loesche 1986). These bacteria are acidogenic since they produce weak organic acids from their metabolism of fermentable carbohydrates. This acid production results in a lower pH, which when below a critical threshold, demineralizes the hard dental tissues.

Recent studies using molecular techniques (polymerase chain reaction (PCR) amplification of DNA) in samples harvested from carious lesions failed to detect these classical bacteria in a significant proportion of cavities, revealing high proportions of other bacteria like *Veillonella* spp., *Capnocytophaga* spp. or *Scardovia wiggsiae* (Simon-Soro et al. 2013; Tanner et al. 2011). These studies have demonstrated that multiple microorganisms acting in a consortium may be the main causative agents (Mira et al. 2017; Sanz et al. 2017).

Traditionally caries management has mainly focused on the restoration of the carious lesion, which despite its limitations, is still the favored method worldwide. However, it results in loss of tooth substance with a possible concomitant harmful effect due to the increased risk for secondary caries or pulp vitality alterations. Currently, the use of modern micro-restorative techniques using adhesive materials have reduced these limitations, mainly allowing the preservation of tooth structure.

To avoid the side effects of restorative therapy, in the past three decades there has been a clear tendency towards preventive approaches in the management of caries. These approaches are mainly based on enhancing dental tissue resistance by the use of fluoridated dentifrices or by community prevention programs (water fluoridation), by reducing the intake of carbohydrates through dietary interventions (reduction of sugar intake), and by dental professional interventions (pit and fissure sealants, dental prophylaxis) (Milgrom et al. 2012; Petersen and Lennon 2004). This holistic approach in caries prevention highlights the multifactorial nature of this disease and also justifies why antimicrobial or immunization strategies targeting a limited number of species have not been effective (Mira 2018).

Current microbiological studies have also shown that there is a symbiotic microflora that provides protection against caries and these studies have prompted the use of pre- and probiotics aimed to promote the establishment of these beneficial bacterial species as a new approach in caries prevention.

Prebiotic Therapy in the Management of Dental Caries

The use of prebiotics aims to selectively boost the growth of symbiotic bacteria. Since carbohydrates are key in the development of the microbial dysbiosis leading to caries, the use of sugar substitutes, especially sugar alcohols (polyols) have been proposed as caries prevention agents. In addition, polyols, as natural sugar substitutes, may provide health benefits by reducing other sugar associated diseases, such as diabetes, obesity and cardiovascular diseases (World Health Organization 2015). The current approach of using polyols for caries prevention has been their topical use such as chewing gum and lozenges with sorbitol, xylitol, or sorbitol/xylitol (Mickenautsch et al. 2007). The efficacy of xylitol, has shown to be moderate in a recent systematic review when used as self-applied caries preventive agent (Janakiram et al. 2017). Its mechanisms of action are related with decreased dental plaque formation, reduced adherence of streptococci, and decreased expression of bacterial genes involved in sucrose metabolism (de Cock et al. 2016). However, long-term use of xylitol may lead to xylitol-resistant *S. mutans*, although these strains seem to be less virulent (Soderling 2009).

Erythritol is also a polyol, although unlike others, is absorbed in the intestine, not systemically metabolized, and excreted unchanged in the urine. Its mechanisms of action seem to be similar to those of xylitol, although it seems that erythritol is more effective than xylitol or sorbitol at reducing dental plaque biomass, maybe due to its lower molecular weight that allows a higher diffusion speed and migration into dental plaque (Makinen et al. 2005). The only published clinical trial comparing candies with xylitol, sorbitol, or erythritol, also demonstrated a lower caries incidence in the erythritol group compared to sorbitol and xylitol through 3 years of follow-up (Honkala et al. 2014).

Arginine is an amino acid and a common natural dietary supplement. Since low pH is one of the driving factors in the development of carious lesions and arginine metabolism via a deiminase pathway produces ammonia, which neutralizes biofilm pH, this amino acid has been studied as an approach for caries prevention (Huang et al. 2016). Over the past 10 years it has been incorporated in toothpaste formulations with or without fluoride, without reported negative side effects. Kraivaphan et al. (2013) demonstrated that a toothpaste containing 1.5% arginine and 1450 ppm of fluoride was able to reduce caries in low and moderate-risk children, and to arrest and reverse carious lesions in children and adults when compared to a regular toothpaste containing 1450 ppm of fluoride alone (Kraivaphan et al. 2013). However, its long-term impact remains to be investigated and more clinical trials are needed.

Other compounds used as prebiotics in oral caries prevention are Bet-methyl-D-galactoside and *N*-acetyl-D-mannosamine, which have proven to stimulate exclusively the growth of beneficial bacteria and inhibit the pathogenic potential of harmful microorganisms (Slomka et al. 2017).

Probiotic Therapy in the Management of Dental Caries

In the last decade, probiotics have been explored as a potential preventive tool against dental caries. Most of the studies have focused on their microbiological impact, especially *S. mutans* reductions (Laleman et al. 2014; Cagetti et al. 2013), with just a few evaluating caries incidence (Näse et al. 2001; Stecksén-Blicks et al. 2009; Petersson et al. 2011; Hasslöf et al. 2013; Taipale et al. 2013; Stensson et al. 2014; Wattanarat et al. 2015). Since probiotics are regularly used for a relatively short period of time and the incidence of caries is a relatively slow process requiring long-term evaluation, the number of studies evaluating its efficacy is relatively low (Näse et al. 2001; Stecksén-Blicks et al. 2009; Petersson et al. 2011; Hasslöf et al. 2013; Taipale et al. 2013; Stensson et al. 2014; Wattanarat et al. 2015; Rodriguez et al. 2016) (Table 2).

As time is important for the demineralization process to occur, most clinical trials have focused on surrogate endpoints such as the capacity of probiotic supplementation to reduce *S. mutans* counts. An important shortcoming of these studies is that they have mainly used a chairside test with questionable validity making comparison of the results difficult. A meta-analysis in a recent systematic review has shown that compared with controls, significantly more patients using the probiotic resulted in lower *S. mutans* counts in saliva ($<10^5$ colony forming units/mL). However, the variety in the probiotics used and the different microbiological techniques used did not allow for a sub-analysis studying the efficacy of the specific probiotic strains (Laleman et al. 2014).

In 2001, Näse et al. were the first to examine the effect of a milk supplemented with a *Lactobacillus* strain, *L. rhamnosus* GG, on its caries-inhibiting ability *in vivo* (Näse et al. 2001). The milk was administered on weekdays in Finnish day-care centers for 7 months to 594 children, between 1 and 6 years old and those using the probiotic milk resulted in a significantly reduced risk of caries. Interestingly, this effect was particularly marked for 3–4-year-old children, reflecting the possibility of a window for the colonization for *L. rhamnosus* GG, although this fact was not studied in this investigation.

Another study in Sweden, reported the effect of milk supplemented with *L. rhamnosus* LB21 and fluoride on caries development in preschool children (Stecksén-Blicks et al. 2009). The study was a double-blinded, randomized clinical trial in which 248 children of 1–5 years of age were randomly assigned to two parallel groups: intake of 150 mL milk supplemented with *L. rhamnosus* LB21 (10^7 colony forming units/mL) and 2.5 mg fluoride per liter for lunch, compared to non-supplemented milk. The mean baseline DMF (decay, missing, filled surfaces) was

Table 2 Studies evaluating caries prevalence as endpoint

Reference	Outcome(s)	Subjects	Strain (concentration)	Delivery format/treatment duration	Groups	Results
Näse et al. (2001)	Caries increment	451 children (1.3–6.8 years)	*Lactobacillus rhamnosus* GG, ATCC 53103 ($5–10 \times 10^5$ CFU/mL)	Milk once daily/5 days a week for 7 months	A: Probiotic B: Placebo	Lower mutans streptococci counts at the end of the study and significantly reduced risk of caries in group A, particularly in the 3–4 year old
Stecksén-Blicks et al. (2009)	Caries increment (dmfs index)	248 children (1–4 years)	*Lactobacillus rhamnosus* LB21 (10^7 CFU/mL)	Milk once daily/21 months	A: Probiotic/ fluoride B: Placebo	Statistically significant difference in caries increment in group A
Petersson et al. (2011)	Root Caries Index (RCI) and Electric Resistance Measurements (ERM)	160 adults (58–84 years)	*Lactobacillus rhamnosus* LB21 (10^7 CFU/mL)	Milk once daily/15 months	A: Placebo B: Fluoride/ probiotic C: Probiotic D: Fluoride	Higher numbers of RCI reversals in groups B, C and D. Mean ECM values increased significantly in groups A, B and C
Hasslöf et al. (2013)	Total caries experience (dmfs + DMFS indexes)	118 children (0–9 years)	*Lactobacillus paracasei* LF19 (10^8 CFU/serving)	Cereals (at least 1 serving/ day) from 4 to 13 months of age	A: Probiotic B: Placebo	No statistically significant differences in caries experience between the two groups at any time point (2–3, 6 and 9 years of age)
Taipale et al. (2013)	Caries increment (ICDAS index)	106 children (4 years)	*Bifidobacterium animalis* subsp. *lactis* BB-12 (10^{10} CFU/mL)	Tablets in slow-release pacifier or spoon twice daily/22–23 months	A: Probiotic B: Xylitol C: Sorbitol	No differences in the occurrence of enamel caries

| Stensson et al. (2014) | Caries prevalence | 113 children (0–9 years) | *Lactobacillus reuteri* ATCC 55730 (10^8 CFU/five drops) | Five drops of oil daily containing living *L. reuteri* derived from breast milk (strain ATCC 55730) during the 4 weeks before the expected date of delivery and continuing until the child was born. From birth, the infant was given five drops orally daily throughout the first year of life (thus 365 days) | A: Probiotic B: Placebo | In group A, statistically significant more children were caries-free in the primary dentition ($n = 49$, 82%), compared with group B ($n = 31$, 58%), and the mean number of initial and manifest approximal caries lesions was statistically significant lower in group A (0.67 ± 1.61 vs 1.53 ± 2.64) |
| Wattanarat et al. (2015) Rodriguez et al. (2016) | Caries increment (ICDAS index) Caries increment | 60 children (13–15 years) 205 children (2–3 years) | *Lactobacillus paracasei* SD1 (7.5×10^8 CFU/g) *Lactobacillus rhamnosus* SP1 (10^7 CFU/mL) | Milk once daily/6 months Milk on weekdays/18 months | A: Probiotic B: Placebo A: Probiotic B: Placebo | The caries increment for the pit and fissure surface was significantly decreased in the group A compared with the group B The percentage of new individuals who developed caries lesions in the control group was significantly higher than that in the probiotic group. At the cavitated lesion level, the increment of new caries lesions was higher in the control group versus the probiotic group |

CFU Colony Forming Units, *RCI* Root Caries Index, *ERM* Electric Resistance Measurements, *DMFS* Decay, Missing, Filled Score, *ICDAS* International Caries Detection and Assessment System

0.5 in the intervention group and 0.6 in the control group; and after 21 months there was a statistically significant difference between the two groups (0.9 and 2.2 respectively). Moreover, children in the test group displayed 60% fewer days with antibiotic therapy and 50% fewer days with otitis media. However, the authors recognized that with this study design, it would be impossible to determine if the positive effect was due to the probiotic, the fluoride, or the combination of both. In 2011, Petersson et al. investigated the effect of a milk with the same strain (*L. rhamnosus* LB21) with or without fluoride particularly on root caries in the elderly. The combination of the probiotic and the fluoride improved the results compared with the use of these products independently (Petersson et al. 2011).

The long-term effect of probiotic administration for caries reduction, however is controversial. Some studies evaluating the caries-related effects several years after probiotic usage did not find differences between the probiotic and control groups (Taipale et al. 2013; Hasslöf et al. 2013). Similarly in another study in Finland aimed to assess the effect of the early administration of probiotics (*B. animalis* subsp. *lactis* BB-12 from the age of 1–2 months to the age of 2 years) and xylitol or sorbitol on the oral colonization of *S. mutans*, did not find differences in the occurrence of caries in 4-year-old children (Taipale et al. 2013). Similar results were reported by Hasslöf et al. (2013) who randomized babies from 4 to 13 months of age to a daily diet of cereals supplemented with *L. paracasei* F19 (LF19) or cereals without the probiotic. At 9-years neither the occurrence of dental caries on deciduous nor permanent teeth, nor the *S. mutans* nor lactobacilli counts had been affected by this intervention and there was no evidence of *L. paracasei* F19 colonization in the oral cavity (as evaluated by polymerase chain reaction, PCR) (Hasslöf et al. 2013).

These findings support the hypothesis that oral probiotics need to be administered continuously in order to exert its beneficial properties, since their effect may disappear after the cessation of its usage.

The only study describing positive long-term results of probiotic administration evaluated the use of daily oral supplementation with the probiotic *L. reuteri* strain ATCC 55730 in mothers during the last month of gestation and in children throughout the first year of life in Sweden (Stensson et al. 2014). This study was a placebo-controlled, multi-center trial involving 113 children. At 9 years of age, children underwent clinical and radiographic examination of the primary dentition, and carious lesions, plaque and gingivitis were recorded. There were statistically significant differences in the number of children that remained caries free with 49 (82%) children in the probiotic group and 31 (58%) in the placebo group. Less approximal caries lesions and fewer sites with gingivitis were observed in the probiotic group. There were no significant differences between the groups with respect to frequency of tooth brushing, plaque and dietary habits, but differences were observed in intake of fluoride supplements, which could have compromised the results. There were no intergroup differences with respect to *L. reuteri* or *S. mutans* counts. According to this study, it seems that daily supplementation with *L. reuteri* from birth throughout the first year of life (the so-called "open window effect") is associated with reduced caries prevalence in the primary dentition at 9 years of age.

These controversial results in the long-term effect of previous probiotic administration during early stages of life may be due to potential confounders that could influence the results (such as oral hygiene, dietary habits, exogenous/endogenous fluoride...) or the selection of the appropriate probiotic strains since *L. paracasei* F19 and *B. animalis* adhere poorly to saliva-coated surfaces and present low co-aggregation rates, preventing them from becoming part of the commensal flora (Haukioja et al. 2006; Lang et al. 2010).

In light of these problems, the use of probiotics in caries prevention has been focused in high caries-risk populations. A recent study performed in 60 school children in Thailand evaluated the daily consumption of 5 g of milk powder containing *L. paracasei* SD1 for 6 months. The significant elevations of HNP1-3 salivary levels at 6 months in the test group indicated a temporal enhancement of the host response; whereas, the significant reductions of *S. mutans* counts also demonstrated a direct probiotic effect as long as the probiotic milk was administered. Moreover, there was a significant impact on caries with a decrease (2.6-fold) in caries incidence for the pit and fissure lesions when compared with the control group (Wattanarat et al. 2015). A similar study in Chile, also focusing on the most vulnerable children with highest risk of dental caries, compared the effect of using a milk supplemented with *L. rhamnosus* SP1 (to attain a final concentration of 10^7 colony forming units/mL) versus standard milk in pre-school children (2–3-year-old). After 10 months of this intervention, the percentage of subjects who developed new caries lesions was significantly lower in the test group (9.7% versus 24.3%), and similar results were reported in regard to cavitated lesions (0.58 versus 1.08) (Rodriguez et al. 2016). The findings from these recently published investigations suggest that the use of milk containing probiotics may be useful in reducing the occurrence of new caries lesions in high caries-risk preschool children.

Recent studies have identified certain streptococci strains more prevalent in caries-free individuals, such as *Streptococcus A12* or *Streptococcus dentisani* (Huang et al. 2016; Lopez-Lopez et al. 2017). Both bacteria seem to inhibit the growth of *S. mutans* and raise the pH within the biofilm. Further studies are needed to evaluate whether these bacteria may be useful as probiotics.

Treatment and Prevention of Periodontal and Peri-implant Diseases

Classification, Pathogenesis and Treatment of Periodontal and Peri-implant Diseases

Periodontal diseases are chronic inflammatory diseases of bacterial etiology affecting the periodontal tissues (Herrera et al. 2018). Among these, plaque-induced gingivitis has been defined as gingival inflammation without clinical attachment loss (American Academy of Periodontology 2000b), and periodontitis as chronic inflammation of the gingival tissues extending into the underlying attachment apparatus, characterized by loss of periodontal attachment (American Academy of Periodontology 2000a).

The currently used 1999 classification of periodontal diseases (Armitage 1999) includes eight main groups: gingival diseases, three types of periodontitis (chronic, aggressive and manifestation of systemic diseases) and four additional periodontal conditions (necrotizing periodontal diseases, abscesses in the periodontium, periodontitis associated with endodontic lesions and developmental or acquired deformities and conditions). More recently, a "World Workshop on the Classification of Periodontal and Peri-implant Diseases and Conditions" took place in November 2017, and a new classification has been introduced in 2018 (Caton et al. 2018), with a new framework for classifying periodontitis, based on stages according to severity and complexity and grades according to the rate of disease progression and presence of demonstrated risk factors.

Bacterial species in the subgingival biofilm are the primary etiological factor in periodontitis with a critical shift in the oral microbiome from periodontal health to disease where the symbiotic flora is replaced with a dysbiotic one (Mira et al. 2017). This microbial imbalance promotes the dysregulation of immuno-inflammatory pathways in the subject's host response leading to persistent local inflammation in cases of gingivitis and chronic inflammation and destruction of the connective tissue attachment and bone in periodontitis (Cekici et al. 2014). In addition, various risk factors influence and modify this interplay (Genco and Borgnakke 2013), including non-modifiable risk factors/indicators (genetic profiles, gender, age and some systemic conditions, such as leukaemia and osteoporosis), and modifiable risk factors/indicators (lifestyle factors, such as smoking and alcohol; metabolic factors, such as obesity, metabolic syndrome and diabetes; dietary factors, such as dietary calcium and vitamin D deficiency; socioeconomic status and stress). Other local factors, such as the amounts of plaque and/or calculus, presence of furcation lesions, enamel pearls, root grooves and concavities, open contacts, malpositioned teeth, and overhanging and/or poorly contoured restorations may also increase the risk for periodontal diseases (Herrera et al. 2018).

There is clear evidence that the prevention and treatment of periodontal diseases should be based on dental biofilm control. In primary and secondary prevention, mechanical control of supragingival biofilm by tooth brushing and interdental cleaning, combined with the use of antiseptic agents for chemical biofilm control have been the main interventions demonstrating long term maintenance of healthy periodontal tissues (Serrano et al. 2015). In the treatment of periodontitis (Graziani et al. 2017), the complete removal of supra- and subgingival biofilms is the main objective, normally achieved by mechanical means, although the adjunctive use of antiseptics and antibiotics in some cases can be considered (Herrera et al. 2002, 2008; Matesanz-Perez et al. 2013). Depending upon the patient and local factors, surgical therapy may also be necessary (Graziani et al. 2017).

In addition to periodontal diseases, inflammation and the destruction of the tissues surrounding dental implants can also occur. These conditions are known as peri-implant diseases and include peri-implant mucositis (similar to gingivitis, characterized by inflammation of the peri-implant tissues without peri-implant bone loss), and peri-implantitis (similar to periodontitis, with both inflammation and peri-implant bone loss).

Prebiotic and Probiotic Therapy in Plaque-Induced Gingivitis

Gingivitis is one of the most common chronic inflammatory diseases affecting mankind with a reported prevalence in the western population of around 75% (Botero et al. 2015). It is characterized by redness, swelling, and frequent bleeding of the gingiva, and it is caused by the accumulation of bacteria in the dento-gingival environment. The inflammation, which is the most remarkable feature, is due to an increase in the amounts of plaque and a shift in the microbiota from gram-positive, facultative microorganisms to predominantly gram-negative, anaerobic organisms (Slots 1992).

Gingivitis is a reversible condition, and can be easily reverted with a combination of personal oral hygiene and plaque control measures in conjunction with professional removal of plaque, calculus and other local contributing factors (Lövdal et al. 1961). However, while it is possible under controlled conditions to remove most of the plaque using a variety of mechanical oral hygiene aids, many patients lack the motivation or skills to attain and maintain a reduced plaque environment compatible with periodontal health for significant periods of time (Sheiham and Netuveli 2002).

To overcome these limitations, antiseptic toothpastes and mouth rinses have been tested as adjuncts to oral hygiene measures by providing additional anti-plaque/anti-gingivitis activity when used daily (Serrano et al. 2015). However, these agents are not used for long periods of time since the advent of secondary effects, such as staining, and taste perturbation is frequent. Other alternative adjunctive treatments, such as the use of probiotics have been proposed.

Prebiotic Therapy in Dental Biofilm-Induced Gingivitis

In functional foods, the most frequently used prebiotics are polyols. In oral research the polyols that have been most investigated are galacto-polysaccharides. This group is used widely as a sweetener in the food industry because it produces low acidogenicity and therefore, it does not increase the levels of *S. mutans* (Makinen 2011).

Some clinical studies in healthy individuals have evaluated the anti-plaque effect of xylitol and have demonstrated a decrease in the levels of plaque assessed through the plaque index (Soderling et al. 1989; Steinberg et al. 1992; Marya et al. 2017). However, the relevant outcome for evaluating the efficacy of prebiotics should be its effect on gingival inflammation. Sharma et al. studied a chewing gum containing sorbitol, maltitol, xylitol and sodium bicarbonate together with once-daily tooth brushing for 60 s in 78 adults with preexisting gingivitis. The control group used breath mints as placebo with the same tooth brushing regimen. After 4 weeks of use they observed that the test group experienced a plaque reduction almost twice (17%) that of the control group (9%). For the gingival index, the test group decreased nearly 10% versus 2% in the control group (Sharma et al. 2001). Keukenmeester et al. tested two different types of chewing gum containing xylitol or maltitol and the effects were compared to the use of a gum base or no gum, five times a day for 3 weeks. This four-group randomized controlled study had a split mouth design, where the subjects did not brush the teeth in the lower jaw to develop experimental gingivitis, while maintaining normal oral hygiene in the upper jaw. The increase in bleeding on marginal probing in the non-brushed jaw was significantly lower in the xylitol and maltitol group compared to the gum base. However, where regular brushing was performed, no effect of chewing gum was observed (Keukenmeester et al. 2014).

Probiotic Therapy in Dental Biofilm-Induced Gingivitis

In 1954, for the first time, a beneficial effect of lactic acid bacteria on inflammatory infections of the oral mucosa was reported (Kragen 1954). Unfortunately, this research line was abandoned in favor of antibiotics. Since the beginning of the twenty-first century, the identification of a so-called "beneficial" oral bacteria and their use in the prevention and treatment of plaque-related periodontal inflammation has undergone a revival. Species like *Lactobacillus* (Stensson et al. 2014; Staab et al. 2009; Hallström et al. 2013; Lee et al. 2015; Iniesta et al. 2012; Montero et al. 2017), *Bifidobacterium* (Kuru et al. 2017), *Pediococcus* (Montero et al. 2017) and *Bacillus* (Alkaya et al. 2017) have been evaluated, although the number of published studies is still low (Table 3).

At least two studies have evaluated the impact of probiotics in the gingival health of children. Karuppaiah et al. (2013) reported a parallel, randomized clinical trial with 216 school children (aged 14–17 years). One week prior to collection of baseline data and probiotic administration, a professional prophylaxis was performed in both groups. Subjects in the test group were asked to eat a curd containing a probiotic in their daily diet for 30 days, while no intervention was programmed in the control group. After 4 weeks, a statistically significant reduction in plaque levels in the test group, when compared with the control group, was observed. However, no differences were detected in the gingival index and no improvement in gingival health could be observed. The methodological quality of the study was low, and the authors did not provide the probiotic strains included in the curd. This low quality in studies assessing probiotics has been highlighted by a systematic review, which evaluated the methodological quality of studies assessing probiotics for periodontal treatment, reporting a high risk of bias (Dhingra 2012).

Similarly, Stensson et al. evaluated the influence of daily probiotic supplementation to mothers during the last month of gestation and to children throughout the first year of life on caries lesions development. They observed significantly fewer sites with gingival inflammation in the probiotic group (oil drops with *L. reuteri*) when compared to the placebo, but without demonstrating differences in the plaque index, suggesting a positive effect of the probiotic on the immune system (Stensson et al. 2014).

Studies using the experimental gingivitis model have evaluated the effects of probiotics on periodontal tissues in young adults, assessing whether the probiotic usage would interfere with plaque formation and/or the development of inflammation (Staab et al. 2009; Hallström et al. 2013; Lee et al. 2015; Kuru et al. 2017). Staab et al. (2009)

Table 3 Studies of probiotic treatment in gingivitis

Reference	Study design	Subjects	Strain (concentration)	Delivery format/treatment duration	Groups	Results
Karuppaiah et al. (2013)	RCT	216 school children (14–17 years)	Not mentioned	Curd/4 weeks	A: Probiotic B: No placebo	No difference in GI between groups. Probiotic group: significant reduction in PI
Stensson et al. (2014)	RCT	113 children (8–9 years)	*Lactobacillus reuteri* ATCC 55730 (10^8 CFU/mL)	Oil drops (5 a day)/ Mother: from 4 weeks before the delivery until the birth of the child; Child: through the first year of life	A: Probiotic B: Placebo	No difference in PI between groups. Probiotic group: significant reduction in gingival bleeding
Staab et al. (2009)	RCT (experimental gingivitis)	50 subjects (22–26 years)	*Lactobacillus casei* Shirota (10^{10} CFU/mL)	Milk (65 mL a day)/8 weeks	A: Probiotic B: No placebo	No differences in PI and papillary bleeding between groups. Probiotic group: elastase activity and MMP-3 amount were significantly lower
Hallström et al. (2013)	RCT crossover (experimental gingivitis)	18 subjects	*Lactobacillus reuteri* DSM 17938 and ATCC PTA 5289 (2×10^8 CFU/mL)	Tablets (2 a day)/3 weeks	A: Probiotic B: Placebo	No differences in PI, GI and bleeding on probing between groups
Lee et al. (2015)	RCT (experimental gingivitis)	30 subjects (19–28 years)	*Lactobacillus brevis* CD2 (10^9 CFU/mL)	Tablets (3 a day)/2 weeks	A: Probiotic B: Placebo	No differences in PI and GI between groups. Probiotics group: significant reduction in bleeding on probing
Kuru et al. (2017)	RCT (experimental gingivitis)	51 subjects (16–26 years) with gingival index ≤ 1	*Bifidobacterium animalis* subsp. *lactis* DN-173010 ($\geq 10^8$ CFU/g)	Yogurt (110 g a day)/28 days	A: Probiotic B: Placebo	Probiotic group: means of PI and GI were significantly lower. Concentration and total amount of IL-1β were significantly lower

(continued)

Table 3 (continued)

Reference	Study design	Subjects	Strain (concentration)	Delivery format/ treatment duration	Groups	Results
Iniesta et al. (2012)	RCT crossover	40 subjects (20–24 years) with gingival index > 1	*Lactobacillus reuteri* DSM 17938 and ATCC PTA 5289 (2×10^8 CFU/mL)	Tablets (1 a day)/28 days	A: Probiotic B: Placebo	No differences in PI and GI between groups. Probiotic group: significant reductions of *Prevotella intermedia* in saliva and *Porphyromonas gingivalis* in subgingival samples
Montero et al. (2017)	RCT	59 subjects (18–44 years) with gingival index > 1.3	*Lactobacillus plantarum* CECT 7481, *Lactobacillus brevis* CECT 7480, and *Pediococcus acidilactici* CECT 8633 (1×10^3 CFU/ mL)	Tablets (2 a day)/6 weeks	A: Probiotic B: Placebo	No differences in PI and GI between groups. Probiotic group: a significantly higher reduction in the number of sites with baseline GI = 3; a statistically significant reduction in *Tannerella forsythia*
Alkaya et al. (2017)	RCT	40 subjects (18–31 years) with generalized gingivitis	*Bacillus subtilis, Bacillus megaterium* and *Bacillus pumilus* ($5–10^7$ CFU/mL)	Toothpaste (twice a day), mouthrinse (once a day) and a probiotic toothbrush cleaner/8 weeks	A: Probiotic B: Placebo	No differences in PI and GI between groups

RCT Randomized Clinical Trial, *GI* Gingival Index, *PI* Plaque Index, *CFU* Colony-Forming Units, *MMP* Metalloproteinase, *IL* Interleukin

evaluated the influence of daily intake of 65 mL of probiotic milk (containing *L. casei* strain *Shirota*, Yakult®, Homsha Co., Tokyo, Japan) in the development of gingivitis. The control group received no treatment. No differences in plaque or gingivitis were observed, but some biomarkers showed significantly better results in the test group (elastase and myeloperoxidase activity, MMP-3 amounts). The data suggested that this strain might have an immune-modulating effect although no clinically relevant effects were observed. Hallström et al. (2013), studied 18 women enrolled in a double-blinded randomized placebo-controlled crossover study that evaluated lozenges containing *L. reuteri*, compared to a placebo. No differences were detected in plaque, gingival or bleeding indices and the microbial composition did not differ between the groups. The only statistically significant difference was the larger increase in gingival crevicular fluid (GCF) volume in the control group. Lee et al. (2015) concluded that *L. brevis* may delay gingivitis development by down-regulating an inflammatory cascade. With 30 healthy young adults (mean age 21.6–22.1 years), after 2 weeks without performing any oral hygiene practice, no differences were observed in plaque or gingival indices, but significant differences between groups were measured for bleeding on probing (BOP) and for the increase in the production of nitric oxide. This may suggest an anti-inflammatory effect that probably did not translate into the gingival index because of the short duration of the study. Kuru et al. (2017) described the first study examining the influence of single-strain *Bifidobacterium* probiotics on gingival health. They studied 51 subjects in a blinded randomized placebo-controlled study and evaluated yogurt containing *B. animalis* subsp. *lactis*, compared to a placebo. After a period of 28 days of yogurt consumption, subjects refrained from any oral hygiene procedure for a 5-day period. Significantly better results for plaque, gingival indices and BOP were observed in the probiotic group. Besides, the probiotic group had a significant reduction in gingival crevicular fluid volume and in total amount and concentration of IL-1β.

Another study in young adults (aged 20–24) with gingivitis was designed as a randomized clinical trial, evaluating the use of tablets containing two strains of *L. reuteri* for 4 weeks (Iniesta et al. 2012). In this study, conducted by our research group, no significant differences were detected for plaque or gingival indices, but some differences in microbiological variables were found, including reductions in total anaerobic, *Lactobacillus* spp. and black-pigmented anaerobic bacteria counts in saliva, as well as *P. gingivalis* and *A. actinomycetemcomitans* counts in subgingival samples. The study also demonstrated that *L. reuteri* was able to colonize both saliva and the subgingival niche in some subjects.

Some studies have also evaluated combinations of probiotic strains. Our research group studied the use of another probiotic containing *L. plantarum*, *L. brevis* and *Pediococcus acidilactici* in gingivitis patients (Montero et al. 2017). No significant differences were reported for mean plaque or gingival indices at baseline or at the end of the follow-up period. However, when evaluating sites with high gingival scores at baseline (gingival index, GI = 3), the probiotic treatment resulted in a significantly higher reduction of these specific sites, to the point that no subjects in the probiotic group presented a mean GI > 1 at 6 weeks, while three subjects in the control group still presented that degree of gingival inflammation. Moreover, a significant reduction of *T. forsythia* was observed in the test group. We suggested that the use of mean GI as the main outcome measurement for assessing the efficacy of new agents for gingivitis management might not be appropriate, as the dilution effect of the most frequently reported event (GI ≤ 1) may mask the positive effect of the agent on sites with clear signs of inflammation (GI ≥ 2).

On the other hand, Alkaya et al. (2017) evaluated some strains of genus bacilli (*Bacillus subtilis, Bacillus megaterium* and *Bacillus pumulus*). Patients received a probiotic containing toothpaste, a probiotic mouth rinse and one box of a probiotic toothbrush cleaner. After brushing, the patients rinsed the toothbrush with water and then place it in a glass with the experimental toothbrush cleaner for 8 weeks. Although plaque and gingivitis indices were significantly reduced, no intergroup differences were found for plaque and gingival indices after the 8 week test period.

Prebiotic and Probiotic Therapy in the Treatment of Periodontitis

The use of prebiotics in the treatment of periodontitis has only been evaluated in preclinical studies (Slomka et al. 2018).

The use of probiotic therapy in the treatment of periodontitis has been evaluated through different randomized clinical trials (RCTs) as an adjunct to scaling and root planing (SRP) (Graziani et al. 2017). The most common study design has been parallel RCTs comparing SRP plus a probiotic versus SRP plus placebo (Ince et al. 2015; Laleman et al. 2015; Morales et al. 2016; Tekce et al. 2015; Teughels et al. 2013; Vicario et al. 2013). Other study designs have used split-mouth approaches (Vivekananda et al. 2010), either evaluating a probiotic together with systemic doxycycline (Shah et al. 2013) or azithromycin (Morales et al. 2018), or by applying the probiotic as a mouth rinse concomitantly with subgingival application (Penala et al. 2016). See Table 4.

Most of these studies have tested a marketed formula containing *L. reuteri* strains DSM17938 and ATCC PTA529 (Ince et al. 2015; Tekce et al. 2015; Teughels et al. 2013; Vivekananda et al. 2010). The results of these studies have been recently included in a systematic review with meta-analysis (Martin-Cabezas et al. 2016), showing a statistically significant short term (from 42 days to 3 months) clinical attachment level (CAL) gain [weighted mean difference (WMD) −0.42 mm] and BOP reduction (WMD −14.66%) for SRP plus probiotic treatment versus SRP alone. When stratifying for probing pocket depth (PPD), statistically significant reductions were observed, particularly for deep pockets (WMD −0.67 mm). This finding could have important clinical implications, as it might reduce the need for surgery in certain patients. Interestingly, Teughels et al. (2013) only found significant differences when considering the "need of surgery" as an outcome measure at the 3 months visit.

The other lactobacilli strain tested in patients with chronic periodontitis (*L. rhamnosus* SP1) have yielded controversial results (Morales et al. 2016, 2018). A different tested approach using indigenous bacteria (such as streptococci species) as probiotics instead of dietary lactobacilli was tested as they were perfectly adapted to the human oral ecology (Teughels et al. 2011). However, the only RCT published evaluating a streptococci containing probiotic formulation (*S. oralis* KJ3,

Table 4 Selected randomized clinical trials assessing probiotics as adjuncts to scaling and root planing (SRP)

Reference	Country	Duration	n	Condition	Design	Control	Test	Probiotic
Vivekananda et al. (2010)	India	42 days	30	ChP	Split mouth	SRP (split-mouth)	Probiotic/placebo	L. reuteri (Prodentis) 21–42 d, 2×
Shah et al. (2013)	India	2 months	30	AgP	Parallel	None	SRP plus probiotic/doxycycline/both	L. brevis (Inersan), 14 d, 2–4 w, 2×
Teughels et al. (2013)	Turkey	12 weeks	30	ChP	Parallel	SRP plus placebo	SRP plus probiotic	L. reuteri (Prodentis) 12 w, 2×
Vicario et al. (2013)	Spain	1 month	20	ChP	Parallel	SRP plus placebo	SRP plus probiotic	L. reuteri (Prodentis) 4 w, 1×
Ince et al. (2015)	Turkey	360 days	30	ChP	Parallel	SRP plus placebo	SRP plus probiotic	L. reuteri, 3 w, 2×
Laleman et al. (2015)	Turkey	12 weeks	48	ChP	Parallel	SRP plus placebo	SRP plus probiotic	S. oralis KJ3, S. uberis KJ2, S. rattus JH145; 12 w, 2×
Tekce et al. (2015)	Turkey	360 days	40	ChP	Parallel	SRP plus placebo	SRP plus probiotic	L. reuteri, 3 w, 2×
Morales et al. (2016)	Chile	1 year	28	ChP	Parallel	SRP plus placebo	SRP plus probiotic	L. rhamnosus SP1, 3 m, 1×
Penala et al. (2016)	India	3 months	32	ChP and halitosis	Parallel	SRP plus placebo	SRP plus probiotic (subgingival and rinse)	L. salivarius and L. reuteri (Unique Biotech)
Morales et al. (2018)	Chile	9 months	47	ChP	Parallel	SRP plus placebo	SRP plus probiotic/azithromycin	L. rhamnosus SP1, 3 m, 1×

L. Lactobacillus; S. Streptococcus

ChP chronic periodontitis, *AgP* aggressive periodontitis, *SRP* scaling and root planing, *d* days, *w* weeks, *m* months, × times per day

Streptococcus uberis KJ2 and *Streptococcus rattus* JH145) as adjunct to SRP in periodontitis patients, failed to detect any clinical or microbiological effect after 12 weeks of administration (Laleman et al. 2015). In a recent systematic review (Matsubara et al. 2016) evaluating studies using probiotics in the treatment of aggressive or chronic periodontitis, with or without adjunctive debridement, 12 papers were identified, concluding that a significant improvement could be observed, although continuous exposure might be necessary.

The critical evaluation of the adjunctive efficacy of probiotics together with SRP must be cautious, since the evaluated RCTs have shown a high degree of heterogeneity with studies showing additional PPD reduction of more than 1 mm (Ince et al. 2015; Tekce et al. 2015) while in others there were no differences (Laleman et al. 2015; Teughels et al. 2013). There is a clear need to define the adequate probiotic formulation, the delivery system, the proper dosage and treatment duration, as well as the appropriate indications. Longer-term studies with larger sample sizes and in different geographical locations are definitively needed to evaluate this interesting approach, which might improve the results of non-surgical periodontal therapy thus reducing the need for surgery without the risks associated with the use of antibiotics.

Prebiotic and Probiotic Therapy in the Treatment of Peri-implant Diseases

With the recent advent of peri-implant diseases, there is a need for developing effective preventive and therapeutic approaches for their control (Schwarz et al. 2015). Noteworthy, the management of peri-implant mucositis is of prime importance as the key preventive treatment for peri-implantitis, a disease for which there are no yet clearly established guidelines for successful treatment.

The use of probiotic therapy in the treatment of peri-implant diseases is limited and only five studies have been identified, three treating patients with peri-implant mucositis, one in patients with peri-implantitis, and one including both (Table 5).

Flichy-Fernandez et al. (2015) were the first to report the effects of *L. reuteri* on the peri-implant health of edentulous patients with peri-implant mucositis (Flichy-Fernandez et al. 2015). The study was a controlled clinical trial (not randomized) with a crossover design and a 6-month wash-out period in edentulous subjects presenting with either peri-implant health or peri-implant mucositis. After probiotic treatment, both groups experienced an improvement in plaque index, probing pocket depth, and IL-1β and IL-8 concentrations. In contrast, these improvements were not observed after the placebo administration. However, no improvement in BOP, which was the main outcome variable was observed in the mucositis group after the probiotic administration. In a randomized clinical trial performed in Sweden, 49 patients with peri-implant mucositis received mechanical debridement and oral hygiene instructions together with a topical oil application (drops containing *L. reuteri* or placebo) followed by twice-daily intake of lozenges (containing *L. reuteri* or placebo) for 3 months (Hallstrom et al. 2016). At the end of the follow-up period

Table 5 Studies evaluating treatment of patients affected by peri-implant diseases with probiotic bacteria

Reference	Study design/follow-up	Disease	Subjects	Strain (concentration)	Delivery format/treatment	Treatment/groups	Results
Flichy-Fernandez et al. (2015)	CCT with a cross-over design/1 month follow-up	Peri-implant mucositis and No peri-implant disease	34 patients/77 implants	*Lactobacillus reuteri* Prodentis 200 million active units (strains ATCC PTA 5289-100 million and DSM 17938-100 million)	One tablet every 24 h during 30 days. All participants received PAPR 1 month prior to the baseline visit	A: No peri-implant disease (22 patients/54 implants) B: Peri-implant mucositis (12 patients/23 implants)	After probiotic treatment both patients with mucositis and peri-implant health improved clinical parameters and cytokine levels. No changes were observed after placebo treatment
Hallstrom et al. (2016)	RCT/12 weeks	Peri-implant mucositis	49 patients	*Lactobacillus reuteri* (strains DSM 17938 and ATCC PTA 5289) (10^8 CFU)	Topical oil application (after PAPR and OHI) plus twice daily intake of lozenges for 3 months	A: Probiotic B: Placebo	No significant differences were displayed between groups for the clinical, microbiological or inflammatory outcome variables
Mongardini et al. (2017)	RCT with a cross-over design/6 weeks follow-up	Experimental peri-implant mucositis	20 patients (39–78 years) contributing with one implant	*Lactobacillus plantarum* and *Lactobacillus brevis*	One tablet per day (before breakfast for 14 days) Following peri-implant mucositis induction, patient underwent PAPR and PDT	A: Placebo B: Probiotic	PAPR and PDT alone or associated with probiotics led to significant reduction in BOP + sites. The adjunctive use of probiotics did not improve the results

(continued)

Table 5 (continued)

Reference	Study design/follow-up	Disease	Subjects	Strain (concentration)	Delivery format/treatment	Treatment/groups	Results
Tada et al. (2018)	RCT, parallel design/6 months	Peri-implantitis	30 patients contributing with one implant	*Lactobacillus reuteri* (strains DSM 17938 and ATCC PTA 5289) (10^8 CFU)	One tablet per day for 6 months Previously, PAPR, OHI and azithromycin (1 each 24 h/3 days) were provided	A: Probiotic B: Placebo	Significant reduction in PPD and mBI in probiotic group versus the placebo. No significant impact on the microbiota
Galofre et al. (2018)	RCT, parallel design/6 months	Peri-implant mucositis and peri-implantitis	44 patients (22 with mucositis and 22 with peri-implantitis)	*Lactobacillus reuteri* (strains DSM 17938 and ATCC PTA 5289) (10^8 CFU)	One tablet, once daily at night after tooth brushing Previously PAPR was performed in mucositis implants and non-surgical mechanical therapy in the peri-implantitis implants	A: Probiotic B: Placebo	The mechanical therapy together with the probiotic administration produced an additional improvement in clinical parameters for both mucositis and peri-implantitis over mechanical therapy alone. A limited effect on peri-implant microbiota was observed

CCT Controlled Clinical Trial, *RCT* Randomized Clinical Trial, *PAPR* Professionally administered plaque removal, *CFU* Colony-Forming Units, *PDT* Photodynamic Therapy, *BOP* Bleeding on probing, *OHI* Oral Hygiene Instructions, *PPD* Probing Pocket Depth, *mBI* marginal Bleeding Index

(3 months), all clinical parameters improved in both groups, but there were no significant differences between the groups. Similarly, there were no significant modifications in the microbiota after the probiotic administration. In this study, the addition of the probiotic did not provide any added benefit to mechanical debridement in the treatment of peri-implant mucositis. Similar results were reported by Mongardini et al. (2017) after an experimental peri-implant mucositis trial with a randomized, crossover, placebo-controlled design (Mongardini et al. 2017). Briefly, after 14 days of undisturbed plaque accumulation due to protection with an acrylic stent during oral hygiene procedures (induction phase), patients received a professionally administered plaque removal (PAPR) together with photodynamic therapy (PDT), and, depending on their allocation, either the application in the peri-implant sulcus of a probiotic powder plus probiotic tablets containing *L. plantarum* and *L. brevis* once per day for 14 days, or saline administration in the sulcus plus placebo tablets. The results showed that the combination of PAPR + PDT either alone or in combination with probiotics significantly reduced the number of BOP-positive sites after 6 weeks; although the adjunctive use of probiotics did not improve the clinical results of PAPR + PDT.

A recent randomized clinical trial evaluated the effects of a probiotic tablet containing two different *L. reuteri* strains after the administration of azithromycin in patients with peri-implantitis (Tada et al. 2018). The selection criterion for subjects was to present at least an implant with mild to moderate peri-implantitis, defined as (1) PPD ≥4 mm and <7 mm, (2) bleeding or suppuration on probing, and (3) marginal bone loss >2 mm assessed from periapical X-rays. Before entering in the study, all subjects received oral hygiene instructions as well as supragingival scaling. At baseline a clinical examination and a microbiological assessment were conducted, followed by the prescription of azithromycin (500 mg/24 h/3 days). One week after, a new clinical examination and bacterial sampling was performed, and patients were randomly allocated to either the test (probiotic tablets for 6 months) or control groups (placebo tablets for 6 months). From a microbiological point of view, significant reductions in total counts and counts of specific periodontal pathogens (*F. nucleatum, P. gingivalis, Prevotella intermedia, A. actinomycetemcomitans, Treponema denticola* and *T. forsythia*) were observed after azithromycin administration; however, these numbers increased again at the 6 month follow-up, without any difference between groups. From a clinical standpoint, however, a significant reduction in PPD occurred in the probiotic group, while no changes were observed in the placebo. Lastly, a recently published investigation mixed cases of peri-implant mucositis and peri-implantitis (Galofre et al. 2018). In a randomized clinical trial, subjects with either mucositis or peri-implantitis were assigned to mechanical debridement plus a container with 30 probiotic lozenges (*L. reuteri*) or mechanical debridement plus a container with 30 placebo lozenges to be used once daily at night, just after dental brushing. In mucositis patients, the authors reported a significant reduction in BOP in the probiotic group compared with the control group. However, *L. reuteri* had a very limited impact on the peri-implant microbiota, as there were no significant reductions in the periodontal pathogens or the total bacterial load due to the probiotic administration.

In the subjects with peri-implantitis, again no effect upon the microflora was observed, although a significant reduction in PPD was observed in the probiotic group when compared to the placebo (0.55 ± 0.37 mm versus 0.20 ± 0.35 mm). Additionally, a higher percentage of subjects in the probiotic group experienced positive results in terms of absence of BOP (45.5% versus 0%).

These limited clinical studies have also resulted in heterogeneous results without clear indication on the most appropriate strains, dosage, delivery format, and treatment duration. These aspects should be addressed by future RCTs before considering providing treatment recommendations.

Halitosis

Oral Halitosis

Halitosis, also called oral malodor or bad breath, is a rather frequent condition among adults and it is defined as foul or offensive odors emanating from the oral cavity independently of its origin. Although several non-oral sites have been related to halitosis, it seems that almost 90% of all the cases have an intraoral cause (Delanghe et al. 1999; Quirynen et al. 2009). The most prevalent causes in adults are: the presence of tongue coating (51%), gingivitis/periodontitis (13%) or a combination of the two (22%) Extra-oral sources have also been identified in approximately 10% of the subjects including: ear–nose–throat (ENT) pathologies, systemic diseases, metabolic or hormonal changes, hepatic or renal insufficiency, bronchial and pulmonary diseases and gastroenterologic conditions (Quirynen et al. 2009).

Prevalence

There are few studies reporting the prevalence of halitosis in broad samples of population. In addition, the results of available studies are difficult to compare, due to the different methodologies and variables used to define a case of halitosis, varying from subjective variables, such as self-reported halitosis to objective measurements such as organoleptic scores or sulfide monitor/chromatograph's readings, and it is well known that subjective variables often cannot be correlated with objective findings.

Nevertheless, the results from different studies worldwide have shown a prevalence in adults ranging from 10 to 39.6% with some geographical and age differences (Miyazaki et al. 1995; Yokoyama et al. 2010; Yaegaki and Sanada 1992; ADA Council on Scientific Affairs 2003; Al-Ansari et al. 2006; Meningaud et al. 1999; de Wit 1966; Liu et al. 2006; Bornstein et al. 2009; Nadanovsky et al. 2007).

Regarding the prevalence in children, only two studies have addressed this subject. Nalçaci et al. reported a prevalence of 14.5% in a population of 628 healthy children aged 7–11 years living in Turkey. Apparently, age and prevalence/severity of dental caries were significantly related with higher oral malodor ratings, while gender, frequency of tooth brushing and regular mouthbreathing were not associated (Nalcaci et al. 2008). Villa et al. studied a group of 101 Italian children aged 6–16 and with a mean age of 11.7 years. Halitosis (VSCs > 100 parts per billion [ppb]) was objectively measured in 37.6% of the patients. Overall, female patients, individuals with dental plaque on more than 25% of the dental surfaces, or patients older than 13 years, were more prone to present halitosis. These results, therefore, suggest that halitosis in the pediatric population is related to poor oral hygiene and might be more common in females and older individuals (Villa et al. 2014).

Etio-pathogenesis

Halitosis occurrence is mainly due to the presence of volatile sulfur compounds (VSCs) in the air expelled through the oral cavity. These compounds are hydrogen sulfide (H_2S), methyl mercaptan (CH_3SH) and dimethyl sulfide ($[CH_3]_2S$) and are generated through the degradation of sulfur-containing protein substrates by gram-negative oral microorganisms, primarily residing on the tongue biofilm, subgingival/supragingival biofilms and other oral areas (Seemann et al. 2014). These substrates are amino acids containing sulfur, such as cysteine, cystine and methionine that can be found free in saliva and crevicular fluid or that are released through the proteolysis of protein substrates (Kleinberg and Westbay 1990).

The main sources of these substrates are desquamated epithelial cells coming from different locations of the oral cavity, pharyngeal mucus, leukocytes that diffuse in locations with certain degree of inflammation, blood cells and to a lesser extent, proteins present in the diet.

The production and release of these VSCs depend on a number of physicochemical local factors (pH, redox potential, concentration of oxygen…), which will modulate both its quality and its quantity. Although VSCs represent 90% of all the malodorous components that contribute to the appearance of bad breath, other components that may contribute to a lesser extent to bad breath have also been identified: these are products that do not contain sulfur, such as volatile aromatic compounds (indole and skatole), organic acids (acetic, propionic) and amines (cadaverine and putrescine) (Goldberg et al. 1994; Greenman et al. 2005; Porter and Scully 2006).

Salivary and Tongue Microbiota of Patients with Intraoral Halitosis

Bacteria play a key role in the production of halitosis. In the absence of bacteria the malodorous compounds are not produced. In different studies, both *in vitro* and *in vivo*, a predominance of gram-negative anaerobic bacteria, mainly asaccharolytic, have been identified (De Boever and Loesche 1995; McNamara et al. 1972), corresponding with species that are normally found in the subgingival niche of patients with periodontitis and that can also be isolated in the tongue biofilm and in the saliva

(Roldan et al. 2003, 2005). Although many bacteria present in subjects with halitosis have not been cultivated yet, several bacterial species capable of producing malodor, such as *Actinomyces* spp., *Veillonella* spp., *Prevotella* spp., *Porphyromonas* spp. and *Fusobacterium* spp. have been identified. While these species appear in greater numbers in patients with halitosis, they have also been isolated in patients without this condition (Goldberg et al. 1997; Kleinberg 1997; Niles 1997; Persson et al. 1990).

It is clear that certain bacteria are more capable of producing malodorous gases than others, and some hypotheses have been proposed about whether some specific bacterial species could be directly associated with intraoral halitosis. In the last decade a bacterial species that is predominantly isolated in patients with halitosis and not in control subjects has been reported, named *Solobacterium moorei* (Kazor et al. 2003; Haraszthy et al. 2007, 2008; Riggio et al. 2008; Vancauwenberghe et al. 2013). However, more studies are needed to clarify what is the role of this bacterial species in the etio-pathogenesis of halitosis.

Therapeutic Approaches to Control Halitosis

In adults, halitosis therapy aims to lower the total numbers of odor-producing bacteria, to reduce the amount of available protein substrates and to neutralize the volatilization of the generated malodorous compounds. Although, twice daily mechanical removal/disruption of the tongue and dental biofilms is a necessary step in the treatment of intraoral halitosis, it is usually not sufficient to control this condition in the long term. The adjunctive everyday use of chemical agents is also normally needed. According to several systematic reviews, the most efficacious chemical formulations combine an antimicrobial agent with zinc salts to precipitate VSCs (Seemann et al. 2014).

Prebiotic Treatment in the Control of Halitosis

Doran and Verran evaluated the efficacy of an inulin mouth rinse to reduce halitosis variables and the proportions of tongue bacteria associated with oral malodour. Thirteen panelists rinsed with either 10% sucrose or a 10% inulin mouth rinse twice daily for 21 days. A reduction in odor levels assessed by Halimeter® and organoleptically was observed immediately after rinsing. The effect was greater by the end of the 21-day regime. No significant differences were observed in total bacterial counts. Immediately after the rinse with either of the carbohydrates, but not the water control, tongue pH levels dropped, returning to baseline scores after 30 min. Thus, there is some indication that the use of inulin mouth rinse can reduce oral malodor by encouraging the growth of acidogenic bacteria and inhibiting acid sensitive asaccharolytic anaerobes. However, the potential cariogenic effect of inulin should be elucidated before its clinical use can be recommended in halitosis patients (Doran and Verran 2007).

Probiotic Treatment in the Control of Halitosis

Halitosis can become, in some cases, a chronic problem and patients have to use specific mouth rinses routinely for long periods of time. Since these products are not always free of side effects, such as staining, and it is clear that re-colonization of halitosis-causing bacteria will occur once the treatment is stopped, and therefore, new therapeutic strategies are needed. Recently, a new interest has emerged in relation to the potential use of probiotics in the control of oral halitosis. In order to prevent the regrowth of odor-causing organisms, pre-emptive colonization of the oral cavity with probiotics might have a potential application as adjunct for both the treatment and prevention of halitosis.

Ideally probiotic therapy for halitosis should be aimed to:

- Elimination or reduction of odor-producing bacteria
- Re-colonization of oral biofilms by bacteria unable to produce malodorous compounds and capable of competing directly with odor-producing bacteria (through the generation of bacteriocins or through alteration of their virulence factors),
- Reduction in the production of VSC and organoleptic values.

Most of the investigations have focused on the bacterial strains *Streptococcus salivarius* K12 (Burton et al. 2005, 2006a; Horz et al. 2007) and *L. salivarius* WB21 (Shimauchi et al. 2008; Mayanagi et al. 2009; Iwamoto et al. 2010; Suzuki et al. 2012, 2014). Other strains that have been studied include *W. cibaria* (Kang et al. 2006b), *L. reuteri* (Keller et al. 2012), *Streptococcus thermophilus* strains HY2, HY3 and HY9012 (Lee and Baek 2014), L. brevis (CD2) (Marchetti et al. 2015), *E. faecium* WB2000 (Suzuki et al. 2016) or a combination of probiotics (*L. salivarius and L. reuteri*) (Penala et al. 2016).

Streptococcus salivarius **K12**

S. salivarius is an early colonizer of the human oral cavity and remains a prominent member of the oropharyngeal tract of healthy humans (Kazor et al. 2003). Some commensal strains of *S. salivarius* have also been used as probiotics in the treatment and prevention of upper respiratory tract infections, since they produce a particularly diverse range of lantibiotic bacteriocins with a broad spectrum against several streptococcal pathogens (Barretto et al. 2012). According to Kazor et al., *S. salivarius* K12, is considered a commensal bacterium of the oral cavity, is an early and dominant colonizer of the tongue microbiota in healthy individuals (Kazor et al. 2003), and it has only a limited capacity to produce VSCs. Besides, it produces two lantibiotic bacteriocins, which have the capability to inhibit or reduce the number of bacteria that produce VSCs (Burton ct al. 2006b; Masdea et al. 2012).

Masdea et al. tested the antimicrobial activity of *S. salivarius* K12 in vitro on different bacteria strains involved in oral malodor (*Solobacterium moorei* CCUG39336 and four clinical *S. moorei* isolates, *Atopobium parvulum* ATCC33793 and *Eubacterium sulci* ATCC35585). The results demonstrated that *S. salivarius* K12 suppressed the growth of all gram-positive bacteria tested, but the extent to which the bacteria were inhibited varied. *E. sulci* ATCC35585 was the most sensitive strain, while all five *S. moorei* isolates were inhibited to a lesser extent (Masdea et al. 2012). Similarly, Moon et al. tested the antimicrobial activity of *S. salivarius* K12 against *P. intermedia in vitro*, and its corresponding effect on VSC levels, and concluded that in concentrations above a certain level (70%), *S. salivarius* K12 showed antibacterial activity against *P. intermedia* and reduced the amount of VSCs produced by *P. intermedia* (Moon et al. 2016).

The use of gum or lozenges containing *S. salivarius* K12 reduced the levels of VSCs among patients diagnosed with halitosis (Burton et al. 2006a). Horz et al. found that *S. salivarius* K12 could be detected at the mucosal membranes for as long as 3 weeks after the use of four lozenges containing *S. salivarius* K12 per day over 3 days (Horz et al. 2007). However, additional studies with larger patient cohorts are needed to confirm the long-term potential of probiotics in preventing and/or treating halitosis.

Lactobacillus salivarius WB21

Lactobacillus species have been used frequently as probiotic bacteria in the oral cavity. The oral consumption of a tablet containing *L. salivarius* WB21 was reported to improve periodontal conditions in healthy smoking volunteers and to reduce the number of the periodontopathic bacterium *T. forsythia* in subgingival plaque (Shimauchi et al. 2008; Mayanagi et al. 2009).

Iwamoto et al. performed an open label pilot study in 20 adult patients with genuine halitosis (nine subjects without periodontitis and 11 subjects with periodontitis), to evaluate the effects of daily administration of three tablets containing 6.7×10^8 CFUs of *L. salivarius* WB21 and 280 mg of xylitol (Wakamoto Pharmaceutical, Tokyo, Japan) on oral malodor and clinical parameters after 2 and 4 weeks. Results showed that all patients were positive for *L. salivarius* DNA in their saliva at 2 weeks. Oral malodor parameters significantly decreased in the subjects without periodontitis at 2 weeks. The organoleptic scores and bleeding on probing significantly decreased in the subjects with periodontitis at 4 weeks (Iwamoto et al. 2010). Suzuki et al., conducted a double-blind, randomized, controlled clinical trial (RCT) using oil drops containing *L. salivarius* WB21 in patients with periodontitis to evaluate their effects on periodontal health and VSC-producing bacteria and found improved BOP compared with the placebo group at 2 weeks, but failed to find a significant reduction of VSC-producing bacteria in the test group compared to placebo (Suzuki et al. 2012). Later, the same group conducted a 14-day, double-blind, placebo-controlled, randomized crossover trial of tablets containing *L. salivarius* WB21 (6.7×10^8 CFU and 280 mg of xylitol) or placebo taken orally by patients with oral malodor. Results showed that organoleptic test scores significantly decreased in both the probiotic and placebo groups compared with baseline scores, but no difference was detected between groups. In contrast, the concentration of VSCs and the average PPD decreased significantly in the probiotic group compared with the placebo. Bacterial quantitative analysis found significantly lower levels of ubiquitous bacteria and *F. nucleatum* in the probiotic group (Suzuki et al. 2014).

Weissella cibaria

Kang et al. isolated *W. cibaria* in healthy children aged between 4 and 7 years, and it was applied in a rinse on the oral cavity of 46 healthy young individuals. Results demonstrated an inhibitory effect on the production of VSCs produced by *F. nucleatum* both *in vitro* and *in vivo*. In young adults, a marked significant reduction in the levels of H_2S and CH_3SH, by approximately 48.2% and 59.4%, respectively, was registered after gargling with *W. cibaria* containing rinse. The possible mechanism in the VSC reduction is the hydrogen peroxide generated by *W. cibaria* that inhibits the proliferation of *F. nucleatum* (Kang et al. 2006b).

Lactobacillus reuteri

Keller et al. studied 28 healthy young adults with self-reported morning breath. The subjects were instructed to chew one gum, twice per day, containing either two strains of probiotic lactobacilli (*L. reuteri* DSM 17938 and *L. reuteri* ATCC PTA 5289 both at the concentration of 1×10^8 CFU) (BioGaia AB; Lund, Sweden) or a placebo. After 14 days of treatment, the organoleptic scores were significantly lower in the probiotic compared with the placebo group. However, assessments of the VSC levels displayed no significant differences between the groups. The authors concluded that the tested probiotic chewing gum may have a slight beneficial effect on oral malodor assessed by organoleptic scores (Keller et al. 2012).

Although the existing research has provided positive results on halitosis parameters, further well-designed studies, including larger populations and long-term results both in safety and effectiveness are needed before any evidence-based conclusions can be entitled and they can be applied in therapeutic protocols for halitosis patients.

Conclusions

Prebiotics and/or probiotics may have a role in the prevention and treatment of relevant oral conditions, such as dental caries, periodontal diseases, peri-implant diseases and halitosis. Both prebiotics and probiotics may be associated with an increase in the number or activity of "beneficial" bacterial species, thus controlling microbiological deleterious shifts.

The current knowledge of the etio-pathogenesis of dental caries suggests that, as a complex disease, the microbial composition alone may be insufficient to predict caries risk, and that treatment strategies aimed to eradicate single species are unlikely to be effective. However, most of the studies evaluating the role of probiotics in caries prevention are focused on surrogate outcome variables, such as the levels of caries-associated bacteria (mostly *S. mutans*), and do not report effects on caries incidence as a true endpoint. Although direct clinical evidence is scarce, it seems that probiotic interventions may be beneficial in children with high caries-risk, while in low caries-risk children this positive effect is doubtful (Wattanarat et al. 2015). It is crucial to define which probiotic strains are the most suitable to colonize the oral cavity, as most long-term studies indicate that dairy strains need to be administered continuously to provide beneficial effects. Further studies are needed to evaluate the inclusion of probiotic strains in products with daily usage, such as toothpastes; or alternatively to test the clinical impact of other probiotics (such as *S. A12* or *S. dentisani*) rather than traditional gut-associated probiotics.

Probiotics, especially those belonging to *Lactobacillus* spp., could have an effect in the treatment and prevention of dental biofilm-induced gingivitis, probably due to their actions against well-known periodontal pathogens. However, most of the studies failed to find relevant clinical effects. The appropriate strains, dosages, delivery format and target populations need to be elucidated through well conducted randomized clinical trials. When used as adjuncts to SRP in the treatment of periodontitis, at least a probiotic formula containing *L. reuteri* strains DSM17938 and ATCC PTA529 have reported better results, when compared to SRP alone, for several surrogate outcome measurements such as CAL, BOP and PPD, with potentially important clinical benefits due to the reduction in the need of surgery. However, these results have been very heterogeneous and these reported positive effects seem to vanish with time, strengthening the need of long-term studies with larger sample sizes in order to evaluate its regular use in clinical practice.

Considering the absence of consensus and the limited results in the management of peri-implant diseases, probiotics may provide a protective role in their prevention, or even a positive effect when used as adjunct treatment. However, evidence is scarce and long-term trials evaluating different strains, dosages, delivery formats and treatment duration should be performed before making any recommendation about its usage.

Halitosis treatment, in general, is mainly focused on the improvement of oral hygiene, including tongue cleaning, and the daily use of specific mouth rinses for long periods of time. However, these products are not always free of side effects, such as staining, and it is clear that a re-colonization of halitosis-causing bacteria will occur after treatment is stopped. Moreover, its use is not recommended for children under 6 years of age and children aged 6–12 years should only use a mouth rinse under close adult supervision. Therefore, the use of probiotics may have an important role in the control of halitosis. Although research is showing some promising results, further well designed clinical trials are still needed in order to recommend its use.

As explained before, probiotics may represent a suitable approach in the management of oral diseases. However, future research needs to identify appropriate strains for each condition, adequate dosages, delivery formats, or the group of individuals who will benefit the most from this therapy. In addition, it could be useful to investigate whether the probiotic effect continues after treatment, since the scientific literature suggests that it will disappear when the patient discontinues its use, since the probiotic application does not seem to induce a definitive shift towards a less pathogenic microbiota. Another issue that needs to be addressed is the removal of the existing biofilm before using the probiotics, since they have difficulties in exerting their beneficial effects on an already matured biofilm (Teughels et al. 2011). Probiotics seem to perform significantly better when the dental biofilm has been previously disrupted; therefore, its contribution to other treatment approaches deserves further investigation.

References

ADA Council on Scientific Affairs. (2003). Malodor. *Journal of the American Dental Association, 134*, 209.

Al-Ansari, J. M., Boodai, H., Al-Sumait, N., Al-Khabbaz, A. K., Al-Shammari, K. F., & Salako, N. (2006). Factors associated with self-reported halitosis in Kuwaiti patients. *Journal of Dentistry, 34*, 444–449. https://doi.org/10.1016/j.jdent.2005.10.002.

Alkaya, B., Laleman, I., Keceli, S., Ozcelik, O., Cenk Haytac, M., & Teughels, W. (2017). Clinical effects of probiotics containing Bacillus species on gingivitis: A pilot randomized controlled trial. *Journal of Periodontal Research, 52*, 497–504. https://doi.org/10.1111/jre.12415.

American Academy of Periodontology. (2000a). Parameter on chronic periodontitis with slight to moderate loss of periodontal support. American Academy of Periodontology. *Journal of Periodontology, 71*, 853–855. https://doi.org/10.1902/jop.2000.71.5-S.853.

American Academy of Periodontology. (2000b). Parameter on plaque-induced gingivitis. American Academy of Periodontology. *Journal of Periodontology, 71*, 851–852. https://doi.org/10.1902/jop.2000.71.5-S.851.

Armitage, G. C. (1999). Development of a classification system for periodontal diseases and conditions. *Annals of Periodontology, 4*, 1–6. https://doi.org/10.1902/annals.1999.4.1.1.

Baca-Castanon, M. L., De la Garza-Ramos, M. A., Alcazar-Pizana, A. G., Grondin, Y., Coronado-Mendoza, A., Sanchez-Najera, R. I., Cardenas-Estrada, E., Medina-De la Garza, C. E., & Escamilla-Garcia, E. (2015). Antimicrobial effect of Lactobacillus reuteri on cariogenic bacteria Streptococcus gordonii, Streptococcus mutans, and periodontal diseases Actinomyces naeslundii and Tannerella forsythia. *Probiotics and Antimicrobial Proteins, 7*, 1–8. https://doi.org/10.1007/s12602-014-9178-y.

Barretto, C., Alvarez-Martin, P., Foata, F., Renault, P., & Berger, B. (2012). Genome sequence of the lantibiotic bacteriocin producer Streptococcus salivarius strain K12. *Journal of Bacteriology, 194*, 5959–5960. https://doi.org/10.1128/JB.01268-12.

Bornstein, M. M., Kislig, K., Hoti, B. B., Seemann, R., & Lussi, A. (2009). Prevalence of halitosis in the population of the city of Bern, Switzerland: A study comparing self-reported and clinical data. *European Journal of Oral Sciences, 117*, 261–267. https://doi.org/10.1111/j.1600-0722.2009.00630.x.

Botero, J. E., Rösing, C. K., Duque, A., Jaramillo, A., & Contreras, A. (2015). Periodontal disease in children and adolescents of Latin America. *Periodontology, 2000*(67), 34–57.

Burton, J. P., Chilcott, C. N., & Tagg, J. R. (2005). The rationale and potential for the reduction of oral malodour using Streptococcus salivarius probiotics. *Oral Diseases, 11*(Suppl. 1), 29–31. https://doi.org/10.1111/j.1601-0825.2005.01084.x.

Burton, J. P., Chilcott, C. N., Moore, C. J., Speiser, G., & Tagg, J. R. (2006a). A preliminary study of the effect of probiotic Streptococcus salivarius K12 on oral malodour parameters. *Journal of Applied Microbiology, 100*, 754–764. https://doi.org/10.1111/j.1365-2672.2006.02837.x.

Burton, J. P., Wescombe, P. A., Moore, C. J., Chilcott, C. N., & Tagg, J. R. (2006b). Safety assessment of the oral cavity probiotic Streptococcus salivarius K12. *Applied and Environmental Microbiology, 72*, 3050–3053. https://doi.org/10.1128/AEM.72.4.3050-3053.2006.

Busarcevic, M., & Dalgalarrondo, M. (2012). Purification and genetic characterisation of the novel bacteriocin LS2 produced by the human oral strain Lactobacillus salivarius BGHO1. *International Journal of Antimicrobial Agents, 40*, 127–134. https://doi.org/10.1016/j.ijantimicag.2012.04.011.

Busarcevic, M., Kojic, M., Dalgalarrondo, M., Chobert, J. M., Haertle, T., & Topisirovic, L. (2008). Purification of bacteriocin LS1 produced by human oral isolate Lactobacillus salivarius BGHO1. *Oral Microbiology and Immunology, 23*, 254–258. https://doi.org/10.1111/j.1399-302X.2007.00420.x.

Cagetti, M. G., Mastroberardino, S., Milia, E., Cocco, F., Lingstrom, P., & Campus, G. (2013). The use of probiotic strains in caries prevention: A systematic review. *Nutrients, 5*, 2530–2550. https://doi.org/10.3390/nu5072530.

Caglar, E., Topcuoglu, N., Cildir, S. K., Sandalli, N., & Kulekci, G. (2009). Oral colonization by Lactobacillus reuteri ATCC 55730 after exposure to probiotics. *International Journal of Paediatric Dentistry, 19*, 377–381. https://doi.org/10.1111/j.1365-263X.2009.00989.x.

Caton, J. G., Armitage, G., Berglundh, T., Chapple, I. L. C., Jepsen, S., Kornman, K. S., Mealey, B.L., Papapanou, P. N., Sanz, M., & Tonetti, M. S. (2018). A new classification scheme for periodontal and peri-implant diseases and conditions – Introduction and key changes from the 1999 classification. *Journal of Clinical Periodontology, 45*, S1–S8

Cekici, A., Kantarci, A., Hasturk, H., & Van Dyke, T. E. (2014). Inflammatory and immune pathways in the pathogenesis of periodontal disease. *Periodontology, 2000*(64), 57–80. https://doi.org/10.1111/prd.12002.

Cosseau, C., Devine, D. A., Dullaghan, E., Gardy, J. L., Chikatamarla, A., Gellatly, S., Yu, L. L., Pistolic, J., Falsafi, R., Tagg, J., & Hancock, R. E. (2008). The commensal Streptococcus salivarius K12 downregulates the innate immune responses of human epithelial cells and promotes host-microbe homeostasis. *Infection and Immunity, 76*, 4163–4175. https://doi.org/10.1128/IAI.00188-08.

Coxam, V. (2007). Current data with inulin-type fructans and calcium, targeting bone health in adults. *Journal of Nutrition, 137*, 2527S–2533S.

De Boever, E. H., & Loesche, W. J. (1995). Assessing the contribution of anaerobic microflora of the tongue to oral malodor. *Journal of the American Dental Association, 126*, 1384–1393.

de Cock, P., Makinen, K., Honkala, E., Saag, M., Kennepohl, E., & Eapen, A. (2016). Erythritol is more effective than xylitol and sorbitol in managing oral health endpoints. *International Journal of Dentistry, 2016*, 9868421. https://doi.org/10.1155/2016/9868421.

de Wit, G. (1966). [Foetor ex ore]. *Nederlands Tijdschrift voor Geneeskunde, 110*, 1689–1692.

Delanghe, G., Bollen, C., & Desloovere, C. (1999). [Halitosis—Foetor ex ore]. *Laryngo-Rhino-Otologie, 78*, 521–524. https://doi.org/10.1055/s-2007-996920.

Delcour, J. A., Aman, P., Courtin, C. M., Hamaker, B. R., & Verbeke, K. (2016). Prebiotics, fermentable dietary fiber, and health claims. *Advances in Nutrition, 7*, 1–4. https://doi.org/10.3945/an.115.010546.

Dhingra, K. (2012). Methodological issues in randomized trials assessing probiotics for periodontal treatment. *Journal of Periodontal Research, 47*, 15–26.

Doran, A. L., & Verran, J. (2007). A clinical study on the effect of the prebiotic inulin in the control of oral malodour. *Microbial Ecology in Health and Disease, 19*, 158–163. https://doi.org/10.4103/2279-042X.179568.

Ericson, D., Hamberg, K., Bratthall, G., Sinkiewicz-Enggren, G., & Ljunggren, L. (2013). Salivary IgA response to probiotic bacteria and mutans streptococci after the use of chewing gum containing Lactobacillus reuteri. *Pathogens and Disease, 68*, 82–87. https://doi.org/10.1111/2049-632x.12048.

FAO/WHO (2001) Report of the Joint FAO/WHO Expert Consultation on Evaluation of Health and Nutritional Properties of Probiotics in Food Including Powder Milk with Live Lactic Acid Bacteria, Córdoba, Argentina, 1-4 October 2001. Food and Agriculture Organization of the United Nations, Geneva.

Fejerskov, O., Nyvad, B., & Kidd, E. (2015). Dental caries, what is it? In *Dental caries: The disease and its clinical management* (3rd ed., pp. 7–10). Oxford: Wiley Blackwell.

Flichy-Fernandez, A. J., Ata-Ali, J., Alegre-Domingo, T., Candel-Marti, E., Ata-Ali, F., Palacio, J. R., & Penarrocha-Diago, M. (2015). The effect of orally administered probiotic Lactobacillus reuteri-containing tablets in peri-implant mucositis: A double-blind randomized controlled trial. *Journal of Periodontal Research, 50*, 775–785. https://doi.org/10.1111/jre.12264.

Frencken, J. E., Sharma, P., Stenhouse, L., Green, D., Laverty, D., & Dietrich, T. (2017). Global epidemiology of dental caries and severe periodontitis—A comprehensive review. *Journal of Clinical Periodontology, 44*(Suppl. 18), S94–S105. https://doi.org/10.1111/jcpe.12677.

Furrie, E., Macfarlane, S., Kennedy, A., Cummings, J. H., Walsh, S. V., O'Neil, D. A., & Macfarlane, G. T. (2005). Synbiotic therapy (Bifidobacterium longum/Synergy 1) initiates resolution of inflammation in patients with active ulcerative colitis: A randomised controlled pilot trial. *Gut, 54*, 242–249. https://doi.org/10.1136/gut.2004.044834.

Galofre, M., Palao, D., Vicario, M., Nart, J., & Violant, D. (2018). Clinical and microbiological evaluation of the effect of Lactobacillus reuteri in the treatment of mucositis and peri-implantitis: A triple-blind randomized clinical trial. *Journal of Periodontal Research, 53*(3), 378–390. https://doi.org/10.1111/jre.12523.

Genco, R. J., & Borgnakke, W. S. (2013). Risk factors for periodontal disease. *Periodontology, 2000*(62), 59–94. https://doi.org/10.1111/j.1600-0757.2012.00457.x.

Gibson, G. R., & Roberfroid, M. B. (1995). Dietary modulation of the human colonic microbiota: Introducing the concept of prebiotics. *Journal of Nutrition, 125*, 1401–1412.

Gibson, G. R., Probert, H. M., Loo, J. V., Rastall, R. A., & Roberfroid, M. B. (2004). Dietary modulation of the human colonic microbiota: Updating the concept of prebiotics. *Nutrition Research Reviews, 17*, 259–275. https://doi.org/10.1079/NRR200479.

Global Burden of Disease Study Collaborators. (2015). Global, regional, and national incidence, prevalence, and years lived with disability for 301 acute and chronic diseases and injuries in 188 countries, 1990-2013: A systematic analysis for the Global Burden of Disease Study 2013. *Lancet, 386*, 743–800. https://doi.org/10.1016/S0140-6736(15)60692-4.

Goldberg, S., Kozlovsky, A., Gordon, D., Gelernter, I., Sintov, A., & Rosenberg, M. (1994). Cadaverine as a putative component of oral malodor. *Journal of Dental Research, 73*, 1168–1172.

Goldberg, S., Cardash, H., Browning, H., 3rd, Sahly, H., & Rosenberg, M. (1997). Isolation of Enterobacteriaceae from the mouth and potential association with malodor. *Journal of Dental Research, 76*, 1770–1775.

Graziani, F., Karapetsa, D., Alonso, B., & Herrera, D. (2017). Nonsurgical and surgical treatment of periodontitis: How many options for one disease? *Periodontology, 2000*(75), 152–188.

Greenman, J., El-Maaytah, M., Duffield, J., Spencer, P., Rosenberg, M., Corry, D., Saad, S., Lenton, P., Majerus, G., & Nachnani, S. (2005). Assessing the relationship between concentrations of malodor compounds and odor scores from judges. *Journal of the American Dental Association, 136*, 749–757.

Gross, E. L., Beall, C. J., Kutsch, S. R., Firestone, N. D., Leys, E. J., & Griffen, A. L. (2012). Beyond Streptococcus mutans: Dental caries onset linked to multiple species by 16S rRNA community analysis. *PLoS One, 7*, e47722.

Hallström, H., Lindgren, S., Yucel-Lindberg, T., Dahlén, G., Renvert, S., & Twetman, S. (2013). Effect of probiotic lozenges on inflammatory reactions and oral biofilm during experimental gingivitis. *Acta Odontologica Scandinavica, 71*, 828–833.

Hallstrom, H., Lindgren, S., Widen, C., Renvert, S., & Twetman, S. (2016). Probiotic supplements and debridement of peri-implant mucositis: A randomized controlled trial. *Acta Odontologica Scandinavica, 74*, 60–66. https://doi.org/10.3109/00016357.2015.1040065.

Haraszthy, V. I., Zambon, J. J., Sreenivasan, P. K., Zambon, M. M., Gerber, D., Rego, R., & Parker, C. (2007). Identification of oral bacterial species associated with halitosis. *Journal of the American Dental Association, 138*, 1113–1120.

Haraszthy, V. I., Gerber, D., Clark, B., Moses, P., Parker, C., Sreenivasan, P. K., & Zambon, J. J. (2008). Characterization and prevalence of Solobacterium moorei associated with oral halitosis. *Journal of Breath Research, 2*, 017002. https://doi.org/10.1088/1752-7155/2/1/017002.

Hasslöf, P., West, C. E., Videhult, F. K., Brandelius, C., & Stecksén-Blicks, C. (2013). Early intervention with probiotic Lactobacillus paracasei F19 has no long-term effect on caries experience. *Caries Research, 47*, 559–565.

Haukioja, A., Yli-Knuuttila, H., Loimaranta, V., Kari, K., Ouwehand, A. C., Meurman, J. H., & Tenovuo, J. (2006). Oral adhesion and survival of probiotic and other lactobacilli and bifidobacteria in vitro. *Oral Microbiology and Immunology, 21*, 326–332.

Haukioja, A., Loimaranta, V., & Tenovuo, J. (2008). Probiotic bacteria affect the composition of salivary pellicle and streptococcal adhesion in vitro. *Oral Microbiology and Immunology, 23*, 336–343. https://doi.org/10.1111/j.1399-302X.2008.00435.x.

Herrera, D., Sanz, M., Jepsen, S., Needleman, I., & Roldan, S. (2002). A systematic review on the effect of systemic antimicrobials as an adjunct to scaling and root planing in periodontitis patients. *Journal of Clinical Periodontology, 29*(Suppl. 3), 136–159; discussion 160–132.

Herrera, D., Alonso, B., Leon, R., Roldan, S., & Sanz, M. (2008). Antimicrobial therapy in periodontitis: The use of systemic antimicrobials against the subgingival biofilm. *Journal of Clinical Periodontology, 35*, 45–66. https://doi.org/10.1111/j.1600-051X.2008.01260.x.

Herrera, D., Meyle, J., Renvert, S., & Lin, J. (2018). White Paper on Prevention and Management of Periodontal Diseases for Oral Health and General Health. FDI World Dental Federation.

Honkala, S., Runnel, R., Saag, M., Olak, J., Nommela, R., Russak, S., Makinen, P. L., Vahlberg, T., Falony, G., Makinen, K., & Honkala, E. (2014). Effect of erythritol and xylitol on dental caries prevention in children. *Caries Research, 48*, 482–490. https://doi.org/10.1159/000358399.

Horz, H. P., Meinelt, A., Houben, B., & Conrads, G. (2007). Distribution and persistence of probiotic Streptococcus salivarius K12 in the human oral cavity as determined by real-time quantitative polymerase chain reaction. *Oral Microbiology and Immunology, 22*, 126–130. https://doi.org/10.1111/j.1399-302X.2007.00334.x.

Huang, X., Palmer, S. R., Ahn, S. J., Richards, V. P., Williams, M. L., Nascimento, M. M., & Burne, R. A. (2016). A highly arginolytic Streptococcus species that potently antagonizes Streptococcus mutans. *Applied and Environmental Microbiology, 82*, 2187–2201. https://doi.org/10.1128/AEM.03887-15.

Human Microbiome Project Consortium. (2012). Structure, function and diversity of the healthy human microbiome. *Nature, 486*, 207–214. https://doi.org/10.1038/nature11234.

Ince, G., Gursoy, H., Ipci, S. D., Cakar, G., Emekli-Alturfan, E., & Yilmaz, S. (2015). Clinical and biochemical evaluation of lozenges containing lactobacillus reuteri as an adjunct to nonsurgical periodontal therapy in chronic periodontitis. *Journal of Periodontology, 86*, 746–754. https://doi.org/10.1902/jop.2015.140612.

Iniesta, M., Herrera, D., Montero, E., Zurbriggen, M., Matos, A. R., Marin, M. J., Sanchez-Beltran, M. C., Llama-Palacio, A., & Sanz, M. (2012). Probiotic effects of orally administered Lactobacillus reuteri-containing tablets on the subgingival and salivary microbiota in patients with gingivitis. A randomized clinical trial. *Journal of Clinical Periodontology, 39*, 736–744. https://doi.org/10.1111/j.1600-051X.2012.01914.x.

Iwamoto, T., Suzuki, N., Tanabe, K., Takeshita, T., & Hirofuji, T. (2010). Effects of probiotic Lactobacillus salivarius WB21 on halitosis and oral health: An open-label pilot trial. *Oral Surgery, Oral Medicine, Oral Pathology, Oral Radiology, and Endodontology, 110*, 201–208. https://doi.org/10.1016/j.tripleo.2010.03.032.

Janakiram, C., Deepan Kumar, C. V., & Joseph, J. (2017). Xylitol in preventing dental caries: A systematic review and meta-analyses. *Journal of Natural Science, Biology and Medicine, 8*, 16–21. https://doi.org/10.4103/0976-9668.198344.

Microbial Manipulation of Dysbiosis: Prebiotics and Probiotics for the Treatment... 229

Jang, H. J., Kang, M. S., Yi, S. H., Hong, J. Y., & Hong, S. P. (2016). Comparative study on the characteristics of Weissella cibaria CMU and probiotic strains for oral care. *Molecules, 21,* E1752. https://doi.org/10.3390/molecules21121752.

Jasberg, H., Soderling, E., Endo, A., Beighton, D., & Haukioja, A. (2016). Bifidobacteria inhibit the growth of Porphyromonas gingivalis but not of Streptococcus mutans in an in vitro biofilm model. *European Journal of Oral Sciences, 124,* 251–258. https://doi.org/10.1111/eos.12266.

Jepsen, S., Blanco, J., Buchalla, W., Carvalho, J. C., Dietrich, T., Dorfer, C., Eaton, K. A., Figuero, E., Frencken, J. E., Graziani, F., Higham, S. M., Kocher, T., Maltz, M., Ortiz-Vigon, A., Schmoeckel, J., Sculean, A., Tenuta, L. M., van der Veen, M. H., & Machiulskiene, V. (2017). Prevention and control of dental caries and periodontal diseases at individual and population level: Consensus report of group 3 of joint EFP/ORCA workshop on the boundaries between caries and periodontal diseases. *Journal of Clinical Periodontology, 44*(Suppl. 18), S85–S93. https://doi.org/10.1111/jcpe.12687.

Kang, M. S., Chung, J., Kim, S. M., Yang, K. H., & Oh, J. S. (2006a). Effect of Weissella cibaria isolates on the formation of Streptococcus mutans biofilm. *Caries Research, 40,* 418–425. https://doi.org/10.1159/000094288.

Kang, M. S., Kim, B. G., Chung, J., Lee, H. C., & Oh, J. S. (2006b). Inhibitory effect of Weissella cibaria isolates on the production of volatile sulphur compounds. *Journal of Clinical Periodontology, 33,* 226–232. https://doi.org/10.1111/j.1600-051X.2006.00893.x.

Kang, M. S., Oh, J. S., Lee, H. C., Lim, H. S., Lee, S. W., Yang, K. H., Choi, N. K., & Kim, S. M. (2011). Inhibitory effect of Lactobacillus reuteri on periodontopathic and cariogenic bacteria. *Journal of Microbiology, 49,* 193–199. https://doi.org/10.1007/s12275-011-0252-9.

Karuppaiah, R. M., Shankar, S., Raj, S. K., Ramesh, K., Prakash, R., & Kruthika, M. (2013). Evaluation of the efficacy of probiotics in plaque reduction and gingival health maintenance among school children—A randomized control trial. *Journal of International Oral Health, 5,* 33–37.

Kassebaum, N. J., Bernabe, E., Dahiya, M., Bhandari, B., Murray, C. J., & Marcenes, W. (2015). Global burden of untreated caries: A systematic review and metaregression. *Journal of Dental Research, 94,* 650–658. https://doi.org/10.1177/0022034515573272.

Kazor, C. E., Mitchell, P. M., Lee, A. M., Stokes, L. N., Loesche, W. J., Dewhirst, F. E., & Paster, B. J. (2003). Diversity of bacterial populations on the tongue dorsa of patients with halitosis and healthy patients. *Journal of Clinical Microbiology, 41,* 558–563.

Keller, M. K., Bardow, A., Jensdottir, T., Lykkeaa, J., & Twetman, S. (2012). Effect of chewing gums containing the probiotic bacterium Lactobacillus reuteri on oral malodour. *Acta Odontologica Scandinavica, 70,* 246–250. https://doi.org/10.3109/00016357.2011.640281.

Keukenmeester, R. S., Slot, D. E., Rosema, N. A., Van Loveren, C., & Van der Weijden, G. A. (2014). Effects of sugar-free chewing gum sweetened with xylitol or maltitol on the development of gingivitis and plaque: A randomized clinical trial. *International Journal of Dental Hygiene, 12,* 238–244. https://doi.org/10.1111/idh.12071.

Kleinberg, I. C. M. (1997). The biological basis of oral malodor formation. In M. Rosenberg (Ed.), *Bad breath: Research perspectives* (pp. 13–39). Tel Aviv, Ramat Aviv: Ramot Publishing, Tel Aviv University.

Kleinberg, I., & Westbay, G. (1990). Oral malodor. *Critical Reviews in Oral Biology and Medicine, 1,* 247–259.

Kolenbrander, P. E. (2011). Multispecies communities: Interspecies interactions influence growth on saliva as sole nutritional source. *International Journal of Oral Science, 3,* 49–54. https://doi.org/10.4248/IJOS11025.

Kragen, H. (1954). [The treatment of inflammatory affections of the oral mucosa with a lactic acid bacterial culture preparation]. *Zahnärztliche Welt, 9,* 306–308.

Kraivaphan, P., Amornchat, C., Triratana, T., Mateo, L. R., Ellwood, R., Cummins, D., DeVizio, W., & Zhang, Y. P. (2013). Two-year caries clinical study of the efficacy of novel dentifrices containing 1.5% arginine, an insoluble calcium compound and 1,450 ppm fluoride. *Caries Research, 47,* 582–590. https://doi.org/10.1159/000353183.

Kumada, M., Motegi, M., Nakao, R., Yonezawa, H., Yamamura, H., Tagami, J., & Senpuku, H. (2009). Inhibiting effects of Enterococcus faecium non-biofilm strain on Streptococcus mutans biofilm formation. *Journal of Microbiology, Immunology and Infection, 42*, 188–196.

Kuru, B. E., Laleman, I., Yalnizoglu, T., Kuru, L., & Teughels, W. (2017). The influence of a bifidobacterium animalis probiotic on gingival health: A randomized controlled clinical trial. *Journal of Periodontology, 88*, 1115–1123. https://doi.org/10.1902/jop.2017.170213.

Laleman, I., Detailleur, V., Slot, D. E., Slomka, V., Quirynen, M., & Teughels, W. (2014). Probiotics reduce mutans streptococci counts in humans: A systematic review and meta-analysis. *Clinical Oral Investigations, 18*, 1539–1552.

Laleman, I., Yilmaz, E., Ozcelik, O., Haytac, C., Pauwels, M., Herrero, E. R., Slomka, V., Quirynen, M., Alkaya, B., & Teughels, W. (2015). The effect of a streptococci containing probiotic in periodontal therapy: A randomized controlled trial. *Journal of Clinical Periodontology, 42*, 1032–1041. https://doi.org/10.1111/jcpe.12464.

Lang, C., Böttner, M., Holz, C., Veen, M., Ryser, M., Reindl, A., Pompejus, M., & Tanzer, J. M. (2010). Specific Lactobacillus/Mutans Streptococcus co-aggregation. *Journal of Dental Research, 89*, 175–179.

Lee, S. H., & Baek, D. H. (2014). Effects of Streptococcus thermophilus on volatile sulfur compounds produced by Porphyromonas gingivalis. *Archives of Oral Biology, 59*, 1205–1210.

Lee, J. K., Kim, S. J., Ko, S. H., Ouwehand, A. C., & Ma, D. S. (2015). Modulation of the host response by probiotic Lactobacillus brevis CD2 in experimental gingivitis. *Oral Diseases, 21*, 705–712. https://doi.org/10.1111/odi.12332.

Liu, X. N., Shinada, K., Chen, X. C., Zhang, B. X., Yaegaki, K., & Kawaguchi, Y. (2006). Oral malodor-related parameters in the Chinese general population. *Journal of Clinical Periodontology, 33*, 31–36. https://doi.org/10.1111/j.1600-051X.2005.00862.x.

Loesche, W. J. (1986). Role of Streptococcus mutans in human dental decay. *Microbiology Reviews, 50*, 353–380.

Lopez-Lopez, A., Camelo-Castillo, A., Ferrer, M. D., Simon-Soro, A., & Mira, A. (2017). Health-associated niche inhabitants as oral probiotics: The case of Streptococcus dentisani. *Frontiers in Microbiology, 8*, 379. https://doi.org/10.3389/fmicb.2017.00379.

Lövdal, A., Ärno, A., Schei, O., & Waerhaug, J. (1961). Combined effect of subgingival scaling and controlled oral hygiene on the incidence of gingivitis. *Acta Odontologica Scandinavica, 19*, 537–555.

Mailander-Sanchez, D., Braunsdorf, C., Grumaz, C., Muller, C., Lorenz, S., Stevens, P., Wagener, J., Hebecker, B., Hube, B., Bracher, F., Sohn, K., & Schaller, M. (2017). Antifungal defense of probiotic Lactobacillus rhamnosus GG is mediated by blocking adhesion and nutrient depletion. *PLoS One, 12*, e0184438. https://doi.org/10.1371/journal.pone.0184438.

Makinen, K. K. (2011). Sugar alcohol sweeteners as alternatives to sugar with special consideration of xylitol. *Medical Principles and Practice, 20*, 303–320. https://doi.org/10.1159/000324534.

Makinen, K. K., Saag, M., Isotupa, K. P., Olak, J., Nommela, R., Soderling, E., & Makinen, P. L. (2005). Similarity of the effects of erythritol and xylitol on some risk factors of dental caries. *Caries Research, 39*, 207–215. https://doi.org/10.1159/000084800.

Marchetti, E., Tecco, S., Santonico, M., Vernile, C., Ciciarelli, D., Tarantino, E., Marzo, G., & Pennazza, G. (2015). Multi-sensor approach for the monitoring of halitosis treatment via Lactobacillus brevis (CD2) containing lozenges—A randomized, double-blind placebo-controlled clinical trial. *Sensors, 15*, 19583–19596. https://doi.org/10.3390/s150819583.

Marsh, P. D. (2018). In sickness and in health-what does the oral microbiome mean to us? An ecological perspective. *Advances in Dental Research, 29*, 60–65. https://doi.org/10.1177/0022034517735295.

Martin-Cabezas, R., Davideau, J. L., Tenenbaum, H., & Huck, O. (2016). Clinical efficacy of probiotics as an adjunctive therapy to non-surgical periodontal treatment of chronic periodontitis: A systematic review and meta-analysis. *Journal of Clinical Periodontology, 43*, 520–530. https://doi.org/10.1111/jcpe.12545.

Marya, C. M., Taneja, P., Nagpal, R., Marya, V., Oberoi, S. S., & Arora, D. (2017). Efficacy of chlorhexidine, xylitol, and chlorhexidine + xylitol against dental plaque, gingivitis, and salivary Streptococcus mutans load: A randomised controlled trial. *Oral Health & Preventive Dentistry, 15*, 529–536. https://doi.org/10.3290/j.ohpd.a39669.

Masdea, L., Kulik, E. M., Hauser-Gerspach, I., Ramseier, A. M., Filippi, A., & Waltimo, T. (2012). Antimicrobial activity of Streptococcus salivarius K12 on bacteria involved in oral malodour. *Archives of Oral Biology, 57*, 1041–1047. https://doi.org/10.1016/j.archoralbio.2012.02.011.

Matesanz-Perez, P., Garcia-Gargallo, M., Figuero, E., Bascones-Martinez, A., Sanz, M., & Herrera, D. (2013). A systematic review on the effects of local antimicrobials as adjuncts to subgingival debridement, compared with subgingival debridement alone, in the treatment of chronic periodontitis. *Journal of Clinical Periodontology, 40*, 227–241. https://doi.org/10.1111/jcpe.12026.

Matsubara, V. H., Bandara, H. M., Ishikawa, K. H., Mayer, M. P., & Samaranayake, L. P. (2016). The role of probiotic bacteria in managing periodontal disease: A systematic review. *Expert Review of Anti-Infective Therapy, 14*, 643–655. https://doi.org/10.1080/14787210.2016.1194 198.

Mayanagi, G., Kimura, M., Nakaya, S., Hirata, H., Sakamoto, M., Benno, Y., & Shimauchi, H. (2009). Probiotic effects of orally administered Lactobacillus salivarius WB21-containing tablets on periodontopathic bacteria: A double-blinded, placebo-controlled, randomized clinical trial. *Journal of Clinical Periodontology, 36*, 506–513. https://doi.org/10.1111/j.1600-051X.2009.01392.x.

McNamara, T. F., Alexander, J. F., & Lee, M. (1972). The role of microorganisms in the production of oral malodor. *Oral Surgery, Oral Medicine, and Oral Pathology, 34*, 41–48.

Mendi, A., Kose, S., Uckan, D., Akca, G., Yilmaz, D., Aral, L., Gultekin, S. E., Eroglu, T., Kilic, E., & Uckan, S. (2016). Lactobacillus rhamnosus could inhibit Porphyromonas gingivalis derived CXCL8 attenuation. *Journal of Applied Oral Science, 24*, 67–75. https://doi.org/10.1590/1678-775720150145.

Meningaud, J. P., Bado, F., Favre, E., Bertrand, J. C., & Guilbert, F. (1999). [Halitosis in 1999]. *Revue de Stomatologie et de Chirurgie Maxillo-Faciale, 100*, 240–244.

Mickenautsch, S., Leal, S. C., Yengopal, V., Bezerra, A. C., & Cruvinel, V. (2007). Sugar-free chewing gum and dental caries: A systematic review. *Journal of Applied Oral Science, 15*, 83–88.

Milgrom, P., Söderling, E. M., Nelson, S., Chi, D. L., & Nakai, Y. (2012). Clinical evidence for polyol efficacy. *Advances in Dental Research, 24*, 112–116.

Mira, A. (2018). Oral microbiome studies: Potential diagnostic and therapeutic implications. *Advances in Dental Research, 29*, 71–77. https://doi.org/10.1177/0022034517737024.

Mira, A., Simon-Soro, A., & Curtis, M. A. (2017). Role of microbial communities in the pathogenesis of periodontal diseases and caries. *Journal of Clinical Periodontology, 44*(Suppl. 18), S23–S38. https://doi.org/10.1111/jcpe.12671.

Miyazaki, H., Sakao, S., Katoh, Y., & Takehara, T. (1995). Correlation between volatile sulphur compounds and certain oral health measurements in the general population. *Journal of Periodontology, 66*, 679–684.

Mongardini, C., Pilloni, A., Farina, R., Di Tanna, G., & Zeza, B. (2017). Adjunctive efficacy of probiotics in the treatment of experimental peri-implant mucositis with mechanical and photodynamic therapy: A randomized, cross-over clinical trial. *Journal of Clinical Periodontology, 44*, 410–417. https://doi.org/10.1111/jcpe.12689.

Montero, E., Iniesta, M., Rodrigo, M., Marin, M. J., Figuero, E., Herrera, D., & Sanz, M. (2017). Clinical and microbiological effects of the adjunctive use of probiotics in the treatment of gingivitis: A randomized controlled clinical trial. *Journal of Clinical Periodontology, 44*, 708–716. https://doi.org/10.1111/jcpe.12752.

Moon, J. E., Moon, Y. M., & Cho, J. W. (2016). The effect of Streptococcus salivarius K12 against Prevotella intermedia on the Reduction of Oral Malodor. *International Journal of Clinical Preventive Dentistry, 12*, 153–161.

Morales, A., Carvajal, P., Silva, N., Hernandez, M., Godoy, C., Rodriguez, G., Cabello, R., Garcia-Sesnich, J., Hoare, A., Diaz, P. I., & Gamonal, J. (2016). Clinical effects of lactobacillus rhamnosus in non-surgical treatment of chronic periodontitis: A randomized placebo-controlled trial with 1-year follow-up. *Journal of Periodontology, 87*, 944–952. https://doi.org/10.1902/jop.2016.150665.

Morales, A., Gandolfo, A., Bravo, J., Carvajal, P., Silva, N., Godoy, C., Garcia-Sesnich, J., Hoare, A., Diaz, P., & Gamonal, J. (2018). Microbiological and clinical effects of probiotics and antibiotics on nonsurgical treatment of chronic periodontitis: A randomized placebo-controlled trial with 9-month follow-up. *Journal of Applied Oral Science, 26*, e20170075. https://doi.org/10.1590/1678-7757-2017-0075.

Moro, G., Arslanoglu, S., Stahl, B., Jelinek, J., Wahn, U., & Boehm, G. (2006). A mixture of prebiotic oligosaccharides reduces the incidence of atopic dermatitis during the first six months of age. *Archives of Disease in Childhood, 91*, 814–819. https://doi.org/10.1136/adc.2006.098251.

Nadanovsky, P., Carvalho, L. B., & Ponce de Leon, A. (2007). Oral malodour and its association with age and sex in a general population in Brazil. *Oral Diseases, 13*, 105–109. https://doi.org/10.1111/j.1601-0825.2006.01257.x.

Nalcaci, R., Dulgergil, T., Oba, A. A., & Gelgor, I. E. (2008). Prevalence of breath malodour in 7-11-year-old children living in Middle Anatolia, Turkey. *Community Dental Health, 25*, 173–177.

Näse, L., Hatakka, K., Savilahti, E., Saxelin, M., Pönkä, A., Poussa, T., Korpela, R., & Meurman, J. H. (2001). Effect of long-term consumption of a probiotic bacterium, Lactobacillus rhamnosus GG, in milk on dental caries and caries risk in children. *Caries Research, 35*, 412–420.

Niles, H. P. G. A. (1997). Advances in mouth odor research. In M. In Rosenberg (Ed.), *Bad breath: Research perspectives* (pp. 13–39). Ramat Aviv: Ramot Publishing, Tel Aviv University.

Nissen, L., Sgorbati, B., Biavati, B., & Belibasakis, G. N. (2014). *Lactobacillus salivarius* and *L. gasseri* down-regulate *Aggregatibacter actinomycetemcomitans* exotoxins expression. *Annals of Microbiology, 64*, 611–617. https://doi.org/10.1007/s13213-013-0694-x.

Pangsomboon, K., Bansal, S., Martin, G. P., Suntinanalert, P., Kaewnopparat, S., & Srichana, T. (2009). Further characterization of a bacteriocin produced by Lactobacillus paracasei HL32. *Journal of Applied Microbiology, 106*, 1928–1940. https://doi.org/10.1111/j.1365-2672.2009.04146.x.

Penala, S., Kalakonda, B., Pathakota, K. R., Jayakumar, A., Koppolu, P., Lakshmi, B. V., Pandey, R., & Mishra, A. (2016). Efficacy of local use of probiotics as an adjunct to scaling and root planing in chronic periodontitis and halitosis: A randomized controlled trial. *Journal of Research in Pharmacy Practice, 5*, 86–93. https://doi.org/10.4103/2279-042x.179568.

Persson, S., Edlund, M.-B., Claesson, R., & Carlsson, J. (1990). The formation of hydrogen sulfide and methyl mercaptan by oral bacteria. *Oral Microbiology and Immunology, 5*, 195–201.

Petersen, P. E., & Lennon, M. A. (2004). Effective use of fluorides for the prevention of dental caries in the 21st century: The WHO approach. *Community Dentistry and Oral Epidemiology, 32*, 319–321.

Petersson, L. G., Magnusson, K., Hakestam, U., Baigi, A., & Twetman, S. (2011). Reversal of primary root caries lesions after daily intake of milk supplemented with fluoride and probiotic lactobacilli in older adults. *Acta Odontologica Scandinavica, 69*, 321–327.

Porter, S. R., & Scully, C. (2006). Oral malodour (halitosis). *British Medical Journal, 333*, 632–635. https://doi.org/10.1136/bmj.38954.631968.AE.

Quirynen, M., Dadamio, J., Van den Velde, S., De Smit, M., Dekeyser, C., Van Tornout, M., & Vandekerckhove, B. (2009). Characteristics of 2000 patients who visited a halitosis clinic. *Journal of Clinical Periodontology, 36*, 970–975. https://doi.org/10.1111/j.1600-051X.2009.01478.x.

Rachmilewitz, D., Katakura, K., Karmeli, F., Hayashi, T., Reinus, C., Rudensky, B., Akira, S., Takeda, K., Lee, J., Takabayashi, K., & Raz, E. (2004). Toll-like receptor 9 signaling mediates the anti-inflammatory effects of probiotics in murine experimental colitis. *Gastroenterology, 126*, 520–528.

Ribeiro, F. C., de Barros, P. P., Rossoni, R. D., Junqueira, J. C., & Jorge, A. O. (2017). Lactobacillus rhamnosus inhibits Candida albicans virulence factors in vitro and modulates immune system in Galleria mellonella. *Journal of Applied Microbiology, 122*, 201–211. https://doi.org/10.1111/jam.13324.

Riccia, D. N., Bizzini, F., Perilli, M. G., Polimeni, A., Trinchieri, V., Amicosante, G., & Cifone, M. G. (2007). Anti-inflammatory effects of Lactobacillus brevis (CD2) on periodontal disease. *Oral Diseases, 13*, 376–385. https://doi.org/10.1111/j.1601-0825.2006.01291.x.

Riggio, M. P., Lennon, A., Rolph, H. J., Hodge, P. J., Donaldson, A., Maxwell, A. J., & Bagg, J. (2008). Molecular identification of bacteria on the tongue dorsum of subjects with and without halitosis. *Oral Diseases, 14*, 251–258. https://doi.org/10.1111/j.1601-0825.2007.01371.x.

Rodriguez, G., Ruiz, B., Faleiros, S., Vistoso, A., Marro, M. L., Sanchez, J., Urzua, I., & Cabello, R. (2016). Probiotic compared with standard milk for high-caries children: A cluster randomized trial. *Journal of Dental Research, 95*, 402–407. https://doi.org/10.1177/0022034515623935.

Roldan, S., Winkel, E. G., Herrera, D., Sanz, M., & Van Winkelhoff, A. J. (2003). The effects of a new mouthrinse containing chlorhexidine, cetylpyridinium chloride and zinc lactate on the microflora of oral halitosis patients: A dual-centre, double-blind placebo-controlled study. *Journal of Clinical Periodontology, 30*, 427–434.

Roldan, S., Herrera, D., O'Connor, A., Gonzalez, I., & Sanz, M. (2005). A combined therapeutic approach to manage oral halitosis: A 3-month prospective case series. *Journal of Periodontology, 76*, 1025–1033. https://doi.org/10.1902/jop.2005.76.6.1025.

Rossoni, R. D., de Barros, P. P., de Alvarenga, J. A., Ribeiro, F. C., Velloso, M. D. S., Fuchs, B. B., Mylonakis, E., Jorge, A. O. C., & Junqueira, J. C. (2018). Antifungal activity of clinical Lactobacillus strains against Candida albicans biofilms: Identification of potential probiotic candidates to prevent oral candidiasis. *Biofouling, 34*, 1–14. https://doi.org/10.1080/0892701 4.2018.1425402.

Salehi, R., Savabi, O., Kazemi, M., Kamali, S., Salehi, A. R., Eslami, G., & Tahmourespour, A. (2014). Effects of Lactobacillus reuteri-derived biosurfactant on the gene expression profile of essential adhesion genes (gtfB, gtfC and ftf) of Streptococcus mutans. *Advanced Biomedical Research, 3*, 169. https://doi.org/10.4103/2277-9175.139134.

Samot, J., & Badet, C. (2013). Antibacterial activity of probiotic candidates for oral health. *Anaerobe, 19*, 34–38. https://doi.org/10.1016/j.anaerobe.2012.11.007.

Sanz, M., Beighton, D., Curtis, M. A., Cury, J. A., Dige, I., Dommisch, H., Ellwood, R., Giacaman, R. A., Herrera, D., Herzberg, M. C., Kononen, E., Marsh, P. D., Meyle, J., Mira, A., Molina, A., Mombelli, A., Quirynen, M., Reynolds, E. C., Shapira, L., & Zaura, E. (2017). Role of microbial biofilms in the maintenance of oral health and in the development of dental caries and periodontal diseases. Consensus report of group 1 of the Joint EFP/ORCA workshop on the boundaries between caries and periodontal disease. *Journal of Clinical Periodontology, 44*(Suppl. 18), S5–S11. https://doi.org/10.1111/jcpe.12682.

Savabi, O., Kazemi, M., Kamali, S., Salehi, A. R., Eslami, G., Tahmourespour, A., & Salehi, R. (2014). Effects of biosurfactant produced by Lactobacillus casei on gtfB, gtfC, and ftf gene expression level in S. mutans by real-time RT-PCR. *Advanced Biomedical Research, 3*, 231. https://doi.org/10.4103/2277-9175.145729.

Schwarz, F., Schmucker, A., & Becker, J. (2015). Efficacy of alternative or adjunctive measures to conventional treatment of peri-implant mucositis and peri-implantitis: A systematic review and meta-analysis. *International Journal of Implant Dentistry, 1*, 22. https://doi.org/10.1186/s40729-015-0023-1.

Seemann, R., Conceicao, M. D., Filippi, A., Greenman, J., Lenton, P., Nachnani, S., Quirynen, M., Roldan, S., Schulze, H., Sterer, N., Tangerman, A., Winkel, E. G., Yaegaki, K., & Rosenberg, M. (2014). Halitosis management by the general dental practitioner—Results of an international consensus workshop. *Journal of Breath Research, 8*, 017101. https://doi.org/10.1088/1752-7155/8/1/017101.

Seifert, S., & Watzl, B. (2007). Inulin and oligofructose: Review of experimental data on immune modulation. *Journal of Nutrition, 137*, 2563S–2567S.

Serrano, J., Escribano, M., Roldan, S., Martin, C., & Herrera, D. (2015). Efficacy of adjunctive anti-plaque chemical agents in managing gingivitis: A systematic review and meta-analysis. *Journal of Clinical Periodontology, 42*(Suppl. 16), S106–S138. https://doi.org/10.1111/jcpe.12331.

Shah, M. P., Gujjari, S. K., & Chandrasekhar, V. S. (2013). Evaluation of the effect of probiotic (inersan(R)) alone, combination of probiotic with doxycycline and doxycycline alone on aggressive periodontitis—A clinical and microbiological study. *Journal of Clinical and Diagnostic Research, 7*, 595–600. https://doi.org/10.7860/jcdr/2013/5225.2834.

Sharma, N. C., Galustians, J. H., & Qaqish, J. G. (2001). An evaluation of a commercial chewing gum in combination with normal toothbrushing for reducing dental plaque and gingivitis. *Compendium of Continuing Education in Dentistry, 22*, 13–17.

Sheiham, A., & Netuveli, G. S. (2002). Periodontal diseases in Europe. *Periodontology, 2000*(29), 104–121.

Shimauchi, H., Mayanagi, G., Nakaya, S., Minamibuchi, M., Ito, Y., Yamaki, K., & Hirata, H. (2008). Improvement of periodontal condition by probiotics with Lactobacillus salivarius WB21: A randomized, double-blind, placebo-controlled study. *Journal of Clinical Periodontology, 35*, 897–905. https://doi.org/10.1111/j.1600-051X.2008.01306.x.

Simon-Soro, A., Belda-Ferre, P., Cabrera-Rubio, R., Alcaraz, L. D., & Mira, A. (2013). A tissue-dependent hypothesis of dental caries. *Caries Research, 47*, 591–600. https://doi.org/10.1159/000351663.

Sliepen, I., Van Essche, M., Loozen, G., Van Eldere, J., Quirynen, M., & Teughels, W. (2009). Interference with Aggregatibacter actinomycetemcomitans: Colonization of epithelial cells under hydrodynamic conditions. *Oral Microbiology and Immunology, 24*, 390–395. https://doi.org/10.1111/j.1399-302X.2009.00531.x.

Slomka, V., Hernandez-Sanabria, E., Herrero, E. R., Zaidel, L., Bernaerts, K., Boon, N., Quirynen, M., & Teughels, W. (2017). Nutritional stimulation of commensal oral bacteria suppresses pathogens: The prebiotic concept. *Journal of Clinical Periodontology, 44*, 344–352. https://doi.org/10.1111/jcpe.12700.

Slomka, V., Herrero, E. R., Boon, N., Bernaerts, K., Trivedi, H. M., Daep, C., Quirynen, M., & Teughels, W. (2018). Oral prebiotics and the influence of environmental conditions in vitro. *Journal of Periodontology, 89*(6), 708–717. https://doi.org/10.1002/JPER.17-0437.

Slots, J. T. M. (1992). *Contemporary oral biology and microbiology.* St Louis: Mosby.

Soderling, E. M. (2009). Xylitol, mutans streptococci, and dental plaque. *Advances in Dental Research, 21*, 74–78. https://doi.org/10.1177/0895937409335642.

Soderling, E., Makinen, K. K., Chen, C. Y., Pape, H. R., Jr., Loesche, W., & Makinen, P. L. (1989). Effect of sorbitol, xylitol, and xylitol/sorbitol chewing gums on dental plaque. *Caries Research, 23*, 378–384. https://doi.org/10.1159/000261212.

Sookkhee, S., Chulasiri, M., & Prachyabrued, W. (2001). Lactic acid bacteria from healthy oral cavity of Thai volunteers: Inhibition of oral pathogens. *Journal of Applied Microbiology, 90*, 172–179.

Staab, B., Eick, S., Knofler, G., & Jentsch, H. (2009). The influence of a probiotic milk drink on the development of gingivitis: A pilot study. *Journal of Clinical Periodontology, 36*, 850–856. https://doi.org/10.1111/j.1600-051X.2009.01459.x.

Stecksén-Blicks, C., Sjöström, I., & Twetman, S. (2009). Effect of long-term consumption of milk supplemented with probiotic lactobacilli and fluoride on dental caries and general health in preschool children: A cluster-randomized study. *Caries Research, 43*, 374–381.

Steinberg, L. M., Odusola, F., & Mandel, I. D. (1992). Remineralizing potential, antiplaque and antigingivitis effects of xylitol and sorbitol sweetened chewing gum. *Clinical Preventive Dentistry, 14*, 31–34.

Stensson, M., Koch, G., Coric, S., Abrahamsson, T. R., Jenmalm, M. C., Birkhed, D., & Wendt, L. K. (2014). Oral administration of Lactobacillus reuteri during the first year of life reduces caries prevalence in the primary dentition at 9 years of age. *Caries Research, 48*, 111–117. https://doi.org/10.1159/000354412.

Surono, I. S., Koestomo, F. P., Novitasari, N., Zakaria, F. R., Yulianasari, & Koesnandar. (2011). Novel probiotic Enterococcus faecium IS-27526 supplementation increased total salivary sIgA level and bodyweight of pre-school children: A pilot study. *Anaerobe, 17*, 496–500. https://doi.org/10.1016/j.anaerobe.2011.06.003.

Suzuki, N., Tanabe, K., Takeshita, T., Yoneda, M., Iwamoto, T., Oshiro, S., Yamashita, Y., & Hirofuji, T. (2012). Effects of oil drops containing Lactobacillus salivarius WB21 on periodontal health and oral microbiota producing volatile sulfur compounds. *Journal of Breath Research, 6*, 017106.

Suzuki, N., Yoneda, M., Tanabe, K., Fujimoto, A., Iha, K., Seno, K., Yamada, K., Iwamoto, T., Masuo, Y., & Hirofuji, T. (2014). Lactobacillus salivarius WB21–containing tablets for the treatment of oral malodor: A double-blind, randomized, placebo-controlled crossover trial. *Oral Surgery, Oral Medicine, Oral Pathology and Oral Radiology, 117*, 462–470.

Suzuki, N., Higuchi, T., Nakajima, M., Fujimoto, A., Morita, H., Yoneda, M., Hanioka, T., & Hirofuji, T. (2016). Inhibitory effect of Enterococcus faecium WB2000 on volatile sulfur compound production by Porphyromonas gingivalis. *International Journal of Dentistry*, 8241681. Epub 2016 Oct 2019.

Tada, H., Masaki, C., Tsuka, S., Mukaibo, T., Kondo, Y., & Hosokawa, R. (2018). The effects of Lactobacillus reuteri probiotics combined with azithromycin on peri-implantitis: A randomized placebo-controlled study. *Journal of Prosthodontic Research, 62*, 89–96. https://doi.org/10.1016/j.jpor.2017.06.006.

Tagg, J. R. (2004). Prevention of streptococcal pharyngitis by anti-Streptococcus pyogenes bacteriocin-like inhibitory substances (BLIS) produced by Streptococcus salivarius. *Indian Journal of Medical Research, 119*(Suppl), 13–16.

Tahmourespour, A., Salehi, R., & Kasra Kermanshahi, R. (2011). Lactobacillus acidophilus-derived biosurfactant effect on GTFB and GTFC expression level in Streptococcus mutans biofilm cells. *Brazilian Journal of Microbiology, 42*, 330–339. https://doi.org/10.1590/s1517-83822011000100042.

Taipale, T., Pienihäkkinen, K., Alanen, P., Jokela, J., & Söderling, E. (2013). Administration of Bifidobacterium animalis subsp. lactis BB-12 in early childhood: A post-trial effect on caries occurrence at four years of age. *Caries Research, 47*, 364–372.

Talarico, T. L., Casas, I. A., Chung, T. C., & Dobrogosz, W. J. (1988). Production and isolation of reuterin, a growth inhibitor produced by Lactobacillus reuteri. *Antimicrobial Agents and Chemotherapy, 32*, 1854–1858.

Tan, Y., Leonhard, M., Moser, D., Ma, S., & Schneider-Stickler, B. (2018). Inhibitory effect of probiotic lactobacilli supernatants on single and mixed non-albicans Candida species biofilm. *Archives of Oral Biology, 85*, 40–45. https://doi.org/10.1016/j.archoralbio.2017.10.002.

Tanner, A. C., Mathney, J. M., Kent, R. L., Chalmers, N. I., Hughes, C. V., Loo, C. Y., Pradhan, N., Kanasi, E., Hwang, J., Dahlan, M. A., Papadopolou, E., & Dewhirst, F. E. (2011). Cultivable anaerobic microbiota of severe early childhood caries. *Journal of Clinical Microbiology, 49*, 1464–1474. https://doi.org/10.1128/jcm.02427-10.

Tao, R., Jurevic, R. J., Coulton, K. K., Tsutsui, M. T., Roberts, M. C., Kimball, J. R., Wells, N., Berndt, J., & Dale, B. A. (2005). Salivary antimicrobial peptide expression and dental caries experience in children. *Antimicrobial Agents and Chemotherapy, 49*, 3883–3888. https://doi.org/10.1128/aac.49.9.3883-3888.2005.

Teanpaisan, R., Piwat, S., & Dahlen, G. (2011). Inhibitory effect of oral Lactobacillus against oral pathogens. *Letters in Applied Microbiology, 53*, 452–459. https://doi.org/10.1111/j.1472-765X.2011.03132.x.

Tekce, M., Ince, G., Gursoy, H., Dirikan Ipci, S., Cakar, G., Kadir, T., & Yilmaz, S. (2015). Clinical and microbiological effects of probiotic lozenges in the treatment of chronic periodontitis: A 1-year follow-up study. *Journal of Clinical Periodontology, 42*, 363–372. https://doi.org/10.1111/jcpe.12387.

Teughels, W., Kinder Haake, S., Sliepen, I., Pauwels, M., Van Eldere, J., Cassiman, J. J., & Quirynen, M. (2007). Bacteria interfere with A. actinomycetemcomitans colonization. *Journal of Dental Research, 86*, 611–617.

Teughels, W., Loozen, G., & Quirynen, M. (2011). Do probiotics offer opportunities to manipulate the periodontal oral microbiota? *Journal of Clinical Periodontology, 38*(Suppl. 11), 159–177.

Teughels, W., Durukan, A., Ozcelik, O., Pauwels, M., Quirynen, M., & Haytac, M. C. (2013). Clinical and microbiological effects of Lactobacillus reuteri probiotics in the treatment of chronic periodontitis: A randomized placebo-controlled study. *Journal of Clinical Periodontology, 40*, 1025–1035. https://doi.org/10.1111/jcpe.12155.

Twetman, S., Derawi, B., Keller, M., Ekstrand, K., Yucel-Lindberg, T., & Stecksen-Blicks, C. (2009). Short-term effect of chewing gums containing probiotic Lactobacillus reuteri on the levels of inflammatory mediators in gingival crevicular fluid. *Acta Odontologica Scandinavica, 67*, 19–24. https://doi.org/10.1080/00016350802516170.

Valdez, R. M., Dos Santos, V. R., Caiaffa, K. S., Danelon, M., Arthur, R. A., Negrini, T. C., Delbem, A. C., & Duque, C. (2016). Comparative in vitro investigation of the cariogenic potential of bifidobacteria. *Archives of Oral Biology, 71*, 97–103. https://doi.org/10.1016/j.archoralbio.2016.07.005.

Vancauwenberghe, F., Dadamio, J., Laleman, I., Van Tornout, M., Teughels, W., Coucke, W., & Quirynen, M. (2013). The role of Solobacterium moorei in oral malodour. *Journal of Breath Research, 7*, 046006.

Vicario, M., Santos, A., Violant, D., Nart, J., & Giner, L. (2013). Clinical changes in periodontal subjects with the probiotic Lactobacillus reuteri Prodentis: A preliminary randomized clinical trial. *Acta Odontologica Scandinavica, 71*, 813–819. https://doi.org/10.3109/00016357.2012.734404.

Villa, A., Zollanvari, A., Alterovitz, G., Cagetti, M. G., Strohmenger, L., & Abati, S. (2014). Prevalence of halitosis in children considering oral hygiene, gender and age. *International Journal of Dental Hygiene, 12*, 208–212. https://doi.org/10.1111/idh.12077.

Vivekananda, M. R., Vandana, K. L., & Bhat, K. G. (2010). Effect of the probiotic Lactobacilli reuteri (Prodentis) in the management of periodontal disease: A preliminary randomized clinical trial. *Journal of Oral Microbiology, 2*. https://doi.org/10.3402/jom.v2i0.5344.

Vuotto, C., Barbanti, F., Mastrantonio, P., & Donelli, G. (2014). Lactobacillus brevis CD2 inhibits Prevotella melaninogenica biofilm. *Oral Diseases, 20*, 668–674. https://doi.org/10.1111/odi.12186.

Walker, G. V., Heng, N. C., Carne, A., Tagg, J. R., & Wescombe, P. A. (2016). Salivaricin E and abundant dextranase activity may contribute to the anti-cariogenic potential of the probiotic candidate Streptococcus salivarius JH. *Microbiology, 162*, 476–486. https://doi.org/10.1099/mic.0.000237.

Wannun, P., Piwat, S., & Teanpaisan, R. (2014). Purification and characterization of bacteriocin produced by oral Lactobacillus paracasei SD1. *Anaerobe, 27*, 17–21. https://doi.org/10.1016/j.anaerobe.2014.03.001.

Wasfi, R., Abd El-Rahman, O. A., Zafer, M. M., & Ashour, H. M. (2018). Probiotic Lactobacillus sp. inhibit growth, biofilm formation and gene expression of caries-inducing Streptococcus mutans. *Journal of Cellular and Molecular Medicine, 22*(3), 1972–1983. https://doi.org/10.1111/jcmm.13496.

Wattanarat, O., Makeudom, A., Sastraruji, T., Piwat, S., Tianviwat, S., Teanpaisan, R., & Krisanaprakornkit, S. (2015). Enhancement of salivary human neutrophil peptide 1-3 levels by probiotic supplementation. *BMC Oral Health, 15*, 19.

World Health Organization. (2015). Guideline: Sugars intake for adults and children. [WWW document].

Yaegaki, K., & Sanada, K. (1992). Volatile sulfur compounds in mouth air from clinically healthy subjects and patients with periodontal disease. *Journal of Periodontal Research, 27*, 233–238.

Yokoyama, S., Ohnuki, M., Shinada, K., Ueno, M., Wright, F. A., & Kawaguchi, Y. (2010). Oral malodor and related factors in Japanese senior high school students. *Journal of School Health, 80*, 346–352. https://doi.org/10.1111/j.1746-1561.2010.00512.x.

Zhang, G., & Rudney, J. D. (2011). Streptococcus cristatus attenuates Fusobacterium nucleatum-induced cytokine expression by influencing pathways converging on nuclear factor-kappaB. *Molecular Oral Microbiology, 26*, 150–163. https://doi.org/10.1111/j.2041-1014.2010.00600.x.

Zhang, G., Chen, R., & Rudney, J. D. (2008). Streptococcus cristatus attenuates Fusobacterium nucleatum-induced interleukin-8 expression in oral epithelial cells. *Journal of Periodontal Research, 43*, 408–416.

Zhao, J. J., Feng, X. P., Zhang, X. L., & Le, K. Y. (2012). Effect of Porphyromonas gingivalis and Lactobacillus acidophilus on secretion of IL1B, IL6, and IL8 by gingival epithelial cells. *Inflammation, 35*, 1330–1337. https://doi.org/10.1007/s10753-012-9446-5.

Part III
The Recipe for Happiness: A Balanced Gut Microbiota

Early Gut Microbiome: A Good Start in Nutrition and Growth May Have Lifelong Lasting Consequences

Amanda L. Thompson

Abstract Initial colonization, establishment, and development of the gut microbiota play an essential role in long-term health. The birthing process is the first major exposure to microorganisms, though there is evidence of pre-natal exposures to low-abundance microorganisms. Different modes of birth facilitate colonization of different microorganisms, which are further modulated by early diet (breast feeding vs formula feeding), environment, and exposure to antibiotics. This chapter discusses mechanisms and consequences of differential colonization and maintenance of the infant microbiota, and their implications for overall health.

Keywords Infant gut microbiome · Microbiota development · Pre-natal development of the microbiome · Mode of birth · Early infant feeding · Complementary infant feeding

Introduction: Development of the Microbiota Development Across the First 1000 Days: Exposures and Health Consequences

Early environmental exposures, such as *in utero* nutrition, delivery mode, antibiotic exposure, and feeding practices, have all been identified as potentially important in shaping long-term vulnerability to obesity and chronic disease (Mueller et al. 2015b; Thompson 2012; Yang and Huffman 2013). At the same time, research has identified the gut microbiome as an important pathway linking environmental and dietary exposures to long-term health (Blaut and Clavel 2007; Vael and Desager 2009; Voreades et al. 2014). The gut microbiome is a complex ecosystem,

---

A. L. Thompson (✉)
Departments of Anthropology and Nutrition, University of North Carolina at Chapel Hill, Chapel Hill, NC, USA
e-mail: althomps@email.unc.edu

© Springer Nature Switzerland AG 2019
M. A. Azcarate-Peril et al. (eds.), *How Fermented Foods Feed a Healthy Gut Microbiota*, https://doi.org/10.1007/978-3-030-28737-5_10

containing a large and diverse number of bacterial populations that vary across individuals and populations and that, once established, remain relatively stable. The importance of these bacterial communities in host nutrition, immune function and metabolism has also become increasingly clear. In the large intestine, gut microbiota help extract nutrients and energy from the diet, maintain barrier function, and balance the innate and acquired immune responses (Hooper et al. 2002; Ley et al. 2005). Disruption of this ecosystem through adverse environmental exposures may contribute to the development of diseases, including inflammatory bowel disease, diabetes, malnutrition, and obesity, through multiple intertwined immune and metabolic pathways (Tilg and Kaser 2011). Thus, increasing attention has turned to examining the development of the microbiome in early life and the factors that may influence the quantity, quality, and timing of this development to promote long term health.

Like many other organ systems, the intestinal microbiome undergoes rapid development in early life. Colonization begins during gestation and intensifies after birth with early exposures, such as maternal weight gain, delivery type, diet and hospital and home environment, playing a major role (Fig. 1). Infants' relatively limited fetal microbiome quickly becomes seeded by their mothers' vaginal, fecal and skin bacteria and the surrounding environment at birth. Predominant microbial groups continue to shift over the first year of life in response to feeding practices, including the type of milk feeding, the introduction of solid foods and the cessation of breastfeeding, illness, and antibiotic use. Though further maturation occurs during childhood, the gut microbiota comes to roughly resemble an adult colonization pattern around age three.

Along with the types of environmental bacteria infants and young children are exposed to, the timing of exposure may also play a critical role in determining health outcomes (Martinez 2014). Early colonizers can modulate host gene expression and gut function, creating a favorable habitat for themselves and preventing colonization by later arriving bacteria (Guarner and Malagelada 2003). Considerable diversity exists between individuals and populations in the composition and timing of these shifts. The diversity and flux of microbes during this period, from gestation

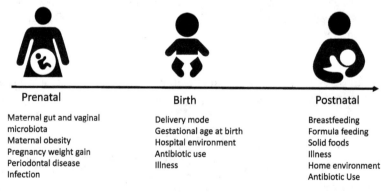

Fig. 1 Environmental factors influencing the development of the gut microbiota across early life

through the first 3 years of life, may be particularly important for health and disease throughout the lifespan since the microbiome develops concurrently with the development of metabolic, immune and cognitive systems and evidence increasingly suggests that these systems interact during this period (Wopereis et al. 2014). Since bacteria differ in their energy extracting and immune regulating capabilities (Blaut and Clavel 2007; Ley et al. 2005), colonization patterns during the first years of life can influence both early growth and immune development and also long-term energy absorption, adipose development, and inflammation. Thus, this period of early life exposures provides a "critical window" during which alterations can have short- and long-term impacts (Arrieta et al. 2014). Since initial colonization may shape final adult colonization, disturbances during this period may alter growth, contribute to the development of immune disorders, such as allergy and infectious disease, and metabolic conditions, like obesity and diabetes, and, potentially, shape brain development, with effects on developmental disorders and cognitive function (Carlson et al. 2018; Wopereis et al. 2014).

Given this potential for early life conditions to influence the long-term trajectory of gut, immune and metabolic health, this chapter examines the maternal, environmental, and dietary factors influencing the development of the gut microbiome across the first years of life and the implications for the development of health and disease. Understanding the salient maternal, dietary and environmental exposures during early life and their sequela is critical for improving long term health; thus, the last section of this chapter will review potential avenues for intervention.

Prenatal Development

Until quite recently, infants were thought to be born with sterile guts that become first exposed to bacteria through the passage through the birth canal or, in the case of infants born by Caesarean-section, through exposure to the hospital environment (Biasucci et al. 2010; Dominguez-Bello et al. 2010). While the massive bacterial inoculation that occurs during birth may be more important for seeding the gut microbiome, prenatal colonization also occurs. The placenta and amniotic fluid of even healthy pregnancies contain a low abundance of bacterial DNA from several common phyla including Firmicutes, Tenericutes, Proteobacteria, Bacteroidetes and Fusobacteria (Aagaard et al. 2014; Arrieta et al. 2014; Meropol and Edwards 2015; Walker et al. 2017). The source of this prenatal gut microbiota remains under debate. Some researchers have described the placental bacteria as more closely resembling the maternal oral microbiota than vaginal microbiota (Aagaard et al. 2014; Mysorekar and Cao 2014). They propose that the oral bacteria translocate hematogenously from the oral cavity to the placenta, as has been previously shown in animal models (Fardini et al. 2010). The similarities between the oral and placental microbiomes are also thought to underlie the association between dental conditions such as periodontal disease and preterm birth (Mysorekar and Cao 2014). Other work, however, has identified maternal gut bacterial DNA in neonatal meconium and cord blood (Ardissone et al. 2014; Walker et al. 2017), suggesting that, instead or in addition, bacteria spread from the maternal gut into the blood stream, to the placenta, and, ultimately, to the fetus as it swallows the nonsterile amniotic fluid (Aagaard et al. 2014; Donnet-Hughes et al. 2010; Walker et al. 2017).

Relatively little is known about whether fetal colonization may be linked to either positive or negative pregnancy outcomes and later health, but prenatal exposures have the potential to shape gut development and function in ways that alter later colonization patterns. Exposure to *Prevotella* and *Gardnerella* have been proposed to contribute to an inflammatory response in newborns while higher levels of *Lactobacillus* have been proposed to suppress inflammation (Rodriguez et al. 2015). Differences in patterns of placental bacterial DNA are seen between infants born at term vs. those born preterm, but not between vaginally and C-section delivered infants suggesting that *in utero* exposures vs. birth conditions shape the placental microbiota (Doyle et al. 2014). Further, maternal health may alter prenatal bacterial exposures as it has been shown that the bacterial content of the meconium differs in neonates born to diabetic mothers (Rodriguez et al. 2015).

Birth Type and the Microbiome

A second and more significant bacterial colonization occurs at birth as the infant is exposed to the vaginal, fecal and skin microbiota of their mother and the bacteria in the surrounding environment. Vaginal delivery is considered critical for establishing a healthy microbiome. As soon as 5 min after birth, differences can be seen in the microbiota of the skin, oral mucosa, an nasopharyngeal aspirate of infants born vaginally and by C-section and differences in the gut microbiota are identifiable within 24 h of birth (Dominguez-Bello et al. 2010). Vaginally delivered infants have bacterial profiles that resemble their mother's pre-delivery vaginal microbiome while infants born by C-section have profiles more like their mother's skin and the hospital environment (Dominguez-Bello et al. 2010). Along with the vaginal microbiota, neonates are also exposed to their mothers' fecal bacteria and this exposure may be beneficial for the development of gut function and immune regulation in early life (Huda et al. 2014). Among the initial bacteria vaginally-delivered infants are exposed to are facultative anaerobic species, such as *Escherichia*, *Staphylococcus*, and *Streptococcus* species, that colonize the gut and change the environment to favor the growth of the obligate anaerobes of the phylum Bacteroidetes and species *Bifidobacterium*, thought to be important for infant health. Differences in the proportion of colonization by these groups can be seen by the third day of life (Biasucci et al. 2010).

C-section delivery may compromise the intergenerational transmission of beneficial maternal bacteria; only 41% of early colonizing bacteria match those in mothers' stool samples compared to 72% in vaginally-delivered infants. Further, these early differences persist through infancy and possibly into childhood. While the gut microbiome of infants born vaginally and by C-section become more similar over the first year of life, the gut microbiome of infants delivered by C-section remain less diverse with a delayed colonization by *Bacteroides* (Arrieta et al. 2014). Differences in colonization by birth type have been shown to persist in 4 month-old Canadian infants (Azad et al. 2013), 12 month-old Swedish infants (Backhed et al. 2015), 24 month-old Finnish toddlers (Backhed et al. 2015), and 7 year-old Swedish children (Salminen et al. 2004). Such differences in the microbiota have been proposed to underlie the higher rates of type 1 diabetes, asthma, and overweight seen in children born by C-section (Belizán et al. 2007).

Although the timing of maternal transmission of bacteria to her infant remains controversial, with some researchers proposing that birth is the most important time for intergenerational exchange (Mueller et al. 2016) and others supporting a later postnatal transmission (Avershina et al. 2014), the maternal microbiome is doubtlessly an important early factor shaping the infant microbiome. Mothers' vaginal and intestinal microbiomes change over the course of pregnancy. The vaginal microbiota become less diverse with an increasing proportion of *Lactobacillus* (Aagaard et al. 2012; Mueller et al. 2015a), an early colonizer important for gut ecology (Backhed et al. 2015). Along with these changes in vaginal microbiota, the maternal intestinal microbiota also shifts to a pattern characterized by increasing proportions of high-energy yielding microbiota with increasing gestational age (Koren et al. 2012; Mueller et al. 2015b). These changes occur independently of maternal weight status, weight gain, gestational diabetes, diet or antibiotic use suggesting that they may be linked to pregnancy-induced alterations in gut motility, immunity and endocrine function (Mueller et al. 2015b) and have been proposed to serve as an adaption to permit greater energy harvest during pregnancy and for the infant in the first days of life (Koren et al. 2012).

Maternal dysbiosis, resulting from obesity, antibiotic use, or other pregnancy conditions, may differentially shape early infant exposures and alter the trajectory of infant microbiome development during the perinatal period. High maternal pre-pregnancy weight and excessive weight gain during pregnancy have been linked to differences in maternal microbiomes, with mothers' who are overweight pre-pregnancy or gain excessive weight during pregnancy having greater proportions of *Bacteroides* in their stool, suggesting that this genus contributes to greater energy storage and weight gain (Collado et al. 2008, 2010). Maternal weight and weight gain further shape patterns of colonization in infant microbiome colonization (Collado et al. 2010). Infants born to mothers with excessive weight gain have lower *Bifidobacterium* and greater *Staphylococcus aureus* at 6 months of age, a pattern associated with greater weight gain in the infants (Collado et al. 2010) and higher risk of inflammatory disease later in life (Bastard et al. 2006; Hotamisligil 2006; Kalliomaki et al. 2001). In a study of Finnish women and their infants, obese mothers or those with excessive weight gain and their infants also had a higher prevalence of *Akkermansia muciniphila*, a species that has previously been linked to overweight, obesity, and metabolic endotoxemia (Cani and Delzenne 2007).

Maternal exposure to antibiotics in late pregnancy or perinatally in response to pregnancy complications or infections may also alter the microbiome. Epidemiological evidence suggests that maternal antibiotic use during pregnancy is associated with greater prevalence of asthma and overweight in childhood (Ajslev et al. 2011; Stensballe et al. 2013), conditions potentially resulting from alterations in the gut or respiratory microbiomes. Maternal antibiotic use around the time of birth has been associated with lower bacterial diversity and lower levels of *Lactobacillus* and *Bifidobacterium* in neonates (Keski-Nisula et al. 2013; Mueller et al. 2015b). Differences in the microbiota have been shown in infants born preterm who tend to be exposed to antibiotics and other medical treatments. Preterm infants have a distinct microbiota with greater proportions of the phyla Proteobacteria and Firmicutes, and the family Enterobacteriaceae compared to term infants (Collado et al. 2015). These differences are linked to greater risk of necrotizing enterocolitis neonatally and with later gut microbiome maturation during infancy (Dogra et al. 2015). Thus, pre- and neonatal exposures may be critical for both the initial colonization and the developmental trajectory of the infant gut microbiome. Disruptions during this period by maternal dysbiosis during pregnancy, premature delivery, perinatal antibiotic use and delivery by C-section may contribute to inadequate or inappropriate colonization and delayed maturation leading to dysbiosis in the infant with consequences for growth, body composition, immune function and metabolism.

Early Infant Feeding and the Gut Microbiome

After this initial colonization, the establishment of the enteric microbiota is further shaped by early milk feeding practices (Table 1). Breast and formula-fed infants tend to differ in their patterns of bacterial colonization. Breastfed infants are generally characterized by lower bacterial diversity and greater proportions of *Bifidobacterium* and *Lactobacillus* (Jost et al. 2014; Thompson et al. 2015; Voreades et al. 2014) while formula-fed infants tend to have a more diverse pattern of colonization with higher proportions of the genera *Bacteroides* and *Clostridium*, and the family *Enterobacteriaceae* (Madan et al. 2016). Although the magnitude of these

Table 1 Microbiota differences associated with infant feeding

	Breastfeeding	Formula	Mixed feeding	Solid foods
Microbiota patterns	↓ Diversity ↑ *Bifidobacterium* and *Lactobacillus*	↑ Diversity ↑ *Bacteroides*, *Clostridium*, and enterobacteria	↑ Diversity ↑ *Bacteroides* and *Clostridium*	↑↑ Diversity *Bacteroides*, *Clostridium*, *Prevotella* and *Xylanibacter*
Exposures	Human Milk Oligosaccharides (HMO), Maternal antibodies and glycoproteins, and breastmilk microbiota	Possible pre- and pro-biotics	Variable HMOs, antibodies, glycoproteins and breastmilk microbiota	Non-digestible plant polysaccharides Animal source protein

differences has decreased in recent years with changes in methodology from culture-dependent to culture-independent techniques and changes in the composition of infant formula to more closely mirror breastmilk (Marques et al. 2010; Vael and Desager 2009), relatively consistent differences persist. A recent study of Australian infants, for example, documented higher abundances of *Bifidobacterium* in breast-fed infants even with the considerable overlap in the relative abundance ranges (Tannock et al. 2013).

Less research has described the pattern of colonization in infants fed both breast-milk and formula, but several studies suggest that considerable differences may exist between exclusively breast fed and those that receive breastmilk in combination with formula. Our research found that American infants receiving formula in addition to breastmilk had greater species diversity with lower levels of Actinobacteria and greater relative abundances of *Bacteroides,* Clostridiales*,* Lachnospiraceae*, Blautia,* and *Faecalibacterium* (Thompson et al. 2015). Similarly, the greater diversity of bacteria seen within a cohort of mixed-fed Swedish infants led to estimates of greater microbiome "maturity" in newborn and 4-month measurements compared to exclusively breast fed infants (Backhed et al. 2015). Overall, then, the current picture suggests that the gut of breastfed infants includes a reduced diversity of bacteria with preferential growth of those, like *Bifidobacterium* and *Lactobacillus* that can utilize milk sugars and human milk oligosaccharides (HMOs). Conversely, the gut of formula fed infants is comprised of a more diverse set of bacteria, even among infants receiving breastmilk along with formula.

These differences in initial colonization may underlie some of the positive health outcomes, such as lower risk of pediatric overweight (Arenz et al. 2004), allergy (Munblit and Verhasselt 2016), and inflammatory conditions (Horta et al. 2015), associated with exclusive breastfeeding. Breastmilk contains multiple bioactive components that promote the colonization and maturation of the infant gut microbiome, contributing to the development of what has been termed the "milk-oriented microbiota" or MOM (Goldsmith et al. 2015; Mueller et al. 2015a; Sela and Mills 2010; Zivkovic et al. 2013). One of the most important groups of bioactive components are the HMOs, a group of over 200 complex sugars that form the third largest component of human milk. HMOs have both direct and indirect effects on the gut microbiota that contribute to the early development of the immune system (Hemarajata and Versalovic 2013; Kunz et al. 2014). First, HMOs may interact directly with pathogenic bacteria, preventing their adhesion to target cells (Zivkovic et al. 2013). Second, HMOs promote the growth of specific microbial communities, including *Bifidobacterium* that utilize the undigested sugars in the gut to produce lactate and short chain fatty acids (SCFA). In turn, SCFAs play an important role in inhibiting the growth of pathogenic bacteria, regulating immune and inflammatory responses, and modulating gut barrier function (Tilg and Kaser 2011). Additional benefits likely come from the presence of maternal antibodies like IgA and glyco-proteins like lactoferrin that prevent the growth of pathogenic bacteria (Gregory and Walker 2013; Meropol and Edwards 2015; Newburg and Walker 2007). Together these beneficial effects lead to greater colonization by MOM and contribute to the stable, relatively uniform gut microbiota with its high proportion of *Bifidobacterium* seen in exclusively breastfed infants (Zivkovic et al. 2011).

In addition to its role as a prebiotic, breastmilk also serves as a source of bacteria, acting as a continuous inoculum until weaning (Wopereis et al. 2014). Human milk contains a diverse and viable bacterial community, that includes an estimated 700 species of *Staphylococcus*, *Streptococcus*, *Bifidobacterium*, and other lactic acid bacteria (Cabrera-Rubio et al. 2012; Marques et al. 2010; Morelli 2008; Reinhardt et al. 2009). The bacterial content of breastmilk varies across lactation, with different predominant taxa seen in colostrum versus late breastmilk (Khodayar-Pardo et al. 2014), in some though not all women (Hunt et al. 2011), and across individual women based on infant gestational age (Khodayar-Pardo et al. 2014), health status (Gomez-Gallego et al. 2016), and delivery type (Cabrera-Rubio et al. 2012). The breastmilk microbiota also seems to be relatively discrete from other sites; principal components analysis of the bacterial groups of colostrum and breastmilk at 1 and 6 months postpartum showed significant clustering of the breastmilk samples away from other measured sites, including skin, vagina, feces and oral epithelium (Cabrera-Rubio et al. 2012). Despite these differences from other sites, the milk microbiome likely arises through a combination of sources including the maternal gut, maternal skin and infant oral microbiota (Jeurink et al. 2013). Sequencing analysis has identified a range of obligate anaerobic bacteria that are shared between maternal feces, breastmilk and neonatal feces (Jost et al. 2014). The presence of several butyrate producing bacteria supports the existence of an entero-mammary pathway that transmits bacteria from the maternal gut into the mammary gland and the milk through the mesenteric lymph nodes (Rodriguez et al. 2015). Hormonal and immunological changes during pregnancy may promote the translocation of these beneficial bacteria into the breast and, subsequently, permit them to influence early colonization patterns and subsequent immune and metabolic development.

The health benefits of breastfeeding have been long established, but examining breastfeeding as a pre- and probiotic modulator of the infant gut microbiome provides a potential mechanism linking early feeding practices to child health. The nutrient, bioactive and bacterial content of breastmilk likely 'optimizes' the proliferation of pioneer bacteria that program the immune system towards homeostasis and prevent inflammation (Walker 2017). *Bacteroides fragilis*, *Bifidobacterium infantis*, and *Lactobacillus acidophilus*, all transmitted through and supported by breastmilk, play unique roles in stimulating the endogenous production of immunoglobulins in the infant gut, activating T-cells, and producing anti-inflammatory factors (Jost et al. 2012; Rautava and Walker 2009). *B. infantis*, in particular, is associated with immune development and response to vaccination. Infants with a higher prevalence of *B. infantis* in their stool at 15 weeks had higher CD4 stimulation responses to oral polio virus, tetanus and tuberculosis vaccines (Huda et al. 2014). Other evidence for the protective effect of the MOM include differences in gene transcription of nearly a dozen genes associated with immune function and virulence between breast- and formula-fed infants (Schwartz et al. 2012) and animal models showing greater T-cell function in milk-fed monkeys that persists across at least the first year of life (Cox et al. 2014).

Complementary Feeding and the Maturation of the Microbiome

The gut microbiota undergoes a rapid change with the cessation of breastfeeding and/or the incorporation of solid foods into the diet. The relative contribution of these processes—the cessation of breastfeeding vs. the introduction of food—remains an open question. Significant differences were seen in the microbiota of 12-month old Swedish infants based on whether they were still receiving breastmilk, leading researchers to conclude that the cessation of breastfeeding shifts the microbial ecology to an adult-like pattern (Backhed et al. 2015). Our work, and that of others, has shown profound shifts in both microbial composition and metabolic function with the introduction of solid foods even in infants still receiving breastmilk (Johnson and Versalovic 2012; Thompson et al. 2015). Similarly, though the microbiota of formula-fed infants is considered more "adult-like" (Harmsen et al. 2000), it also undergoes shifts with the introduction of solids that, in our study, were even more dramatic than those seen in breastmilk infants (Thompson et al. 2015). In any case, the inclusion of new foods into the diet and the cessation of breastfeeding appear to be more important than the *timing* of solid food introduction (Backhed et al. 2015; Laursen et al. 2016). The age of first introduction to solid foods, ranging from 3 to 6 months, was not associated with patterns of colonization at 9 months of age in two large cohorts of Danish infants (Laursen et al. 2016).

Early complementary diets doubtlessly shape the colonization of the gut microbiota as infants are exposed to new foods that provide new substrates for the survival and growth of bacteria not previously supported by breastmilk or formula (Fallani et al. 2011; Parrett and Edwards 1997). Animal models (Reinhardt et al. 2009) and human studies (Backhed et al. 2015; Koenig et al. 2011; Thompson et al. 2015) document a shift in the predominant phyla as the high-fat breastmilk diet is replaced by or supplemented with a carbohydrate-rich weaning diet. Predominant bacterial communities shift from a microbiota dominated by *Lactobacillus, Bifidobacterium,* and the family Enterobacteriaceae to one dominated by *Clostridium* and *Bacteroides* species (Bergstrom et al. 2014). Further shifts are seen in complexity and species function across the first years of life as diets become more diverse (Laursen et al. 2016). Microbiota that are more capable of digesting complex sugars and starches and processing higher protein foods become more prevalent and the microbiome comes to resemble the adults' in structure and function (Avershina et al. 2014; Backhed et al. 2015; Meropol and Edwards 2015). The exact age at which the microbiome comes to resemble a stable adult pattern is unclear, but early colonization patterns continue to shape this maturation. A longitudinal study of European infants has documented the persistent effects of early postnatal conditions, including geographic location, milk feeding, and delivery type, on the maturation of microbiota in the second half of infancy, suggesting that early exposures may shape longer term kinetics (Fallani et al. 2011).

While few human studies have examined the dynamics of microbiota composition in relation to the types and sequence of new foods added into infant diets, limited evidence suggests that the macronutrient content of the diet is also important. Differences in predominant microbiota are associated with differences in carbohydrate vs. protein intakes across populations and individuals. High carbohydrate intake was associated with greater prevalence of Bacteroidetes groups, particularly *Prevotella* and *Xylanibacter*, in cross-national comparisons of infants and young children from Italy and Burkina Faso (De Filippo et al. 2010). More recently, a study of 5-month old American infants randomized to receive meats vs. micronutrient-fortified cereals as the primary complementary food until 9 months found differences in the prevalence of key bacteria, including decreased *Lactobacillus* and increased *Bacteroides* in infants fed iron-fortified cereals (Krebs et al. 2013). Higher fiber intake was associated with a greater prevalence of Prevotellaceae, Veionellaceae, Fusobacteriaceae, Eubacteriaceae, and Pasteurellaceae in Danish infants, while higher protein intake was associated with a decreased prevalence of Bifidobacteriaceae (Laursen et al. 2016). These patterns were proposed to result from the increasing diversity and protein content of a diet that more closely resembled family foods, rather than any specific food or food group. Yet data from slightly older children, aged 2–3 years, has shown that specific intakes of dairy and plant-based foods (fruit, vegetables, soy, nuts and pulses) were associated with distinct microbiota profiles (Smith-Brown et al. 2016). In this cohort of Australian children, higher dairy intake, particularly of yogurts, was associated with lower species diversity and richness and greater prevalence of Firmicutes, while vegetable intakes were associated with a greater abundance of *Lachnospira*.

That the early diet should play an important role in shaping the developing microbiome is not surprising. Diet is a well-known factor in shaping adult gut microbiota composition and metabolic function (Flint et al. 2015; Ussar et al. 2015). Shifts in predominant microbiota with the introduction of solid foods are accompanied by changes in the metabolic capacity and function of the intestine (Reinhardt et al. 2009), with implications for the development of gut health, inflammation, and obesity. Bacteria differ in their energy-extracting capabilities, so differences in early bacterial colonization due to different diets may lead to differences in available energy and contribute to infant weight gain and subsequent obesity risk. A cohort study of Danish infants demonstrated that change in BMI from 9 to 18 months was positively correlated with the abundance of Firmicutes, particularly in *Clostridium leptum and Enterobacter hallii*, butyrate-producing bacteria that may increase host energy harvest (Bergstrom et al. 2014). Longitudinal studies examining infant diet, microbiota, and weight gain across infancy and childhood are rare; however, preliminary evidence suggests that the composition of the microbiome may be associated with weight gain and adipose deposition in infancy (Thompson 2012). The relative abundance of *Bacteroides* at 3 and 52 weeks of age was positively associated with BMIz in 1 and 3 year olds Belgian infants, controlling for infant feeding and a host of other potential confounders, while *Staphylococcus* at 3 and 26 weeks of age was inversely associated with BMIz at these later ages (Vael and Desager 2009). Similarly, overweight at age 7 was related to microbiota colonization at 6 and 12 months of age in a sample of Finnish children (Kalliomaki et al. 2008).

Implications for Long Term Health

As the above sections describe, early life exposures, beginning prenatally and continuing through the first years of life, are critical for establishing the infant gut microbiome. Prenatally, the fetus may be exposed to maternal bacteria through the placenta, an exposure that may differ according to maternal diet, weight status, and health and may have implications for birth outcomes. At birth, the neonate is exposed to an even greater range of bacteria from their mothers and the external environment. Vaginally-born infants receive an inoculum of potential beneficial *Bacteroides* and *Lactobacillus* through exposure to their mothers' vaginal and fecal microbiota, while the acquisition of these bacteria appears to be diminished or delayed in infants delivered by C-section. Early colonization is further enhanced by breastfeeding, which, unlike formula, provides infants with both beneficial bacteria and bioactive factors that promote their proliferation. The predominant bacterial groups shift for both breast and formula fed infants with the introduction of solid foods, which leads to an increase in species diversity, an increase in the relative abundance of Bacteroidetes and Firmicutes, and a decrease in the relative abundance of *Bifidobacterium*. As more and varied foods are added to the diet in the second and third years of life, the microbiota becomes more stable and the enterotypes responsive to habitual diet form (Bergstrom et al. 2014).

Disruptions to this process, through exposure to maternal overweight or health-related dysbiosis, C-section delivery, exclusive formula feeding, and prolonged and/or recurrent antibiotic exposure, have the potential to alter colonization with long-term impacts on health (Table 2). While few prospective studies have traced early exposures to differences in microbiota colonization and then to later developmental and health outcomes, a good deal of research shows that these pathways are likely to be important. Delivery by Caesarean section, for example, has been established as a risk factor for the development of allergy in epidemiological studies (Bager et al. 2008; West 2014) and linked to differences in gut microbiota that persist into childhood (Azad et al. 2013; Backhed et al. 2015; Salminen et al. 2004). Similarly, retrospective studies have established an association between antibiotic use in infancy and early childhood and an increased risk of inflammatory conditions, like IBD (Kronman et al. 2012), obesity (Trasande et al. 2013) and Type 2 diabetes (Bailey et al. 2014). Shorter term studies have documented differences in microbiota composition predating the development of these conditions. Infants with higher numbers of *E. coli* were more likely to develop eczema in the first 2 years of life than children with lower abundance (Penders et al. 2007). Children at high risk for developing allergy had a greater relative abundance of *Enterococcus* at 4 months of age compared to a control group (Vebo et al. 2011). Other studies have found lower rates of *Bifidobacterium* and *Lactobacillus* in infants who later develop allergy (Sjogren et al. 2009; Wang et al. 2008).

Again, while few studies have established the long-term links between infant feeding practices, microbiota and long-term risk of obesity, intervention studies in humans and animal models support the importance of the gut microbiome in energy

Table 2 Risk factors for altered microbiota development and long term health impacts

Risk factors	Microbiota impact in infant	Health consequences	References
Prenatal			
Maternal obesity	↓ *Bacteroides, Prevotella,* and *Bifidobacterium* ↑ *Clostridium*	↑ Child overweight and obesity ↑ Weight gain ↑ Inflammation	Collado et al. (2010); Cani and Delzenne (2007)
High pregnancy weight	↓ *Bacteroides, Prevotella,* and *Bifidobacterium* ↑ *Clostridium*		
Antibiotic use	↓ Diversity ↓ *Bifidobacterium* and *Lactobacillus*	↑ Child overweight ↑ Asthma	Keski-Nisula et al. (2013); Mueller et al. (2015b)
Birth			
Preterm birth	↑ Proteobacteria, Firmicutes, Enterobacteriaceae, *Staphylococcus, Propionibacterium,* and *Corynebacterium*	↑ NEC ↑ Inflammation	Collado et al. (2015); Dogra et al. (2015)
Caesarian section	↑ *Staphylococcus, Propionibacterium,* and *Corynebacterium* ↓ *Lactobacillus, Prevotella,* and *Sneathia* Delayed *Bacteroides*	↑ Child overweight ↑ Allergy ↑ Asthma	Arrieta et al. (2014); Belizán et al. (2007)
Postnatal			
Formula feeding	↑ Diversity ↑ *Bacteroides, Clostridium,* enterobacteria ↓ *Bifidobacterium* and *Lactobacillus*	↑ Child overweight/ obesity ↑ Inflammation	Arenz et al. (2004); Munblit and Verhasselt (2016); Horta et al. (2015)
Prolonged/ habitual antibiotic use	↓ Diversity ↓ *Actinobacteria* ↑ *Bacteroidetes* and *Proteobacteria*	↑ Inflammatory Bowel Disease ↑ Overweight/ obesity ↑ Type 2 Diabetes	Korpela et al. (2016); Kronman et al. (2012); Trasande et al. (2013) Bailey et al. (2014)

harvest and adiposity. Differences in the gut microbiota of obese and non-obese adults have documented (Ley et al. 2005) as have shifts in predominant bacteria with weight loss (Turnbaugh et al. 2006). Germ-free rodent studies have further supported causality between bacterial colonization and weight regulation; gut microbiota transfer from obese mice to germ free mice resulted in higher weight gain compared to genetically-identical mice receiving gut microbiota from lean mice (Backhed et al. 2007). Similar findings have been documented for undernutrition. The relative abundance of Actinobacteria differed between twin pairs discordant for kwashiorkor, a form of severe protein-energy malnutrition, and fecal transplants from the malnourished twins into germ-free mice resulted in significant weight loss (Smith et al. 2013).

Diet helps to regulate these energy modulating effects, providing nutrients for the maintenance and growth of commensal bacterial communities and acting as a selective factor for differential bacterial colonization (Blaut and Clavel 2007). High fat or carbohydrate diets, for example, modulate dominant bacteria populations in the gut. Through multiple intertwined metabolic and inflammatory pathways, including fermentation of non-digestible carbohydrates into short chain fatty acids, bacteria can have downstream effects on inflammation and the development of insulin resistance and type 2-diabetes (Cani and Delzenne 2007). Such metabolic processes may increase energy extraction through the up-regulation of carbohydrate and lipid-utilizing genes, providing almost 10% additional energy from the same quantity of food (Ley 2010). Alterations in the gut microbiota associated with early feeding patterns, therefore, may influence the absorption and storage of energy early in life, itself a risk factor for the development of later obesity.

At the same time, the importance of early exposures in shaping the microbiome also presents opportunities for intervention. The intergenerational transmission of the microbiome from mothers to infants prenatally, at birth, and postnatally through breastmilk provide opportunities to intervene in maternal diet and health to influence both maternal and child health. The importance of maternal weight status and weight gain during pregnancy for early colonization, for example, suggests that interventions targeted at weight reduction before pregnancy or promoting healthy levels of weight gain during pregnancy may improve mothers' microbiota and lead to the transmission of healthier initial bacterial inocula to their infants. Other approaches to improve initial colonization include swabbing the mouths of infants born by C-section with the vaginal and fecal microbiota of their mothers, which has been shown to shift initial colonization in a small sample of infants (Dominguez-Bello et al. 2010).

The use of prebiotics, non-digestible food ingredients that stimulate the growth or metabolism of beneficial bacteria, or probiotics, live bacteria administered in food or supplements (chapter "Fermented Vegetables as Vectors for Relocation of Microbial Diversity from the Environment to the Human Gut"), may also improve initial patterns of colonization both pre- and postnatally (Sohn and Underwood 2017). Dietary interventions or supplementation aimed at pregnant women have the potential to influence *in utero* exposures and improve long-term health outcomes (Luoto et al. 2011). Probiotic supplementation of mothers from 35 weeks gestation to when their infants were 6 months of age, for example, was shown to reduce the development of eczema in a sample of infants at high risk for the development of eczema (Wickens et al. 2008). Direct supplementation to infants also has the potential to shape health outcomes, though these strategies have had mixed success (Gritz and Bhandari 2015). The incorporation of pre- and probiotics into infant formula may help improve colonization by *Bifidobacterium* to levels like those found in breastfed infants (Baglatzi et al. 2016; Vandenplas et al. 2015) for infants whose mothers are unable to exclusively breastfeed and may explain the lack of pronounced differences between these groups in recent studies. Direct probiotic supplementation of infants with *Lactobacillus* and *Bifidobacterium* strains appears to reduce the risk of necrotizing enterocolitis (NEC) in preterm infants in many studies

(Gritz and Bhandari 2015) may reduce the risk of atopy in high-risk infants (Wickens et al. 2008). While questions remain about the efficacy of providing pre- and probiotics to healthy infants, these dietary and microbe-modulating therapies may have the potential to directly and indirectly alter the composition of the microbiota to promote healthy growth and development.

Conclusion

As this chapter describes, substantial evidence points to the importance of exposures across early life in shaping the gut microbiome and its subsequent metabolic and immunological function. Thus, promoting optimal microbial development through limiting unnecessary medical intervention, supporting breastfeeding and increasing access to healthy diets for mothers and infants is important for shaping both short-term growth and development and long-term health. Increased attention to the mechanisms underlying the intergenerational transmission and development of the microbiota in early life will help to identify opportunities for intervention and improved health across the lifespan.

References

Aagaard, K., Riehle, K., Ma, J., Segata, N., Mistretta, T. A., Coarfa, C., Raza, S., Rosenbaum, S., Van den Veyver, I., Milosavljevic, A., et al. (2012). A metagenomic approach to characterization of the vaginal microbiome signature in pregnancy. *PLoS One, 7*(6), e36466.

Aagaard, K., Ma, J., Antony, K. M., Ganu, R., Petrosino, J., & Versalovic, J. (2014). The placenta harbors a unique microbiome. *Science Translational Medicine, 6*(237), 237ra265.

Ajslev, T., Andersen, C., Gamborg, M., Sørensen, T., & Jess, T. (2011). Childhood overweight after establishment of the gut microbiota: The role of delivery mode, pre-pregnancy weight and early administration of antibiotics. *International Journal of Obesity, 35*(4), 522–529.

Ardissone, A. N., de la Cruz, D. M., Davis-Richardson, A. G., Rechcigl, K. T., Li, N., Drew, J. C., Murgas-Torrazza, R., Sharma, R., Hudak, M. L., Triplett, E. W., et al. (2014). Meconium microbiome analysis identifies bacteria correlated with premature birth. *PLoS One, 9*(3), e90784.

Arenz, S., Ruckerl, R., Koletzko, B., & von Kries, R. (2004). Breast-feeding and childhood obesity—A systematic review. *International Journal of Obesity and Related Metabolic Disorders, 28*(10), 1247–1256.

Arrieta, M. C., Stiemsma, L. T., Amenyogbe, N., Brown, E. M., & Finlay, B. (2014). The intestinal microbiome in early life: Health and disease. *Frontiers in Immunology, 5*, 427.

Avershina, E., Storro, O., Oien, T., Johnsen, R., Pope, P., & Rudi, K. (2014). Major faecal microbiota shifts in composition and diversity with age in a geographically restricted cohort of mothers and their children. *FEMS Microbiology Ecology, 87*(1), 280–290.

Azad, M. B., Konya, T., Maughan, H., Guttman, D. S., Field, C. J., Chari, R. S., Sears, M. R., Becker, A. B., Scott, J. A., & Kozyrskyj, A. L. (2013). Gut microbiota of healthy Canadian infants: Profiles by mode of delivery and infant diet at 4 months. *Canadian Medical Association Journal, 185*(5), 385–394.

Backhed, F., Manchester, J. K., Semenkovich, C. F., & Gordon, J. I. (2007). Mechanisms underlying the resistance to diet-induced obesity in germ-free mice. *Proceedings of the National Academy of Sciences of the United States of America, 104*(3), 979–984.

Backhed, F., Roswall, J., Peng, Y., Feng, Q., Jia, H., Kovatcheva-Datchary, P., Li, Y., Xia, Y., Xie, H., Zhong, H., et al. (2015). Dynamics and stabilization of the human gut microbiome during the first year of life. *Cell Host & Microbe, 17*(6), 852.

Bager, P., Wohlfahrt, J., & Westergaard, T. (2008). Caesarean delivery and risk of atopy and allergic disease: Meta-analyses. *Clinical and Experimental Allergy, 38*(4), 634–642.

Baglatzi, L., Gavrili, S., Stamouli, K., Zachaki, S., Favre, L., Pecquet, S., Benyacoub, J., & Costalos, C. (2016). Effect of infant formula containing a low dose of the probiotic Bifidobacterium lactis CNCM I-3446 on immune and gut functions in C-section delivered babies: A pilot study. *Clinical Medicine Insights: Pediatrics, 10*, 11.

Bailey, L. C., Forrest, C. B., Zhang, P., Richards, T. M., Livshits, A., & DeRusso, P. A. (2014). Association of antibiotics in infancy with early childhood obesity. *JAMA Pediatrics, 168*(11), 1063–1069.

Bastard, J. P., Maachi, M., Lagathu, C., Kim, M. J., Caron, M., Vidal, H., Capeau, J., & Feve, B. (2006). Recent advances in the relationship between obesity, inflammation, and insulin resistance. *European Cytokine Network, 17*(1), 4–12.

Belizán, J. M., Althabe, F., & Cafferata, M. L. (2007). Health consequences of the increasing caesarean section rates. *Epidemiology, 18*(4), 485–486.

Bergstrom, A., Skov, T. H., Bahl, M. I., Roager, H. M., Christensen, L. B., Ejlerskov, K. T., Molgaard, C., Michaelsen, K. F., & Licht, T. R. (2014). Establishment of intestinal microbiota during early life: A longitudinal, explorative study of a large cohort of Danish infants. *Applied and Environmental Microbiology, 80*(9), 2889–2900.

Biasucci, G., Rubini, M., Riboni, S., Morelli, L., Bessi, E., & Retetangos, C. (2010). Mode of delivery affects the bacterial community in the newborn gut. *Early Human Development, 86*(Suppl. 1), 13–15.

Blaut, M., & Clavel, T. (2007). Metabolic diversity of the intestinal microbiota: Implications for health and disease. *The Journal of Nutrition, 137*(3 Suppl. 2), 751S–755S.

Cabrera-Rubio, R., Collado, M. C., Laitinen, K., Salminen, S., Isolauri, E., & Mira, A. (2012). The human milk microbiome changes over lactation and is shaped by maternal weight and mode of delivery. *The American Journal of Clinical Nutrition, 96*(3), 544–551.

Cani, P. D., & Delzenne, N. M. (2007). Gut microflora as a target for energy and metabolic homeostasis. *Current Opinion in Clinical Nutrition and Metabolic Care, 10*(6), 729–734.

Carlson, A. L., Xia, K., Azcarate-Peril, M. A., Goldman, B. D., Ahn, M., Styner, M. A., Thompson, A. L., Geng, X., Gilmore, J. H., & Knickmeyer, R. C. (2018). Infant gut microbiome associated with cognitive development. *Biological Psychiatry, 83*(2), 148–159.

Collado, M. C., Isolauri, E., Laitinen, K., & Salminen, S. (2008). Distinct composition of gut microbiota during pregnancy in overweight and normal-weight women. *The American Journal of Clinical Nutrition, 88*(4), 894–899.

Collado, M. C., Isolauri, E., Laitinen, K., & Salminen, S. (2010). Effect of mother's weight on infant's microbiota acquisition, composition, and activity during early infancy: A prospective follow-up study initiated in early pregnancy. *The American Journal of Clinical Nutrition, 92*(5), 1023–1030.

Collado, M. C., Cernada, M., Neu, J., Perez-Martinez, G., Gormaz, M., & Vento, M. (2015). Factors influencing gastrointestinal tract and microbiota immune interaction in preterm infants. *Pediatric Research, 77*(6), 726–731.

Cox, L. M., Yamanishi, S., Sohn, J., Alekseyenko, A. V., Leung, J. M., Cho, I., Kim, S. G., Li, H., Gao, Z., Mahana, D., et al. (2014). Altering the intestinal microbiota during a critical developmental window has lasting metabolic consequences. *Cell, 158*(4), 705–721.

De Filippo, C., Cavalieri, D., Di Paola, M., Ramazzotti, M., Poullet, J. B., Massart, S., Collini, S., Pieraccini, G., & Lionetti, P. (2010). Impact of diet in shaping gut microbiota revealed by a comparative study in children from Europe and rural Africa. *Proceedings of the National Academy of Sciences of the United States of America, 107*(33), 14691–14696.

Dogra, S., Sakwinska, O., Soh, S. E., Ngom-Bru, C., Bruck, W. M., Berger, B., Brussow, H., Karnani, N., Lee, Y. S., Yap, F., et al. (2015). Rate of establishing the gut microbiota in infancy has consequences for future health. *Gut Microbes, 6*(5), 321–325.

Dominguez-Bello, M. G., Costello, E. K., Contreras, M., Magris, M., Hidalgo, G., Fierer, N., & Knight, R. (2010). Delivery mode shapes the acquisition and structure of the initial microbiota across multiple body habitats in newborns. *Proceedings of the National Academy of Sciences of the United States of America, 107*(26), 11971–11975.

Donnet-Hughes, A., Perez, P. F., Dore, J., Leclerc, M., Levenez, F., Benyacoub, J., Serrant, P., Segura-Roggero, I., & Schiffrin, E. J. (2010). Potential role of the intestinal microbiota of the mother in neonatal immune education. *The Proceedings of the Nutrition Society, 69*(3), 407–415.

Doyle, R. M., Alber, D. G., Jones, H. E., Harris, K., Fitzgerald, F., Peebles, D., & Klein, N. (2014). Term and preterm labour are associated with distinct microbial community structures in placental membranes which are independent of mode of delivery. *Placenta, 35*(12), 1099–1101.

Fallani, M., Amarri, S., Uusijarvi, A., Adam, R., Khanna, S., Aguilera, M., Gil, A., Vieites, J. M., Norin, E., Young, D., et al. (2011). Determinants of the human infant intestinal microbiota after the introduction of first complementary foods in infant samples from five European centres. *Microbiology, 157*(Pt 5), 1385–1392.

Fardini, Y., Chung, P., Dumm, R., Joshi, N., & Han, Y. W. (2010). Transmission of diverse oral bacteria to murine placenta: Evidence for the oral microbiome as a potential source of intrauterine infection. *Infection and Immunity, 78*(4), 1789–1796.

Flint, H. J., Duncan, S. H., Scott, K. P., & Louis, P. (2015). Links between diet, gut microbiota composition and gut metabolism. *Proceedings of the Nutrition Society, 74*(1), 13–22.

Goldsmith, F., O'Sullivan, A., Smilowitz, J. T., & Freeman, S. L. (2015). Lactation and intestinal microbiota: How early diet shapes the infant gut. *Journal of Mammary Gland Biology and Neoplasia, 20*(3–4), 149–158.

Gomez-Gallego, C., Garcia-Mantrana, I., Salminen, S., & Collado, M. C. (2016). The human milk microbiome and factors influencing its composition and activity. *Seminars in Fetal and Neonatal Medicine, 21*(6), 400–405. Elsevier.

Gregory, K. E., & Walker, W. A. (2013). Immunologic factors in human milk and disease prevention in the preterm infant. *Current Pediatrics Reports, 1*(4), 222–228.

Gritz, E. C., & Bhandari, V. (2015). The human neonatal gut microbiome: A brief review. *Frontiers in Pediatrics, 3*, 17.

Guarner, F., & Malagelada, J. R. (2003). Gut flora in health and disease. *Lancet, 361*(9356), 512–519.

Harmsen, H. J., Wildeboer-Veloo, A. C., Raangs, G. C., Wagendorp, A. A., Klijn, N., Bindels, J. G., & Welling, G. W. (2000). Analysis of intestinal flora development in breast-fed and formula-fed infants by using molecular identification and detection methods. *Journal of Pediatric Gastroenterology and Nutrition, 30*(1), 61–67.

Hemarajata, P., & Versalovic, J. (2013). Effects of probiotics on gut microbiota: Mechanisms of intestinal immunomodulation and neuromodulation. *Therapeutic Advances in Gastroenterology, 6*(1), 39–51.

Hooper, L. V., Midtvedt, T., & Gordon, J. I. (2002). How host-microbial interactions shape the nutrient environment of the mammalian intestine. *Annual Review of Nutrition, 22*, 283–307.

Horta, B. L., Loret de Mola, C., & Victora, C. G. (2015). Long-term consequences of breastfeeding on cholesterol, obesity, systolic blood pressure and type 2 diabetes: A systematic review and meta-analysis. *Acta Paediatrica, 104*(S467), 30–37.

Hotamisligil, G. S. (2006). Inflammation and metabolic disorders. *Nature, 444*(7121), 860–867.

Huda, M. N., Lewis, Z., Kalanetra, K. M., Rashid, M., Ahmad, S. M., Raqib, R., Qadri, F., Underwood, M. A., Mills, D. A., & Stephensen, C. B. (2014). Stool microbiota and vaccine responses of infants. *Pediatrics, 134*(2), e362–e372.

Hunt, K. M., Foster, J. A., Forney, L. J., Schutte, U. M., Beck, D. L., Abdo, Z., Fox, L. K., Williams, J. E., McGuire, M. K., & McGuire, M. A. (2011). Characterization of the diversity and temporal stability of bacterial communities in human milk. *PLoS One, 6*(6), e21313.

Jeurink, P. V., van Bergenhenegouwen, J., Jimenez, E., Knippels, L. M., Fernandez, L., Garssen, J., Knol, J., Rodriguez, J. M., & Martin, R. (2013). Human milk: A source of more life than we imagine. *Beneficial Microbes, 4*(1), 17–30.

Johnson, C. L., & Versalovic, J. (2012). The human microbiome and its potential importance to pediatrics. *Pediatrics, 129*(5), 950–960.

Jost, T., Lacroix, C., Braegger, C. P., & Chassard, C. (2012). New insights in gut microbiota establishment in healthy breast fed neonates. *PLoS One, 7*(8), e44595.

Jost, T., Lacroix, C., Braegger, C. P., Rochat, F., & Chassard, C. (2014). Vertical mother-neonate transfer of maternal gut bacteria via breastfeeding. *Environmental Microbiology, 16*(9), 2891–2904.

Kalliomaki, M., Kirjavainen, P., Eerola, E., Kero, P., Salminen, S., & Isolauri, E. (2001). Distinct patterns of neonatal gut microflora in infants in whom atopy was and was not developing. *The Journal of Allergy and Clinical Immunology, 107*(1), 129–134.

Kalliomaki, M., Collado, M. C., Salminen, S., & Isolauri, E. (2008). Early differences in fecal microbiota composition in children may predict overweight. *The American Journal of Clinical Nutrition, 87*(3), 534–538.

Keski-Nisula, L., Kyynäräinen, H. R., Kärkkäinen, U., Karhukorpi, J., Heinonen, S., & Pekkanen, J. (2013). Maternal intrapartum antibiotics and decreased vertical transmission of Lactobacillus to neonates during birth. *Acta Paediatrica, 102*(5), 480–485.

Khodayar-Pardo, P., Mira-Pascual, L., Collado, M. C., & Martinez-Costa, C. (2014). Impact of lactation stage, gestational age and mode of delivery on breast milk microbiota. *Journal of Perinatology, 34*(8), 599–605.

Koenig, J. E., Spor, A., Scalfone, N., Fricker, A. D., Stombaugh, J., Knight, R., Angenent, L. T., & Ley, R. E. (2011). Succession of microbial consortia in the developing infant gut microbiome. *Proceedings of the National Academy of Sciences of the United States of America, 108*(Suppl. 1), 4578–4585.

Koren, O., Goodrich, J. K., Cullender, T. C., Spor, A., Laitinen, K., Backhed, H. K., Gonzalez, A., Werner, J. J., Angenent, L. T., Knight, R., et al. (2012). Host remodeling of the gut microbiome and metabolic changes during pregnancy. *Cell, 150*(3), 470–480.

Korpela, K., Salonen, A.,Virta, L.J., Kekkonen, R.A., Forslund, K., Bork, P., & de Vos, W. (2016). Intestinal microbiome is related to lifetime antibiotic use in Finnish pre-school children. *Nature Communications 7*(1), 10410.

Krebs, N. F., Sherlock, L. G., Westcott, J., Culbertson, D., Hambidge, K. M., Feazel, L. M., Robertson, C. E., & Frank, D. N. (2013). Effects of different complementary feeding regimens on iron status and enteric microbiota in breastfed infants. *The Journal of Pediatrics, 163*(2), 416–423.

Kronman, M. P., Zaoutis, T. E., Haynes, K., Feng, R., & Coffin, S. E. (2012). Antibiotic exposure and IBD development among children: A population-based cohort study. *Pediatrics, 130*(4), e794–e803.

Kunz, C., Kuntz, S., Rudloff ,S. (2014). Bioactivity of human milk oligosaccharides, In: Moreno FJ, Sanz MLE, editors. *Food Oligosaccharides: Production, Analysis and Bioactivity*. Chichester: Wiley-Blackwell, p. 5–20.

Laursen, M. F., Andersen, L. B., Michaelsen, K. F., Molgaard, C., Trolle, E., Bahl, M. I., & Licht, T. R. (2016). Infant gut microbiota development is driven by transition to family foods independent of maternal obesity. *mSphere, 1*(1), e00069-15.

Ley, R. E. (2010). Obesity and the human microbiome. *Current Opinion in Gastroenterology, 26*(1), 5–11.

Ley, R. E., Backhed, F., Turnbaugh, P., Lozupone, C. A., Knight, R. D., & Gordon, J. I. (2005). Obesity alters gut microbial ecology. *Proceedings of the National Academy of Sciences of the United States of America, 102*(31), 11070–11075.

Luoto, R., Kalliomaki, M., Laitinen, K., Delzenne, N. M., Cani, P. D., Salminen, S., & Isolauri, E. (2011). Initial dietary and microbiological environments deviate in normal-weight compared to overweight children at 10 years of age. *Journal of Pediatric Gastroenterology and Nutrition, 52*(1), 90–95.

Madan, J. C., Hoen, A. G., Lundgren, S. N., Farzan, S. F., Cottingham, K. L., Morrison, H. G., Sogin, M. L., Li, H., Moore, J. H., & Karagas, M. R. (2016). Association of cesarean delivery and formula supplementation with the intestinal microbiome of 6-week-old infants. *JAMA Pediatrics, 170*(3), 212–219.

Marques, T. M., Wall, R., Ross, R. P., Fitzgerald, G. F., Ryan, C. A., & Stanton, C. (2010). Programming infant gut microbiota: Influence of dietary and environmental factors. *Current Opinion in Biotechnology, 21*(2), 149–156.

Martinez, F. D. (2014). The human microbiome. Early life determinant of health outcomes. *Annals of the American Thoracic Society, 11*(Suppl. 1), S7–S12.

Meropol, S. B., & Edwards, A. (2015). Development of the infant intestinal microbiome: A bird's eye view of a complex process. *Birth Defects Research. Part C, Embryo Today, 105*(4), 228–239.

Morelli, L. (2008). Postnatal development of intestinal microflora as influenced by infant nutrition. *The Journal of Nutrition, 138*(9), 1791S–1795S.

Mueller, N. T., Bakacs, E., Combellick, J., Grigoryan, Z., & Dominguez-Bello, M. G. (2015a). The infant microbiome development: Mom matters. *Trends in Molecular Medicine, 21*(2), 109–117.

Mueller, N. T., Whyatt, R., Hoepner, L., Oberfield, S., Dominguez-Bello, M. G., Widen, E., Hassoun, A., Perera, F., & Rundle, A. (2015b). Prenatal exposure to antibiotics, cesarean section and risk of childhood obesity. *International Journal of Obesity, 39*(4), 665–670.

Mueller, N. T., Shin, H., Pizoni, A., Werlang, I. C., Matte, U., Goldani, M. Z., Goldani, H. A., & Dominguez-Bello, M. G. (2016). Birth mode-dependent association between pre-pregnancy maternal weight status and the neonatal intestinal microbiome. *Scientific Reports, 6*, 23133.

Munblit, D., & Verhasselt, V. (2016). Allergy prevention by breastfeeding: Possible mechanisms and evidence from human cohorts. *Current Opinion in Allergy and Clinical Immunology, 16*(5), 427–433.

Mysorekar, I. U., & Cao, B. (2014). Microbiome in parturition and preterm birth. *Seminars in Reproductive Medicine, 32*(1), 50–55.

Newburg, D. S., & Walker, W. A. (2007). Protection of the neonate by the innate immune system of developing gut and of human milk. *Pediatric Research, 61*(1), 2–8.

Parrett, A. M., & Edwards, C. A. (1997). In vitro fermentation of carbohydrate by breast fed and formula fed infants. *Archives of Disease in Childhood, 76*(3), 249–253.

Penders, J., Thijs, C., van den Brandt, P. A., Kummeling, I., Snijders, B., Stelma, F., Adams, H., van Ree, R., & Stobberingh, E. E. (2007). Gut microbiota composition and development of atopic manifestations in infancy: The KOALA Birth Cohort Study. *Gut, 56*(5), 661–667.

Rautava, S., & Walker, W. A. (2009). Academy of Breastfeeding Medicine founder's lecture 2008: Breastfeeding—An extrauterine link between mother and child. *Breastfeeding Medicine, 4*(1), 3–10.

Reinhardt, C., Reigstad, C. S., & Backhed, F. (2009). Intestinal microbiota during infancy and its implications for obesity. *Journal of Pediatric Gastroenterology and Nutrition, 48*(3), 249–256.

Rodriguez, J. M., Murphy, K., Stanton, C., Ross, R. P., Kober, O. I., Juge, N., Avershina, E., Rudi, K., Narbad, A., Jenmalm, M. C., et al. (2015). The composition of the gut microbiota throughout life, with an emphasis on early life. *Microbial Ecology in Health and Disease, 26*, 26050.

Salminen, S., Gibson, G., McCartney, A., & Isolauri, E. (2004). Influence of mode of delivery on gut microbiota composition in seven year old children. *Gut, 53*(9), 1388–1389.

Schwartz, S., Friedberg, I., Ivanov, I. V., Davidson, L. A., Goldsby, J. S., Dahl, D. B., Herman, D., Wang, M., Donovan, S. M., & Chapkin, R. S. (2012). A metagenomic study of diet-dependent interaction between gut microbiota and host in infants reveals differences in immune response. *Genome Biology, 13*(4), r32.

Sela, D. A., & Mills, D. A. (2010). Nursing our microbiota: Molecular linkages between bifidobacteria and milk oligosaccharides. *Trends in Microbiology, 18*(7), 298–307.

Sjogren, Y. M., Jenmalm, M. C., Bottcher, M. F., Bjorksten, B., & Sverremark-Ekstrom, E. (2009). Altered early infant gut microbiota in children developing allergy up to 5 years of age. *Clinical and Experimental Allergy, 39*(4), 518–526.

Smith, M. I., Yatsunenko, T., Manary, M. J., Trehan, I., Mkakosya, R., Cheng, J., Kau, A. L., Rich, S. S., Concannon, P., Mychaleckyj, J. C., et al. (2013). Gut microbiomes of Malawian twin pairs discordant for kwashiorkor. *Science, 339*(6119), 548–554.

Smith-Brown, P., Morrison, M., Krause, L., & Davies, P. S. (2016). Dairy and plant based food intakes are associated with altered faecal microbiota in 2 to 3 year old Australian children. *Scientific Reports, 6,* 32385.

Sohn, K., & Underwood, M. A. (2017). Prenatal and postnatal administration of prebiotics and probiotics. *Seminars in Fetal and Neonatal Medicine, 22*(5), 284–289. Elsevier.

Stensballe, L. G., Simonsen, J., Jensen, S. M., Bønnelykke, K., & Bisgaard, H. (2013). Use of antibiotics during pregnancy increases the risk of asthma in early childhood. *The Journal of Pediatrics, 162*(4), 832–838.e833.

Tannock, G. W., Lawley, B., Munro, K., Gowri Pathmanathan, S., Zhou, S. J., Makrides, M., Gibson, R. A., Sullivan, T., Prosser, C. G., Lowry, D., et al. (2013). Comparison of the compositions of the stool microbiotas of infants fed goat milk formula, cow milk-based formula, or breast milk. *Applied and Environmental Microbiology, 79*(9), 3040–3048.

Thompson, A. L. (2012). Developmental origins of obesity: Early feeding environments, infant growth, and the intestinal microbiome. *American Journal of Human Biology, 24*(3), 350–360.

Thompson, A. L., Monteagudo-Mera, A., Cadenas, M. B., Lampl, M. L., & Azcarate-Peril, M. A. (2015). Milk- and solid-feeding practices and daycare attendance are associated with differences in bacterial diversity, predominant communities, and metabolic and immune function of the infant gut microbiome. *Frontiers in Cellular and Infection Microbiology, 5,* 3.

Tilg, H., & Kaser, A. (2011). Gut microbiome, obesity, and metabolic dysfunction. *The Journal of Clinical Investigation, 121*(6), 2126–2132.

Trasande, L., Blustein, J., Liu, M., Corwin, E., Cox, L. M., & Blaser, M. J. (2013). Infant antibiotic exposures and early-life body mass. *International Journal of Obesity, 37*(1), 16–23.

Turnbaugh, P. J., Ley, R. E., Mahowald, M. A., Magrini, V., Mardis, E. R., & Gordon, J. I. (2006). An obesity-associated gut microbiome with increased capacity for energy harvest. *Nature, 444*(7122), 1027–1031.

Ussar, S., Griffin, N. W., Bezy, O., Fujisaka, S., Vienberg, S., Softic, S., Deng, L., Bry, L., Gordon, J. I., & Kahn, C. R. (2015). Interactions between gut microbiota, host genetics and diet modulate the predisposition to obesity and metabolic syndrome. *Cell Metabolism, 22*(3), 516–530.

Vael, C., & Desager, K. (2009). The importance of the development of the intestinal microbiota in infancy. *Current Opinion in Pediatrics, 21*(6), 794–800.

Vandenplas, Y., Zakharova, I., & Dmitrieva, Y. (2015). Oligosaccharides in infant formula: More evidence to validate the role of prebiotics. *British Journal of Nutrition, 113*(9), 1339–1344.

Vebo, H. C., Sekelja, M., Nestestog, R., Storro, O., Johnsen, R., Oien, T., & Rudi, K. (2011). Temporal development of the infant gut microbiota in immunoglobulin E-sensitized and non-sensitized children determined by the GA-map infant array. *Clinical and Vaccine Immunology, 18*(8), 1326–1335.

Voreades, N., Kozil, A., & Weir, T. L. (2014). Diet and the development of the human intestinal microbiome. *Frontiers in Microbiology, 5,* 494.

Walker, W. A. (2017). The importance of appropriate initial bacterial colonization of the intestine in newborn, child, and adult health. *Pediatric Research, 82*(3), 387–395.

Walker, R. W., Clemente, J. C., Peter, I., & Loos, R. J. F. (2017). The prenatal gut microbiome: Are we colonized with bacteria in utero? *Pediatric Obesity, 12*(Suppl. 1), 3–17.

Wang, M., Karlsson, C., Olsson, C., Adlerberth, I., Wold, A. E., Strachan, D. P., Martricardi, P. M., Aberg, N., Perkin, M. R., Tripodi, S., et al. (2008). Reduced diversity in the early fecal microbiota of infants with atopic eczema. *The Journal of Allergy and Clinical Immunology, 121*(1), 129–134.

West, C. E. (2014). Gut microbiota and allergic disease: New findings. *Current Opinion in Clinical Nutrition and Metabolic Care, 17*(3), 261–266.

Wickens, K., Black, P. N., Stanley, T. V., Mitchell, E., Fitzharris, P., Tannock, G. W., Purdie, G., Crane, J., & Probiotic Study Group. (2008). A differential effect of 2 probiotics in the prevention of eczema and atopy: A double-blind, randomized, placebo-controlled trial. *The Journal of Allergy and Clinical Immunology, 122*(4), 788–794.

Wopereis, H., Oozeer, R., Knipping, K., Belzer, C., & Knol, J. (2014). The first thousand days—Intestinal microbiology of early life: Establishing a symbiosis. *Pediatric Allergy and Immunology, 25*(5), 428–438.

Yang, Z., & Huffman, S. L. (2013). Nutrition in pregnancy and early childhood and associations with obesity in developing countries. *Maternal & Child Nutrition, 9*(S1), 105–119.

Zivkovic, A. M., German, J. B., Lebrilla, C. B., & Mills, D. A. (2011). Human milk glycobiome and its impact on the infant gastrointestinal microbiota. *Proceedings of the National Academy of Sciences of the United States of America, 108*(Suppl. 1), 4653–4658.

Zivkovic, A. M., Lewis, Z. T., German, J. B., & Mills, D. A. (2013). Establishment of a milk-oriented microbiota (MOM) in early life: How babies meet their MOMs. *Functional Food Reviews, 5*(1), 3–12.

"We Are What We Eat": How Diet Impacts the Gut Microbiota in Adulthood

Taojun Wang, Dominique I. M. Roest, Hauke Smidt, and Erwin G. Zoetendal

Abstract The important role of the microbes residing in our gut, collectively called the microbiota, in human health is widely acknowledged. There are numerous factors that have an impact on the microbiota in the gut of which diet is considered a crucial one. In this chapter we highlight our current knowledge on the ecology of the microbiota in adults and how it is affected by diet. We summarize observations from different cross-sectional and intervention studies that focused on the impact of diet on microbiota composition and activity. Special attention is paid to which microbial metabolites can be produced in the gut; how these are affected by different dietary components such as carbohydrates, fat, and proteins; and how these are associated to human health. Finally, we provide recommendations for future intervention studies in order to improve our understanding of the complex interplay between microbes, diet, and ourselves.

Keywords Adult microbiome · Western diet · Traditional diet · Microbial stability · Resilience of the gut microbiome · Intestinal microbial metabolites

Introduction

The gut microbiota evolves with age from infant to adult (Yatsunenko et al. 2012; O'Toole and Jeffery 2015). In adulthood, the gut microbiota reaches its highest diversity, compared with infants and the elderly (Cheng et al. 2016; Lynch and Pedersen 2016; An et al. 2018). Furthermore, composition of the gut microbiota is considered stable over time, although a long term study showed that significant alterations may also happen during adulthood (Rajilić-Stojanović et al. 2013). Firmicutes, Bacteroidetes, and Actinobacteria are in general the predominant phyla

---

T. Wang · D. I. M. Roest · H. Smidt · E. G. Zoetendal (✉)
Laboratory of Microbiology, Wageningen University & Research, Wageningen, Netherlands
e-mail: erwin.zoetendal@wur.nl

© Springer Nature Switzerland AG 2019
M. A. Azcarate-Peril et al. (eds.), *How Fermented Foods Feed a Healthy Gut Microbiota*, https://doi.org/10.1007/978-3-030-28737-5_11

within the gut microbiota; however, the composition is host-specific, and each individual has a unique gut microbiota composition (Zoetendal and de Vos 2014). In contrast, the functional capacity encoded by the gut metagenome is more similar between subjects (Qin et al. 2010; The Human Microbiome Project Consortium 2012). Given the fact that we are all unique as human beings but our bodies function in a highly similar fashion, this is not very surprising.

A number of factors have been associated with the composition of the gut microbiota. These include the genetic background, environment, health status, use of antibiotics, and diet (Maslowski and Mackay 2011; Goodrich et al. 2014; Falony et al. 2016). The famous expression "you are what you eat" emphasizes that diet is essential. However, we have become more aware of the significance of the fact that our daily diet not only feeds us but also provides substrates for the gut microbiota. Long-term diet is the dietary habit kept for several years or decades with relatively stable dietary components, and it has been shown to contribute to shaping gut microbial composition (Wu et al. 2011). By contrast, short-term dietary changes have in general a minor impact on microbial composition, but may change overall activity patterns (David et al. 2014).

Typical long-term dietary patterns, such as consumption of Western or traditional diets, have been associated with specific gut microbial profiles. For example, Western diets are rich in protein and fat, a diet that is markedly different from African diets, which are traditionally rich in dietary fibres, and these differences are reflected in the microbiota composition (Yatsunenko et al. 2012). It was shown that even short-term consumption of a Western diet resulted in a microbiota-mediated increase in colorectal cancer (CRC) associated risk factors, while the opposite was observed with consumption of a traditional South African diet (O'Keefe et al. 2015). Other examples of the importance of diet with respect to human health are the association between irritable bowel syndrome (IBS) symptoms and fermentable oligosaccharides, disaccharides, monosaccharides and polyols (FODMAP) diets (Halmos et al. 2014b) as well as the associations of, among others, obesity, inflammatory bowel diseases (IBD), malnutrition, and type 2 diabetes with the gut microbiota and diets (Ley et al. 2006; Frank et al. 2007; Larsen et al. 2010; Kau et al. 2011).

Overall, diet is one of the most important factors associated with human health and diseases, which can exert its effects either directly or indirectly via the gut microbiota. In this chapter, we provide an overview of our insight into the adult gut microbiota and its association with diet.

The Adult Gut Microbiota

The gut microbiota is individual and niche-specific Human beings are recognized as "superorganisms" composed both of human and microbial cells. It is estimated that the total number of bacterial (approximately 3.9×10^{13}) is similar to that of human cells (Sender et al. 2016a). Collectively, the gut microbiota comprises more than 1000 microbial species, harbouring approximately ten million nonredundant genes, which is 2–3 orders of magnitude larger than the human gene complement (Qin et al. 2010; Li et al. 2014).

As indicated before, the gut microbiota in the healthy adult is dominated by Firmicutes and Bacteroidetes as well as Actinobacteria at the phylum level, with lower proportions of other phyla such as Proteobacteria and Verrucomicrobia (Rajilić-Stojanović et al. 2007; Zhernakova et al. 2016). However, the microbial composition varies along the intestinal tract due to variation in environmental conditions (Zoetendal and de Vos 2014). The small intestine consisting of duodenum, jejunum and ileum is considered a harsh environment for the gut microbiota due to the relatively high oxygen level, presence of bile acids and digestive enzymes, and short transit time. As a result, microbial numbers are only a fraction of the total gut microbiota with 10^3–10^4 per millilitre in duodenum and jejunum, increasing to 10^8 in the ileum (Booijink et al. 2010; Sender et al. 2016b). Since it is difficult to access the small intestine for sampling, studies on its residing microbiota are limited. It has been shown that the microbial composition in the small intestine is different from that in the colon, and even different when comparing proximal and distal parts of the small intestine (Ou et al. 2009; Zoetendal et al. 2012). Microbiota profiling of ileal effluent samples of ileostomists as well as samples taken at proximal sites in the small intestine demonstrated that among others, members of the genera *Streptococcus* and *Veillonella* are predominant (Booijink et al. 2010; Zoetendal et al. 2012). In contrast, samples from the distal small intestine resemble those of the colon, which could be due to colonic reflux (Zoetendal et al. 2012).

The colon is considered the main fermentation vessel of the human body and is the most densely populated organ with 10^{11} cells per millilitre and approximately 400 mL in total. The high density and biomass levels are attributed to the extended transit time and more suitable fermentation conditions in comparison to the small intestine (Zoetendal and de Vos 2014; Sender et al. 2016a). Most studies in the past decades used faecal samples to characterize the gut microbiota in the colon due to their accessibility. Although a variety of studies have demonstrated that the gut microbiota is individual-specific, population-level studies observed that the gut microbiota is characterized by certain conserved patterns at a higher organizational level, termed enterotypes (Arumugam et al. 2011). It is evident that such a stratification of the microbiota may improve our understanding of gut microbial ecology (Costea et al. 2018); however, there is an ongoing debate regarding the identification and stability of the enterotypes. Meta-analyses of microbial composition data indicated that *Faecalibacterium prausnitzii*, *Oscillospira guillermondii* and *Ruminococcus obeum* are the top three taxa shared by adults (Shetty et al. 2017). Of note is that the faecal microbiota only resembles that of the distal colon, and that differences have been observed between lumen and mucosa-associated microbiota (Zoetendal et al. 2002; Lepage et al. 2005). A study characterizing the microbiota associated with human rectal biopsies and mucosal swab samples indicated a higher proportion of Proteobacteria and Actinobacteria compared to faecal samples. Taxa belonging to these phyla are often described to primarily metabolize peptones and amino acids in the mucus layer, reflecting the selective pressure and adaption to substrates most readily available in this niche (Albenberg et al. 2014; Jones et al. 2018). Furthermore, *Akkermansia muciniphila*, an abundant mucin degrader, is thought to act as the gatekeeper of the mucosa maintaining the stability of our gut microbial ecosystem (Derrien et al. 2004; de Vos 2017; Geerlings et al. 2018). Overall, it is evident that the differences in microbiota composition between subjects and intestinal locations makes it challenging to comprehensively study the role of the microbiota and how it is impacted by diet.

Microbial diversity, stability and resilience The adult gut microbiota remains relatively stable over time, suggesting that it is resilient to environmental perturbations. One hypothesis is that the microbes residing in microhabitats like the colonic crypts, the appendix or the mucus layer serve as reservoirs of microbial diversity and can replenish the gut microbiota after perturbations (Donaldson et al. 2016). Moreover, a deep phylogenetic analysis of gut microbial data in Western adults demonstrated that a limited number of gut microbial taxa show bimodal distributions, being either highly abundant or nearly absent in most individuals. Such bistability may also play a role in maintaining the homeostasis of the gut ecosystem (Lahti et al. 2014).

Despite the resilience of the gut ecosystem, perturbations exceeding its capacity, like the use of antibiotics and extreme dietary changes can disrupt gut microbial composition and function, converting the healthy stable state of gut microbiota into a degraded stable state (Lozupone et al. 2012). Accumulating evidence demonstrated that the human gut microbiota composition differs between healthy subjects and those suffering from disorders like obesity, IBD and CRC (Turnbaugh et al. 2006; Frank et al. 2007; Wang et al. 2012). It has been postulated that these disorders may be induced by the consumption of a Western diet via diet-microbiota interactions (Zoetendal and de Vos 2014). However, whether a different microbiota composition is the cause or the result of disorders and diseases remains unknown in most cases. On the other hand, it was pointed out that compromised individuals have generally a lower gut microbial diversity compared to healthy individuals (Lozupone et al. 2012; Zoetendal and de Vos 2014; Menni et al. 2018). Therefore, homeostasis of healthy individuals is hypothesized to be associated with a highly diverse and resilient microbiota (Zoetendal and de Vos 2014). In line with this, obese individuals with a low microbial richness are at increased risk for developing insulin resistance, dyslipidaemia and a more pronounced inflammatory phenotype in comparison with those with higher microbial richness (Ley et al. 2006; Le Chatelier et al. 2013).

The transference of microbes from healthy donors to patients (faecal microbiota transplantation, FMT) has been tested as a therapy for a variety of health disorders. The most successful application of FMT has been without doubt in patients suffering from recurrent *Clostridium difficile* infections (Van Nood et al. 2013). FMT has also been used in individuals suffering from ulcerative colitis (UC) and IBD. Studies have shown that the treatment can relieve symptoms or eradicate disease after one or more transplants (Moayyedi et al. 2015; Vermeire et al. 2016), indicating that some microbiomes contain health promoting aspects and can pass on their beneficial

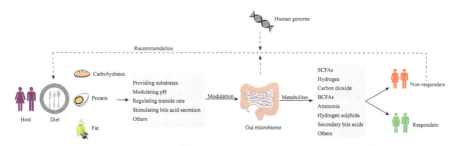

Fig. 1 Targets for improving the modulation of gut microbiota as a basis for individualized dietary recommendations. Diet modulates gut microbiota via many factors. Gut microbes convert dietary components into a broad range of small molecule metabolites. However, variable responsiveness to a given diet is seen in different individuals who can be stratified into responders and non-responders. Accordingly, precision diets based on information related to individualized gut microbiome and human genome data is recommended with the ambition to change non-responders into responders

effects to others. This may be explained by the fact that a more diverse and more resilient microbiota can help compromised individuals to rebuild a stable personalized microbiota restoring intestinal homeostasis. This hypothesis is supported by the recent discovery that donor microbiota richness determines FMT success in IBD (Vermeire et al. 2016). However, not all individuals respond to the FMT. For example studies in UC patients found that FMT induces remission in some individuals, but not all (Moayyedi et al. 2015; Rossen et al. 2015). Moreover, one donor faeces does not suit all individuals. Analysis of the microbiota of responders and non-responders may help elucidating the mechanisms underlying FMT success (Fuentes et al. 2017).

Microbial metabolites Dietary components escaping the digestion and absorption in the small intestine include mostly complex carbohydrates, such as dietary fibre, but also certain proteins and peptides. These are subsequently fermented by the gut microbiota, generating specific metabolites (as shown in Fig. 1) including short-chain fatty acids (SCFAs) and gasses such as hydrogen and carbon dioxide (Nicholson et al. 2012; Louis et al. 2014; Donia and Fischbach 2015). Besides diet, the mucin secreted by goblet cells provides a substrate for gut microbes (Atuma et al. 2001; Derrien et al. 2004).

SCFAs, notably acetate, propionate and butyrate, produced in the intestine can reach as far as the lungs through circulation, and cross the blood-brain barrier. The combined concentration of the three SCFAs acetate, propionate and butyrate is approximately 50–150 mM with a typical ratio of 3:1:1 in the colon (Louis et al. 2014). Butyrate is the main energy source for the epithelial cells and is locally consumed resulting in a lower concentration in the systemic circulation. Propionate that is absorbed from the lumen is transported via the portal vein and subsequently metabolized by the liver. Acetate remains in relatively high concentrations in peripheral blood circulation (Cummings et al. 1987; Louis et al. 2014). Butyrate is formed during fermentation using acetyl-CoA as the starting point via the phosphotransbu-tyrylase/butyrate kinase or butyryl-CoA:acetate CoA-transferase route, which allows butyrate producers to utilize sugars, lactate and acetate as well as amino acids as substrates (Vital et al. 2014; Koh et al. 2016). A wide variety of microbes are able to produce butyrate, including e.g. *Faecalibacterium prausnitzii*, *Anaerobutyricum hallii* (recently renamed from *Eubacterium hallii*), *Roseburia intestinalis*, and *Intestinimonas* strain AF211 (Bui et al. 2015; Koh et al. 2016). Propionate is generated predominantly via the succinate pathway in the colon (Louis et al. 2014), but the acrylate and propanediol pathways have also been described for gut microbes (Koh et al. 2016). Like butyrate, propionate can also be produced by a number of bacterial taxa, which include *Bacteroides* spp. and *Veillonella* spp., among others (Louis et al. 2014). As for acetate, it can be produced via acetyl-CoA or via the Wood-Ljungdahl pathway (Koh et al. 2016), in which acetate is produced via reductive acetogenesis. Acetate can be produced by almost all intestinal microbes. When fermentable carbohydrates are limited, which is typical for Western diets, the gut microbiota switches to utilize less favourable sources such as amino acids. This may result in increased concentrations of branched-chain fatty acids (BCFAs), amines, ammonia, and phenolic compounds, which are considered detrimental (Louis et al. 2014; Koh et al. 2016).

Besides organic acids, fermentation also leads to production of gasses. Hydrogen is one of the main gasses that is produced during microbial fermentation in the gut. The three most common pathways for hydrogen production include the reoxidation of reduced pyridine and flavin nucleotides, the metabolism of formate generated by the cleavage of pyruvate, and the activity of pyruvate:ferredoxin oxidoreductase and hydrogenase (Carbonero et al. 2012). Hydrogen producers include, among others, strains of *Ruminococcus* spp., *Roseburia* spp., *Clostridium* spp., and *Bacteroides* spp. (Carbonero et al. 2012). Hydrogen accumulation by hydrogenogenic microbes increases the partial pressure in the gut, limiting further microbial fermentation ther-modynamically. In turn, hydrogen can be removed from the intestine via flatus and can also be transferred into the blood with subsequent excretion via the lungs. Microbial disposal of hydrogen in the gut by hydrogenotrophic microbes has also been described and includes reductive acetogenesis, methanogenesis and sulphate-reduction, which are performed by archaea or bacteria that use hydrogen as electron donor for anaerobic respiration (Stams and Plugge 2009; Nakamura et al. 2010; Carbonero et al. 2012). Reductive acetogenic bacteria like *Blautia hydrogenotro-phica* and *Marvinbryantia formatexigens* can utilize hydrogen and carbon dioxide or formate which results in the generation of acetate (Rey et al. 2010). It is estimated that almost one third of acetate in the gut ecosystem is derived from this activity (Miller and Wolin 1996). The most common methanogenic archaeal species isolated from the human colon is *Methanobrevibacter smithii* that converts hydrogen and carbon dioxide to methane. It is estimated that one third of the Western population carries approximately 10^9 CFU/g of methanogens in stool, while individuals who are non-methane producers harbour less than 10^4 CFU/g (Pochart et al. 1992; Levitt et al. 2006). Remarkably, in Africans approximately 80% of the population harbours high numbers of intestinal methanogens (Segal et al. 1988). Whether this is related to the differences in diets remains speculative. Sulphidogenic bacteria can generate the toxic component hydrogen sulfide from inorganic and organic sulphur containing compounds. Many use sulphate and/or sulphite and hydrogen as the electron acceptor and donor, respectively, although it should be noted that sulphide can also be produced through fermentation of sulphur-containing organic compounds (Feng et al. 2017). Hydrogen sulfide was reported to be associated with IBS, IBD and CRC (Carbonero et al. 2012). Sources of intestinal sulphur include host-derived mucin and taurine as well as dietary sulphur-containing components including amino acids.

A vast number of metabolites from dietary sources are produced by the gut microbiota. Some of them are beneficial, but some detrimental (Louis et al. 2014; Donia and Fischbach 2015). They play a critical role in the connection between the gut microbiota and human health and diseases. Therefore, clarifying how metabolites affect human health and diseases and how diets are associated with metabolite production can be very meaningful but challenging in the future.

The Impact of Diet on the Gut Microbiota

Dietary modulation of the gut microbiota Diet is considered a crucial determinant of gut microbial composition, and a variety of factors may explain why diet acts as the ecological driving force, which include being the energy source of the microbes, affecting the pH of the lumen, regulating gut transit rate and stimulating the secretion of bile acids and other components (Fig. 1). First, dietary components are used as energy source, conferring selective growth advantages to specific microbial taxa. Typical Western diets with high fat and protein content are markedly different from non-Western diets, such as traditional African diets which are rich in complex carbohydrates. Hence, the finding that *Prevotella* is abundantly detected in Africans with non-Western diet, suggests that this taxon plays a significant role in extracting energy from undigested plant-derived carbohydrates (De Filippo et al. 2010, 2017; Ou et al. 2013). Additionally, the conversion of dietary components by the gut microbiota can lead to changes in pH, caused by metabolites such as SCFAs that can decrease pH in the intestinal lumen. It has been observed that pH values vary from 5.5 in the caecum to 6.5 in the descending colon (Cummings et al. 1987). An *in vitro* study using continuous flow fermenters indicated that the Bacteroidetes outcompeted Firmicutes and dominated in the system at pH 6.5, but could not persist at pH 5.5, demonstrating the critical role of pH in shaping gut microbial composition (Duncan et al. 2009). Another factor that can be a link between diet and microbiota is intestinal transit time (ITT). ITT has an impact on microbial competition since a faster transit favours the survival of fast growing organisms over slow growing ones. A murine model study showed that ITT was accelerated by high dietary fibre intake (Kashyap et al. 2013), and the explanation for a faster ITT is that dietary fibre, and especially incompletely fermented fibre, attracts water, increasing digesta mass, which shortens the transit time (Conlon and Bird 2014). Last but not least, secretion of digestive enzymes and other components like bile acids into the intestinal lumen are stimulated by diet intake and is a selective force affecting residing microbes. Bile acids in the colon can act as antibacterial components selecting resistant microorganisms, indirectly making diet a selective force on gut microbial composition and function (Jones et al. 2008; Ridlon et al. 2014). Overall, studying the impact of diet on our microbiota can be very complex due to a myriad of actions.

A range of dietary components have been investigated for their impact on human health through gut microbial functionality (Table 1). The studies can be roughly divided into two classes. The first class examined the effects of long-term dietary patterns on gut microbiota by cross-sectional comparative analysis of individuals that consistently consumed a specific diet such as Western diet, traditional diet, vegetarian or vegan diets. The second class comprises short-term intervention studies that investigated the impact of diet on gut microbiota by switching from one dietary pattern to another for a certain period of time. These studies allowed to determine whether a specific diet or increased intake of certain macronutrients can lead to variations in gut microbiota composition and/or activity.

Several human short-term intervention studies have indicated that the microbiota can quickly adapt to a change in dietary pattern, which may not only lead to drastic microbiota activity changes, but also to compositional changes (Wu et al. 2011; David et al. 2014; O'Keefe et al. 2015), while the gut microbiota reverted to its original structure when the dietary intervention stopped (David et al. 2014). Long-term dietary patterns have been linked to distinct gut microbial communities (Wu et al. 2011). Major differences in microbiota composition, notably within the Bacteroidetes phylum (*Prevotella* in native Africans and *Bacteroides* in African Americans) have been observed when comparing the gut microbiota of Africans consuming a native diet to that of African Americans consuming a Western diet (Ou et al. 2013). A recent study showed that, as a consequence of modernization, a shift occurred from ancient microbial communities with a higher capacity to degrade complex carbohydrates, to microbial taxa that are more suited to metabolize protein and fat (De Filippo et al. 2017). In line with this observation, a mouse model study indicated that long-term adaptation to a Western diet over generations may result in progressive loss of gut microbial species and diversity which cannot be fully recovered by the reintroduction of traditional diets (Sonnenburg et al. 2016). This is a worrying scenario as Westernization of our diet, associated with increased risk of gastrointestinal diseases, is increasing world-wide.

Responsiveness to dietary interventions appears to be host dependent, although a number of studies have identified dietary components that may have a universal effect on the gut microbiota. A study with obese men indicated that responsiveness of the gut microbiota to dietary components, and the generated fermentation products, differed drastically between subjects, suggesting that individuals could be stratified into responders and non-responders based on their gut microbial dynamics (Salonen et al. 2014). Similarly, a dietary intervention study using a 3-day window of consumption of barley kernel-based bread showed that certain individuals gained improvement in glucose metabolism with a corresponding increase of *Prevotella copri* abundance while others did not (Kovatcheva-Datchary et al.

Table 1 Overview of diet intervention studies in individuals

Diet	Cohort	Time	Analysis method	Gut microbial response	References
Inulin/Oligofructose (50%/50%, 16 g/day)	30 obese women	3 months	Human Intestinal Tract Chip (HITChip); qPCR	*Bifidobacterium* ↑, *Faecalibacterium prausnitzii* ↑, *Bacteroides intestinalis* ↓, *Bacteroides vulgatus* ↓, *Propionibacterium* ↓	Dewulf et al. (2013)
Inulin (12 g/day)	44 healthy subjects	8 weeks	Sequencing (16S rRNA gene)	*Bilophila* ↓	Vandeputte et al. (2017)
Very long-chain inulin (VLCI) 10 g/day	32 healthy adults	6 weeks	Fluorescent in situ hybridization (FISH)	bifidobacteria ↑, lactobacilli ↑, *Atopobium* ↑, *Bacteroides* ↓, *Prevotella* ↓	Costabile et al. (2010)
Galacto-oligosaccharides (GOS) supplementation (15 g/day)	44 obese prediabetic individuals	12 weeks	HITChip	*Bifidobacterium* ↑	Canfora et al. (2017)
Xylooligosaccharides-(XOS-) enriched rice porridge	20 healthy subjects	6 weeks	Medium selection	*Lactobacillus* spp. ↑, *Bifidobacterium* spp. ↑, *Clostridium perfringens* ↓	Lin et al. (2016)
Fructooligosaccharides (FOS); Galactooligosaccharides (GOS) (16 g/day)	35 adults	14 days	318 V2 chip	FOS: *Bifidobacterium* ↑, *Phascolarctobacterium* ↓; GOS: *Bifidobacterium* ↑, *Ruminococcus* ↓	Liu et al. (2017)
Partially hydrolysed guar gum (PHGG) and fructooligosaccharides (FOS) in biscuits	31 healthy subjects	6 weeks	FISH	bifidobacteria ↑	Tuohy et al. (2001)
Arabinoxylan and resistant starch type 2	19 adults	8 weeks	Sequencing (16S rRNA gene)	Bacterial diversity ↓, *Bifidobacterium* ↑	Hald et al. (2016)
Resistant starch (RS) or non-starch polysaccharides (NSPs) supplementation; weight-loss (WL) diet	14 obese males	3 weeks	HITChip; qPCR	RS diet: *Ruminococcaceae* ↑ NSP diet: *Lachnospiraceae* ↑ WL diet: bifidobacteria ↓	Salonen et al. (2014)

(continued)

Table 1 (continued)

Diet	Cohort	Time	Analysis method	Gut microbial response	References
10 g or 40 g dietary fibre per day	19 healthy adults	6 weeks	Pyrosequencing (16S rRNA gene); qPCR	High dietary fibre: microbiota richness ↑, *Prevotella* ↑, *Coprococcus* ↑	Tap et al. (2015)
Barley kernel-based bread (BKB)	39 subjects	3 days	Metagenomics; qPCR	BKB: *Prevotella/Bacteroides* ↑, *Prevotella copri* ↑	Kovatcheva-Datchary et al. (2015)
High resistant starch (RS) diet; Weight loss (WL) diet	14 overweight men	10 weeks	Sequencing (16S rRNA gene); qPCR; Denaturing Gradient Gel Glectrophoresis (DGGE)	RS diet: *Ruminococcus bromii* ↑, Uncultured *Oscillibacter* group ↑, *Eubacterium rectale* ↑ WL diets: Uncultured *Oscillibacter* group ↑, *Eubacterium rectale* ↓, *Collinsella aerofaciens* ↓	Walker et al. (2011)
Vegetable/fruit juices	20 healthy adults	3 days	Sequencing (16S rRNA gene)	Cyanobacteria ↑, Firmicutes ↓, Proteobacteria ↓, Bacteroidetes ↑,	Henning et al. (2017)
High-vegetable/low-protein diet (HV/LP)	20 healthy adults	12 months	Sequencing (16S rRNA gene)	HV/LP: *Lachnospiraceae* ↑	Saresella et al. (2017)
Diet rich in whole grain (WG) Diet rich in red meat (RM)	20 healthy adults	10 weeks	DGGE (16S rRNA gene)	WG: *Collinsella aerofaciens* ↑ RM: *Clostridium* spp. ↑	Foerster et al. (2014)
Whole-grain wheat (WGW); whole-grain rye (WGR); refined wheat (RW)	70 healthy adults	6 weeks	Sequencing (16S rRNA gene)	No effects	Vuholm et al. (2017)
Whole-grain foods; fibre-rich rye breads; refined wheat breads	51 Finnish individuals with metabolic syndrome	12 weeks	HITChip; qPCR	Refined wheat breads: *Bacteroides vulgatus* ↓, *B. plebeius* ↓, *Prevotella tannerae* ↓, *Collinsella* ↑, *Clostridium* clusters IV and XI ↑	Lappi et al. (2013)
Whole grains, traditional Chinese medicinal foods, and prebiotics (WTP diet)	93 obese volunteers	9 weeks	Pyrosequencing (16S rRNA) gene	*Bifidobacteriaceae* ↑, *Enterobacteriaceae* ↓, *Desulfovibrionaceae* ↓	Xiao et al. (2014)

Whole-grain barley (WGB); Brown rice (BR); BR + WGB (60 g/g)	28 healthy humans	4 weeks	Pyrosequencing (16S rRNA gene)	All treatments: microbial diversity ↑, Firmicutes/Bacteroidetes ratio ↑, *Blautia* spp.↑; WGB: *Roseburia* spp.↑, *Bifidobacterium* spp.↑, *Dialister* spp.↑	Martinez et al. (2013)
Whole grain diet (WG); Refined grain diet (RG)	60 Danish adults	8 weeks	Shotgun metagenomics; Sequencing (16S rRNA gene)	No major changes of the gut microbiome WG: *Faecalibacterium prausnitzii* ↑, *Prevotella copri* ↑, and *Bacteroides thetaiotaomicron* ↓	Roager et al. (2017)
Whole-grain (WG) wheat; wheat bran	31 individuals	3 weeks	FISH	WG: bifidobacteria ↑, lactobacilli ↑	Costabile et al. (2008)
WG diet; Refined grain (RG) diet	17 healthy subjects	2 weeks	qPCR	WG: *Clostridium leptum* ↑	Ross et al. (2011)
FODMAPs (low 3.5 g/day vs. high 23.7 g/day)	27 IBS patients; 6 healthy subjects	21 days	qPCR; DGGE	Low FODMAPs: microbial diversity ↑, total bacterial abundance ↓ High FODMAPs: *Ruminococcus torques* ↓ *Clostridium* cluster XIVa ↑, *Akkermansia muciniphila* ↑	Halmos et al. (2014a)
FODMAPs-restricted diet	41 IBS patients	4 weeks	FISH	bifidobacteria ↓	(Staudacher et al. 2012)
African style diet (high fibre); Western-style diet (high fat)	20 healthy African Americans; 20 rural Africans	2 weeks	HITChip	Western style diet: *Eubacterium rectale* ↓, *Clostridium symbiosum* et rel ↓, *Oscillospira guillermondii* ↓	O'Keefe et al. (2015)
High-fat/low-fibre diet; Low-fat/high-fibre diet	98 healthy individuals	10 days	Sequencing (16S rRNA gene)	High-fat: Bacteroidetes ↑, Actinobacteria↑ Low-fat: Bacteroidetes ↓, Actinobacteria ↓	Wu et al. (2011)

(continued)

Table 1 (continued)

Diet	Cohort	Time	Analysis method	Gut microbial response	References
High saturated fat diet (HS); high monounsaturated fat diet (HM)	88 subjects at increased metabolic syndrome risk	28 weeks	FISH	HS: *Faecalibacterium prausnitzii* ↑ HM: total bacteria ↓	Fava et al. (2013)
High-fat diet with additional whipping cream	24 healthy men	7 days	Sequencing (16S rRNA gene)	*Bacteroidaceae* ↓, *Betaproteobacteria* ↑	Ott et al. (2018)
High-protein/moderate-carbohydrate diet (HPMC); High-protein/low-carbohydrate diet (HPLC)	17 obese men	8 weeks	FISH	HPMC and HPLC: bacterial numbers ↓ HPLC: *Roseburia/Eubacterium rectale* ↓, *Bacteroides* spp. ↓	Russell et al. (2011)
High protein (HP) diet; low protein (LP) diet	20 healthy subjects	4 weeks	DGGE	No effects	Windey et al. (2012a)
Gluten-free diet (GFD)	10 healthy subjects	1 month	Fluorescent in situ hybridization (FISH); qPCR	*Clostridium lituseburense* ↓, *Faecalibacterium prausnitzii* ↓, *Bifidobacterium* ↓, *Lactobacillus* ↓, *Bifidobacterium longum* ↓, *Enterobacteriaceae* ↑, *Escherichia coli* ↑	De Palma et al. (2009)
GFD	21 healthy subjects	4 weeks	Pyrosequencing (16S rRNA gene)	*Veillonellaceae* ↓	Bonder et al. (2016)

↑: increase of the microbial abundance; ↓ decrease of the microbial abundance

2015). Comparative analyses between responders and non-responders are essential for the identification of signature microbes that could predict physiological changes in the host (Korpela et al. 2014). A recent study validated a comprehensive computational platform, the "community and system-level interactive optimization" (CASINO) toolbox, which was capable of predicting faecal and blood metabolomics data in a dietary interventional study with 45 obese and overweight individuals, thereby providing a powerful tool to determine diet-induced metabolic changes of the gut microbiome (Shoaie et al. 2015). Similarly, a different study successfully predicted personalized postprandial glycaemic responses to real-life meals by integrating microbial and host datasets using machine-learning algorithms (Zeevi et al. 2015). Predictive studies and algorithms may lead to the design of personalized dietary interventions, which could prospectively convert non-responders into responders using human genome and gut microbiome information (Fig. 1) (Bashiardes et al. 2018).

Gut microbial interactions with dietary macronutrients The main macronutrients in diet are carbohydrates, fat and protein. As described earlier, after escaping digestion and absorption by the human digestive system, nutrients can be utilized by the gut microbiota, resulting in the generation of a range of metabolites via microbial fermentation and cross-feeding. Composition of the macronutrients thus exerts a considerable influence on gut microbiota composition, activity, and human physiology (Fig. 1).

(a) **Carbohydrates**: Microbiota-accessible carbohydrates are fermented by the resident microbes resulting in the production of SCFAs, notably acetate, propionate and butyrate, and gasses such as hydrogen and carbon dioxide. Manipulating the gut microbial composition and activity by specific dietary carbohydrates, notably non-digestible fibre, is increasingly accepted as a promising approach to benefit human health (see the chapter "Beneficial Modulation of the Gut Microbiome: Probiotics and Prebiotics" for more in depth discussion about prebiotics and their impact on human health). A study by Tap and colleagues (Tap et al. 2015) using 19 healthy volunteers indicated that increased dietary fibre restored richness and stability of the gut microbiota in adults. Moreover, a double-blind, placebo-controlled cross-over study in individuals given a 3-week oral administration of the prebiotic inulin showed that inulin increased the numbers of bifidobacteria and lactobacilli in faecal samples (Costabile et al. 2010). Not only do the dietary fibres alter gut microbiota composition, but also contribute to human physiology. For example, a 4-week randomized cross-over trial in healthy individuals found that intake of whole grains increased gut microbial diversity, and that this was also associated with the reduction of the postprandial glucose peak and immunological improvements (Martinez et al. 2013). However, a dietary component-induced change of the microbiota not always results in an altered physiological response. For example, a recent intervention with galacto-oligosaccharides in prediabetic individuals resulted in increased relative abundance of bifidobacteria but did not improve insulin sensitivity or host energy metabolism (Canfora et al. 2017). It has to be realized that carbohydrate fermentation does not always lead to improvements in the host. For example, FODMAPs, which are usually digested and absorbed in the small intestine, may cause a quick increase of glycaemic concentrations. When FODMAPs reach the colon, they can be rapidly fermented, leading to a mass production of hydrogen, an associated factor underlying susceptibility in irritable bowel syndrome (IBS) patients (Halmos et al. 2014b). Hence, restriction of FODMAP diets may reduce gastrointestinal symptoms in IBS patients (Staudacher et al. 2012; Halmos et al. 2014b).

As indicated above, microbial fermentation of carbohydrates results mainly in the production of SCFAs. However, there is no general consensus on the link between diets and SCFAs since many conflicting results have been published. A positive correlation was observed between diets rich in fibre and high levels of SCFAs (De Filippis et al. 2016). In contrast, another study showed unexpected reductions in faecal SCFAs concentrations in vegans compared with omnivores (Reiss et al. 2016), and a trial in which a whole grain-based diet was compared to a red meat-based diet did not show differences in SCFA levels (Foerster et al. 2014). Reasons for observed variations between studies could be that a minimum intake of dietary fibre is needed to induce measurable alterations in SCFA production, and that SCFA production occurs but is not measurable in faecal samples due to intestinal absorption and subsequent microbial interactions. Other reasons could include the type of carbohydrate used in an intervention, the intestinal location where a certain carbohydrate fermentation takes place as well as the metabolic state of the individual. Overall, these intervention studies confirmed the complex interactions between carbohydrates and gut microbiota, and thus more mechanistic studies are needed to further elucidate how carbohydrates affect gut microbial composition and metabolites, and subsequently human physiology.

(b) Fat: Dietary fat is believed to be degraded and absorbed in the small intestine. Thus, the colonic microbiota is not expected to have significant interactions with this macronutrient directly. Primary bile acids, produced from host cholesterol in the liver and then conjugated with taurine and glycine, are secreted into the small intestine to solubilize lipids, facilitating the digestion and absorption of fat. Secretion of bile acids is stimulated by consumption of high-fat diets, and excessive amounts are metabolized by bile salt hydrolases (BSH) from intestinal microorganisms into secondary bile acids, such as deoxycholic acid and lithocholic acid (Begley et al. 2005; Ridlon et al. 2006). The accumulation of secondary bile acids, which are biologically active, may contribute to the development of gallstones, CRC, and other diseases (Ridlon et al. 2006; Azcárate-Peril et al. 2011; Ou et al. 2012). Bile acids can also regulate gut microbial composition due to their antibacterial properties. Research in rats indicated that Firmicutes and Proteobacteria are generally more resistant to bile salts compared to Bacteroidetes (Islam et al. 2011). Although these observations have not been consistently reported (Duncan et al. 2008; Schwiertz et al. 2010), differences in bacterial resistance to bile acids may in part explain why obese individuals harbour more Firmicutes but less Bacteroidetes (Ley et al. 2006).

Many studies have described the effects of high fat diets on host metabolism and the gut microbiota. However, human studies do not show a similar consistency.

A 2-week dietary exchange study in which rural Africans were fed a high-fat diet and African Americans a low-fat diet demonstrated that the switch from a high-fat diet to a high-complex carbohydrate diet increased butyrate production and suppressed the synthesis of secondary bile acids (O'Keefe et al. 2015). In addition, an intervention study in subjects at increased risk for development of metabolic syndrome indicated that a diet high in mono-unsaturated fat did not affect gut microbial composition, but rather reduced the number of total bacteria, whereas a diet high in saturated fat increased the abundance of *Faecalibacterium prausnitzii* and SCFAs concentrations (Fava et al. 2013). Therefore, attention should be paid to both the quantity and quality of the dietary fat consumed and how this affects microbiota composition and activity.

(c) **Protein**: Protein may also escape digestion and absorption in the small intestine, reaching the colon where it is exposed to the colonic microbiota. When carbohydrates become limited substrates for the gut microbiota, fermentation of proteins occurs in the distal colon. Protein fermentation may result in the production of metabolites consisting of sulphur compounds, N-Nitroso compounds, ammonia, heterocyclic amines and organic acids, which are potentially toxic and detrimental to human health (Nyangale et al. 2012; Windey et al. 2012b).

Both, short-term and long-term dietary intervention studies in humans indicated that *Bacteroides* is strongly associated with an animal-based diet high in fat and proteins (Wu et al. 2011; David et al. 2014). *In vitro* fermentation studies revealed that the predominant proteolytic bacteria belong to the genera *Bacteroides* and *Propionibacterium* as well as species from to the genera *Streptococcus*, *Clostridium*, *Bacillus* and *Staphylococcus*, as determined by culturing (Macfarlane et al. 1986). More recent culture-independent studies using the TIM-2 dynamic colon model indicated that a high-protein diet enriches the genus *Bacteroides* with a concomitant increase in BCFAs, such as *iso*-butyrate and *iso*-valerate (Hermes 2016). In line with these observations, a 4-week dietary intervention study in obese individuals indicated that a diet high in proteins reduced the abundance of members of the *Roseburia/Eubacterium rectale* group with a significant decrease in beneficial metabolites like butyrate but concomitant increase in the concentration of metabolites such as BCFAs in comparison with the maintenance diet (Russell et al. 2011). Due to the production of potentially detrimental protein-derived metabolites, it has been assumed that the intake of high protein is associated with several human diseases including CRC, IBD and cardiovascular diseases (Jantchou et al. 2010; De Filippis et al. 2016; O'Keefe et al. 2015). However, we still lack the evidence explaining the relationship of the protein-derived metabolites and human diseases. Therefore, more studies should be carried out to discover the underlying causality.

Explaining the impact of a dietary change mechanistically is challenging. In intervention studies, diets are often calorically matched and as a result, the increase of one component is compensated by the decrease in other. For example, typical Western diets contain lower amounts of fibre and therefore concomitantly the amounts of fat and protein are higher. As a consequence, it remains speculative if the observations in comparative analyses between diets are a direct effect of the carbohydrate, fat or protein content, or a combination of these. Moreover, it could also be that the effect of diet is indirect via bile or other host secretions as a result of the dietary intake. It is evident that there is an urgent need of dedicated studies to unravel the mechanisms underlying the observations in dietary intervention studies.

Conclusions

Cross-sectional and interventional studies have investigated differences of long-term and short-term dietary changes. Comparative analyses indicated that long-term consumption of traditional versus Western diets are associated with major differences in microbiota composition. On the other hand, short-term dietary interventions, even those that included major dietary changes, had a limited impact on composition but could have significant impact on metabolite production. It is evident that diet is a crucial determinant in the ecology of the gut microbiota, with concomitant effects on human metabolism and physiology. Nevertheless, our knowledge of how diets affect the microbiota in adulthood and what impact this has on human physiology is still in its infancy, mainly because the mechanisms underlying observations involve many factors often dependent on each other. In addition, comparative analyses between studies are hampered by the facts that the setup, the choice of approaches to study the gut microbiota, background of the study participants, duration of the study, and choice of diet components and their amounts differ between studies. Moreover, due to the individuality of microbiota composition, responses towards the same intervention can be drastically different between subjects. This may explain why contrasting observations with similar diets or dietary components have been reported.

In order to improve our understanding of how diet impacts gut microbiota and human physiology (Fig. 2), prospective studies should be performed to determine whether the gut microbiota is altered after the dietary interventions, which can help to form new hypotheses based on new observations. Then, the generated hypotheses should be tested and validated using microbial consortia inlcuding defined (synthetic) communities, cell lines and intestinal organoids, animal models or other effecitive ways to improve or form new concepts of diet-microbiota interactions (Elzinga et al. 2019; Shetty et al. 2019). After that, individualized precision interventional studies can be carried out to further evaluate the new concepts. Such microbial research triangle is essential to elucidate whether the gut microbiota is causally linked to host metabolism in humans (Fig. 2). Besides this, different approaches with the ambition to predict diet-induced metabolic changes of the human gut microbiome (Shoaie et al. 2015) or the personalized postprandial glycaemic response to real-life meals have been reported (Zeevi et al. 2015). These are very promising approaches to move from observations towards subject-specific predictions on the effect of diets on the gut microbiota and human physiology and enable microbiota-focused precision nutrition in the future.

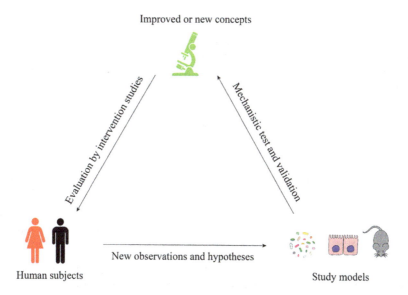

Fig. 2 Strategies for investigating the mechanisms of diet-microbiota interactions. Performed prospective studies that observe altered gut microbiota composition and/or activity after the dietary interventions in individuals can be used to generate new observations and hypotheses. Subsequently, relevant experimental models, such as microbial consortia including (defined) synthetic communities, cell lines, intestinal organoids and animal models can be used to test and validate the underlying mechanisms to improve or generate new concepts. These can be further evaluated by using precision dietary interventions

Acknowledgements The research by Taojun Wang is financially supported by the China Scholarship Council (File No. 201600090211).

References

Albenberg, L., Esipova, T. V., Judge, C. P., Bittinger, K., Chen, J., Laughlin, A., Grunberg, S., Baldassano, R. N., Lewis, J. D., & Li, H. (2014). Correlation between intraluminal oxygen gradient and radial partitioning of intestinal microbiota. *Gastroenterology, 147*(5), 1055–1063. e1058.

An, R., Wilms, E., Masclee, A. A., Smidt, H., Zoetendal, E. G., & Jonkers, D. (2018). Age-dependent changes in GI physiology and microbiota: Time to reconsider? *Gut, 67*(12), 2213–2222.

Arumugam, M., Raes, J., Pelletier, E., Le Paslier, D., Yamada, T., Mende, D. R., Fernandes, G. R., Tap, J., Bruls, T., Batto, J.-M., Bertalan, M., Borruel, N., Casellas, F., Fernandez, L., Gautier, L., Hansen, T., Hattori, M., Hayashi, T., Kleerebezem, M., Kurokawa, K., Leclerc, M., Levenez, F., Manichanh, C., Nielsen, H. B., Nielsen, T., Pons, N., Poulain, J., Qin, J., Sicheritz-Ponten, T., Tims, S., Torrents, D., Ugarte, E., Zoetendal, E. G., Wang, J., Guarner, F., Pedersen, O., de Vos, W. M., Brunak, S., Dore, J., Weissenbach, J., Ehrlich, S. D., Bork, P., & Meta, H. I. T. C. (2011). Enterotypes of the human gut microbiome. *Nature, 473*(7346), 174–180.

Atuma, C., Strugala, V., Allen, A., & Holm, L. (2001). The adherent gastrointestinal mucus gel layer: Thickness and physical state in vivo. *American Journal of Physiology-Gastrointestinal and Liver Physiology, 280*(5), G922–G929.

Azcárate-Peril, M. A., Sikes, M., & Bruno-Bárcena, J. M. (2011). The intestinal microbiota, gastrointestinal environment and colorectal cancer: A putative role for probiotics in prevention of colorectal cancer? *American Journal of Physiology-Gastrointestinal and Liver Physiology, 301*(3), G401–G424.

Bashiardes, S., Godneva, A., Elinav, E., & Segal, E. (2018). Towards utilization of the human genome and microbiome for personalized nutrition. *Current Opinion in Biotechnology, 51*, 57–63.

Begley, M., Gahan, C. G., & Hill, C. (2005). The interaction between bacteria and bile. *FEMS Microbiology Reviews, 29*(4), 625–651.

Bonder, M. J., Tigchelaar, E. F., Cai, X., Trynka, G., Cenit, M. C., Hrdlickova, B., Zhong, H., Vatanen, T., Gevers, D., Wijmenga, C., Wang, Y., & Zhernakova, A. (2016). The influence of a short-term gluten-free diet on the human gut microbiome. *Genome Medicine, 8*(1), 45.

Booijink, C. C., El-Aidy, S., Rajilić-Stojanović, M., Heilig, H. G., Troost, F. J., Smidt, H., Kleerebezem, M., De Vos, W. M., & Zoetendal, E. G. (2010). High temporal and inter-individual variation detected in the human ileal microbiota. *Environmental Microbiology, 12*(12), 3213–3227.

Bui, T. P. N., Ritari, J., Boeren, S., De Waard, P., Plugge, C. M., & De Vos, W. M. (2015). Production of butyrate from lysine and the Amadori product fructoselysine by a human gut commensal. *Nature Communications, 6*, 10062.

Canfora, E. E., van der Beek, C. M., Hermes, G. D., Goossens, G. H., Jocken, J. W., Holst, J. J., van Eijk, H. M., Venema, K., Smidt, H., & Zoetendal, E. G. (2017). Supplementation of diet with galacto-oligosaccharides increases bifidobacteria, but not insulin sensitivity, in obese prediabetic individuals. *Gastroenterology, 153*(1), 87–97.

Carbonero, F., Benefiel, A. C., & Gaskins, H. R. (2012). Contributions of the microbial hydrogen economy to colonic homeostasis. *Nature Reviews. Gastroenterology & Hepatology, 9*(9), 504–518.

Cheng, J., Ringel-Kulka, T., Heikamp-de Jong, I., Ringel, Y., Carroll, I., De Vos, W. M., Salojärvi, J., & Satokari, R. (2016). Discordant temporal development of bacterial phyla and the emergence of core in the fecal microbiota of young children. *The ISME Journal, 10*(4), 1002–1014.

Conlon, M. A., & Bird, A. R. (2014). The impact of diet and lifestyle on gut microbiota and human health. *Nutrients, 7*(1), 17–44.

Costabile, A., Klinder, A., Fava, F., Napolitano, A., Fogliano, V., Leonard, C., Gibson, G. R., & Tuohy, K. M. (2008). Whole-grain wheat breakfast cereal has a prebiotic effect on the human gut microbiota: A double-blind, placebo-controlled, crossover study. *British Journal of Nutrition, 99*(1), 110–120.

Costabile, A., Kolida, S., Klinder, A., Gietl, E., Bäuerlein, M., Frohberg, C., Landschütze, V., & Gibson, G. R. (2010). A double-blind, placebo-controlled, cross-over study to establish the bifidogenic effect of a very-long-chain inulin extracted from globe artichoke (Cynara scolymus) in healthy human subjects. *British Journal of Nutrition, 104*(7), 1007–1017.

Costea, P. I., Hildebrand, F., Manimozhiyan, A., Bäckhed, F., Blaser, M. J., Bushman, F. D., De Vos, W. M., Ehrlich, S. D., Fraser, C. M., & Hattori, M. (2018). Enterotypes in the landscape of gut microbial community composition. *Nature Microbiology, 3*(1), 8.

Cummings, J., Pomare, E., Branch, W., Naylor, C., & Macfarlane, G. (1987). Short chain fatty acids in human large intestine, portal, hepatic and venous blood. *Gut, 28*(10), 1221–1227.

David, L. A., Maurice, C. F., Carmody, R. N., Gootenberg, D. B., Button, J. E., Wolfe, B. E., Ling, A. V., Devlin, A. S., Varma, Y., Fischbach, M. A., Biddinger, S. B., Dutton, R. J., & Turnbaugh, P. J. (2014). Diet rapidly and reproducibly alters the human gut microbiome. *Nature, 505*(7484), 559.

De Filippis, F., Pellegrini, N., Vannini, L., Jeffery, I. B., La Storia, A., Laghi, L., Serrazanetti, D. I., Di Cagno, R., Ferrocino, I., & Lazzi, C. (2016). High-level adherence to a Mediterranean diet beneficially impacts the gut microbiota and associated metabolome. *Gut, 65*(11), 1812–1821.

De Filippo, C., Cavalieri, D., Di Paola, M., Ramazzotti, M., Poullet, J. B., Massart, S., Collini, S., Pieraccini, G., & Lionetti, P. (2010). Impact of diet in shaping gut microbiota revealed by a comparative study in children from Europe and rural Africa. *Proceedings of the National Academy of Sciences, 107*(33), 14691–14696.

De Filippo, C., Di Paola, M., Ramazzotti, M., Albanese, D., Pieraccini, G., Banci, E., Miglietta, F., Cavalieri, D., & Lionetti, P. (2017). Diet, environments, and gut microbiota. a preliminary investigation in children living in rural and urban burkina faso and italy. *Frontiers in Microbiology, 8*, 1979.

De Palma, G., Nadal, I., Carmen Collado, M., & Sanz, Y. (2009). Effects of a gluten-free diet on gut microbiota and immune function in healthy adult human subjects. *British Journal of Nutrition, 102*(8), 1154–1160.

de Vos, W. M. (2017). Microbe profile: Akkermansia muciniphila: A conserved intestinal symbiont that acts as the gatekeeper of our mucosa. *Microbiology, 163*(5), 646–648.

Derrien, M., Vaughan, E. E., Plugge, C. M., & de Vos, W. M. (2004). Akkermansia muciniphila gen. nov., sp. nov., a human intestinal mucin-degrading bacterium. *International Journal of Systematic and Evolutionary Microbiology, 54*(5), 1469–1476.

Dewulf, E. M., Cani, P. D., Claus, S. P., Fuentes, S., Puylaert, P. G., Neyrinck, A. M., Bindels, L. B., de Vos, W. M., Gibson, G. R., & Thissen, J.-P. (2013). Insight into the prebiotic concept: Lessons from an exploratory, double blind intervention study with inulin-type fructans in obese women. *Gut, 62*(8), 1112–1121.

Donaldson, G. P., Lee, S. M., & Mazmanian, S. K. (2016). Gut biogeography of the bacterial microbiota. *Nature Reviews Microbiology, 14*(1), 20–32.

Donia, M. S., & Fischbach, M. A. (2015). Small molecules from the human microbiota. *Science, 349*(6246), 1254766.

Duncan, S. H., Lobley, G., Holtrop, G., Ince, J., Johnstone, A., Louis, P., & Flint, H. J. (2008). Human colonic microbiota associated with diet, obesity and weight loss. *International Journal of Obesity, 32*(11), 1720.

Duncan, S. H., Louis, P., Thomson, J. M., & Flint, H. J. (2009). The role of pH in determining the species composition of the human colonic microbiota. *Environmental Microbiology, 11*(8), 2112–2122.

Elzinga, J., van der Oost, J., de Vos, W. M., & Smidt, H. (2019). The use of defined microbial communities to model host-microbe interactions in the human gut. *Microbiology and Molecular Biology Reviews, 83*(2), e00054–e00018.

Falony, G., Joossens, M., Vieira-Silva, S., Wang, J., Darzi, Y., Faust, K., Kurilshikov, A., Bonder, M. J., Valles-Colomer, M., & Vandeputte, D. (2016). Population-level analysis of gut microbiome variation. *Science, 352*(6285), 560–564.

Fava, F., Gitau, R., Griffin, B., Gibson, G., Tuohy, K., & Lovegrove, J. (2013). The type and quantity of dietary fat and carbohydrate alter faecal microbiome and short-chain fatty acid excretion in a metabolic syndrome 'at-risk' population. *International Journal of Obesity, 37*(2), 216.

Feng, Y., Stams, A. J., De Vos, W. M., & Sánchez-Andrea, I. (2017). Enrichment of sulfidogenic bacteria from the human intestinal tract. *FEMS Microbiology Letters, 364*(4), fnx028.

Foerster, J., Maskarinec, G., Reichardt, N., Tett, A., Narbad, A., Blaut, M., & Boeing, H. (2014). The influence of whole grain products and red meat on intestinal microbiota composition in normal weight adults: A randomized crossover intervention trial. *PLoS One, 9*(10), e109606.

Frank, D. N., Amand, A. L. S., Feldman, R. A., Boedeker, E. C., Harpaz, N., & Pace, N. R. (2007). Molecular-phylogenetic characterization of microbial community imbalances in human inflammatory bowel diseases. *Proceedings of the National Academy of Sciences, 104*(34), 13780–13785.

Fuentes, S., Rossen, N. G., van der Spek, M. J., Hartman, J. H., Huuskonen, L., Korpela, K., Salojärvi, J., Aalvink, S., de Vos, W. M., & D'Haens, G. R. (2017). Microbial shifts and signatures of long-term remission in ulcerative colitis after faecal microbiota transplantation. *The ISME Journal, 11*(8), 1877.

Geerlings, S., Kostopoulos, I., de Vos, W., & Belzer, C. (2018). Akkermansia muciniphila in the human gastrointestinal tract: When, where, and how? *Microorganisms, 6*(3), 75.

Goodrich, J. K., Waters, J. L., Poole, A. C., Sutter, J. L., Koren, O., Blekhman, R., Beaumont, M., Van Treuren, W., Knight, R., & Bell, J. T. (2014). Human genetics shape the gut microbiome. *Cell, 159*(4), 789–799.

Hald, S., Schioldan, A. G., Moore, M. E., Dige, A., Lærke, H. N., Agnholt, J., Knudsen, K. E. B., Hermansen, K., Marco, M. L., & Gregersen, S. (2016). Effects of arabinoxylan and resistant starch on intestinal microbiota and short-chain fatty acids in subjects with metabolic syndrome: A randomised crossover study. *PLoS One, 11*(7), e0159223.

Halmos, E. P., Christophersen, C. T., Bird, A. R., Shepherd, S. J., Gibson, P. R., & Muir, J. G. (2014a). Diets that differ in their FODMAP content alter the colonic luminal microenvironment. *Gut, 64*(1), 93–100.

Halmos, E. P., Power, V. A., Shepherd, S. J., Gibson, P. R., & Muir, J. G. (2014b). A diet low in FODMAPs reduces symptoms of irritable bowel syndrome. *Gastroenterology, 146*(1), 67–75. e65.

Henning, S. M., Yang, J., Shao, P., Lee, R.-P., Huang, J., Ly, A., Hsu, M., Lu, Q.-Y., Thames, G., & Heber, D. (2017). Health benefit of vegetable/fruit juice-based diet: Role of microbiome. *Scientific Reports, 7*(1), 2167.

Hermes, G. (2016). *Mining the human intestinal microbiota for biomarkers associated with metabolic disorders*. Ph.D. thesis, Wageningen University.

Islam, K. S., Fukiya, S., Hagio, M., Fujii, N., Ishizuka, S., Ooka, T., Ogura, Y., Hayashi, T., & Yokota, A. (2011). Bile acid is a host factor that regulates the composition of the cecal microbiota in rats. *Gastroenterology, 141*(5), 1773–1781.

Jantchou, P., Morois, S., Clavel-Chapelon, F., Boutron-Ruault, M.-C., & Carbonnel, F. (2010). Animal protein intake and risk of inflammatory bowel disease: The E3N prospective study. *The American Journal of Gastroenterology, 105*(10), 2195.

Jones, B. V., Begley, M., Hill, C., Gahan, C. G., & Marchesi, J. R. (2008). Functional and comparative metagenomic analysis of bile salt hydrolase activity in the human gut microbiome. *Proceedings of the National Academy of Sciences, 105*(36), 13580–13585.

Jones, R. B., Zhu, X. Z., Moan, E., Murff, H. J., Ness, R. M., Seidner, D. L., Sun, S., Yu, C., Dai, Q., Fodor, A. A., Azcarate-Peril, M. A., & Shrubsole, M. J. (2018). Inter-niche and inter-individual variation in gut microbial community assessment using stool, rectal swab, and mucosal samples. *Scientific Reports, 8*(1), 4139.

Kashyap, P. C., Marcobal, A., Ursell, L. K., Larauche, M., Duboc, H., Earle, K. A., Sonnenburg, E. D., Ferreyra, J. A., Higginbottom, S. K., & Million, M. (2013). Complex interactions among diet, gastrointestinal transit, and gut microbiota in humanized mice. *Gastroenterology, 144*(5), 967–977.

Kau, A. L., Ahern, P. P., Griffin, N. W., Goodman, A. L., & Gordon, J. I. (2011). Human nutrition, the gut microbiome and the immune system. *Nature, 474*(7351), 327–336.

Koh, A., De Vadder, F., Kovatcheva-Datchary, P., & Bäckhed, F. (2016). From dietary fiber to host physiology: Short-chain fatty acids as key bacterial metabolites. *Cell, 165*(6), 1332–1345.

Korpela, K., Flint, H. J., Johnstone, A. M., Lappi, J., Poutanen, K., Dewulf, E., Delzenne, N., De Vos, W. M., & Salonen, A. (2014). Gut microbiota signatures predict host and microbiota responses to dietary interventions in obese individuals. *PLoS One, 9*(3), e90702.

Kovatcheva-Datchary, P., Nilsson, A., Akrami, R., Lee, Y. S., De Vadder, F., Arora, T., Hallen, A., Martens, E., Bjorck, I., & Backhed, F. (2015). Dietary fiber-induced improvement in glucose metabolism is associated with increased abundance of Prevotella. *Cell Metabolism, 22*(6), 971–982.

Lahti, L., Salojärvi, J., Salonen, A., Scheffer, M., & De Vos, W. M. (2014). Tipping elements in the human intestinal ecosystem. *Nature Communications, 5*, 4344.

Lappi, J., Salojärvi, J., Kolehmainen, M., Mykkänen, H., Poutanen, K., de Vos, W. M., & Salonen, A. (2013). Intake of whole-grain and fiber-rich rye bread versus refined wheat bread does not differentiate intestinal microbiota composition in Finnish adults with metabolic syndrome. *The Journal of Nutrition, 143*(5), 648–655.

Larsen, N., Vogensen, F. K., van den Berg, F. W., Nielsen, D. S., Andreasen, A. S., Pedersen, B. K., Al-Soud, W. A., Sørensen, S. J., Hansen, L. H., & Jakobsen, M. (2010). Gut microbiota in human adults with type 2 diabetes differs from non-diabetic adults. *PLoS One, 5*(2), e9085.

Le Chatelier, E., Nielsen, T., Qin, J., Prifti, E., Hildebrand, F., Falony, G., Almeida, M., Arumugam, M., Batto, J.-M., & Kennedy, S. (2013). Richness of human gut microbiome correlates with metabolic markers. *Nature, 500*(7464), 541.

Lepage, P., Seksik, P., Sutren, M., de la Cochetiere, M.-F., Jian, R., Marteau, P., & Doré, J. (2005). Biodiversity of the mucosa-associated microbiota is stable along the distal digestive tract in healthy individuals and patients with IBD. *Inflammatory Bowel Diseases, 11*(5), 473–480.

Levitt, M. D., Furne, J. K., Kuskowski, M., & Ruddy, J. (2006). Stability of human methanogenic flora over 35 years and a review of insights obtained from breath methane measurements. *Clinical Gastroenterology and Hepatology, 4*(2), 123–129.

Ley, R. E., Turnbaugh, P. J., Klein, S., & Gordon, J. I. (2006). Microbial ecology: Human gut microbes associated with obesity. *Nature, 444*(7122), 1022–1023.

Li, J., Jia, H., Cai, X., Zhong, H., Feng, Q., Sunagawa, S., Arumugam, M., Kultima, J. R., Prifti, E., & Nielsen, T. (2014). An integrated catalog of reference genes in the human gut microbiome. *Nature Biotechnology, 32*(8), 834.

Lin, S. H., Chou, L. M., Chien, Y. W., Chang, J. S., & Lin, C. I. (2016). Prebiotic effects of xylooligosaccharides on the improvement of microbiota balance in human subjects. *Gastroenterology Research and Practice, 2016*(2016), 5789232.

Liu, F., Li, P., Chen, M., Luo, Y., Prabhakar, M., Zheng, H., He, Y., Qi, Q., Long, H., & Zhang, Y. (2017). Fructooligosaccharide (FOS) and galactooligosaccharide (GOS) increase bifidobacterium but reduce butyrate producing bacteria with adverse glycemic metabolism in healthy young population. *Scientific Reports, 7*(1), 11789.

Louis, P., Hold, G. L., & Flint, H. J. (2014). The gut microbiota, bacterial metabolites and colorectal cancer. *Nature Reviews Microbiology, 12*(10), 661.

Lozupone, C. A., Stombaugh, J. I., Gordon, J. I., Jansson, J. K., & Knight, R. (2012). Diversity, stability and resilience of the human gut microbiota. *Nature, 489*(7415), 220–230.

Lynch, S. V., & Pedersen, O. (2016). The human intestinal microbiome in health and disease. *New England Journal of Medicine, 375*(24), 2369–2379.

Macfarlane, G., Cummings, J., & Allison, C. (1986). Protein degradation by human intestinal bacteria. *Microbiology, 132*(6), 1647–1656.

Martinez, I., Lattimer, J. M., Hubach, K. L., Case, J. A., Yang, J., Weber, C. G., Louk, J. A., Rose, D. J., Kyureghian, G., Peterson, D. A., Haub, M. D., & Walter, J. (2013). Gut microbiome composition is linked to whole grain-induced immunological improvements. *ISME Journal, 7*(2), 269–280.

Maslowski, K. M., & Mackay, C. R. (2011). Diet, gut microbiota and immune responses. *Nature Immunology, 12*(1), 5–9.

Menni, C., Lin, C., Cecelja, M., Mangino, M., Matey-Hernandez, M. L., Keehn, L., Mohney, R. P., Steves, C. J., Spector, T. D., Kuo, C.-F., Chowienczyk, P., & Valdes, A. M. (2018). Gut microbial diversity is associated with lower arterial stiffness in women. *European Heart Journal, 39*(25), 2390–2397.

Miller, T. L., & Wolin, M. J. (1996). Pathways of acetate, propionate, and butyrate formation by the human fecal microbial flora. *Applied and Environmental Microbiology, 62*(5), 1589–1592.

Moayyedi, P., Surette, M. G., Kim, P. T., Libertucci, J., Wolfe, M., Onischi, C., Armstrong, D., Marshall, J. K., Kassam, Z., & Reinisch, W. (2015). Fecal microbiota transplantation induces remission in patients with active ulcerative colitis in a randomized controlled trial. *Gastroenterology, 149*(1), 102–109. e106.

Nakamura, N., Lin, H. C., McSweeney, C. S., Mackie, R. I., & Gaskins, H. R. (2010). Mechanisms of microbial hydrogen disposal in the human colon and implications for health and disease. *Annual Review of Food Science and Technology, 1*, 363–395.

Nicholson, J. K., Holmes, E., Kinross, J., Burcelin, R., Gibson, G., Jia, W., & Pettersson, S. (2012). Host-gut microbiota metabolic interactions. *Science, 336*(6086), 1262–1267.

Nyangale, E. P., Mottram, D. S., & Gibson, G. R. (2012). Gut microbial activity, implications for health and disease: The potential role of metabolite analysis. *Journal of Proteome Research, 11*(12), 5573–5585.

O'Keefe, S. J., Li, J. V., Lahti, L., Ou, J., Carbonero, F., Mohammed, K., Posma, J. M., Kinross, J., Wahl, E., & Ruder, E. (2015). Fat, fibre and cancer risk in African Americans and rural Africans. *Nature Communications, 6*, 6342.

O'Toole, P. W., & Jeffery, I. B. (2015). Gut microbiota and aging. *Science, 350*(6265), 1214–1215.

Ott, B., Skurk, T., Lagkouvardos, L., Fischer, S., Büttner, J., Lichtenegger, M., Clavel, T., Lechner, A., Rychlik, M., & Haller, D. (2018). Short-term overfeeding with dairy cream does not modify gut permeability, the fecal microbiota, or glucose metabolism in young healthy men. *The Journal of Nutrition, 148*(1), 77–85.

Ou, G., Hedberg, M., Hörstedt, P., Baranov, V., Forsberg, G., Drobni, M., Sandström, O., Wai, S. N., Johansson, I., & Hammarström, M.-L. (2009). Proximal small intestinal microbiota and identification of rod-shaped bacteria associated with childhood celiac disease. *The American Journal of Gastroenterology, 104*, 3058–3067.

Ou, J., DeLany, J. P., Zhang, M., Sharma, S., & O'Keefe, S. J. (2012). Association between low colonic short-chain fatty acids and high bile acids in high colon cancer risk populations. *Nutrition and Cancer, 64*(1), 34–40.

Ou, J., Carbonero, F., Zoetendal, E. G., DeLany, J. P., Wang, M., Newton, K., Gaskins, H. R., & O'keefe, S. J. (2013). Diet, microbiota, and microbial metabolites in colon cancer risk in rural Africans and African Americans. *The American Journal of Clinical Nutrition, 98*(1), 111–120.

Pochart, P., Dore, J., Lemann, F., Goderel, I., & Rambaud, J. C. (1992). Interrelations between populations of methanogenic archaea and sulfate-reducing bacteria in the human colon. *FEMS Microbiology Letters, 98*(1–3), 225–228.

Qin, J. J., Li, R. Q., Raes, J., Arumugam, M., Burgdorf, K. S., Manichanh, C., Nielsen, T., Pons, N., Levenez, F., Yamada, T., Mende, D. R., Li, J. H., Xu, J. M., Li, S. C., Li, D. F., Cao, J. J., Wang, B., Liang, H. Q., Zheng, H. S., Xie, Y. L., Tap, J., Lepage, P., Bertalan, M., Batto, J. M., Hansen, T., Le Paslier, D., Linneberg, A., Nielsen, H. B., Pelletier, E., Renault, P., Sicheritz-Ponten, T., Turner, K., Zhu, H. M., Yu, C., Li, S. T., Jian, M., Zhou, Y., Li, Y. R., Zhang, X. Q., Li, S. G., Qin, N., Yang, H. M., Wang, J., Brunak, S., Dore, J., Guarner, F., Kristiansen, K., Pedersen, O., Parkhill, J., Weissenbach, J., Bork, P., Ehrlich, S. D., Wang, J., & Consortium, M. (2010). A human gut microbial gene catalogue established by metagenomic sequencing. *Nature, 464*(7285), 59–U70.

Rajilić-Stojanović, M., Smidt, H., & De Vos, W. M. (2007). Diversity of the human gastrointestinal tract microbiota revisited. *Environmental Microbiology, 9*(9), 2125–2136.

Rajilić-Stojanović, M., Heilig, H. G., Tims, S., Zoetendal, E. G., & Vos, W. M. (2013). Long-term monitoring of the human intestinal microbiota composition. *Environmental Microbiology, 15*(4), 1146–1159.

Reiss, A., Jacobi, M., Rusch, K., & Schwiertz, A. (2016). Association of dietary type with fecal microbiota and short chain fatty acids in vegans and omnivores. *The Journal International Society of Microbiota, 2*, 1.

Rey, F. E., Faith, J. J., Bain, J., Muehlbauer, M. J., Stevens, R. D., Newgard, C. B., & Gordon, J. I. (2010). Dissecting the in vivo metabolic potential of two human gut acetogens. *Journal of Biological Chemistry, 285*(29), 22082–22090.

Ridlon, J. M., Kang, D.-J., & Hylemon, P. B. (2006). Bile salt biotransformations by human intestinal bacteria. *Journal of Lipid Research, 47*(2), 241–259.

Ridlon, J. M., Kang, D. J., Hylemon, P. B., & Bajaj, J. S. (2014). Bile acids and the gut microbiome. *Current Opinion in Gastroenterology, 30*(3), 332.

Roager, H. M., Vogt, J. K., Kristensen, M., Hansen, L. B. S., Ibrügger, S., Mærkedahl, R. B., Bahl, M. I., Lind, M. V., Nielsen, R. L., & Frøkiær, H. (2017). Whole grain-rich diet reduces body weight and systemic low-grade inflammation without inducing major changes of the gut microbiome: A randomised cross-over trial. *Gut, 68*(1), 83–93.

Ross, A. B., Bruce, S. J., Blondel-Lubrano, A., Oguey-Araymon, S., Beaumont, M., Bourgeois, A., Nielsen-Moennoz, C., Vigo, M., Fay, L.-B., & Kochhar, S. (2011). A whole-grain cereal-rich diet increases plasma betaine, and tends to decrease total and LDL-cholesterol compared with a refined-grain diet in healthy subjects. *British Journal of Nutrition, 105*(10), 1492–1502.

Rossen, N. G., Fuentes, S., van der Spek, M. J., Tijssen, J. G., Hartman, J. H., Duflou, A., Löwenberg, M., van den Brink, G. R., Mathus-Vliegen, E. M., & de Vos, W. M. (2015). Findings from a randomized controlled trial of fecal transplantation for patients with ulcerative colitis. *Gastroenterology, 149*(1), 110–118. e114.

Russell, W. R., Gratz, S. W., Duncan, S. H., Holtrop, G., Ince, J., Scobbie, L., Duncan, G., Johnstone, A. M., Lobley, G. E., & Wallace, R. J. (2011). High-protein, reduced-carbohydrate weight-loss diets promote metabolite profiles likely to be detrimental to colonic health. *The American Journal of Clinical Nutrition, 93*(5), 1062–1072.

Salonen, A., Lahti, L., Salojarvi, J., Holtrop, G., Korpela, K., Duncan, S. H., Date, P., Farquharson, F., Johnstone, A. M., Lobley, G. E., Louis, P., Flint, H. J., & de Vos, W. M. (2014). Impact of diet and individual variation on intestinal microbiota composition and fermentation products in obese men. *ISME Journal, 8*(11), 2218–2230.

Saresella, M., Mendozzi, L., Rossi, V., Mazzali, F., Piancone, F., LaRosa, F., Marventano, I., Caputo, D., Felis, G. E., & Clerici, M. (2017). Immunological and clinical effect of diet modulation of the gut microbiome in multiple sclerosis patients: A pilot study. *Frontiers in Immunology, 8*, 1391.

Schwiertz, A., Taras, D., Schaefer, K., Beijer, S., Bos, N. A., Donus, C., & Hardt, P. D. (2010). Microbiota and SCFA in lean and overweight healthy subjects. *Obesity, 18*(1), 190–195.

Segal, I., Walker, A., Lord, S., & Cummings, J. (1988). Breath methane and large bowel cancer risk in contrasting African populations. *Gut, 29*(5), 608–613.

Sender, R., Fuchs, S., & Milo, R. (2016a). Are we really vastly outnumbered? Revisiting the ratio of bacterial to host cells in humans. *Cell, 164*(3), 337–340.

Sender, R., Fuchs, S., & Milo, R. (2016b). Revised estimates for the number of human and bacteria cells in the body. *PLoS Biology, 14*(8), e1002533.

Shetty, S. A., Hugenholtz, F., Lahti, L., Smidt, H., & de Vos, W. M. (2017). Intestinal microbiome landscaping: Insight in community assemblage and implications for microbial modulation strategies. *FEMS Microbiology Reviews, 41*(2), 182–199.

Shetty, S. A., Smidt, H., & de Vos, W. M. (2019). Reconstructing functional networks in the human intestinal tract using synthetic microbiomes. *Current Opinion in Biotechnology, 58*, 146–154.

Shoaie, S., Ghaffari, P., Kovatcheva-Datchary, P., Mardinoglu, A., Sen, P., Pujos-Guillot, E., de Wouters, T., Juste, C., Rizkalla, S., & Chilloux, J. (2015). Quantifying diet-induced metabolic changes of the human gut microbiome. *Cell Metabolism, 22*(2), 320–331.

Sonnenburg, E. D., Smits, S. A., Tikhonov, M., Higginbottom, S. K., Wingreen, N. S., & Sonnenburg, J. L. (2016). Diet-induced extinctions in the gut microbiota compound over generations. *Nature, 529*(7585), 212–U208.

Stams, A. J., & Plugge, C. M. (2009). Electron transfer in syntrophic communities of anaerobic bacteria and archaea. *Nature Reviews Microbiology, 7*(8), 568.

Staudacher, H. M., Lomer, M. C., Anderson, J. L., Barrett, J. S., Muir, J. G., Irving, P. M., & Whelan, K. (2012). Fermentable carbohydrate restriction reduces luminal bifidobacteria and gastrointestinal symptoms in patients with irritable bowel syndrome. *The Journal of Nutrition, 142*(8), 1510–1518.

Tap, J., Furet, J. P., Bensaada, M., Philippe, C., Roth, H., Rabot, S., Lakhdari, O., Lombard, V., Henrissat, B., & Corthier, G. (2015). Gut microbiota richness promotes its stability upon increased dietary fibre intake in healthy adults. *Environmental Microbiology, 17*(12), 4954–4964.

The Human Microbiome Project Consortium. (2012). Structure, function and diversity of the healthy human microbiome. *Nature, 486*(7402), 207–214.

Tuohy, K., Kolida, S., Lustenberger, A., & Gibson, G. R. (2001). The prebiotic effects of biscuits containing partially hydrolysed guar gum and fructo-oligosaccharides—A human volunteer study. *British Journal of Nutrition, 86*(3), 341–348.

Turnbaugh, P. J., Ley, R. E., Mahowald, M. A., Magrini, V., Mardis, E. R., & Gordon, J. I. (2006). An obesity-associated gut microbiome with increased capacity for energy harvest. *Nature, 444*(7122), 1027–1031.

Van Nood, E., Vrieze, A., Nieuwdorp, M., Fuentes, S., Zoetendal, E. G., de Vos, W. M., Visser, C. E., Kuijper, E. J., Bartelsman, J. F., & Tijssen, J. G. (2013). Duodenal infusion of donor feces for recurrent Clostridium difficile. *New England Journal of Medicine, 368*(5), 407–415.

Vandeputte, D., Falony, G., Vieira-Silva, S., Wang, J., Sailer, M., Theis, S., Verbeke, K., & Raes, J. (2017). Prebiotic inulin-type fructans induce specific changes in the human gut microbiota. *Gut, 66*(11), 1968–1974.

Vermeire, S., Joossens, M., Verbeke, K., Wang, J., Machiels, K., Sabino, J., Ferrante, M., Van Assche, G., Rutgeerts, P., & Raes, J. (2016). Donor species richness determines faecal microbiota transplantation success in inflammatory bowel disease. *Journal of Crohn's and Colitis, 10*(4), 387–394.

Vital, M., Howe, A. C., & Tiedje, J. M. (2014). Revealing the bacterial butyrate synthesis pathways by analyzing (meta) genomic data. *MBio, 5*(2), e00889–e00814.

Vuholm, S., Nielsen, D. S., Iversen, K. N., Suhr, J., Westermann, P., Krych, L., Andersen, J. R., & Kristensen, M. (2017). Whole-grain rye and wheat affect some markers of gut health without altering the fecal microbiota in healthy overweight adults: A 6-week randomized trial. *The Journal of Nutrition, 147*(11), 2067–2075.

Walker, A. W., Ince, J., Duncan, S. H., Webster, L. M., Holtrop, G., Ze, X., Brown, D., Stares, M. D., Scott, P., & Bergerat, A. (2011). Dominant and diet-responsive groups of bacteria within the human colonic microbiota. *The ISME Journal, 5*(2), 220–230.

Wang, T., Cai, G., Qiu, Y., Fei, N., Zhang, M., Pang, X., Jia, W., Cai, S., & Zhao, L. (2012). Structural segregation of gut microbiota between colorectal cancer patients and healthy volunteers. *The ISME Journal, 6*(2), 320–329.

Windey, K., De Preter, V., Louat, T., Schuit, F., Herman, J., Vansant, G., & Verbeke, K. (2012a). Modulation of protein fermentation does not affect fecal water toxicity: A randomized crossover study in healthy subjects. *PLoS One, 7*(12), e52387.

Windey, K., De Preter, V., & Verbeke, K. (2012b). Relevance of protein fermentation to gut health. *Molecular Nutrition & Food Research, 56*(1), 184–196.

Wu, G. D., Chen, J., Hoffmann, C., Bittinger, K., Chen, Y.-Y., Keilbaugh, S. A., Bewtra, M., Knights, D., Walters, W. A., & Knight, R. (2011). Linking long-term dietary patterns with gut microbial enterotypes. *Science, 334*(6052), 105–108.

Xiao, S., Fei, N., Pang, X., Shen, J., Wang, L., Zhang, B., Zhang, M., Zhang, X., Zhang, C., & Li, M. (2014). A gut microbiota-targeted dietary intervention for amelioration of chronic inflammation underlying metabolic syndrome. *FEMS Microbiology Ecology, 87*(2), 357–367.

Yatsunenko, T., Rey, F. E., Manary, M. J., Trehan, I., Dominguez-Bello, M. G., Contreras, M., Magris, M., Hidalgo, G., Baldassano, R. N., Anokhin, A. P., Heath, A. C., Warner, B., Reeder, J., Kuczynski, J., Caporaso, J. G., Lozupone, C. A., Lauber, C., Clemente, J. C., Knights, D., Knight, R., & Gordon, J. I. (2012). Human gut microbiome viewed across age and geography. *Nature, 486*(7402), 222.

Zeevi, D., Korem, T., Zmora, N., Israeli, D., Rothschild, D., Weinberger, A., Ben-Yacov, O., Lador, D., Avnit-Sagi, T., Lotan-Pompan, M., Suez, J., Mahdi, J. A., Matot, E., Malka, G., Kosower, N., Rein, M., Zilberman-Schapira, G., Dohnalova, L., Pevsner-Fischer, M., Bikovsky, R., Halpern, Z., Elinav, E., & Segal, E. (2015). Personalized nutrition by prediction of glycemic responses. *Cell, 163*(5), 1079–1094.

Zhernakova, A., Kurilshikov, A., Bonder, M. J., Tigchelaar, E. F., Schirmer, M., Vatanen, T., Mujagic, Z., Vila, A. V., Falony, G., & Vieira-Silva, S. (2016). Population-based metagenomics analysis reveals markers for gut microbiome composition and diversity. *Science, 352*(6285), 565–569.

Zoetendal, E. G., & de Vos, W. M. (2014). Effect of diet on the intestinal microbiota and its activity. *Current Opinion in Gastroenterology, 30*(2), 189–195.

Zoetendal, E. G., von Wright, A., Vilpponen-Salmela, T., Ben-Amor, K., Akkermans, A. D., & de Vos, W. M. (2002). Mucosa-associated bacteria in the human gastrointestinal tract are uniformly distributed along the colon and differ from the community recovered from feces. *Applied and Environmental Microbiology, 68*(7), 3401–3407.

Zoetendal, E. G., Raes, J., Van Den Bogert, B., Arumugam, M., Booijink, C. C., Troost, F. J., Bork, P., Wels, M., De Vos, W. M., & Kleerebezem, M. (2012). The human small intestinal microbiota is driven by rapid uptake and conversion of simple carbohydrates. *The ISME Journal, 6*(7), 1415–1426.

The Aging Gut Microbiota

Erin S. Keebaugh, Leslie D. Williams, and William W. Ja

Abstract Researchers have detailed changes in host–intestinal microbe homeostasis in elderly humans, but it is not clear whether gut microbiota influence these changes, or if maintaining intestinal homeostasis would support overall health with age. Insight into age-related changes in hosts and their microbiota has been gained by studying vertebrate models such as mice, rats, and African turquoise killifish, and invertebrates, including *Drosophila melanogaster* and *Caenorhabditis elegans*. Studies using aged, germ-free models show that intestinal microbiota do not initiate all age-related pathologies, suggesting that host-specific changes may be a factor in declining host–intestinal microbe homeostasis with age. Although it is not clear how model-based host–intestinal microbe research applies to the elderly, understanding the interplay between aging hosts and gut microbiota will be critical toward the design of therapeutic interventions. Since research on aging microbiota systems is an emerging field, further developments may come through attempts to translate model findings to humans.

Keywords Aging microbiome · Inflammaging · Intestinal permeability · Healthy aging · Age-associated dysbiosis · Model organisms

Introduction

With a growing population of longer-living people, the promotion of healthy aging is an increasingly urgent task. Our intestinal microbiota has gained attention because of the notable changes in host–microbe homeostasis in aged hosts, though there is great difficulty in distinguishing the physiological changes associated with the

---

Erin S. Keebaugh and William W. Ja share senior authorship.

E. S. Keebaugh · L. D. Williams · W. W. Ja (✉)
Department of Neuroscience, Center on Aging, The Scripps Research Institute,
Jupiter, FL, USA
e-mail: WJa@scripps.edu

© Springer Nature Switzerland AG 2019
M. A. Azcarate-Peril et al. (eds.), *How Fermented Foods Feed a Healthy Gut Microbiota*, https://doi.org/10.1007/978-3-030-28737-5_12

aging host from that of microbe-driven pathologies. Being able to make those distinctions will be of clinical importance, especially in the promotion of healthy aging.

The gut is a highly complex organ system with a tremendous surface area. It not only serves as a barrier against luminal macromolecules and microbes, but it is also involved in immune function, digestion, and nutrient assimilation. Like any other organ system, gut aging is accompanied by physiological changes that lead to impaired function, with ultimately far-reaching consequences on health. For example, age-related changes in intestinal transit time (Woodmansey 2007) and in gut function can impact nutritional intake and absorption (Lovat 1996), potentially exacerbating diet-related influences on intestinal and organismal physiology.

Evidence from invertebrate and vertebrate model organisms suggests that gut barrier integrity is compromised with age (Tran and Greenwood-Van Meerveld 2013; Tricoire and Rera 2015; Dambroise et al. 2016; Gelino et al. 2016; Rera et al. 2018); in mice and the invertebrate model *Drosophila melanogaster*, this change in intestinal permeability is thought to allow the translocation of bacteria or bacterial products from the lumen into circulation (Li et al. 2016; Thevaranjan et al. 2017). The host is then thought to mount an inflammatory response against the leaked microbial signatures. Although it is not yet known if a 'leaky gut' is a natural occurrence in the elderly, aged humans do show increased inflammation. Further, the ability to resolve inflammation may be impaired with advancing age (Sarkar and Fisher 2006). When the homeostatic balance of the immune system is no longer in check, an age-associated inflammatory state, dubbed 'inflammaging', may ensue, resulting in a chronic, low state of inflammation (Franceschi et al. 2000, 2007). Chronic, low-grade inflammation may contribute to a range of comorbidities, accentuating the aging phenotype.

There are documented examples of age-associated effects on components of the intestinal barrier and immune system in humans, but what about the gut microbiota? Here, we overview studies showing that aging is accompanied by changes in the composition of the gut microbiota, but the extent to which these changes are causes or consequences of aging gut physiology remains uncertain.

What Is a 'Healthy' Gut Microbiota?

The gut was believed to be sterile up until birth, although recent studies point to the highly debated possibility of *in utero* colonization (Jimenez et al. 2005; Rautava et al. 2012; Collado et al. 2016; de Goffau et al. 2019; Martin et al. 2016; Perez-Munoz et al. 2017). As delineated earlier in this book, mode of delivery and feeding influence the early intestinal microbiota composition, with possible consequences for immune system development (Hallstrom et al. 2004; Rutayisire et al. 2016). The early life microbiota may be an important factor in health outcomes, as Cesarean births are sometimes associated with higher incidences of immune-related disorders in early life (Negele et al. 2004). Indeed, microbes have been found to influence the developing immune system and have an impact on mucosal and systemic immune tissues (Macpherson and Harris 2004; Malamitsi-Puchner et al. 2005; Hooper et al. 2012; Tamburini et al. 2016).

With the introduction of solid foods, the composition of gut microbiota is diversified and at around age 3 begins to resemble that of the adult, after which differences resulting from the mode of delivery and breast or formula feeding are less pronounced (Koenig et al. 2011; Yatsunenko et al. 2012; Duncan and Flint 2013). The gut microbiota continues to diversify with age up through adulthood, although there are individual differences in this maturation process (Odamaki et al. 2016). Microbes are housed throughout the intestinal tract, continually increasing in abundance and found in the highest amount in the large intestine (O'Hara and Shanahan 2006). The precise composition of microbiota in the adult gut varies between individuals and even among siblings, although there is thought to be a core, shared microbiota amongst different people (Qin et al. 2010; Yatsunenko et al. 2012). It is thought that greater than 20% of the observed interindividual variation in microbiota composition is due to diet and other environmental factors as opposed to host genetics (Rothschild et al. 2018). And though the gut microbiota is responsive to environmental perturbations, the microbial communities inhabiting an individual remain relatively stable (Costello et al. 2009).

The characterization of a healthy microbiota might assist in the diagnosis and intervention of health conditions associated with alterations in gut microbes. Identifying what constitutes a healthy microbiota, however, has proven difficult despite a number of population-scale studies that have set out to do so (Human Microbiome Project Consortium 2012). The microbial composition of subjects ranging in age and geography have been measured to establish common microbial features (Turnbaugh et al. 2007) and have generally focused on searching for taxa abundant in healthy guts (Lloyd-Price et al. 2016). *Faecalibacterium prausnitzii*, a member of Firmicutes, is one of the most abundant species in a healthy intestine and has anti-inflammatory properties; a loss in abundance has been observed in various intestinal disorders such as irritable bowel disease (Mueller et al. 2006; Sokol et al. 2008; Miquel et al. 2013; Cao et al. 2014). Not all microbes are ubiquitous across humans, and a low prevalence of certain microbial groups does not indicate the lack of functionality. Lesser-represented groups in some Western populations, such as methanogenic archaea, are important despite their low relative abundance; methanogens are useful for energy harvest from ingested food (Walker 2007).

Attempts to identify imbalances that reflect disease states are complicated by interindividual diversity and by the existence of a range of possible 'healthy' microbiota configurations (Lloyd-Price et al. 2016). An 'unhealthy' gut may be defined by a disproportionate amount of pathogenic bacteria, for example, *Clostridium difficile*, or when disease phenotypes manifest in the host as a result of an unbalanced intestinal microbiota, or intestinal dysbiosis (Bien et al. 2013; Henderson and Nibali 2016). Alternatively, attempts to characterize a 'healthy' microbiome can use a metagenomic approach that focuses on the functionality of genes present (Lozupone et al. 2012; Rosen and Palm 2017). An analysis of fecal metagenomes (the genetic material isolated from fecal samples) of humans from different countries found that 12 genetic biomarkers correlate with increasing age, including an elevation of digestive enzymes that degrade starch (Arumugam et al. 2011). These findings point to a potential use of microbial biomarkers for the detection of an 'aging' microbiota (Arumugam et al. 2011).

A high degree of diversity in an ecosystem is sometimes considered better for adaptation to environmental stresses. Similarly, the diverse gut microbiota is somewhat malleable, responding to dietary changes to the potential benefit of the host (David et al. 2014; Biagi et al. 2017). A system of checks and balances may render a diverse microbiota ecosystem less susceptible to disease (Candela et al. 2012). Therefore, instead of defining a healthy microbiota by a set of taxa known to support health, it may be more informative to identify a set of general characteristics such as microbe diversity, stability, and plasticity (Backhed et al. 2012; Lloyd-Price et al. 2016).

The Elderly Gut Microbiota

Though diversity and stability of the gut microbiome increases with age beginning at birth and throughout early life (Palmer et al. 2007; Koenig et al. 2011; Yatsunenko et al. 2012; Rodriguez et al. 2015), the general trajectory for the aging gut is a loss of biodiversity (Woodmansey 2007; Biagi et al. 2016), compromised stability, and greater individual variation (Claesson et al. 2011, 2012). Although increased inter-individual variation makes it difficult to make generalizations, there are some broad trends and commonalities that are worth mentioning. A study focused on humans ranging in age from adulthood to centenarians and beyond identified a core set of shared microbes that were found to change in abundance over time. The dominant core microbiota was mostly comprised of Ruminococcaceae, Lachnospiraceae, and Bacteroidaceae (Biagi et al. 2016). While this dominant core shrank in representation in increasingly aged humans, subdominant groups increased in abundance (Biagi et al. 2016). Across multiple studies, facultative anaerobes, streptococci, staphylococci, enterococci, and enterobacteria were among the microbial groups elevated with age (Candela et al. 2014). However, the same intestinal microbiota shifts are not always common across studies (Magrone and Jirillo 2013).

Studies on aging humans are often focused on distinct populations and are therefore not always broadly applicable. As such, there are differences found in studies across various groups of aging humans, and the conflicting observations are at least partially due to differences in diet and other environmental factors between cohorts (Magrone and Jirillo 2013). Further, the experimental design, microbe sampling method (Biagi et al. 2012), and targeted populations in clinical trials can lead to varying results between studies and lower the translatability of datasets. That said, it is possible to identify similarities across distinct human populations. A study focused on multiple European countries found that enterobacteria levels were increased in the elderly sampled across countries; between-cohort differences, however, were also noted in the *Bifidobacterium* group within the same study (Mueller et al. 2006). For these reasons, we focus largely on one dataset that used a unique study design to capture the interplay between gut microbiota and the environment of aging humans.

The ELDERMET project was designed to identify links between health and gut microbiota structure within elderly Irish subjects. The ongoing project has been enacted in phases; one study tracked the elderly across different living situations and care facilities to provide insight into the interaction between health status, diet, lifestyle, and microbiota composition in aged humans. Subjects were categorized as one of the following: individuals living in the community, making out-patient hospital visits, receiving short term care (<6 weeks) for rehabilitation, or residing long-term in residential care. The study identified several shifts in fecal microbiota associated with residence, which also closely associated with diet (Claesson et al. 2012). The extent to which diet is a controllable factor to modulate age-related disease remains a major line of current research to determine if specialized diets can delay the onset of age-related illness.

Most of the long-stay subjects reported a diet that was moderate to high in fat and low in fiber, and their fecal metabolites revealed higher levels of glucose, glycine, and lipids. Microbiota from long-term care residents was composed of a higher proportion of Bacteroidetes over Firmicutes and was associated with *Parabacteroides*, *Eubacterium*, *Anaerotruncus*, *Lactonifactor*, and *Coprobacillus* genera. In contrast, the majority of community dwellers reported diets classified as low to moderate in fat, and high in fiber. Community dwellers had higher levels of the metabolites glutarate, butyrate, acetate, propionate, and valerate. Further, their microbiota showed more abundant *Coprococcus* and *Roseburia* at the genus level, and Lachnospiraceae was among the most prominent of associated families. Those that ate a diet classified as low fat/high fiber not only had the most diverse diet but also had the most diverse intestinal microbiota (Claesson et al. 2012). It is possible that those living at home had more exposure to a variety of foods. These findings may indicate that care facilities can benefit from diversified food menus that promote intestinal microbiota diversity.

The shifts observed in elderly gut microbiota were reflective of changes in health as measured by a number of indices, including mental state, inflammatory markers, and functional independence. A loss of certain community-associated microbes was correlated with increased measures of frailty (Claesson et al. 2012). Frailty can be a useful indicator of health deficit, and studies have shown that reduced microbiota diversity is associated with increased frailty (Jackson et al. 2016). There are reported differences in fecal microbiota composition between elderly persons with low and high frailty scores. For example, Lactobacilli, *Bacteroides/Prevotella*, and *F. prausnitzii* were decreased, while Enterobacteriaceae were increased in the high-frailty subjects (van Tongeren et al. 2005). These microbial changes associated with frailty may represent diagnostic targets to monitor as individuals age.

Longitudinal studies following humans across their lifespan are not readily attainable; these studies are more feasible in shorter-lived animal models. Some human studies are semi-longitudinal over a brief portion of the human lifespan, however most are cross-sectional, whereby representative groups are sampled at one point in time. Although longitudinal studies may be ideal to record age-related trajectories, cross-sectional studies have provided insight into broad differences across age, health, and lifestyle cohorts. Numerous studies not covered here detected shifts in the intestinal microbiota of aged humans (Hopkins et al. 2001; Hopkins and Macfarlane 2002; Hayashi et al. 2003; Woodmansey et al. 2004; Mariat et al. 2009; Biagi et al. 2010; Rampelli et al. 2013; Odamaki et al. 2016; Buford 2017), including other ELDERMET consortium studies (Claesson et al. 2011; Jeffery et al. 2016). It is worth reiterating that the variation in gut microbiota composition and the specificity of human studies can complicate attempts to reveal common associations between microbes and host age; the degree to which observed changes are due to dietary or lifestyle factors, or are part of the natural aging process, is not always clear. Future research may benefit from an integrative approach to reveal how environmental factors impact a broader range of human populations.

Age-Related Changes in the Host–Microbiota System

Existing studies have detected age-related changes in the composition of gut microbial populations (Buford 2017), leading to an interest in detailing causative factors. One proposed causal factor is a change in nutrition in older adults (Lu and Wang 2018), which can be driven by natural processes, age-related illnesses, or behavioral and lifestyle changes (Nagpal et al. 2018; Riaz Rajoka et al. 2018). Beyond changes in physiologic systems, external factors including the environment and how an individual responds to their environment can also influence the intestinal microbiota in an age-dependent manner. Elderly humans show a higher threshold for sweet, salty, sour, and bitter tastants, indicating that taste is altered with age (Fukunaga et al. 2005), potentially contributing to changes in food intake. Since changes in nutrition correlate with fecal microbiota composition in healthy (Wu et al. 2011) and elderly humans (Claesson et al. 2012), the factors that alter dietary intake with age may interact to influence intestinal microbiota during the aging process. It is not yet clear if specific diets can prevent aging-related microbial changes. However, it was proposed that achieving optimal levels of protein, fiber, and fat may support intestinal and immune health in the elderly (Clements and Carding 2018).

In addition to changes in nutrition, antibiotic use in elderly patients can reduce or eradicate certain microbial species in fecal microbiota (Bartosch et al. 2004). The rising antibiotic use in residential care facilities (Lim et al. 2014) and in older United States residents (Lee et al. 2013, 2014) represents an increasingly influential factor on intestinal microbes. Additionally, living conditions can impact how intestinal microbiota respond to antibiotic treatment (Jeffery et al. 2016), making it difficult to perform controlled analyses on human populations. Given the myriad of factors that influence intestinal microbiota in humans, it follows that laboratory models are commonly used for a more controlled approach to researching aging host–microbe systems. The use of model systems has produced some of the most informative data to date on age-associated changes in hosts and their intestinal microbes (Maynard and Weinkove 2018).

Humans and model organisms experience age-associated changes in systemic and intestinal immunity (Man et al. 2014); models are useful to study aging host–microbe systems because altered immune regulation can impact microbial symbionts, and vice versa. Advancing studies use models to pursue a century-old hypothesis generated by Elie Metchnikoff: maleffects of old age stem from changes in intestinal microbiota and a restoration of host–microbe homeostasis can improve age-related illnesses (Metchnikoff 1908). To test this idea, researchers have begun detailing aging guts to determine if altered host–microbe homeostasis underlies broader age-related maladies.

One commonality of interest is the increase in chronic, systemic inflammation with age (i.e., inflammaging). An age-related increase in inflammation is seen in a range of organisms from insects (Rera et al. 2012; Clark et al. 2015; Li et al. 2016) to mice (Conley et al. 2016) and humans. The inflammation status of elderly humans is thought to be an indicator of disease and mortality risk (Franceschi and Campisi 2014). Although associative changes are known to occur along with inflammaging, the definitive cause of age-related inflammation remains mostly unknown. As a starting point for investigations of age-related inflammation, some research has focused on the innermost layer of the intestine, the mucosa.

The mucus layer coats the inner lining of the intestinal tract and comes into contact with luminal microbes; deterioration of this interface may be a source for the homeostatic breakdown between host and microbes. In rodents, the density of the colonic mucus layer varies both vertically and longitudinally, and this viscosity gradient can impact the distribution of colonic microbes (Swidsinski et al. 2007b). Although the mucus layer thickness varies across species, many animals (Varum et al. 2012) including humans have two colonic mucus layers (Matsuo et al. 1997) with a relatively dense inner layer adjacent to epithelial cell surfaces, and a less-dense outer layer exposed to the intestinal luminal contents. The inner layer is expected to be absent of microbes, whereas the outer layer is colonized with microbes, suggesting that microbes are typically partitioned from the epithelium by the dense inner mucosal layer (Johansson et al. 2008, 2011).

Since the mucus layer covering the intestinal epithelium acts as a barrier for those epithelial cells (Johansson 2014), a malfunctioning mucus layer is associated with translocation of bacteria into intestinal crypts and an increase in intestinal inflammation (Johansson et al. 2008; Johansson 2014). The mucus layer of the mouse colon declines with age (van Beek et al. 2016). Aged mice show a diminishing mucus layer and bacterial translocation into the mucus or even into the intestinal epithelium. These changes are associated with a change in microbiota composition and activation of the intestinal immune response (Elderman et al. 2017).

As in other organisms, the human colonic mucus layers largely prevent contact between intestinal microbes and the epithelium. An investigation of normal and inflamed colons from human subjects found an association with decreased mucus layer thickness and increased inflammation, as well as a migration of bacteria into the mucosa (Swidsinski et al. 2007a). Further, upon some intestinal insults, a diminished mucus layer is associated with increased intestinal epithelial permeability in rats (Qin et al. 2011; Fishman et al. 2013). Thus, once this protective mucosal layer is diminished with age or with illness, the cellular barrier of the intestine may be compromised and an associated increase in inflammation can occur.

It has been proposed that a compromised intestinal barrier function upon age may allow gut microbes, or microbial products, to leak into non-tolerant areas (Franceschi and Campisi 2014); a translocation of microbial signatures then sparks a subsequent inflammatory response against exogenous products that hosts encounter in circulation. Indeed, aged mice show increased intestinal permeability—specifically, the colonic region shows higher paracellular permeability, indicating that the increased intestinal permeability is due to compromised passage between intercellular spaces (Thevaranjan et al. 2017). Aged mice also show higher levels of a bacterial cell-wall product called muramyl dipeptide outside of the intestinal lumen, which may indicate that displaced microbes or microbial products are indeed circulating systemically (Thevaranjan et al. 2017).

Mice null for TNF, a pro-inflammatory cytokine, have been used to determine how inflammation influences these age-associated changes. Aged TNF mutants do not show heightened systemic inflammation, compromised intestinal barrier function, or increased microbial signatures in circulation (Thevaranjan et al. 2017). Further, these mutants appear to have less prominent age-associated microbial adjustments, and anti-TNF therapy has the capacity to modulate microbial diversity (Thevaranjan et al. 2017). These results indicate that TNF-mediated inflammation may influence aging phenotypes related to altered gut barrier function and microbial composition.

Aged germ-free mice also lack some of the aforementioned age-related symptoms perhaps because of the absence of intestinal microbiota. Old germ-free mice do not show a decline in intestinal barrier function or increased systemic inflammation. Co-housing germ-free mice with young or old conventional mice exposed germ-free mice to a conventional, or 'standard', microbiota. Young germ-free mice exposed to aged donor mice demonstrate an increase in intestinal permeability and systemic inflammation (Thevaranjan et al. 2017). These results are consistent with a model suggesting that microbiota from aged individuals can drive these intestinal and systemic symptoms. However, no causal changes within the microbiota from aged mice have been identified. Thus, specific dysbiotic changes in the microbiota composition or quantity remain unknown. Interestingly, aged germ-free mice show increased TNF when exposed to microbes from both young and old mice, suggesting that older mice may also possess sensitivities to intestinal microbiota that are absent in younger mice (Thevaranjan et al. 2017). Sensitivity to microbiota upon age could potentially compound age-related symptoms.

In a similar study, fecal microbiota were transferred from young or old conventional mice into young, germ-free mice by oral gavage. This process exposed formerly germ-free mice to conventional youthful or aged mouse gut microbiota. After 4 weeks, recipients of 'old' microbes showed systemic immune activation along with an upregulation of several immune pathways in the small intestine (Fransen et al. 2017). These changes were not detected in young mice, or recipients of 'young' microbiota (Fransen et al. 2017). A bioinformatics analysis suggested that

The Aging Gut Microbiota 293

lipopolysaccharides, molecules found on the outer membrane of some bacteria, induce the immune modulatory effect of 'old' microbiota (Fransen et al. 2017).

Further, cell culture-based tests indicated that sera from recipients of 'old' microbiota, but not from 'young' microbiota, may contain immune-stimulatory factors, a proxy measurement for microbial signatures (Fransen et al. 2017). This is consistent with a model in which the transfer of 'old' microbiota into young mice may lead to translocation of immune-activating bacterial moieties systemically. More definitive tests should be performed, however, since sera from conventional aged mice show no increased signatures of bacterial components when compared to young conventional mice. Changes in a few groups of bacteria including decreases in *Akkermansia* and increases in TM7 and Proteobacteria are associated with older mice or recipients of 'old' microbiota. These changes in microbiota composition are dynamic over a month-long period and it is not known if shifts in any of these groups are causative to the observed age-related outcomes (Fransen et al. 2017).

Many of the current studies on the interaction between microbiota and the aging host are associative. Further research is necessary to approach the status of clear, causal evidence. Overall, these studies show that age-related changes occur in both the host and microbiota across a range of animals, and that these changes are associated with negative health outcomes. As a result, there is great interest in finding ways to prevent or delay ailments of old age by treating both the host and intestinal microbiota.

Preventing Age-Related Deterioration by Genetically Manipulating the Host

Studies on *D. melanogaster* were some of the first to provide detailed information on the homeostatic changes in intestinal, commensal, and host physiology in aged animals. Some of the benefits of the fly model are its genetic tractability, relatively short lifespan (typically ranging from 30 to 80 days), and the similarities between mammalian and fly intestinal biology (Buchon et al. 2013; Marianes and Spradling 2013). These benefits allow rapid studies on the connection between host–commensal physiology, and organismal aging and longevity. A properly functioning intestinal barrier is influential on longevity in *D. melanogaster*. Further, the general status of fly–microbe homeostasis is indicative of intestinal barrier integrity and host mortality (Rera et al. 2012; Clark et al. 2015; Li et al. 2016). As flies age, they demonstrate changes in the configuration and numbers of intestinal microbes, and flies also show diminished barrier function (Guo et al. 2014; Clark et al. 2015). A recent focus on the etiology of intestinal and commensal maleffects has identified genetic manipulations that can impede or lessen these age-related breakdowns in host–microbe homeostasis, ultimately extending life.

The fly intestine normally comprises ten or more compartments (Buchon et al. 2013, Marianes and Spradling 2013) that are involved in the localization of luminal microbes along the intestinal tract (Li et al. 2016). An acidic region of the gut is formed by the 'copper cells'. pH alterations of the acidic region can modulate microbiota levels, suggesting this distinct compartment has a regulatory role over gut microbes (Overend et al. 2016). When the acidic region is intact, microbes are most commonly housed within the anterior gut. When the copper cell region is genetically ablated, however, the quantity of luminal microbes increases throughout the intestinal tract (Li et al. 2016). As these changes occur, systemic inflammation elevates as measured by the activity of conserved pathways that control inflammation in mammals (Li et al. 2016). It is possible that this inflammatory response occurs as the fly responds to translocated microbial factors, similar to what was suggested in the mouse model (Fransen et al. 2017; Thevaranjan et al. 2017).

The copper cell region of aging flies undergoes metaplastic changes as copper cells are replaced with cell types typically found in other intestinal compartments; these changes are also demonstrated by germ-free animals (Li et al. 2016). Aged flies, even when germ-free, experience changes in septate junction protein localization that may negatively impact intestinal barrier integrity (Byri et al. 2015; Resnik-Docampo et al. 2017, 2018; Salazar et al. 2018). The occurrence of these cellular alterations in germ-free flies indicates that intestinal microbiota do not initiate these age-related intestinal pathologies. This is consistent with the possibility that age-related changes in the copper cell region and intercellular junctions may drive changes in host–microbe homeostasis. Importantly, these results suggest that genetic manipulations of the aging fly host could help pinpoint the onset of age-related malfunctions in fly–microbe homeostasis.

To determine the etiology of age-related intestinal pathologies, researchers selectively focused on the JAK/Stat pathway, which can control inflammatory-like responses against infection and is deregulated with age (Guo et al. 2014). Further analysis detected heightened JAK/Stat pathway activity in the intestine of aged flies. Importantly, intestinal JAK/Stat activation causes metaplastic changes in the copper cell region, comprised of both mis-differentiated and trans-differentiated cells (Li et al. 2016). These results indicate that JAK/Stat activation can impair intestinal partitions, one of the hallmarks of declining host–microbe homeostasis in the fly.

In further supportive studies, knocking-down JAK/Stat activity in the intestinal copper cell region counteracts negative health parameters in aged flies; these flies harbor lower counts of intestinal bacteria and showed cellular characteristics of younger flies (Li et al. 2016). Interestingly, these animals also have an increased lifespan even when germ-free, suggesting that JAK/Stat misregulation in the copper cell region generates negative health outcomes in aging flies, and inhibiting JAK/Stat activity reverses some of those symptoms to extend life. Further, the longer lifespan and intestinal compartment preservation in axenic flies with decreased JAK/Stat signaling adds evidence that changes in intestinal microbiota alone do not explain all of the ailments of age. Further studies are required to determine why intestinal decompartmentalization drives negative effects with age, and how these effects may influence host–microbe homeostasis.

Although there are some noted similarities between models and humans, such as late-life shifts in microbiota (Claesson et al. 2011; Guo et al. 2014; Clark et al.

2015; Conley et al. 2016; Li et al. 2016; Fransen et al. 2017; Thevaranjan et al. 2017) and increased inflammation (Franceschi and Campisi 2014; Li et al. 2016; Thevaranjan et al. 2017), whether the remaining age-related disturbances occur in humans is unknown. Therefore, while there are similarities between the age-related pathologies across mice and flies, the implication of these findings for humans is not definitive. Still, model systems have contributed valuable insight into potential mechanisms leading to age-related inflammation and intestinal decline. Continued work may enhance the translational power of models. Recent *Drosophila* studies suggest that microbes can influence host nutritional status or act as a nutritional resource (Ridley et al. 2012; Broderick et al. 2014; Wong et al. 2014; Chaston et al. 2016) to impact fly lifespan under certain conditions (Yamada et al. 2015; Bing et al. 2018; Keebaugh et al. 2018). It may be interesting to consider passive versus active microbial effects in an aging fly model, potentially by differentiating between microbes that stably colonize the fly intestine (Obadia et al. 2017; Pais et al. 2018) versus those that pass through during meals. Ultimately, modern studies with various models may provide a deeper understanding of the physiological alterations influencing host–microbe homeostasis and whether these changes impact longevity. Future research might also investigate how interventions targeting the aging intestine can influence host–microbiota outcomes.

Treating Age-Related Symptoms with Probiotics

Some of the intestinal microbial species that decrease in aged humans can be beneficial for preventing inflammatory responses. Whether compositional changes impacting these species influence increased age-related inflammation remains unknown (Rehman 2012). Research suggests that the immunomodulatory effect of some microbial strains is impacted by aging (You and Yaqoob 2012). Because of the potential link between microbiota, aging, and immune regulation, there is an interest in treating aging symptoms by promoting beneficial microbes.

Live microbes that promote health benefits when adequately consumed are generally referred to as probiotics (see chapter on "Probiotics and Prebiotics"). There are different ways probiotics can be administered, including as foods or as supplements, and probiotics can have a range of beneficial effects on hosts (Hill et al. 2014). Fermented foods have also been found to have beneficial effects in older adults (Turchet et al. 2003; Beausoleil et al. 2007; Fukushima et al. 2007; Hickson et al. 2007; Guillemard et al. 2010) although not all fermented foods can be considered probiotics given their unquantified amount of microbes. Further, it is not always straightforward to differentiate the benefits of microbes within fermented foods with those associated with the food item itself (Hill et al. 2014). That said, studies have found that dietary supplementation or fermented drinks with quantified levels of *Bifidobacterium* can increase the levels of these microbes in fecal samples (Ahmed et al. 2007; Lahtinen et al. 2009), and is correlated with an increase in cellular immune function (Gill et al. 2001) and potentially beneficial shifts in pro- and anti-inflammatory cytokines in elderly subjects (Ouwehand et al. 2008).

Although some of the individual symptoms of age may be treated with probiotics, there is no current probiotic or fermented food regimen to prevent the suite of aforementioned age-related pathologies in host–microbe homeostasis. Most of the evidence for age-related probiotic treatments is largely produced in mice or in accelerated-aging mouse models. There is some evidence for specific microbial strains that can improve the intestinal permeability and longevity of aged mice. Middle-aged mice gavaged with *Bifidobacterium animalis* strain LKM512 three times a week show decreased colon permeability and improved survival rates over the 11-month dosing period (Matsumoto et al. 2011). Further, mice show suppressed systemic and colonic inflammation at 45 weeks of treatment (Matsumoto et al. 2011). Currently, the relevance of these findings in humans is unknown.

Many studies, including some using human subjects, focus on the bacterial species *Lactobacillus plantarum. L. plantarum* is a fermentative lactic acid bacterium that is found in a variety of food products and in the intestines of multiple animals (Ahrne et al. 1998; de Vries et al. 2006). A study testing the adherence capacity of different *L. plantarum* strains found that a majority of tested strains have the capacity to bind to a human-derived colonic cell line via what appears to be a mannose-specific mechanism, suggesting that some *L. plantarum* strains adhere to mannose-containing receptors within the intestine (Ahrne et al. 1998). Certain strains of *L. plantarum* from a fermented diet can survive the gastrointestinal tract and become associated with the intestinal mucosa in both healthy (Johansson et al. 1993) and ill patients (Klarin et al. 2005), although the capacity for *L. plantarum* to colonize the human intestinal tract varies (Johansson et al. 1993; Vesa et al. 2000). Since constant exposure is required for persistence of some strains (Vesa et al. 2000), recent attempts to identify 'persisting' *L. plantarum* strains are focusing on strains derived from healthy human guts as opposed to other sources (Suryavanshi et al. 2017). Such strains that are sustained within the intestine may be more suited for probiotic applications. To date, various *L. plantarum* strains have been tested for probiotic effects in human trials on patients harboring a diverse range of illnesses (Darby and Jones 2017); of potential interest for the aging population, *L. plantarum* strain 299v has the potential to attenuate systemic inflammation in ill patients (McNaught et al. 2005; Jones et al. 2013).

Recently, a mouse model of accelerated aging was used to test for the effects microbes have on aging intestines, since little is known about the impact of specific microbial strains on the aging gut. Accelerated-aging mutant mice and their wild-type littermates were exposed to *L. plantarum* strain WCFS1 by gavage three times per week for a 10-week period. The WCFS1 strain impedes the thinning of the colonic mucus barrier, which is a normal occurrence in the accelerated-aging mutants. Interestingly, there are no noted effects of WCFS1 supplementation in wild-type littermates, suggesting that the beneficial effects of this strain may be specific to aged animals (van Beek et al. 2016). As *L. plantarum* strain WCFS1 has a sequenced and annotated genome, it may provide a powerful system to investigate the mechanisms underlying beneficial effects in aging mice (Kleerebezem et al. 2003).

Conclusion and Future Directions

The expanse of recent gut microbiota research details a complex relationship between hosts and their associated microbes. It is increasingly evident that the maintenance of intestinal homeostasis may contribute to the overall health status of aged individuals. With a growing population of elderly people, understanding how an aging microbiota might accelerate or slow the pathophysiology of aging is of particular interest and may lead to novel therapeutics or dietary interventions that can restore intestinal homeostasis and support health.

It is currently unclear the degree to which aging gut physiology is a cause or consequence of the microbiota shifts accompanying age. Numerous studies on the elderly have detected changes in gut microbiota composition as well as increased levels of inflammation, but it is not known whether microbiota drive inflammaging. Although we do not yet understand the underlying etiological mechanisms in their entirety, we know that there are a number of factors that may compromise our homeostatic relationship with gut microbes, possibly tipping the scale toward a dysbiotic ecology. Although the appealing idea to enterically treat the suite of age-related gut and microbe alterations has no current support in humans, studies have demonstrated that certain microbes may have the ability to modify the host phenotype in ways that pertain to host health.

Diet is a somewhat controllable factor by which to manipulate gut microbiota, and a diverse, healthy diet is associated with a diverse gut microbiota. Modern approaches may help in the development of dietary interventions for aging humans. Researchers are investigating long-lived models of 'healthy aging' to identify lifestyle and dietary habits that might support the maintenance of microbial diversity and health with age (Kong et al. 2016, 2018; Franceschi et al. 2018), and considering biological markers of aging as opposed to chronological age to better understand the interaction of diet, aging, and the microbiota (Kim and Jazwinski 2018). These studies, in combination with longitudinal approaches (Santoro et al. 2018) and new genome-scale metabolic modeling methods (Kumar et al. 2016), may eventually reveal how physiological changes upon age impact nutritional intake and microbiota composition and reveal nutritional means by which aging humans can maintain health.

Researchers are responding to the mounting knowledge on aging intestinal microbiota with attempts to develop food-based or probiotic treatments. A downstream initiative from the ELDERMET studies, referred to as ELDERFOOD, is identifying food ingredients that support a healthy microbiota and overall health in the elderly. As researchers continue to catalog specific functions performed by particular microbial strains, we may see an increase in targeted therapeutic probiotics. Fermented foods are another abundant source of microbes, some of which are part of traditional diets. Future studies may focus on aging human subjects to infer beneficial effects of specific microbial strains or fermented foods. However, mechanistic investigations into age-related changes are likely to be restricted to genetically tractable model organisms.

Most of the aforementioned treatment-focused studies rely on model organisms, and they would not be possible without the prior progress made by aging model research. Fly and mouse research provided premier details on the interrelated, age-associated changes in host intestines, microbiota, and systemic immune regulation. Although research has found correlative changes between intestinal microbiota and age, causal roles that distinct microbial strains play in age-related changes have not yet been detailed. Subsequent work using models may focus on identifying specific dysbiotic changes that influence, or are characteristic of, age-related pathologies. Identifying specific dysbiotic shifts across animals may help to identify health- or age-associated microbes that may ultimately support direct probiotic developments.

There is still more to come from research pertaining to aging and gut microbiota across animal systems. The nematode *Caenorhabditis elegans* has been used to identify pro-longevity variants in *Escherichia coli* mutant libraries (Han et al. 2017); downstream efforts from this study may aid in the development of pro-longevity probiotics. *C. elegans* research has also demonstrated that intestinal microbes can influence drug efficiency (Garcia-Gonzalez et al. 2017; Scott et al. 2017), and recent studies in mice indicate that the microbiome can contribute greatly to drug metabolism (Zimmermann et al. 2019). Future work on modeling host–gut microbe–drug interactions may be important for aging humans because of the increasing polypharmacy observed with age (Charlesworth et al. 2015).

Aged African turquoise killifish lose gut microbe diversity during aging and live longer when colonized with microbiota from younger fish (Smith et al. 2017). This suggests that negative changes occur in killifish microbiota with age, and restoring microbiota to a more youthful state is beneficial to older fish. Model organisms have unique attributes and limitations (Douglas 2018). Although innate differences in gut anatomy or microbiota partitioning may interfere with translating findings from study organisms to humans (Nguyen et al. 2015; Keebaugh and Ja 2016), animal models will continue to be valued for their use in uncovering molecular mechanisms and in developing host- or microbe-targeted interventions.

Researchers have only scraped the surface in terms of aging microbiota research. In particular, microbial populations outside of the intestine are lesser-studied and may have significance for aged humans. For example, it has been suggested that toxic proteases from *Porphyromonas gingivalis,* a bacterium associated with periodontal disease, are found in higher levels in the brains of Alzheimer's patients; small-molecule inhibitors of those proteases reduced Alzheimer's-like disease pathology in the mouse brain and are now being tested in human trials (Dominy et al. 2019). Further interesting developments may come as researchers continue to compile and analyze data across species, and attempt to translate findings from model organisms to the human system.

References

Ahmed, M., Prasad, J., Gill, H., Stevenson, L., & Gopal, P. (2007). Impact of consumption of different levels of *Bifidobacterium lactis* HN019 on the intestinal microflora of elderly human subjects. *The Journal of Nutrition, Health & Aging, 11*(1), 26–31.

Ahrne, S., Nobaek, S., Jeppsson, B., Adlerberth, I., Wold, A. E., & Molin, G. (1998). The normal Lactobacillus flora of healthy human rectal and oral mucosa. *Journal of Applied Microbiology, 85*(1), 88–94.

Arumugam, M., Raes, J., Pelletier, E., Le Paslier, D., Yamada, T., Mende, D. R., Fernandes, G. R., Tap, J., Bruls, T., Batto, J. M., Bertalan, M., Borruel, N., Casellas, F., Fernandez, L., Gautier, L., Hansen, T., Hattori, M., Hayashi, T., Kleerebezem, M., Kurokawa, K., Leclerc, M., Levenez, F., Manichanh, C., Nielsen, H. B., Nielsen, T., Pons, N., Poulain, J., Qin, J., Sicheritz-Ponten, T., Tims, S., Torrents, D., Ugarte, E., Zoetendal, E. G., Wang, J., Guarner, F., Pedersen, O., de Vos, W. M., Brunak, S., Dore, J., Meta, H. I. T. C., Antolin, M., Artiguenave, F., Blottiere, H. M., Almeida, M., Brechot, C., Cara, C., Chervaux, C., Cultrone, A., Delorme, C., Denariaz, G., Dervyn, R., Foerstner, K. U., Friss, C., van de Guchte, M., Guedon, E., Haimet, F., Huber, W., van Hylckama-Vlieg, J., Jamet, A., Juste, C., Kaci, G., Knol, J., Lakhdari, O., Layec, S., Le Roux, K., Maguin, E., Merieux, A., Melo Minardi, R., M'Rini, C., Muller, J., Oozeer, R., Parkhill, J., Renault, P., Rescigno, M., Sanchez, N., Sunagawa, S., Torrejon, A., Turner, K., Vandemeulebrouck, G., Varela, E., Winogradsky, Y., Zeller, G., Weissenbach, J., Ehrlich, S. D., & Bork, P. (2011). Enterotypes of the human gut microbiome. *Nature, 473*(7346), 174–180.

Backhed, F., Fraser, C. M., Ringel, Y., Sanders, M. E., Sartor, R. B., Sherman, P. M., Versalovic, J., Young, V., & Finlay, B. B. (2012). Defining a healthy human gut microbiome: Current concepts, future directions, and clinical applications. *Cell Host & Microbe, 12*(5), 611–622.

Bartosch, S., Fite, A., Macfarlane, G. T., & McMurdo, M. E. (2004). Characterization of bacterial communities in feces from healthy elderly volunteers and hospitalized elderly patients by using real-time PCR and effects of antibiotic treatment on the fecal microbiota. *Applied and Environmental Microbiology, 70*(6), 3575–3581.

Beausoleil, M., Fortier, N., Guenette, S., L'Ecuyer, A., Savoie, M., Franco, M., Lachaine, J., & Weiss, K. (2007). Effect of a fermented milk combining *Lactobacillus acidophilus* CL1285 and *Lactobacillus casei* in the prevention of antibiotic-associated diarrhea: A randomized, double-blind, placebo-controlled trial. *Canadian Journal of Gastroenterology, 21*(11), 732–736.

Biagi, E., Nylund, L., Candela, M., Ostan, R., Bucci, L., Pini, E., Nikkila, J., Monti, D., Satokari, R., Franceschi, C., Brigidi, P., & De Vos, W. (2010). Through ageing, and beyond: Gut microbiota and inflammatory status in seniors and centenarians. *PLoS One, 5*(5), e10667.

Biagi, E., Candela, M., Fairweather-Tait, S., Franceschi, C., & Brigidi, P. (2012). Ageing of the human metaorganism: The microbial counterpart. *Age, 34*(1), 247–267.

Biagi, E., Franceschi, C., Rampelli, S., Severgnini, M., Ostan, R., Turroni, S., Consolandi, C., Quercia, S., Scurti, M., Monti, D., Capri, M., Brigidi, P., & Candela, M. (2016). Gut microbiota and extreme longevity. *Current Biology, 26*(11), 1480–1485.

Biagi, E., Rampelli, S., Turroni, S., Quercia, S., Candela, M., & Brigidi, P. (2017). The gut microbiota of centenarians: Signatures of longevity in the gut microbiota profile. *Mechanisms of Ageing and Development, 165*(Pt B), 180–184.

Bien, J., Palagani, V., & Bozko, P. (2013). The intestinal microbiota dysbiosis and *Clostridium difficile* infection: Is there a relationship with inflammatory bowel disease? *Therapeutic Advances in Gastroenterology, 6*(1), 53–68.

Bing, X., Gerlach, J., Loeb, G., & Buchon, N. (2018). Nutrient-dependent impact of microbes on *Drosophila suzukii* development. *MBio, 9*(2), e02199.

Broderick, N. A., Buchon, N., & Lemaitre, B. (2014). Microbiota-induced changes in drosophila melanogaster host gene expression and gut morphology. *MBio, 5*(3), e01117–e01114.

Buchon, N., Osman, D., David, F. P., Fang, H. Y., Boquete, J. P., Deplancke, B., & Lemaitre, B. (2013). Morphological and molecular characterization of adult midgut compartmentalization in Drosophila. *Cell Reports, 3*(5), 1725–1738.

Buford, T. W. (2017). (Dis)Trust your gut: The gut microbiome in age-related inflammation, health, and disease. *Microbiome, 5*(1), 80.

Byri, S., Misra, T., Syed, Z. A., Batz, T., Shah, J., Boril, L., Glashauser, J., Aegerter-Wilmsen, T., Matzat, T., Moussian, B., Uv, A., & Luschnig, S. (2015). The triple-repeat protein anakonda controls epithelial tricellular junction formation in *Drosophila*. *Developmental Cell, 33*(5), 535–548.

Candela, M., Biagi, E., Maccaferri, S., Turroni, S., & Brigidi, P. (2012). Intestinal microbiota is a plastic factor responding to environmental changes. *Trends in Microbiology, 20*(8), 385–391.

Candela, M., Biagi, E., Brigidi, P., O'Toole, P. W., & De Vos, W. M. (2014). Maintenance of a healthy trajectory of the intestinal microbiome during aging: A dietary approach. *Mechanisms of Ageing and Development, 136–137*, 70–75.

Cao, Y., Shen, J., & Ran, Z. H. (2014). Association between *Faecalibacterium prausnitzii* reduction and inflammatory bowel disease: A meta-analysis and systematic review of the literature. *Gastroenterology Research and Practice, 2014*, 872725.

Charlesworth, C. J., Smit, E., Lee, D. S. H., Alramadhan, F., & Odden, M. C. (2015). Polypharmacy among adults aged 65 years and older in the United States: 1988-2010. *Journals of Gerontology Series A-Biological Sciences and Medical Sciences, 70*(8), 989–995.

Chaston, J. M., Dobson, A. J., Newell, P. D., & Douglas, A. E. (2016). Host genetic control of the microbiota mediates the *Drosophila* nutritional phenotype. *Applied and Environmental Microbiology, 82*(2), 671–679.

Claesson, M. J., Cusack, S., O'Sullivan, O., Greene-Diniz, R., de Weerd, H., Flannery, E., Marchesi, J. R., Falush, D., Dinan, T., Fitzgerald, G., Stanton, C., van Sinderen, D., O'Connor, M., Harnedy, N., O'Connor, K., Henry, C., O'Mahony, D., Fitzgerald, A. P., Shanahan, F., Twomey, C., Hill, C., Ross, R. P., & O'Toole, P. W. (2011). Composition, variability, and temporal stability of the intestinal microbiota of the elderly. *Proceedings of the National Academy of Sciences of the United States of America, 108*(Suppl 1), 4586–4591.

Claesson, M. J., Jeffery, I. B., Conde, S., Power, S. E., O'Connor, E. M., Cusack, S., Harris, H. M. B., Coakley, M., Lakshminarayanan, B., O'Sullivan, O., Fitzgerald, G. F., Deane, J., O'Connor, M., Harnedy, N., O'Connor, K., O'Mahony, D., van Sinderen, D., Wallace, M., Brennan, L., Stanton, C., Marchesi, J. R., Fitzgerald, A. P., Shanahan, F., Hill, C., Ross, R. P., & O'Toole, P. W. (2012). Gut microbiota composition correlates with diet and health in the elderly. *Nature, 488*(7410), 178–184.

Clark, R. I., Salazar, A., Yamada, R., Fitz-Gibbon, S., Morselli, M., Alcaraz, J., Rana, A., Rera, M., Pellegrini, M., Ja, W. W., & Walker, D. W. (2015). Distinct shifts in microbiota composition during *Drosophila* aging impair intestinal function and drive mortality. *Cell Reports, 12*(10), 1656–1667.

Clements, S. J., & Carding, S. R. (2018). Diet, the intestinal microbiota, and immune health in aging. *Critical Reviews in Food Science and Nutrition, 58*(4), 651–661.

Collado, M. C., Rautava, S., Aakko, J., Isolauri, E., & Salminen, S. (2016). Human gut colonisation may be initiated in utero by distinct microbial communities in the placenta and amniotic fluid. *Scientific Reports, 6*, 23129.

Conley, M. N., Wong, C. P., Duyck, K. M., Hord, N., Ho, E., & Sharpton, T. J. (2016). Aging and serum MCP-1 are associated with gut microbiome composition in a murine model. *PeerJ, 4*, e1854.

Costello, E. K., Lauber, C. L., Hamady, M., Fierer, N., Gordon, J. I., & Knight, R. (2009). Bacterial community variation in human body habitats across space and time. *Science, 326*(5960), 1694–1697.

de Vries, M. C., Vaughan, E. E., Kleerebezem, M., & de Vos, W. M. (2006). *Lactobacillus plantarum*- survival, functional and potential probiotic properties in the human intestinal tract. *International Dairy Journal, 16*(9), 1018–1028.

de Goffau, M. C., Lager, S., Sovio, U., Gaccioli, F., Cook, E., Peacock, S. J., Parkhill, J., Charnock-Jones, D. S., Smith, G. C. S. (2019). Human placenta has no microbiome but can contain potential pathogens. *Nature, 572*(7769), 329–334.

Dambroise, E., Monnier, L., Ruisheng, L., Aguilaniu, H., Joly, J. S., Tricoire, H., & Rera, M. (2016). Two phases of aging separated by the Smurf transition as a public path to death. *Scientific Reports, 6*, 23523.

Darby, T. M., & Jones, R. M. (2017). Beneficial influences of *Lactobacillus plantarum* on human health and disease. In Y. Ringel & W. A. Walker (Eds.), *The microbiota in gastrointestinal pathophysiology* (pp. 109–117). Boston: Academic.

The Aging Gut Microbiota 301

David, L. A., Maurice, C. F., Carmody, R. N., Gootenberg, D. B., Button, J. E., Wolfe, B. E., Ling, A. V., Devlin, A. S., Varma, Y., Fischbach, M. A., Biddinger, S. B., Dutton, R. J., & Turnbaugh, P. J. (2014). Diet rapidly and reproducibly alters the human gut microbiome. *Nature, 505*(7484), 559–563.

Dominy, S. S., Lynch, C., Ermini, F., Benedyk, M., Marczyk, A., Konradi, A., Nguyen, M., Haditsch, U., Raha, D., Griffin, C., Holsinger, L. J., Arastu-Kapur, S., Kaba, S., Lee, A., Ryder, M. I., Potempa, B., Mydel, P., Hellvard, A., Adamowicz, K., Hasturk, H., Walker, G. D., Reynolds, E. C., Faull, R. L. M., Curtis, M. A., Dragunow, M., & Potempa, J. (2019). Porphyromonas gingivalis in Alzheimer's disease brains: Evidence for disease causation and treatment with small-molecule inhibitors. *Science Advances, 5*(1), eaau3333.

Douglas, A. E. (2018). Which experimental systems should we use for human microbiome science? *PLoS Biology, 16*(3), e2005245. https://doi.org/10.1371/journal.pbio.2005245.

Duncan, S. H., & Flint, H. J. (2013). Probiotics and prebiotics and health in ageing populations. *Maturitas, 75*(1), 44–50.

Elderman, M., Sovran, B., Hugenholtz, F., Graversen, K., Huijskes, M., Houtsma, E., Belzer, C., Boekschoten, M., de Vos, P., Dekker, J., Wells, J., & Faas, M. (2017). The effect of age on the intestinal mucus thickness, microbiota composition and immunity in relation to sex in mice. *PLoS One, 12*(9), e0184274.

Fishman, J. E., Levy, G., Alli, V., Sheth, S., Lu, Q., & Deitch, E. A. (2013). Oxidative modification of the intestinal mucus layer is a critical but unrecognized component of trauma hemorrhagic shock-induced gut barrier failure. *American Journal of Physiology. Gastrointestinal and Liver Physiology, 304*(1), G57–G63.

Franceschi, C., & Campisi, J. (2014). Chronic inflammation (inflammaging) and its potential contribution to age-associated diseases. *The Journals of Gerontology. Series A, Biological Sciences and Medical Sciences, 69*(Suppl 1), S4–S9.

Franceschi, C., Bonafe, M., Valensin, S., Olivieri, F., De Luca, M., Ottaviani, E., & De Benedictis, G. (2000). Inflamm-aging. An evolutionary perspective on immunosenescence. *Annals of the New York Academy of Sciences, 908*, 244–254.

Franceschi, C., Capri, M., Monti, D., Giunta, S., Olivieri, F., Sevini, F., Panouraia, M. P., Invidia, L., Celani, L., Scurti, M., Cevenini, E., Castellani, G. C., & Salvioli, S. (2007). Inflammaging and anti-inflammaging: A systemic perspective on aging and longevity emerged from studies in humans. *Mechanisms of Ageing and Development, 128*(1), 92–105.

Franceschi, C., Ostan, R., & Santoro, A. (2018). Nutrition and Inflammation: Are centenarians similar to individuals on calorie-restricted diets? *Annual Review of Nutrition, 38*, 329–356.

Fransen, F., van Beek, A. A., Borghuis, T., Aidy, S. E., Hugenholtz, F., van der Gaast-de Jongh, C., Savelkoul, H. F. J., De Jonge, M. I., Boekschoten, M. V., Smidt, H., Faas, M. M., & de Vos, P. (2017). Aged gut microbiota contributes to systemical inflammaging after transfer to germ-free mice. *Frontiers in Immunology, 8*, 1385.

Fukunaga, A., Uematsu, H., & Sugimoto, K. (2005). Influences of aging on taste perception and oral somatic sensation. *The Journals of Gerontology: Series A, 60*(1), 109–113.

Fukushima, Y., Miyaguchi, S., Yamano, T., Kaburagi, T., Iino, H., Ushida, K., & Sato, K. (2007). Improvement of nutritional status and incidence of infection in hospitalised, enterally fed elderly by feeding of fermented milk containing probiotic *Lactobacillus johnsonii* La1 (NCC533). *The British Journal of Nutrition, 98*(5), 969–977.

Garcia-Gonzalez, A. P., Ritter, A. D., Shrestha, S., Andersen, E. C., Yilmaz, L. S., & Walhout, A. J. M. (2017). Bacterial Metabolism Affects the *C. elegans* Response to Cancer Chemotherapeutics. *Cell, 169*(3), 431–441. e438.

Gelino, S., Chang, J. T., Kumsta, C., She, X., Davis, A., Nguyen, C., Panowski, S., & Hansen, M. (2016). Intestinal autophagy improves healthspan and longevity in *C. elegans* during dietary restriction. *PLOS Genetics, 12*(7), e1006135.

Gill, H. S., Rutherfurd, K. J., Cross, M. L., & Gopal, P. K. (2001). Enhancement of immunity in the elderly by dietary supplementation with the probiotic Bifidobacterium lactis HN019. *The American Journal of Clinical Nutrition, 74*(6), 833–839.

Guillemard, E., Tondu, F., Lacoin, F., & Schrezenmeir, J. (2010). Consumption of a fermented dairy product containing the probiotic *Lactobacillus casei* DN-114001 reduces the duration of respiratory infections in the elderly in a randomised controlled trial. *The British Journal of Nutrition, 103*(1), 58–68.

Guo, L., Karpac, J., Tran, S. L., & Jasper, H. (2014). PGRP-SC2 promotes gut immune homeostasis to limit commensal dysbiosis and extend lifespan. *Cell, 156*(1–2), 109–122.

Hallstrom, M., Eerola, E., Vuento, R., Janas, M., & Tammela, O. (2004). Effects of mode of delivery and necrotising enterocolitis on the intestinal microflora in preterm infants. *European Journal of Clinical Microbiology & Infectious Diseases, 23*(6), 463–470.

Han, B., Sivaramakrishnan, P., Lin, C. C. J., Neve, I. A. A., He, J. Q., Tay, L. W. R., Sowa, J. N., Sizovs, A., Du, G. W., Wang, J., Herman, C., & Wang, M. C. (2017). Microbial genetic composition tunes host longevity. *Cell, 169*(7), 1249–1262.

Hayashi, H., Sakamoto, M., Kitahara, M., & Benno, Y. (2003). Molecular analysis of fecal microbiota in elderly individuals using 16S rDNA library and T-RFLP. *Microbiology and Immunology, 47*(8), 557–570.

Henderson, B., & Nibali, L. (2016). *The human microbiota and chronic disease: Dysbiosis as a cause of human pathology*. Hoboken, NJ: Wiley Blackwell.

Hickson, M., D'Souza, A. L., Muthu, N., Rogers, T. R., Want, S., Rajkumar, C., & Bulpitt, C. J. (2007). Use of probiotic Lactobacillus preparation to prevent diarrhoea associated with antibiotics: Randomised double blind placebo controlled trial. *BMJ, 335*(7610), 80–83.

Hill, C., Guarner, F., Reid, G., Gibson, G. R., Merenstein, D. J., Pot, B., Morelli, L., Canani, R. B., Flint, H. J., Salminen, S., Calder, P. C., & Sanders, M. E. (2014). Expert consensus document. The International Scientific Association for Probiotics and Prebiotics consensus statement on the scope and appropriate use of the term probiotic. *Nature Reviews. Gastroenterology & Hepatology, 11*(8), 506–514.

Hooper, L. V., Littman, D. R., & Macpherson, A. J. (2012). Interactions between the microbiota and the immune system. *Science, 336*(6086), 1268–1273.

Hopkins, M. J., & Macfarlane, G. T. (2002). Changes in predominant bacterial populations in human faeces with age and with *Clostridium difficile* infection. *Journal of Medical Microbiology, 51*(5), 448–454.

Hopkins, M. J., Sharp, R., & Macfarlane, G. T. (2001). Age and disease related changes in intestinal bacterial populations assessed by cell culture, 16S rRNA abundance, and community cellular fatty acid profiles. *Gut, 48*(2), 198–205.

Human Microbiome Project Consortium (2012). Structure, function and diversity of the healthy human microbiome. *Nature, 486*(7402), 207–214.

Jackson, M. A., Jeffery, I. B., Beaumont, M., Bell, J. T., Clark, A. G., Ley, R. E., O'Toole, P. W., Spector, T. D., & Steves, C. J. (2016). Signatures of early frailty in the gut microbiota. *Genome Medicine, 8*(1), 8.

Jeffery, I. B., Lynch, D. B., & O'Toole, P. W. (2016). Composition and temporal stability of the gut microbiota in older persons. *The ISME Journal, 10*(1), 170–182.

Jimenez, E., Fernandez, L., Marin, M. L., Martin, R., Odriozola, J. M., Nueno-Palop, C., Narbad, A., Olivares, M., Xaus, J., & Rodriguez, J. M. (2005). Isolation of commensal bacteria from umbilical cord blood of healthy neonates born by cesarean section. *Current Microbiology, 51*(4), 270–274.

Johansson, M. E. (2014). Mucus layers in inflammatory bowel disease. *Inflammatory Bowel Diseases, 20*(11), 2124–2131.

Johansson, M. L., Molin, G., Jeppsson, B., Nobaek, S., Ahrne, S., & Bengmark, S. (1993). Administration of different Lactobacillus strains in fermented oatmeal soup: In vivo colonization of human intestinal mucosa and effect on the indigenous flora. *Applied and Environmental Microbiology, 59*(1), 15–20.

Johansson, M. E. V., Phillipson, M., Petersson, J., Velcich, A., Holm, L., & Hansson, G. C. (2008). The inner of the two Muc2 mucin-dependent mucus layers in colon is devoid of bacteria. *Proceedings of the National Academy of Sciences of the United States of America, 105*(39), 15064–15069.

Johansson, M. E. V., Larsson, J. M. H., & Hansson, G. C. (2011). The two mucus layers of colon are organized by the MUC2 mucin, whereas the outer layer is a legislator of host-microbial interactions. *Proceedings of the National Academy of Sciences of the United States of America, 108*, 4659–4665.

Jones, C., Badger, S. A., Regan, M., Clements, B. W., Diamond, T., Parks, R. W., & Taylor, M. A. (2013). Modulation of gut barrier function in patients with obstructive jaundice using probiotic LP299v. *European Journal of Gastroenterology & Hepatology, 25*(12), 1424–1430.

Keebaugh, E. S., & Ja, W. W. (2016). Microbes without borders: Decompartmentalization of the aging gut. *Cell Host & Microbe, 19*(2), 133–135.

Keebaugh, E. S., Yamada, R., Obadia, B., Ludington, W. B., & Ja, W. W. (2018). Microbial quantity impacts *Drosophila* nutrition, development, and lifespan. *iScience, 4*, 247–259.

Kim, S., & Jazwinski, S. M. (2018). The gut microbiota and healthy aging: A mini-review. *Gerontology, 64*(6), 513–520.

Klarin, B., Johansson, M. L., Molin, G., Larsson, A., & Jeppsson, B. (2005). Adhesion of the probiotic bacterium *Lactobacillus plantarum* 299v onto the gut mucosa in critically ill patients: A randomised open trial. *Critical Care, 9*(3), R285–R293.

Kleerebezem, M., Boekhorst, J., van Kranenburg, R., Molenaar, D., Kuipers, O. P., Leer, R., Tarchini, R., Peters, S. A., Sandbrink, H. M., Fiers, M. W. E. J., Stiekema, W., Lankhorst, R. M. K., Bron, P. A., Hoffer, S. M., Groot, M. N. N., Kerkhoven, R., de Vries, M., Ursing, B., de Vos, W. M., & Siezen, R. J. (2003). Complete genome sequence of *Lactobacillus plantarum* WCFS1. *Proceedings of the National Academy of Sciences of the United States of America, 100*(4), 1990–1995.

Koenig, J. E., Spor, A., Scalfone, N., Fricker, A. D., Stombaugh, J., Knight, R., Angenent, L. T., & Ley, R. E. (2011). Succession of microbial consortia in the developing infant gut microbiome. *Proceedings of the National Academy of Sciences of the United States of America, 108*(Suppl 1), 4578–4585.

Kong, F., Hua, Y., Zeng, B., Ning, R., Li, Y., & Zhao, J. (2016). Gut microbiota signatures of longevity. *Current Biology, 26*(18), R832–R833.

Kong, F., Deng, F., Li, Y., & Zhao, J. (2018). Identification of gut microbiome signatures associated with longevity provides a promising modulation target for healthy aging. *Gut Microbes, 10*(2), 210–215.

Kumar, M., Babaei, P., Ji, B., & Nielsen, J. (2016). Human gut microbiota and healthy aging: Recent developments and future prospective. *Nutrition and Healthy Aging, 4*(1), 3–16.

Lahtinen, S. J., Tammela, L., Korpela, J., Parhiala, R., Ahokoski, H., Mykkanen, H., & Salminen, S. J. (2009). Probiotics modulate the bifidobacterium microbiota of elderly nursing home residents. *Age (Dordrecht, Netherlands), 31*(1), 59–66.

Lee, G. C., Daniels, K., Lawson, K. A., Attridge, R. T., Lewis, J., & Frei, C. R. (2013). Age-based outpatient antibiotic prescribing in the United States from 2000 to 2010. *Value in Health, 16*(3), A78–A78.

Lee, G. C., Reveles, K. R., Attridge, R. T., Lawson, K. A., Mansi, I. A., Lewis, J. S., & Frei, C. R. (2014). Outpatient antibiotic prescribing in the United States: 2000 to 2010. *BMC Medicine, 12*, 96.

Li, H., Qi, Y., & Jasper, H. (2016). Preventing age-related decline of gut compartmentalization limits microbiota dysbiosis and extends lifespan. *Cell Host & Microbe, 19*(2), 240–253.

Lim, C. J., Kong, D. C. M., & Stuart, R. L. (2014). Reducing inappropriate antibiotic prescribing in the residential care setting: Current perspectives. *Clinical Interventions in Aging, 9*, 165–177.

Lloyd-Price, J., Abu-Ali, G., & Huttenhower, C. (2016). The healthy human microbiome. *Genome Medicine, 8*(1), 51.

Lovat, L. B. (1996). Age related changes in gut physiology and nutritional status. *Gut, 38*(3), 306–309.

Lozupone, C. A., Stombaugh, J. I., Gordon, J. I., Jansson, J. K., & Knight, R. (2012). Diversity, stability and resilience of the human gut microbiota. *Nature, 489*(7415), 220–230.

Lu, M., & Wang, Z. (2018). Linking gut microbiota to aging process: A new target for anti-aging. *Food Science and Human Wellness, 7*(2), 111–119.

Macpherson, A. J., & Harris, N. L. (2004). Interactions between commensal intestinal bacteria and the immune system. *Nature Reviews. Immunology, 4*(6), 478–485.

Magrone, T., & Jirillo, E. (2013). The interaction between gut microbiota and age-related changes in immune function and inflammation. *Immunity & Ageing, 10*(1), 31.

Malamitsi-Puchner, A., Protonotariou, E., Boutsikou, T., Makrakis, E., Sarandakou, A., & Creatsas, G. (2005). The influence of the mode of delivery on circulating cytokine concentrations in the perinatal period. *Early Human Development, 81*(4), 387–392.

Man, A. L., Gicheva, N., & Nicoletti, C. (2014). The impact of ageing on the intestinal epithelial barrier and immune system. *Cellular Immunology, 289*(1–2), 112–118.

Marianes, A., & Spradling, A. C. (2013). Physiological and stem cell compartmentalization within the *Drosophila* midgut. *eLife, 2*, e00886.

Mariat, D., Firmesse, O., Levenez, F., Guimaraes, V. D., Sokol, H., Dore, J., Corthier, G., & Furet, J. P. (2009). The Firmicutes/Bacteroidetes ratio of the human microbiota changes with age. *BMC Microbiology, 9*, 123.

Martin, R., Makino, H., Yavuz, A. C., Ben-Amor, K., Roelofs, M., Ishikawa, E., Kubota, H., Swinkels, S., Sakai, T., Oishi, K., Kushiro, A., & Knol, J. (2016). Early-life events, including mode of delivery and type of feeding, siblings and gender, shape the developing gut microbiota. *PLoS One, 11*(6), e0158498.

Matsumoto, M., Kurihara, S., Kibe, R., Ashida, H., & Benno, Y. (2011). Longevity in mice Is promoted by probiotic-induced suppression of colonic senescence dependent on upregulation of gut bacterial polyamine production. *PLoS One, 6*(8), e23652.

Matsuo, K., Ota, H., Akamatsu, T., Sugiyama, A., & Katsuyama, T. (1997). Histochemistry of the surface mucous gel layer of the human colon. *Gut, 40*(6), 782–789.

Maynard, C., & Weinkove, D. (2018). The gut microbiota and ageing. *Sub-Cellular Biochemistry, 90*, 351–371.

McNaught, C. E., Woodcock, N. P., Anderson, A. D., & MacFie, J. (2005). A prospective randomised trial of probiotics in critically ill patients. *Clinical Nutrition, 24*(2), 211–219.

Metchnikoff, E. (1908). *The prolongation of life: Optimistic studies*. New York and London: GP Putnam's Sons.

Miquel, S., Martin, R., Rossi, O., Bermudez-Humaran, L. G., Chatel, J. M., Sokol, H., Thomas, M., Wells, J. M., & Langella, P. (2013). *Faecalibacterium prausnitzii* and human intestinal health. *Current Opinion in Microbiology, 16*(3), 255–261.

Mueller, S., Saunier, K., Hanisch, C., Norin, E., Alm, L., Midtvedt, T., Cresci, A., Silvi, S., Orpianesi, C., Verdenelli, M. C., Clavel, T., Koebnick, C., Zunft, H.-J. F., Doré, J., & Blaut, M. (2006). Differences in fecal microbiota in different European study populations in relation to age, gender, and country: A cross-sectional study. *Applied and Environmental Microbiology, 72*(2), 1027–1033.

Nagpal, R., Mainali, R., Ahmadi, S., Wang, S., Singh, R., Kavanagh, K., Kitzman, D. W., Kushugulova, A., Marotta, F., & Yadav, H. (2018). Gut microbiome and aging: Physiological and mechanistic insights. *Nutrition and Healthy Aging, 4*(4), 267–285.

Negele, K., Heinrich, J., Borte, M., von Berg, A., Schaaf, B., Lehmann, I., Wichmann, H. E., Bolte, G., & L. S. Group. (2004). Mode of delivery and development of atopic disease during the first 2 years of life. *Pediatric Allergy and Immunology, 15*(1), 48–54.

Nguyen, T. L. A., Vieira-Silva, S., Liston, A., & Raes, J. (2015). How informative is the mouse for human gut microbiota research? *Disease Models & Mechanisms, 8*(1), 1–16.

O'Hara, A. M., & Shanahan, F. (2006). The gut flora as a forgotten organ. *EMBO Reports, 7*(7), 688–693.

Obadia, B., Guvener, Z. T., Zhang, V., Ceja-Navarro, J. A., Brodie, E. L., Ja, W. W., & Ludington, W. B. (2017). Probabilistic invasion underlies natural gut microbiome stability. *Current Biology, 27*(13), 1999–2006. e1998.

Odamaki, T., Kato, K., Sugahara, H., Hashikura, N., Takahashi, S., Xiao, J. Z., Abe, F., & Osawa, R. (2016). Age-related changes in gut microbiota composition from newborn to centenarian: A cross-sectional study. *BMC Microbiology, 16*, 90.

Ouwehand, A. C., Bergsma, N., Parhiala, R., Lahtinen, S., Gueimonde, M., Finne-Soveri, H., Strandberg, T., Pitkala, K., & Salminen, S. (2008). Bifidobacterium microbiota and parameters of immune function in elderly subjects. *FEMS Immunology and Medical Microbiology, 53*(1), 18–25.

Overend, G., Luo, Y., Henderson, L., Douglas, A. E., Davies, S. A., & Dow, J. A. (2016). Molecular mechanism and functional significance of acid generation in the *Drosophila* midgut. *Scientific Reports, 6*, 27242.

Pais, I. S., Valente, R. S., Sporniak, M., & Teixeira, L. (2018). *Drosophila melanogaster* establishes a species-specific mutualistic interaction with stable gut-colonizing bacteria. *PLoS Biology, 16*(7), e2005710.

Palmer, C., Bik, E. M., DiGiulio, D. B., Relman, D. A., & Brown, P. O. (2007). Development of the human infant intestinal microbiota. *PLoS Biology, 5*(7), e177.

Perez-Munoz, M. E., Arrieta, M. C., Ramer-Tait, A. E., & Walter, J. (2017). A critical assessment of the "sterile womb" and "in utero colonization" hypotheses: Implications for research on the pioneer infant microbiome. *Microbiome, 5*(1), 48.

Qin, J. J., Li, R. Q., Raes, J., Arumugam, M., Burgdorf, K. S., Manichanh, C., Nielsen, T., Pons, N., Levenez, F., Yamada, T., Mende, D. R., Li, J. H., Xu, J. M., Li, S. C., Li, D. F., Cao, J. J., Wang, B., Liang, H. Q., Zheng, H. S., Xie, Y. L., Tap, J., Lepage, P., Bertalan, M., Batto, J. M., Hansen, T., Le Paslier, D., Linneberg, A., Nielsen, H. B., Pelletier, E., Renault, P., Sicheritz-Ponten, T., Turner, K., Zhu, H. M., Yu, C., Li, S. T., Jian, M., Zhou, Y., Li, Y. R., Zhang, X. Q., Li, S. G., Qin, N., Yang, H. M., Wang, J., Brunak, S., Dore, J., Guarner, F., Kristiansen, K., Pedersen, O., Parkhill, J., Weissenbach, J., Bork, P., Ehrlich, S. D., Wang, J., & MetaHIT Consortium (2010). A human gut microbial gene catalogue established by metagenomic sequencing. *Nature, 464*(7285), 59–65.

Qin, X. F., Sheth, S. U., Sharpe, S. M., Dong, W., Lu, Q., Xu, D. Z., & Deitch, E. A. (2011). The mucus layer Is critical in protecting against ischemia-reperfusion-mediated gut injury and in the restitution of gut barrier function. *Shock, 35*(3), 275–281.

Rampelli, S., Candela, M., Turroni, S., Biagi, E., Collino, S., Franceschi, C., O'Toole, P. W., & Brigidi, P. (2013). Functional metagenomic profiling of intestinal microbiome in extreme ageing. *Aging (Albany NY), 5*(12), 902–912.

Rautava, S., Luoto, R., Salminen, S., & Isolauri, E. (2012). Microbial contact during pregnancy, intestinal colonization and human disease. *Nature Reviews. Gastroenterology & Hepatology, 9*(10), 565–576.

Rehman, T. (2012). Role of the gut microbiota in age-related chronic inflammation. *Endocrine, Metabolic & Immune Disorders Drug Targets, 12*(4), 361–367.

Rera, M., Clark, R. I., & Walker, D. W. (2012). Intestinal barrier dysfunction links metabolic and inflammatory markers of aging to death in *Drosophila. Proceedings of the National Academy of Sciences of the United States of America, 109*(52), 21528–21533.

Rera, M., Vallot, C., & Lefrancois, C. (2018). The Smurf transition: New insights on ageing from end-of-life studies in animal models. *Current Opinion in Oncology, 30*(1), 38–44.

Resnik-Docampo, M., Koehler, C. L., Clark, R. I., Schinaman, J. M., Sauer, V., Wong, D. M., Lewis, S., D'Alterio, C., Walker, D. W., & Jones, D. L. (2017). Tricellular junctions regulate intestinal stem cell behaviour to maintain homeostasis. *Nature Cell Biology, 19*(1), 52–59.

Resnik-Docampo, M., Sauer, V., Schinaman, J. M., Clark, R. I., Walker, D. W., & Jones, D. L. (2018). Keeping it tight: The relationship between bacterial dysbiosis, septate junctions, and the intestinal barrier in *Drosophila. Fly (Austin), 12*(1), 34–40.

Riaz Rajoka, M. S., Zhao, H., Li, N., Lu, Y., Lian, Z., Shao, D., Jin, M., Li, Q., Zhao, L., & Shi, J. (2018). Origination, change, and modulation of geriatric disease-related gut microbiota during life. *Applied Microbiology and Biotechnology, 102*(19), 8275–8289.

Ridley, E. V., Wong, A. C., Westmiller, S., & Douglas, A. E. (2012). Impact of the resident microbiota on the nutritional phenotype of *Drosophila melanogaster. PLoS One, 7*(5), e36765.

Rodriguez, J. M., Murphy, K., Stanton, C., Ross, R. P., Kober, O. I., Juge, N., Avershina, E., Rudi, K., Narbad, A., Jenmalm, M. C., Marchesi, J. R., & Collado, M. C. (2015). The composition of the gut microbiota throughout life, with an emphasis on early life. *Microbial Ecology in Health and Disease, 26*, 26050.

Rosen, C. E., & Palm, N. W. (2017). Functional classification of the gut microbiota: The key to cracking the microbiota composition code: Functional classifications of the gut microbiota reveal previously hidden contributions of indigenous gut bacteria to human health and disease. *BioEssays, 39*(12), 1700032.

Rothschild, D., Weissbrod, O., Barkan, E., Kurilshikov, A., Korem, T., Zeevi, D., Costea, P. I., Godneva, A., Kalka, I. N., Bar, N., Shilo, S., Lador, D., Vila, A. V., Zmora, N., Pevsner-Fischer, M., Israeli, D., Kosower, N., Malka, G., Wolf, B. C., Avnit-Sagi, T., Lotan-Pompan, M., Weinberger, A., Halpern, Z., Carmi, S., Fu, J., Wijmenga, C., Zhernakova, A., Elinav, E., & Segal, E. (2018). Environment dominates over host genetics in shaping human gut microbiota. *Nature, 555*, 210–215.

Rutayisire, E., Huang, K., Liu, Y., & Tao, F. (2016). The mode of delivery affects the diversity and colonization pattern of the gut microbiota during the first year of infants' life: A systematic review. *BMC Gastroenterology, 16*(1), 86.

Salazar, A. M., Resnik-Docampo, M., Ulgherait, M., Clark, R. I., Shirasu-Hiza, M., Jones, D. L., & Walker, D. W. (2018). Intestinal snakeskin limits microbial dysbiosis during aging and promotes longevity. *iScience, 9*, 229–243.

Santoro, A., Ostan, R., Candela, M., Biagi, E., Brigidi, P., Capri, M., & Franceschi, C. (2018). Gut microbiota changes in the extreme decades of human life: A focus on centenarians. *Cellular and Molecular Life Sciences, 75*(1), 129–148.

Sarkar, D., & Fisher, P. B. (2006). Molecular mechanisms of aging-associated inflammation. *Cancer Letters, 236*(1), 13–23.

Scott, T. A., Quintaneiro, L. M., Norvaisas, P., Lui, P. P., Wilson, M. P., Leung, K. Y., Herrera-Dominguez, L., Sudiwala, S., Pessia, A., Clayton, P. T., Bryson, K., Velagapudi, V., Mills, P. B., Typas, A., Greene, N. D. E., & Cabreiro, F. (2017). Host-Microbe Co-metabolism Dictates Cancer Drug Efficacy in *C. elegans*. *Cell, 169*(3), 442–456. e418.

Smith, P., Willemsen, D., Popkes, M., Metge, F., Gandiwa, E., Reichard, M., et al. (2017). Regulation of life span by the gut microbiota in the short-lived African turquoise killifish. *Elife. 6*.

Sokol, H., Pigneur, B., Watterlot, L., Lakhdari, O., Bermúdez-Humarán, L. G., Gratadoux, J.-J., Blugeon, S., Bridonneau, C., Furet, J.-P., Corthier, G., Grangette, C., Vasquez, N., Pochart, P., Trugnan, G., Thomas, G., Blottière, H. M., Doré, J., Marteau, P., Seksik, P., & Langella, P. (2008). *Faecalibacterium prausnitzii* is an anti-inflammatory commensal bacterium identified by gut microbiota analysis of Crohn disease patients. *Proceedings of the National Academy of Sciences of the United States of America, 105*(43), 16731–16736.

Suryavanshi, M. V., Paul, D., Doijad, S. P., Bhute, S. S., Hingamire, T. B., Gune, R. P., & Shouche, Y. S. (2017). Draft genome sequence of *Lactobacillus plantarum* strains E2C2 and E2C5 isolated from human stool culture. *Standards in Genomic Sciences, 12*, 15.

Swidsinski, A., Loening-Baucke, V., Theissig, F., Engelhardt, H., Bengmark, S., Koch, S., Lochs, H., & Dorffel, Y. (2007a). Comparative study of the intestinal mucus barrier in normal and inflamed colon. *Gut, 56*(3), 343–350.

Swidsinski, A., Sydora, B. C., Doerffel, Y., Loening-Baucke, V., Vaneechoutte, M., Lupicki, M., Scholze, J., Lochs, H., & Dieleman, L. A. (2007b). Viscosity gradient within the mucus layer determines the mucosal barrier function and the spatial organization of the intestinal microbiota. *Inflammatory Bowel Diseases, 13*(8), 963–970.

Tamburini, S., Shen, N., Wu, H. C., & Clemente, J. C. (2016). The microbiome in early life: Implications for health outcomes. *Nature Medicine, 22*(7), 713–722.

Thevaranjan, N., Puchta, A., Schulz, C., Naidoo, A., Szamosi, J. C., Verschoor, C. P., Loukov, D., Schenck, L. P., Jury, J., Foley, K. P., Schertzer, J. D., Larche, M. J., Davidson, D. J., Verdu, E. F., Surette, M. G., & Bowdish, D. M. E. (2017). Age-associated microbial dysbiosis promotes intestinal permeability, systemic inflammation, and macrophage dysfunction. *Cell Host & Microbe, 21*(4), 455–466. e454.

Tran, L., & Greenwood-Van Meerveld, B. (2013). Age-Associated Remodeling of the Intestinal Epithelial Barrier. *Journals of Gerontology Series A-Biological Sciences and Medical Sciences, 68*(9), 1045–1056.

Tricoire, H., & Rera, M. (2015). A new, discontinuous 2 phases of aging model: Lessons from *Drosophila melanogaster*. *PLoS One, 10*(11), e0141920.

Turchet, P., Laurenzano, M., Auboiron, S., & Antoine, J. M. (2003). Effect of fermented milk containing the probiotic *Lactobacillus casei* DN-114001 on winter infections in free-living elderly subjects: A randomised, controlled pilot study. *The Journal of Nutrition, Health & Aging, 7*(2), 75–77.

Turnbaugh, P. J., Ley, R. E., Hamady, M., Fraser-Liggett, C. M., Knight, R., & Gordon, J. I. (2007). The human microbiome project. *Nature, 449*(7164), 804–810.

van Beek, A. A., Sovran, B., Hugenholtz, F., Meijer, B., Hoogerland, J. A., Mihailova, V., van der Ploeg, C., Belzer, C., Boekschoten, M. V., Hoeijmakers, J. H., Vermeij, W. P., de Vos, P., Wells, J. M., Leenen, P. J., Nicoletti, C., Hendriks, R. W., & Savelkoul, H. F. (2016). Supplementation with *Lactobacillus plantarum* WCFS1 prevents decline of mucus barrier in colon of accelerated aging *Ercc1(-/Delta7)* mice. *Frontiers in Immunology, 7*, 408.

van Tongeren, S. P., Slaets, J. P., Harmsen, H. J., & Welling, G. W. (2005). Fecal microbiota composition and frailty. *Applied and Environmental Microbiology, 71*(10), 6438–6442.

Varum, F. J. O., Veiga, F., Sousa, J. S., & Basit, A. W. (2012). Mucus thickness in the gastrointestinal tract of laboratory animals. *The Journal of Pharmacy and Pharmacology, 64*(2), 218–227.

Vesa, T., Pochart, P., & Marteau, P. (2000). Pharmacokinetics of *Lactobacillus plantarum* NCIMB 8826, *Lactobacillus fermentum* KLD, and *Lactococcus lactis* MG 1363 in the human gastrointestinal tract. *Alimentary Pharmacology & Therapeutics, 14*(6), 823–828.

Walker, A. (2007). Genome watch—Say hello to our little friends. *Nature Reviews. Microbiology, 5*(8), 572–573.

Wong, A. C., Dobson, A. J., & Douglas, A. E. (2014). Gut microbiota dictates the metabolic response of *Drosophila* to diet. *The Journal of Experimental Biology, 217*(Pt 11), 1894–1901.

Woodmansey, E. J. (2007). Intestinal bacteria and ageing. *Journal of Applied Microbiology, 102*(5), 1178–1186.

Woodmansey, E. J., McMurdo, M. E., Macfarlane, G. T., & Macfarlane, S. (2004). Comparison of compositions and metabolic activities of fecal microbiotas in young adults and in antibiotic-treated and non-antibiotic-treated elderly subjects. *Applied and Environmental Microbiology, 70*(10), 6113–6122.

Wu, G. D., Chen, J., Hoffmann, C., Bittinger, K., Chen, Y. Y., Keilbaugh, S. A., Bewtra, M., Knights, D., Walters, W. A., Knight, R., Sinha, R., Gilroy, E., Gupta, K., Baldassano, R., Nessel, L., Li, H., Bushman, F. D., & Lewis, J. D. (2011). Linking long-term dietary patterns with gut microbial enterotypes. *Science, 334*(6052), 105–108.

Yamada, R., Deshpande, S. A., Bruce, K. D., Mak, E. M., & Ja, W. W. (2015). Microbes promote amino acid harvest to rescue undernutrition in *Drosophila*. *Cell Rep, 10*(6), 865–872.

Yatsunenko, T., Rey, F. E., Manary, M. J., Trehan, I., Dominguez-Bello, M. G., Contreras, M., Magris, M., Hidalgo, G., Baldassano, R. N., Anokhin, A. P., Heath, A. C., Warner, B., Reeder, J., Kuczynski, J., Caporaso, J. G., Lozupone, C. A., Lauber, C., Clemente, J. C., Knights, D., Knight, R., & Gordon, J. I. (2012). Human gut microbiome viewed across age and geography. *Nature, 486*(7402), 222–227.

You, J. L., & Yaqoob, P. (2012). Evidence of immunomodulatory effects of a novel probiotic, *Bifidobacterium longum bv. infantis* CCUG 52486. *FEMS Immunology and Medical Microbiology, 66*(3), 353–362.

Zimmermann, M., Zimmermann-Kogadeeva, M., Wegmann, R., & Goodman, A. L. (2019). Separating host and microbiome contributions to drug pharmacokinetics and toxicity. *Science, 363*(6427), eaat9931.

Beneficial Modulation of the Gut Microbiome: Probiotics and Prebiotics

M. Andrea Azcarate-Peril

Abstract The gut microbiota plays a critical role in the overall health of its host. Benefits derived from bacterial members of the gut microbiota can influence host growth, immune response, pathogen colonization, and intestinal physiology. Use of probiotics, prebiotics, and synbiotics are emerging as effective mechanisms to selectively modulate composition and function of the gut microbiota. This chapter introduces the concept of probiotics and prebiotics from a historic perspective, and attempts to answer the fundamental questions of the impact of probiotics and prebiotics on microbiome composition in health versus disease states, colonization of the human gut by probiotics (is it necessary?), and how the food or product matrix impact probiotic delivery and effect. The conclusion of this chapter focuses on the next generation of probiotics: novel species and bacterial consortia.

Keywords Probiotics · Prebiotics · Microbiota modulation · *Bifidobacterium* · *Lactobacillus* · Next generation probiotics

Introduction: The Origin and Evolution of the Concept and Definition of Probiotics and Prebiotics

Ilya Metchnikoff in his classic book, "Prolongation of Life: Optimistic Studies" (Metchnikoff and Mitchell 2004), posed a number of provocative questions to which scientists were only recently able to provide answers. He contemplated a potential causality relationship between the relatively short life of mammals, compared to birds and lower vertebrates, and their over developed large intestine. Metchnikoff hypothesized in the book that the large intestine "has been increased in mammals to make it possible for these animals to run long distances without having to stand still

---

M. A. Azcarate-Peril (✉)
Division of Gastroenterology and Hepatology, Department of Medicine, University of North Carolina at Chapel Hill, Chapel Hill, NC, USA
e-mail: azcarate@med.unc.edu

© Springer Nature Switzerland AG 2019
M. A. Azcarate-Peril et al. (eds.), *How Fermented Foods Feed a Healthy Gut Microbiota*, https://doi.org/10.1007/978-3-030-28737-5_13

for defecation". In other words, although the large intestine provided an advantage to animals in cases of emergency, its only role was to accumulate waste matter and hence could become a "nidus for microbes which produce fermentations and putrefaction harmful to the organism". Furthermore, he proposed that accumulation of microorganisms for relatively long periods of time, for example in cases of constipation, led to auto-intoxication (a concept widely accepted at the time) and that microbes were the cause of senility. It is in the following chapters of his book that the concept of modulation or manipulation of the gut microbiota arose as an approach to prolong human life. Although these chapters included a thorough review of methods that attempt to "completely disinfect the intestine", like the use of β-naphthol, naphthaline, camphor, and purgatives, Metchnikoff eventually reaches the conclusion that if the lactic acid in acidified foods, like sour milk, can prevent the putrefaction of for example meat, they could do the same within the gastrointestinal tract. Citing previous studies, he eventually reached the conclusion that "intestinal putrefaction is to be combated not by lactic acid itself but by the introduction into the organism of cultures of the lactic bacilli" thus originating the concept of probiotics as beneficial modulators of the gut microbiota. Metchnikoff did not use the term "probiotics", this expression was actually introduced in the 1950s by researchers from Germany and Sweden, and then by Lilly and Stillwell (1965) to refer to substances produced by protozoa during their logarithmic phase of growth that prolonged the logarithmic phase of other species. A 2003 letter to the editor of the British Journal of Nutrition by Hamilton-Miller, Gibson, and Bruck nicely detailed the derivation and early uses of the term "probiotics" (Hamilton-Miller et al. 2003). Currently probiotics are defined as "live microorganisms that, when administered in adequate amounts, confer a health benefit on the host" (Bindels et al. 2015). The International Scientific Association for Probiotics and Prebiotics (ISAPP) in 2013 confirmed this definition stated by the FAO/WHO Expert Panel (Hill et al. 2014) and previously by Reid and collaborators (Reid et al. 2003).

The concept and definition of prebiotics is more recent with a delineation of the concept by the study of Rettger and Cheplin (1921). At the time of their publication "Treatise on the Transformation of the Intestinal Flora: with Special Reference to the Implantation of *Bacillus acidophilus*", the fact that diet had a profound influence on the composition and predominant bacteria of the gastrointestinal tract was well established; however, the authors propose that certain carbohydrates, specifically dextrose and lactose in their study, were capable of simplifying the gut microbiota and "encourage a non-putrefactive flora". A formal definition of prebiotics was introduced in 1995, establishing that prebiotics are "nondigestible food ingredients that beneficially affect the host by selectively stimulating the growth and/or activity of one or a limited number of bacterial species already resident in the colon" (Gibson and Roberfroid 1995). The term "selectively" was challenged by the group of Jens Walter (Bindels et al. 2015) concluding that only two types of dietary oligosaccharides (inulin and trans-GOS) fulfil this criteria for classification as a prebiotic. Based on this and other questions, which included the restriction to carbohydrates and the gut, the authors propose that prebiotics could be defined as "a nondigestible compound that, through its metabolization by microorganisms in the gut, modulates composition and/or activity of the gut microbiota, thus conferring a beneficial physiological effect on the host". This broader definition allowed to include new compounds to the category including pectin, arabinoxylan, whole grains and various dietary fibers. However, in 2017 the ISAPP expert panel conveyed that in fact the criterion of *selective utilization* is the one that distinguishes prebiotics from a range of substances hence introducing the following (and most current) definition of prebiotics as "a substrate that is selective utilized by host microorganisms conferring a health benefit".

The one criterion shared by the definitions of probiotics and prebiotics is their ability of conferring a health benefit to the host. This criterion undoubtedly leads to the concept of synbiotics, defined as "a mixture of probiotics and prebiotics that beneficially affects the host by improving the survival and implantation of live microbial dietary supplements in the gastrointestinal tract, by selectively stimulating the growth and/or by activating the metabolism of one or a limited number of health-promoting bacteria, and thus improving host welfare" (Gibson and Roberfroid 1995). The definition implies a synergistic relationship between the prebiotic (selectively favors the probiotic strain or mixture of strains) and the probiotic (selected to metabolize the prebiotic compound). However, synergism also occurs between the prebiotic and the intestinal probiome (autochthonous beneficial bacteria).

Do Probiotics Actually Alter the Composition and Functionality of the Gut Microbiota?

Before we were able to rapidly determine how an intervention modified the composition of the gut (fecal) microbiome, it was assumed that probiotics temporarily modified abundances of bacteria in the gastrointestinal tract, and through those modifications, impacted gut functionality. A clear proof of modification of the gut microbiome composition was lacking until fairly recently. However, modifications to gut functionality where identified in the 1970s, for example in the study by Goldin and Gorbach (1977) which showed that *L. acidophilus* significantly lowered the activity of fecal nitroreductase and azoreductase in meat-eating rats. Figure 1 lists the proven impacts of probiotics of the *Bifidobacterium* and *Lactobacillus* genera. This figure does not intend to represent an exhaustive literature search but instead to serve as confirmation of previous assumptions on probiotics effects on composition, functionality, and disease or disease prevention.

Although there is a consensus that probiotics must be able to survive transit through the gut, and several studies have focused on the mechanisms used by probiotic bacteria to survive the acidity of the stomach, as well as pH shifts and bile acids in the small intestine and colon (Azcarate-Peril et al. 2004; Buck et al. 2006; Pfeiler et al. 2006, 2007; Oozeer et al. 2006; Bruno-Barcena et al. 2010; Watson et al. 2008), research has clearly shown that probiotics are not capable of persistent

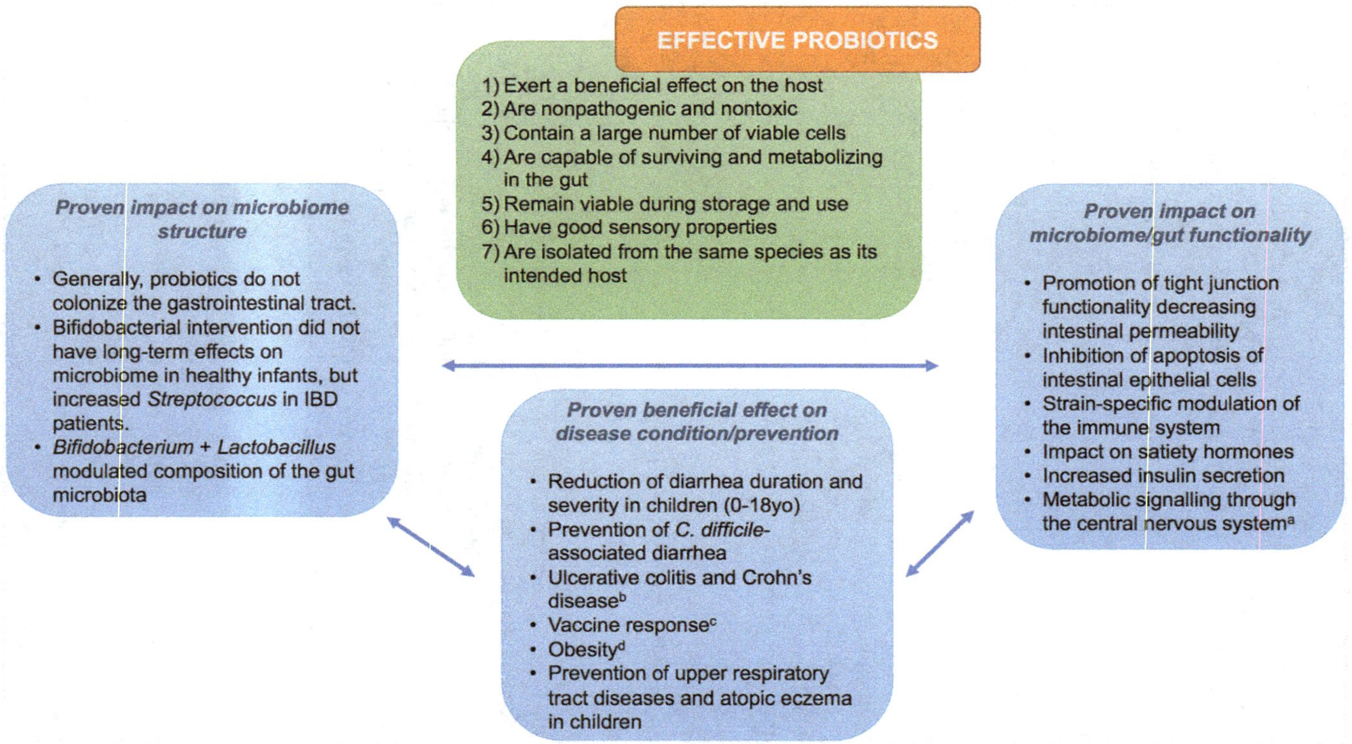

EFFECTIVE PROBIOTICS

1) Exert a beneficial effect on the host
2) Are nonpathogenic and nontoxic
3) Contain a large number of viable cells
4) Are capable of surviving and metabolizing in the gut
5) Remain viable during storage and use
6) Have good sensory properties
7) Are isolated from the same species as its intended host

Proven impact on microbiome/gut functionality

- Promotion of tight junction functionality decreasing intestinal permeability
- Inhibition of apoptosis of intestinal epithelial cells
- Strain-specific modulation of the immune system
- Impact on satiety hormones
- Increased insulin secretion
- Metabolic signalling through the central nervous system[a]

Proven beneficial effect on disease condition/prevention

- Reduction of diarrhea duration and severity in children (0–18yo)
- Prevention of C. difficile-associated diarrhea
- Ulcerative colitis and Crohn's disease[b]
- Vaccine response[c]
- Obesity[d]
- Prevention of upper respiratory tract diseases and atopic eczema in children

Proven impact on microbiome structure

- Generally, probiotics do not colonize the gastrointestinal tract. Bifidobacterial intervention did not have long-term effects on microbiome in healthy infants, but increased *Streptococcus* in IBD patients.
- *Bifidobacterium* + *Lactobacillus* modulated composition of the gut microbiota

Fig. 1 Characteristics of effective probiotics and their impact on gut microbiome structure and functionality. The figure was constructed based on included references (Collins and Gibson 1999; Goldenberg et al. 2015; Allen et al. 2010; Ganji-Arjenaki and Rafieian-Kopaei 2018; Zimmermann and Curtis 2018; Crovesy et al. 2017; Liu et al. 2013; Kalliomaki et al. 2001; Karczewski et al. 2010; Yan et al. 2007, 2013; Yan and Polk 2002; Forsten et al. 2013; Simon et al. 2015; Westfall et al. 2017; Bazanella et al. 2017; Staudacher et al. 2017; Seo et al. 2017; Toscano et al. 2017a, b). [a]The review by Westfall et al. (2017) contains a very comprehensive table of host metabolites regulated by probiotic and commensal bacteria. [b]Data on IBD is conflicting. The review by Goldenberg et al. (2015) reported no beneficial impact of probiotics on IBD conditions, conversely the review by Ganji-Arjenaki and Rafieian-Kopaei (2018) reported the opposite. [c]Of 26 studies included in Zimmermann and Curtis (2018), 50% reported beneficial effect of probiotics on vaccine response. [d]Strain dependent (Crovesy et al. 2017)

colonization of the adult GI tract (Alander et al. 1999; Bezkorovainy 2001). On the other hand, colonization of the infant gut by probiotics is still under debate. In the UK, a randomized controlled trial of the probiotic *Bifidobacterium breve* BBG-001 in preterm babies to prevent sepsis, necrotizing enterocolitis and death showed that, although there were no evident benefits for the primary outcomes, the probiotic strain persisted for up to 2 weeks in stools of treated infants after the end of the intervention (Costeloe et al. 2016). In another study, breastfed infants that received *B. infantis* EVC001 until postnatal day 28, maintained significantly higher abundances of fecal *B. infantis* for 30 days after supplementation compared to control infants (Frese et al. 2017). Conversely, in a study conducted on healthy infants receiving a standard whey-based formula containing a total of 10^7 colony-forming units (CFU)/g of *B. bifidum*, *B. breve*, *B. longum*, and *B. longum* subspecies *infantis*), long-term colonization (24 months) of the supplemented *Bifidobacterium* strains was not detected (Bazanella et al. 2017).

Still to this day there is no consensus regarding effectiveness of probiotics in general. Most probably because it is not correct to include all probiotics in one general category. Can we really compare efficacy of probiotic interventions for one specific disease or disease prevention when different strains and doses were used? For example in the review of efficacy of probiotics on vaccine response, Zimmerman and Curtis included a total of 26 studies, involving 3812 participants and 40 different probiotic strains on the efficacy of 17 different vaccines (Zimmermann and Curtis 2018). The review acknowledged that the large variation in the reported effect of probiotics was probably due to the substantial variation between studies in the choice of probiotics, strain, dose, viability, purity, and duration and timing of administration. Importantly nevertheless and despite this broad variability, the review concluded that probiotics increased responses to influenza vaccination in elderly people, whom have lower seroconversion rates to influenza vaccination compared to younger people. We also highlighted the probiotic effect variation on colon cancer prevention in our earlier review (Azcarate-Peril et al. 2011).

We can conclude from this section that some affirmations can be made of probiotics in general. Although assumed for decades, a number of recent research studies have demonstrated that in fact probiotics modify gut microbiota composition. However, those modifications are short lived and depend on the continuous supply of the beneficial strain(s). A similar conclusion can be made regarding impact on functionality. In other words, our gut microbiota needs continuous replenishment of beneficial bacteria to stay healthy. A healthy, balanced microbiota will be then more resilient and resistant to disease. Conclusions about effectiveness are harder to make due to the variation in probiotic strains and disease conditions.

Who Is Enhanced by Prebiotics in the Colon?

The high complexity of the microbial populations residing in the human gastrointestinal tract adds another confounding factor to prebiotic and probiotic research. Hence, trying to discern the effect of one specific compound on a bacterial population of 500–700 species (and millions of different strains), which varies from individual to individual, is impacted by the host's genotype, diet, and other environmental factors can be a daunting task. Nevertheless, we recently reviewed the extensive evidence on the bacterial targets of the most studied and widely used prebiotics (galacto-oligosaccharides [GOS] and fructo-oligosaccharides [FOS]) (Bruno-Barcena and Azcarate-Peril 2015).

The bifidogenic effect of GOS has been historically documented in infants fed formula containing β (1–4) GOS (Scalabrin et al. 2012), or β (1–4) GOS plus FOS (in a 9:1 ratio) (Salvini et al. 2011; Holscher et al. 2012; Bruzzese et al. 2009). GOS is generally metabolized by β-galactosidases (β-Gal, EC 3.2.1.23), also known as lactases (Campbell et al. 2005) because β-Gal enzymes are also responsible for the hydrolysis of terminal non-reducing β-D-galactose residue of the disaccharide lactose (4-O-β-galactopyranosyl-D-glucopyranose). In fact, lactose human milk oligosaccharides (HMOs) represent the highest proportion of carbohydrates in the breast milk of mammals being the first and most natural source of nutrients for the newborn. It is recognized that one of the most abundant taxa in breastfed babies is *Bifidobacterium* (Thompson et al. 2015; Tannock et al. 2013) and hence the most extensively characterized β-Gal enzymes are those from *Bifidobacterium* species. *B. bifidum* and some strains of *B. longum* subsp. *longum* have a dedicated pathway for degrading type I HMOs, which involves liberation of lacto-N-biose type I (LNB) and galacto-N-biose type I (GNB) from their natural substrates by extracellular enzymes [endo-α-N-acetylgalactosaminidase (Fujita et al. 2005) and/or lacto-N-biosidase (Wada et al. 2008)], transport and subsequent cleavage by the lacto-N-biose phosphorylase LnpA. The products of this process are α-galactosyl phosphate, which enters glycolysis, and N-acetylhexosamines, which enter the aminosugar metabolic cycle (Nishimoto and Kitaoka 2007). Strains of *B. longum* subsp. *infantis* characterized so far have shown no presence of lacto-N-biosidase homologs (LoCascio et al. 2010). A study by Yoshida et al. (2012) showed that *B. longum* subsp. *infantis* can directly incorporate lacto-N-tetraose (LNT) and hydrolyze it *via* a specific β-Gal enzyme. The authors identified two different β-Gal enzymes (Bga42A and Bga2A) responsible for the degradation of type-1 and type-2 HMOs respectively.

The bacterial enzymatic machinery necessary for hydrolysis of β-GOS in the colon is quite ubiquitous. β-Gal enzymes have been identified in the phyla Proteobacteria, Firmicutes, Actinobacteria, the CFB (Bacteroidetes-Chlorobi-Fibrobacteres) group, Verrucomicrobia, and Spirochaetes, and also in Victivallaceae, Thermotogales, Chloroflexi, Acidobacteriales, and over 350 more taxa. β-Galactosidases have been identified in over 20 species of *Lactobacillus*, including *L. pentosus* (Maischberger et al. 2010), *L. sakei* (Iqbal et al. 2011), *L. delbrueckii* subsp. *bulgaricus* (Nguyen et al. 2012), *L. plantarum* (Iqbal et al. 2010), *L. reuteri* (Nguyen et al. 2006) and *L. acidophilus* (Nguyen et al. 2007). However, the dedicated pathways and enzymes involved in GOS degradation by β-Galactosidases in lactobacilli have not been extensively characterized.

Although the impact of prebiotics on beneficial bacteria of the gastrointestinal tract is well documented, we cannot assume that prebiotics will not increase abundance of other taxa, especially considering that β-galactosidases are widely distributed in the gut microbiome. The study by Davis et al. (2011) done in 18 healthy adult human volunteers, whom received (1–4) GOS during 16 weeks showed that only a few taxa other than bifidobacteria, were impacted by GOS. Statistically significant decreases were observed for the family Bacteroidaceae and the genus *Bacteroides* while abundance of *Coprococcus comes* and *Faecalibacterium prausnitzii* was significantly increased at doses of 5 and 10 g/day. We further demonstrated that specific bifidobacteria (*B. longum, B. adolescentis, B. catenulatum,* and *B. breve*) increased in lactose-intolerant adults receiving purified GOS (Azcarate Peril et al. 2013, 2017). In addition, our study showed that GOS also enhanced other beneficial taxa like *Faecalibacterium prausnitzii, Lactobacillus,* Christensenellaceae, *Collinsella, Prevotella,* and *Catenibacterium* (Azcarate-Peril et al. 2017; Monteagudo-Mera et al. 2016). It remains to be investigated the reasons why, although the enzymes that target GOS are widely distributed, their impact appears to be limited to mostly beneficial members of the gut microbiota.

Health and Disease: Different Circumstances, Different Impact

Foods containing probiotics and prebiotics have been consumed by humans for millennia, mostly fermented foods (yogurt, kimchi, fermented vegetables) and vegetables containing high levels of prebiotic compounds like inulin (asparagus, onion) or FOS (garlic). Is it possible to quantify their effect on healthy individuals? Probably not. A recent opinion article in Scientific American states that "although certain bacteria help treat some gut disorders, they have no known benefits for healthy people" (Jabr 2017). The author mainly refers to food supplements, deemed as another "nutritional craze". Although the point of view of this particular article is understandable, in terms of manufacturers' attempts to manipulate consumers into purchasing their products, it is clear that the consumption of probiotics and prebiotics is essentially what keeps healthy people healthy! Given the complexity of the issue, no long-term study has attempted to compare the overall health status of persons consuming or not beneficial bacteria and/or prebiotics. However, a probiotic study on 68 healthy adults living in the Helsinki area showed that specific strains of probiotics (in this case *L. rhamnosus* GG and *P. freudenreichii* ssp. *shermanii* JS) have the ability to induce anti-inflammatory and cytokine responses and may have a moderate anti-inflammatory effect shown as a decrease in serum highly sensitive CRP (hsCRP) levels (Kekkonen et al. 2008).

In essence, the probiotics in food are not considered treatment or cure, and hence cannot be used to make health claims. Moreover, is unclear which biomarkers can serve as indicators for interventions to prevent diseases in healthy individuals. There could be however, an agreement on pre-disease states, in which alterations of specific physiologic parameters in otherwise healthy individuals could predispose to particular ailments. We find one example in the study by Takahashi and collaborators

(Takahashi et al. 2016) that showed that consumption of *Bifidobacterium animalis* ssp. *lactis* GCL2505 during 12 weeks by overweight or mildly obese, but otherwise healthy Japanese subjects resulted in a measured decrease in visceral fat area, which can be a key factor associated with metabolic disorders, including metabolic syndrome. It is clear that more studies are needed in healthy subjects to determine a "health baseline" and to identify biomarkers impacted by beneficial modulators of the gut microbiota.

Does It Matter How the Probiotics Are Delivered to Our Gut?

Nearly 10,000 years ago, before we even started using the term 'probiotics', beneficial bacteria were delivered to our gastrointestinal tract in food matrices, as fermented foodstuffs. As products of non-controlled fermentative processes, they probably had food safety issues, but microbial diversity in those products was undoubtedly high and contributed to an overall healthier gut diversity. There are clear challenges to the delivery of probiotics including protecting the microorganisms from the harsh gastric environment (or selecting highly resistant strains) to reach the colon in adequate amounts of viable bacteria. Immobilization (the process of attaching a cell or entrapping it within a suitable inert material called a matrix) and encapsulation (the process of forming a continuous coating around an inner matrix that is wholly contained within the capsule wall as a core of encapsulated material) are methods used to increase viability of probiotics in food and nutritional supplements (Sipailiene and Petraityte 2017). Today, carefully characterized probiotic strains selected for their industrial robustness are delivered via conventional and non-conventional food matrices and as dietary supplements (Fig. 2).

Decades of research have shown that processing, delivery, and passage through the gastrointestinal tract elicit rapid gene expression responses from probiotic bacteria (Azcarate-Peril et al. 2009; Desmond et al. 2004; Sheehan et al. 2007), and hence are likely to influence functionality by inducing changes to cell composition, generation of end-products like organic acids, and bioactives like bacteriocins and small peptides (Tripathi and Giri 2014; Ananta et al. 2004; Sanders and Marco 2010). Moreover, the delivery matrix has an effect on viability of strains during shelf life and passage through the upper gastrointestinal tract. Finally, delivery of probiotics in a food matrix could induce synergy with other active ingredients, including fiber and prebiotics (Sanders and Marco 2010).

Not many studies have focused in comparative analyses of probiotic efficacy in regard to the delivery matrix. The efficacy of *L. rhamnosus* GG on duration of diarrhea in children was evaluated in fermented milk or as a freeze-dried powder (Isolauri et al. 1991). Duration of diarrhea episodes was shorter in the fermented milk and freeze-dried powder groups compared with the pasteurized yogurt group, suggesting that, in this study, the delivery matrix was not critical for probiotic function. Conversely, efficacy of *Lactobacillus casei* BL23 in a dextran sulfate sodium (DSS)-induced murine model of ulcerative colitis showed that the strain protected

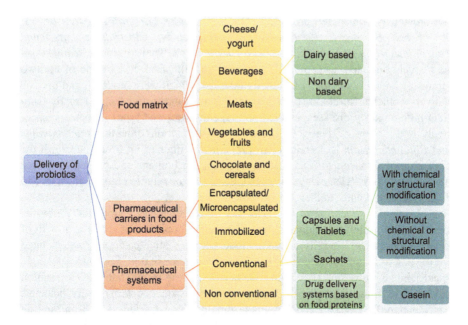

Fig. 2 Methods and vehicles currently used for the delivery of probiotics. Based on the following references (Sipailiene and Petraityte 2017; Phillips et al. 2006; Govender et al. 2014; Dalli et al. 2017)

against the development of colitis when ingested in milk but not in a nutrient-free buffer simulating a nutritional supplement (Lee et al. 2015). Earlier studies in animal models showed comparable results. Administration of *L. rhamnosus* GG to a rat model of arthritis in water or milk conferred minor preventative effects, whereas animals that received plain yogurt or yogurt containing the probiotics were more protected in the prevention of either adjuvant or tropomyosin arthritis (Baharav et al. 2004). Similarly, significantly fewer adenomas were detected in the small intestine of mice fed micro capsulated *L. acidophilus* in yogurt compared with mice fed the same strain in saline solution (Urbanska et al. 2009).

The Next Generation of Beneficial Modulators of the Gut Microbiota

Traditional probiotics draw from a very narrow spectrum of bacterial species, the species that are more robust and resilient, and presumably, the species in which the highest percent of strains share probiotic features, making them Generally Regarded as Safe (GRAS). The explosion of microbiome research however has uncovered a number of potentially beneficial bacteria beyond *Lactobacillus* and *Bifidobacterium* species. A recent review (O'Toole et al. 2017) listed species of probiotic potential not previously considered probiotics. The list includes *Bacteroides xylanisolvens*, *B. ovatus*, *B. dorei*, *B. fragilis*, *B. acidifaciens*, *Clostridium butyricum*, *Faecalibacterium prausnitzii*, and *Lactococcus lactis*. Their probiotic features can be natively encoded, or provided by genetic modification, like in one of the two strains of *Lactococcus lactis* included in the list, which were modified and used as delivery vehicles for elafin (peptidase inhibitor 3) and trefoil factor 1 or IL-10 to treat inflammatory diseases, and autoimmune diseases and type1 diabetes, respectively. At this point, evidence of probiotic potential of strains other than *Lactobacillus* and *Bifidobacterium* is correlative or preclinical, either in vitro or mice, with the exception of *B. xylanisolvens*, shown to be safe in humans and proven to induce TFα-specific IgM (Ulsemer et al. 2016), *C. butyricum*, studied mostly in Asia, and *L. lactis* (Robert and Steidler 2014).

Akkermansia muciniphila has also recently emerged as a potential probiotic, based on the notion that humans make their own prebiotics to enhance growth of beneficial bacteria. Literally meaning "mucin-loving", this bacterium was isolated for the first time in 2004 (Derrien et al. 2004), being one of the most abundant single species in the human intestine with abundances ranging from 0.5 to 5%. Correlation studies have shown that this bacterium is inversely associated with obesity, diabetes, cardiometabolic diseases and low-grade inflammation (reviewed by Cani and de Vos (2017)). Furthermore, animal experiments demonstrated that administration of *A. muciniphila* improved the metabolism of obese and diabetic mice (Plovier et al. 2017) and protected against atherosclerosis by preventing metabolic endotoxemia-induced inflammation in Apoe$^{-/-}$ mice (Li et al. 2016). Conversely, Seregin et al. (2017a) showed that loss of the innate immune receptor NLRP6 in mice resulted in impaired production of interleukin-18 (IL-18) and increased susceptibility to epithelial-induced injury. Subsequently, they demonstrated that NLRP6 was important for suppressing the development of spontaneous colitis in Il10$^{-/-}$ mice and that NLRP6 deficiency resulted in the enrichment of *A. muciniphila*. In fact, *A. muciniphila* was sufficient for promoting intestinal inflammation in both specific-pathogen-free and germ-free Il10$^{-/-}$ mice suggesting that this organism can have a detrimental effect in genetically susceptible hosts (Seregin et al. 2017b). Further studies will contribute to establish (or not) *A. muciniphila* as a probiotic microorganism.

The success of fecal microbiota transplantation (FMT) for treatment of *C. difficile* infections hints that the future of beneficial modulation of the gut microbiota lies not on isolated strains but in bacterial consortia capable of restoring complete metabolic pathways potentially affected by disease. Microbiome therapeutics using characterized microbial consortia are being investigated by multiple groups as more controlled and/ or safer alternatives to FMT. One example is RePOOPulate (Petrof et al. 2013), a consortium of 33 bacterial strains developed by extensively culturing the microbial diversity from the stool of a healthy, 41-year-old female donor. The mix contains species of *Lactobacillus*, *Eubacterium*, *Dorea*, *Clostridium*, *Faecalibacterium*, *Roseburia*, *Ruminococcus*, and *Streptococcus*. Although treatment of two *C. difficile* patients was reported successful in the study, this mixture has not been tested in a larger cohort and it is not commercially available.

Conclusions

A recent review by Puebla-Barragan and Reid (2019) provided an extraordinary summary of a 45-year evolution of probiotics and probiotics research from being completely ignored to "snake oil sold from the back of covered wagons", as defined in 1999 by then the President of the American Society of Microbiology (Atlas 1999), to the present day where omics tools have clearly demonstrated that beneficial modulation of the gut microbiota translates into host health and well-being. Unfortunately, the National Institutes of Health (NIH) and other funding agencies are still focused on curing rather than preventing disease, slowing down the advancement of probiotics, prebiotics, and nutrition research. It will be up to consumers to guide research for healthy future generations.

References

Alander, M., Satokari, R., Korpela, R., Saxelin, M., Vilpponen-Salmela, T., Mattila-Sandholm, T., & von Wright, A. (1999). Persistence of colonization of human colonic mucosa by a probiotic strain, Lactobacillus rhamnosus GG, after oral consumption. *Applied and Environmental Microbiology, 65*(1), 351–354.

Allen, S. J., Martinez, E. G., Gregorio, G. V., & Dans, L. F. (2010). Probiotics for treating acute infectious diarrhoea. *Cochrane Database of Systematic Reviews, 11*, CD003048.

Ananta, E., Birkeland, S.-E., Corcoran, B. M., Fitzgerald, G., Hinz, S., Klijn, A., Matto, J., Mercenier, A., Nilsson, U., Saarela, C., Stanton, C., Stahl, U., Suomalainen, T., Vincken, J.-P., Virkajarvi, I., Voragen, F., Wesenfeld, J., Wouters, R., & Knorr, D. (2004). Processing effects on the nutritional advancement of probiotics and prebiotics. *Microbial Ecology in Health and Disease, 16*(2–3), 113–124.

Atlas, R. M. (1999). Probiotics—Snake oil for the new millennium? *Environmental Microbiology, 1*(5), 377–382.

Azcarate Peril, M. A., Savaiano, D. A., Ritter, A. J., & Klaenhammer, T. (2013). Microbiome alterations of lactose intolerant individuals in response to dietary intervention with galactooligosaccharides may help negate symptoms of lactose intolerance. *Gastroenterology, 144*(5), S-893.

Azcarate-Peril, M. A., Altermann, E., Hoover-Fitzula, R. L., Cano, R. J., & Klaenhammer, T. R. (2004). Identification and inactivation of genetic loci involved with Lactobacillus acidophilus acid tolerance. *Applied and Environmental Microbiology, 70*(9), 5315–5322.

Azcarate-Peril, M. A., Tallon, R., & Klaenhammer, T. R. (2009). Temporal gene expression and probiotic attributes of Lactobacillus acidophilus during growth in milk. *Journal of Dairy Science, 92*(3), 870–886.

Azcarate-Peril, M. A., Sikes, M., & Bruno-Barcena, J. M. (2011). The intestinal microbiota, gastrointestinal environment and colorectal cancer: A putative role for probiotics in prevention of colorectal cancer? *American Journal of Physiology. Gastrointestinal and Liver Physiology, 301*(3), G401–G424.

Azcarate-Peril, M. A., Ritter, A. J., Savaiano, D., Monteagudo-Mera, A., Anderson, C., Magness, S. T., & Klaenhammer, T. R. (2017). Impact of short-chain galactooligosaccharides on the gut microbiome of lactose-intolerant individuals. *Proceedings of the National Academy of Sciences of the United States of America, 114*(3), E367–E375.

Baharav, E., Mor, F., Halpern, M., & Weinberger, A. (2004). Lactobacillus GG bacteria ameliorate arthritis in Lewis rats. *The Journal of Nutrition, 134*(8), 1964–1969.

Bazanella, M., Maier, T. V., Clavel, T., Lagkouvardos, I., Lucio, M., Maldonado-Gomez, M. X., Autran, C., Walter, J., Bode, L., Schmitt-Kopplin, P., & Haller, D. (2017). Randomized controlled trial on the impact of early-life intervention with bifidobacteria on the healthy infant fecal microbiota and metabolome. *The American Journal of Clinical Nutrition, 106*(5), 1274–1286.

Bezkorovainy, A. (2001). Probiotics: Determinants of survival and growth in the gut. *The American Journal of Clinical Nutrition, 73*(2 Suppl), 399S–405S.

Bindels, L. B., Delzenne, N. M., Cani, P. D., & Walter, J. (2015). Towards a more comprehensive concept for prebiotics. *Nat Rev Gastroenterol Hepatol, 12*(5), 303–310.

Bruno-Barcena, J. M., & Azcarate-Peril, M. A. (2015). Galacto-oligosaccharides and colorectal cancer: Feeding our intestinal probiome. *Journal of Functional Foods, 12*, 92–108.

Bruno-Barcena, J. M., Azcarate-Peril, M. A., & Hassan, H. M. (2010). Role of antioxidant enzymes in bacterial resistance to organic acids. *Applied and Environmental Microbiology, 76*(9), 2747–2753.

Bruzzese, E., Volpicelli, M., Squeglia, V., Bruzzese, D., Salvini, F., Bisceglia, M., Lionetti, P., Cinquetti, M., Iacono, G., Amarri, S., & Guarino, A. (2009). A formula containing galacto- and fructo-oligosaccharides prevents intestinal and extra-intestinal infections: An observational study. *Clinical Nutrition, 28*(2), 156–161.

Buck, B. L., Azcarate-Peril, M. A., Altermann, E., & Klaenhammer, T. *Methods and compositions to modulate adhesion and stress tolerance in bacteria.*Google Patents; 2006.

Campbell, A. K., Waud, J. P., & Matthews, S. B. (2005). The molecular basis of lactose intolerance. *Science Progress, 88*(Pt 3), 157–202.

Cani, P. D., & de Vos, W. M. (2017). Next-generation beneficial microbes: The case of Akkermansia muciniphila. *Frontiers in Microbiology, 8*, 1765.

Collins, M. D., & Gibson, G. R. (1999). Probiotics, prebiotics, and synbiotics: Approaches for modulating the microbial ecology of the gut. *The American Journal of Clinical Nutrition, 69*(5), 1052S–1057S.

Costeloe, K., Bowler, U., Brocklehurst, P., Hardy, P., Heal, P., Juszczak, E., King, A., Panton, N., Stacey, F., Whiley, A., Wilks, M., & Millar, M. R. (2016). A randomised controlled trial of the probiotic Bifidobacterium breve BBG-001 in preterm babies to prevent sepsis, necrotising enterocolitis and death: The Probiotics in Preterm infantS (PiPS) trial. *Health Technology Assessment, 20*(66), 1–194.

Crovesy, L., Ostrowski, M., Ferreira, D., Rosado, E. L., & Soares-Mota, M. (2017). Effect of Lactobacillus on body weight and body fat in overweight subjects: A systematic review of randomized controlled clinical trials. *International Journal of Obesity, 41*(11), 1607–1614.

Dalli, S. S., Uprety, K., & Rakshit, S. K. (2017). Industrial production of active probiotics for food enrichment. In Y. H. Roos & Y. D. Livney (Eds.), *Engineering foods for bioactives stability and delivery*. New York: Springer.

Davis, L. M., Martinez, I., Walter, J., Goin, C., & Hutkins, R. W. (2011). Barcoded pyrosequencing reveals that consumption of galactooligosaccharides results in a highly specific bifidogenic response in humans. *PLoS One, 6*(9), e25200.

Derrien, M., Vaughan, E. E., Plugge, C. M., & de Vos, W. M. (2004). Akkermansia muciniphila gen. nov., sp. nov., a human intestinal mucin-degrading bacterium. *International Journal of Systematic and Evolutionary Microbiology, 54*(Pt 5), 1469–1476.

Desmond, C., Fitzgerald, G. F., Stanton, C., & Ross, R. P. (2004). Improved stress tolerance of GroESL-overproducing Lactococcus lactis and probiotic Lactobacillus paracasei NFBC 338. *Applied and Environmental Microbiology, 70*(10), 5929–5936.

Forssten, S. D., Korczynska, M. Z., Zwijsen, R. M., Noordman, W. H., Madetoja, M., & Ouwehand, A. C. (2013). Changes in satiety hormone concentrations and feed intake in rats in response to lactic acid bacteria. *Appetite, 71*, 16–21.

Frese, S. A., Hutton, A. A., Contreras, L. N., Shaw, C. A., Palumbo, M. C., Casaburi, G., Xu, G., Davis, J. C. C., Lebrilla, C. B., Henrick, B. M., Freeman, S. L., Barile, D., German, J. B., Mills, D. A., Smilowitz, J. T., & Underwood, M. A. (2017). Persistence of supplemented Bifidobacterium longum subsp. infantis EVC001 in breastfed infants. *mSphere, 2*(6), e00501–e00517.

Fujita, K., Oura, F., Nagamine, N., Katayama, T., Hiratake, J., Sakata, K., Kumagai, H., & Yamamoto, K. (2005). Identification and molecular cloning of a novel glycoside hydrolase family of core 1 type O-glycan-specific endo-alpha-N-acetylgalactosaminidase from Bifidobacterium longum. *The Journal of Biological Chemistry, 280*(45), 37415–37422.

Ganji-Arjenaki, M., & Rafieian-Kopaei, M. (2018). Probiotics are a good choice in remission of inflammatory bowel diseases: A meta analysis and systematic review. *Journal of Cellular Physiology, 233*(3), 2091–2103.

Gibson, G. R., & Roberfroid, M. B. (1995). Dietary modulation of the human colonic microbiota—Introducing the concept of prebiotics. *Journal of Nutrition, 125*(6), 1401–1412.

Goldenberg, J. Z., Lytvyn, L., Steurich, J., Parkin, P., Mahant, S., & Johnston, B. C. (2015). Probiotics for the prevention of pediatric antibiotic-associated diarrhea. *Cochrane Database of Systematic Reviews, 12*, CD004827.

Goldin, B., & Gorbach, S. L. (1977). Alterations in fecal microflora enzymes related to diet, age, lactobacillus supplements, and dimethylhydrazine. *Cancer, 40*(5 Suppl), 2421–2426.

Govender, M., Choonara, Y. E., Kumar, P., du Toit, L. C., van Vuuren, S., & Pillay, V. (2014). A review of the advancements in probiotic delivery: Conventional vs. non-conventional formulations for intestinal flora supplementation. *AAPS PharmSciTech, 15*(1), 29–43.

Hamilton-Miller, J. M., Gibson, G. R., & Bruck, W. (2003). Some insights into the derivation and early uses of the word 'probiotic'. *The British Journal of Nutrition, 90*(4), 845.

Hill, C., Guarner, F., Reid, G., Gibson, G. R., Merenstein, D. J., Pot, B., Morelli, L., Canani, R. B., Flint, H. J., Salminen, S., Calder, P. C., & Sanders, M. E. (2014). Expert consensus document. The International Scientific Association for Probiotics and Prebiotics consensus statement on the scope and appropriate use of the term probiotic. *Nature Reviews. Gastroenterology & Hepatology, 11*(8), 506–514.

Holscher, H. D., Faust, K. L., Czerkies, L. A., Litov, R., Ziegler, E. E., Lessin, H., Hatch, T., Sun, S., & Tappenden, K. A. (2012). Effects of prebiotic-containing infant formula on gastrointestinal tolerance and fecal microbiota in a randomized controlled trial. *JPEN Journal of Parenteral and Enteral Nutrition, 36*(1 Suppl), 95S–105S.

Iqbal, S., Nguyen, T. H., Nguyen, T. T., Maischberger, T., & Haltrich, D. (2010). beta-Galactosidase from Lactobacillus plantarum WCFS1: Biochemical characterization and formation of prebiotic galacto-oligosaccharides. *Carbohydrate Research, 345*(10), 1408–1416.

Iqbal, S., Nguyen, T. H., Nguyen, H. A., Maischberger, T., Kittl, R., & Haltrich, D. (2011). Characterization of a heterodimeric GH2 beta-galactosidase from Lactobacillus sakei Lb790 and formation of prebiotic galacto-oligosaccharides. *Journal of Agricultural and Food Chemistry, 59*(8), 3803–3811.

Isolauri, E., Juntunen, M., Rautanen, T., Sillanaukee, P., & Koivula, T. (1991). A human Lactobacillus strain (Lactobacillus casei sp strain GG) promotes recovery from acute diarrhea in children. *Pediatrics, 88*(1), 90–97.

Jabr, F. Do probiotics really work? *Scientific American.* 2017.

Kalliomaki, M., Salminen, S., Arvilommi, H., Kero, P., Koskinen, P., & Isolauri, E. (2001). Probiotics in primary prevention of atopic disease: A randomised placebo-controlled trial. *Lancet, 357*(9262), 1076–1079.

Karczewski, J., Troost, F. J., Konings, I., Dekker, J., Kleerebezem, M., Brummer, R. J., & Wells, J. M. (2010). Regulation of human epithelial tight junction proteins by Lactobacillus plantarum in vivo and protective effects on the epithelial barrier. *American Journal of Physiology. Gastrointestinal and Liver Physiology, 298*(6), G851–G859.

Kekkonen, R. A., Lummela, N., Karjalainen, H., Latvala, S., Tynkkynen, S., Jarvenpaa, S., Kautiainen, H., Julkunen, I., Vapaatalo, H., & Korpela, R. (2008). Probiotic intervention has strain-specific anti-inflammatory effects in healthy adults. *World Journal of Gastroenterology, 14*(13), 2029–2036.

Lee, B., Yin, X., Griffey, S. M., & Marco, M. L. (2015). Attenuation of colitis by Lactobacillus casei BL23 is dependent on the dairy delivery matrix. *Applied and Environmental Microbiology, 81*(18), 6425–6435.

Li, J., Lin, S., Vanhoutte, P. M., Woo, C. W., & Xu, A. (2016). Akkermansia muciniphila protects against atherosclerosis by preventing metabolic endotoxemia-induced inflammation in Apoe$^{-/-}$ mice. *Circulation, 133*(24), 2434–2446.

Lilly, D. M., & Stillwell, R. H. (1965). Probiotics: Growth-promoting factors produced by microorganisms. *Science, 147*(3659), 747–748.

Liu, S., Hu, P., Du, X., Zhou, T., & Pei, X. (2013). Lactobacillus rhamnosus GG supplementation for preventing respiratory infections in children: A meta-analysis of randomized, placebo-controlled trials. *Indian Pediatrics, 50*(4), 377–381.

LoCascio, R. G., Desai, P., Sela, D. A., Weimer, B., & Mills, D. A. (2010). Broad conservation of milk utilization genes in Bifidobacterium longum subsp. infantis as revealed by comparative genomic hybridization. *Applied and Environmental Microbiology, 76*(22), 7373–7381.

Maischberger, T., Leitner, E., Nitisinprasert, S., Juajun, O., Yamabhai, M., Nguyen, T. H., & Haltrich, D. (2010). Beta-galactosidase from Lactobacillus pentosus: Purification, characterization and formation of galacto-oligosaccharides. *Biotechnology Journal, 5*(8), 838–847.

Metchnikoff, I. I., & Mitchell, P. (2004). *Prolongation of life: Optimistic studies*. New York: Springer.

Monteagudo-Mera, A., Arthur, J. C., Jobin, C., Keku, T. O., Bruno Barcena, J. M., & Azcarate-Peril, M. A. (2016). High purity galacto-oligosaccharides enhance specific Bifidobacterium species and their metabolic activity in the mouse gut microbiome. *Beneficial Microbes, 3*, 1–18.

Nguyen, T. H., Splechtna, B., Steinbock, M., Kneifel, W., Lettner, H. P., Kulbe, K. D., & Haltrich, D. (2006). Purification and characterization of two novel beta-galactosidases from Lactobacillus reuteri. *Journal of Agricultural and Food Chemistry, 54*(14), 4989–4998.

Nguyen, T. H., Splechtna, B., Krasteva, S., Kneifel, W., Kulbe, K. D., Divne, C., & Haltrich, D. (2007). Characterization and molecular cloning of a heterodimeric beta-galactosidase from the probiotic strain Lactobacillus acidophilus R22. *FEMS Microbiology Letters, 269*(1), 136–144.

Nguyen, T. T., Nguyen, H. A., Arreola, S. L., Mlynek, G., Djinovic-Carugo, K., Mathiesen, G., Nguyen, T. H., & Haltrich, D. (2012). Homodimeric beta-galactosidase from Lactobacillus delbrueckii subsp. bulgaricus DSM 20081: Expression in Lactobacillus plantarum and biochemical characterization. *Journal of Agricultural and Food Chemistry, 60*(7), 1713–1721.

Nishimoto, M., & Kitaoka, M. (2007). Identification of N-acetylhexosamine 1-kinase in the complete lacto-N-biose I/galacto-N-biose metabolic pathway in Bifidobacterium longum. *Applied and Environmental Microbiology, 73*(20), 6444–6449.

O'Toole, P. W., Marchesi, J. R., & Hill, C. (2017). Next-generation probiotics: The spectrum from probiotics to live biotherapeutics. *Nature Microbiology, 2*, 17057.

Oozeer, R., Leplingard, A., Mater, D. D., Mogenet, A., Michelin, R., Seksek, I., Marteau, P., Dore, J., Bresson, J. L., & Corthier, G. (2006). Survival of Lactobacillus casei in the human digestive tract after consumption of fermented milk. *Applied and Environmental Microbiology, 72*(8), 5615–5617.

Petrof, E. O., Gloor, G. B., Vanner, S. J., Weese, S. J., Carter, D., Daigneault, M. C., Brown, E. M., Schroeter, K., & Allen-Vercoe, E. (2013). Stool substitute transplant therapy for the eradication of Clostridium difficile infection: 'RePOOPulating' the gut. *Microbiome, 1*(1), 3.

Pfeiler, E. A., Azcarate-Peril, M. A., & Klaenhammer, T. R. (2006). Characterization of a two-component regulatory system implicated in the bile tolerance of Lactobacillus acidophilus NCFM. *Journal of Animal Science, 84*, 181–181.

Pfeiler, E. A., Azcarate-Peril, M. A., & Klaenhammer, T. R. (2007). Characterization of a novel bile-inducible operon encoding a two-component regulatory system in Lactobacillus acidophilus. *Journal of Bacteriology, 189*(13), 4624–4634.

Phillips, M., Kailasapathy, K., & Tran, L. (2006). Viability of commercial probiotic cultures (L. acidophilus, Bifidobacterium sp., L. casei, L. paracasei and L. rhamnosus) in cheddar cheese. *International Journal of Food Microbiology, 108*(2), 276–280.

Plovier, H., Everard, A., Druart, C., Depommier, C., Van Hul, M., Geurts, L., Chilloux, J., Ottman, N., Duparc, T., Lichtenstein, L., Myridakis, A., Delzenne, N. M., Klievink, J., Bhattacharjee, A., van der Ark, K. C., Aalvink, S., Martinez, L. O., Dumas, M. E., Maiter, D., Loumaye,

A., Hermans, M. P., Thissen, J. P., Belzer, C., de Vos, W. M., & Cani, P. D. (2017). A purified membrane protein from Akkermansia muciniphila or the pasteurized bacterium improves metabolism in obese and diabetic mice. *Nature Medicine, 23*(1), 107–113.

Puebla-Barragan, S., & Reid, G. (2019). Forty-five-year evolution of probiotic therapy. *Microbial Cell, 6*(4), 184–196.

Reid, G., Sanders, M. E., Gaskins, H. R., Gibson, G. R., Mercenier, A., Rastall, R., Roberfroid, M., Rowland, I., Cherbut, C., & Klaenhammer, T. R. (2003). New scientific paradigms for probiotics and prebiotics. *Journal of Clinical Gastroenterology, 37*(2), 105–118.

Rettger, L. F., & Cheplin, H. A. (1921). *Treatise on the transformation of the intestinal flora: With special reference to the implantation of Bacillus acidophilus* (Vol. 13). Yale University Press.

Robert, S., & Steidler, L. (2014). Recombinant Lactococcus lactis can make the difference in antigen-specific immune tolerance induction, the Type 1 Diabetes case. *Microbial Cell Factories, 13*(Suppl 1), S11.

Salvini, F., Riva, E., Salvatici, E., Boehm, G., Jelinek, J., Banderali, G., & Giovannini, M. (2011). A specific prebiotic mixture added to starting infant formula has long-lasting bifidogenic effects. *The Journal of Nutrition, 141*(7), 1335–1339.

Sanders, M. E., & Marco, M. L. (2010). Food formats for effective delivery of probiotics. *Annual Review of Food Science and Technology, 1*, 65–85.

Scalabrin, D. M., Mitmesser, S. H., Welling, G. W., Harris, C. L., Marunycz, J. D., Walker, D. C., Bos, N. A., Tolkko, S., Salminen, S., & Vanderhoof, J. A. (2012). New prebiotic blend of polydextrose and galacto-oligosaccharides has a bifidogenic effect in young infants. *Journal of Pediatric Gastroenterology and Nutrition, 54*(3), 343–352.

Seo, M., Heo, J., Yoon, J., Kim, S. Y., Kang, Y. M., Yu, J., Cho, S., & Kim, H. (2017). Methanobrevibacter attenuation via probiotic intervention reduces flatulence in adult human: A non-randomised paired-design clinical trial of efficacy. *PLoS One, 12*(9), e0184547.

Seregin, S. S., Golovchenko, N., Schaf, B., Chen, J., Eaton, K. A., & Chen, G. Y. (2017a). NLRP6 function in inflammatory monocytes reduces susceptibility to chemically induced intestinal injury. *Mucosal Immunology, 10*(2), 434–445.

Seregin, S. S., Golovchenko, N., Schaf, B., Chen, J., Pudlo, N. A., Mitchell, J., Baxter, N. T., Zhao, L., Schloss, P. D., Martens, E. C., Eaton, K. A., & Chen, G. Y. (2017b). NLRP6 protects Il10(-/-) mice from colitis by limiting colonization of Akkermansia muciniphila. *Cell Reports, 19*(4), 733–745.

Sheehan, V. M., Sleator, R. D., Hill, C., & Fitzgerald, G. F. (2007). Improving gastric transit, gastrointestinal persistence and therapeutic efficacy of the probiotic strain Bifidobacterium breve UCC2003. *Microbiology, 153*(Pt 10), 3563–3571.

Simon, M. C., Strassburger, K., Nowotny, B., Kolb, H., Nowotny, P., Burkart, V., Zivehe, F., Hwang, J. H., Stehle, P., Pacini, G., Hartmann, B., Holst, J. J., MacKenzie, C., Bindels, L. B., Martinez, I., Walter, J., Henrich, B., Schloot, N. C., & Roden, M. (2015). Intake of Lactobacillus reuteri improves incretin and insulin secretion in glucose-tolerant humans: A proof of concept. *Diabetes Care, 38*(10), 1827–1834.

Sipailiene, A., & Petraityte, S. (2017). Encapsulation of probiotics: Proper selection of the probiotic strain and the influence of encapsulation technology and materials on the viability of encapsulated microorganisms. *Probiotics Antimicrob Proteins, 10*(1), 1–10.

Staudacher, H. M., Lomer, M. C. E., Farquharson, F. M., Louis, P., Fava, F., Franciosi, E., Scholz, M., Tuohy, K. M., Lindsay, J. O., Irving, P. M., & Whelan, K. (2017). A diet low in FODMAPs reduces symptoms in patients with irritable bowel syndrome and a probiotic restores bifidobacterium species: A randomized controlled trial. *Gastroenterology, 153*(4), 936–947.

Takahashi, S., Anzawa, D., Takami, K., Ishizuka, A., Mawatari, T., Kamikado, K., Sugimura, H., & Nishijima, T. (2016). Effect of Bifidobacterium animalis ssp. lactis GCL2505 on visceral fat accumulation in healthy Japanese adults: A randomized controlled trial. *Bioscience of Microbiota, Food and Health, 35*(4), 163–171.

Tannock, G. W., Lawley, B., Munro, K., Gowri Pathmanathan, S., Zhou, S. J., Makrides, M., Gibson, R. A., Sullivan, T., Prosser, C. G., Lowry, D., & Hodgkinson, A. J. (2013). Comparison of the compositions of the stool microbiotas of infants fed goat milk formula, cow milk-based formula, or breast milk. *Applied and Environmental Microbiology, 79*(9), 3040–3048.

Thompson, A. L., Monteagudo-Mera, A., Cadenas, M. B., Lampl, M. L., & Azcarate-Peril, M. A. (2015). Milk- and solid-feeding practices and daycare attendance are associated with differences in bacterial diversity, predominant communities, and metabolic and immune function of the infant gut microbiome. *Frontiers in Cellular and Infection Microbiology, 5*, 3.

Toscano, M., De Grandi, R., Miniello, V. L., Mattina, R., & Drago, L. (2017a). Ability of Lactobacillus kefiri LKF01 (DSM32079) to colonize the intestinal environment and modify the gut microbiota composition of healthy individuals. *Digestive and Liver Disease, 49*(3), 261–267.

Toscano, M., De Grandi, R., Stronati, L., De Vecchi, E., & Drago, L. (2017b). Effect of Lactobacillus rhamnosus HN001 and Bifidobacterium longum BB536 on the healthy gut microbiota composition at phyla and species level: A preliminary study. *World Journal of Gastroenterology, 23*(15), 2696–2704.

Tripathi, M. K., & Giri, S. K. (2014). Probiotic functional foods: Survival of probiotics during processing and storage. *Journal of Functional Foods, 9*, 225–241.

Ulsemer, P., Toutounian, K., Kressel, G., Goletz, C., Schmidt, J., Karsten, U., Hahn, A., & Goletz, S. (2016). Impact of oral consumption of heat-treated Bacteroides xylanisolvens DSM 23964 on the level of natural TFalpha-specific antibodies in human adults. *Beneficial Microbes, 7*(4), 485–500.

Urbanska, A. M., Bhathena, J., Martoni, C., & Prakash, S. (2009). Estimation of the potential antitumor activity of microencapsulated Lactobacillus acidophilus yogurt formulation in the attenuation of tumorigenesis in Apc(Min/+) mice. *Digestive Diseases and Sciences, 54*(2), 264–273.

Wada, J., Ando, T., Kiyohara, M., Ashida, H., Kitaoka, M., Yamaguchi, M., Kumagai, H., Katayama, T., & Yamamoto, K. (2008). Bifidobacterium bifidum lacto-N-biosidase, a critical enzyme for the degradation of human milk oligosaccharides with a type 1 structure. *Applied and Environmental Microbiology, 74*(13), 3996–4004.

Watson, D., Sleator, R. D., Hill, C., & Gahan, C. G. (2008). Enhancing bile tolerance improves survival and persistence of Bifidobacterium and Lactococcus in the murine gastrointestinal tract. *BMC Microbiology, 8*, 176.

Westfall, S., Lomis, N., Kahouli, I., Dia, S. Y., Singh, S. P., & Prakash, S. (2017). Microbiome, probiotics and neurodegenerative diseases: Deciphering the gut brain axis. *Cellular and Molecular Life Sciences, 74*(20), 3769–3787.

Yan, F., & Polk, D. B. (2002). Probiotic bacterium prevents cytokine-induced apoptosis in intestinal epithelial cells. *The Journal of Biological Chemistry, 277*(52), 50959–50965.

Yan, F., Cao, H., Cover, T. L., Whitehead, R., Washington, M. K., & Polk, D. B. (2007). Soluble proteins produced by probiotic bacteria regulate intestinal epithelial cell survival and growth. *Gastroenterology, 132*(2), 562–575.

Yan, F., Liu, L., Dempsey, P. J., Tsai, Y. H., Raines, E. W., Wilson, C. L., Cao, H., Cao, Z., Liu, L., & Polk, D. B. (2013). A Lactobacillus rhamnosus GG-derived soluble protein, p40, stimulates ligand release from intestinal epithelial cells to transactivate epidermal growth factor receptor. *The Journal of Biological Chemistry, 288*(42), 30742–30751.

Yoshida, E., Sakurama, H., Kiyohara, M., Nakajima, M., Kitaoka, M., Ashida, H., Hirose, J., Katayama, T., Yamamoto, K., & Kumagai, H. (2012). Bifidobacterium longum subsp. infantis uses two different beta-galactosidases for selectively degrading type-1 and type-2 human milk oligosaccharides. *Glycobiology, 22*(3), 361–368.

Zimmermann, P., & Curtis, N. (2018). The influence of probiotics on vaccine responses—A systematic review. *Vaccine, 36*(2), 207–213.

The Disappearing Microbiota: Diseases of the Western Civilization

Emiliano Salvucci

Abstract The human being is a superorganism composed of human cells and its associated microbiota. Humans did not emerge alone along evolution but in coexistence and intricate metabolic integration with microorganisms. The microorganisms that co-evolve and co-live with humans are called the microbiota. The human gut microbiota is a dynamic taxonomically complex community that participates in several processes related to normal function of the host-microbiota superorganism, maintaining the health status. Changes to the social aspects of the Western civilization and technological developments impacted on the evolutionary host-microbes' association. As a consequence of the disruption to this equilibrium, immunological, endocrine, metabolic and neurological alterations have arisen. Maternal diet, lifestyle, mode of delivery, administration of antibiotics to the mother during pregnancy, early nutrition (breastfeeding or formula) and treatment with antibiotics in newborns are crucial factors that affect microbiota structure. Microbiota and epigenome are involved in the reduced or increased risk to develop different microbiome-associated diseases in adult life.

Keywords Human-microbes superorganism · Western diseases · Western diet · Low diversity · Antibiotics · Epigenetics

Introduction

The purpose of this chapter is to analyze the emergence of diseases specific to the Western civilization and their potential relationship with the depletion of the human-associated microbiota. This evolutionary perspective takes into consideration the fact that integration of systems is a common pattern of life. The human being is a superorganism composed of human cells and organs, and its associated microbiota. Humans did not emerge alone along evolution but in coexistence and intricate metabolic integration with microorganisms. Changes to the social aspects of the Western civilization

E. Salvucci (✉)
Instituto de Ciencia y Tecnología de Alimentos Córdoba (ICYTAC), Consejo Nacional de Investigaciones Científicas y Técnicas (CONICET), Córdoba, Argentina

© Springer Nature Switzerland AG 2019
M. A. Azcarate-Peril et al. (eds.), *How Fermented Foods Feed a Healthy Gut Microbiota*, https://doi.org/10.1007/978-3-030-28737-5_14

and technological developments impacted the evolutionary host-microbes' association. The relatively recent alteration of the coexistence between host and microbiota has led to the emergence of diseases related to an over reactive immune system. A better knowledge of the human superorganism will yield approaches to restore or modulate the microbiota to treat or alleviate symptoms of these emergent diseases.

Origin of the Human-Microbes Superorganism

Maturana and Varela (Varela et al. 1974) created a new term to explain life: autopoiesis. Literally, it means self-production. It is the common trait of all living organisms. Cells and organisms produce a myriad of molecules and create a net of functionality to maintain their distinctiveness, cohesiveness, and relative autonomy. This net of components maintains cell self-production (Varela et al. 1974; Zelený 1981). According to this perspective, it is not correct to state that the organism neither adapts to the milieu nor the milieu selects changes in the organisms. The environment and the niche do not pre-exist to individuals. Organism and milieu change concomitantly. Maturana calls 'ontogenic drift' the process in which each organism is part of the niche where others organisms change and change it, in reciprocal building of their niches. The diversity of species is the result of reproduction and maintenance of autopoiesis and the systemic conservation of their organization in relation with other organisms or lineages (Maturana-Romesin and Mpodozis 2000).

There is an intrinsic force, inseparable of life, that causes system evolution: drastic changes result in collapse of the system or its re-organization (an evolutive step). This should not be confused with vitalism. The inevitable and inseparable propulsion of life to evolve is not extern to the organism but a characteristic that defines life. So there is nothing outside life that defines what is fit and what is not. Under this perspective, there is no selector (because it implies a teleological thinking) and natural selection is what an observer sees in the differential reproduction of two lineages of organism in different historical moments. The process behind the observed "selection" remains hidden. That is, the conservation of phenotypes in a natural drift (ontogenic and phylogenetic drift) is inseparable from autopoiesis (Maturana-Romesin and Mpodozis 2000). The properties of life imply evolution.

There are crucial mechanisms in evolution that involve the maintenance of this autopoiesis and they imply integration. Main steps of evolution are based on integration processes. A crucial mechanism is symbiogenesis. According to Margulis and Sagan (2003) symbiosis is simply the living together of organisms that are different from each other and symbiogenesis is the origin of evolutionary novelty via symbiosis (Margulis 2010; Margulis and Sagan 2003). As an example, the emergence of the eukaryotic cell involved the integration of pre-existent bacterial cells. Moreover, there is strong evidence that mitochondria and chloroplasts are evolutionary structures originated from different ancestors, with the mitochondria evolving from *proto-Rickettsiales*, *proto-Rhizobiales*, proto-alphaproteobacteria and current alphaproteobacterial species (Georgiades and Raoult 2012). The symbiogenetic theory was developed by Lynn Margulis and, independent and previously, by

Ivan Wallin, Paul Poitier, Konstantin Merezhkovski and Boris Kozo-Polianski (Kozo-Polyansky and Margulis 2010; Brucker and Bordenstein 2012; Wallin 1927; Symbiosis as a Source of Evolutionary Innovation 2017).

Consistently, the nuclear structure of eukaryotes is another example of an evolutionary integration step (Bell 2001; Forterre 2006). The role of viruses in the emergence of nucleus demonstrate the importance of those microooorganisms in evolution, based on their ability to integrate into the nucleic acids, modifying genomic behaviour and regulation. This integrative process is associated with the emergence of new structures (Feschotte and Gilbert 2012; Belyi et al. 2010; Ding and Lipshitz 1994; Gifford et al. 2008; Jamain et al. 2001; Medstrand and Mager 1998; Villarreal and DeFilippis 2000). For example, virus-related genes are involved in the process of placentation in mammals (Mallet et al. 2004). Likewise, retrotransposons participate in the regulation of genes related to the histocompatibility in humans, other mammals and invertebrates (Ding and Lipshitz 1994; McDonald et al. 1997; Kidwell and Lisch 2000).

A crucial evolutionary mechanism is horizontal gene transfer (HGT). The high prevalence of HGT events in Bacteria, Archaea and Eukaryotes has resulted in the mosaicism of genomes. It is difficult to identify a single common ancestor for the gene repertoire of any organism. HGT is not only extensive and directional but also ongoing, and acquired genes are related to metabolism and biogenesis with crucial value in evolution (Deschamps et al. 2014; Fuchsman et al. 2017; Craig et al. 2009; Hotopp 2011). HGT has a central importance in evolution of microorganism and metazoan hosts. As examples, endogenous viral elements from different families (Bornaviridae, Filoviridae, Bunyaviridae, Flaviviridae, Parvoviridae, Hapadnaviridae) are part of animal genomes including primates (Holmes 2011; Katzourakis and Gifford 2010; Gilbert et al. 2010). Human genome sequencing has demonstrated the high content of bacterium- and virus-related genes including retrotransposons (Mallet et al. 2004; Hotopp 2011; Gilbert et al. 2010; Gifford et al. 2008; Mi et al. 2000). This reveals that the genomic structure is the result of a dynamic equilibrium between genetic and cellular processes. The primary structure of the DNA is the result of the continuous feedback between an organism and the rest of living beings and the environment in a net of mutual building (Belyi et al. 2010; Gifford et al. 2008; Casjens 2003; Merhej and Raoult 2012). Evolution of species including metazoan is described like a rhizome by some researchers, pointing out different and mixed origins of genomic sequences in species (Georgiades and Raoult 2011, 2012). Ramulu et al. (2012) have emphasized that many proposals have emerged replacing the tree-like pattern, that consider a the existence of a common ancestor of organisms, with more complex models such as the "reticulate evolution", "synthesis of life", "web of life" or "network of life" (Ramulu et al. 2012).

The mechanisms of integration and symbiopoiesis are presently at work. It is possible to understand that co-evolution of humans and microorganisms results in an intricate metabolism and homeostasis, and this is not possible to maintain if any part of this symbiosis is altered or depleted. The microorganisms that co-evolve and co-live with humans are called the microbiota. In fact, the human body contains more bacterial cells than human cells. Recently, it has been stated that human:microorganisms cells ratio is 1,3:1, but that takes into account bacterial cells

(Sender et al. 2016), but not eukaryotic components of the microbiota (Parfrey et al. 2011). All organisms emerge in relation with their environment and the surrounding organisms that coevolve with them as a unity. Humans arose with the microorganisms that defined their metabolism, their systems, and their structure.

The human gut microbiota is a dynamic taxonomically complex community that participates in several processes related to normal function of the host-microbiota superorganism, maintaining the health status (Kau et al. 2011; Lederberg and McCray 2001; Salvucci 2016). These include vitamin production, digestion and utilization of carbohydrates and lipids, energy homeostasis, tryptophan metabolism regulation, integrity of intestinal barrier and angiogenesis (Arrieta et al. 2014; van der Meulen et al. 2016; Evans et al. 2013). As an example, large primates are incapable of vitamin C synthesis. As a consequence, humans have an evolved dependence on fruit and vegetables since our genes for vitamin synthesis were lost (Rook 2011).

Short-chain fatty acids (SCFAs), mainly acetic, propionic and butyric acids are products of bacterial fermentation in the intestinal tract. Cross-feeding interactions between bifidobacteria and butyrate-producing colon bacteria, such as *Faecalibacterium prausnitzii* (clostridial cluster IV) and *Anaerostipes, Eubacterium*, and *Roseburia* species (clostridial cluster XIVa) result in an enhancement of butyrate production. SCFAs are a source of ATP for intestinal epithelial cells (IECs) and modulate IECs and leukocytes development, survival and function through activation of G protein coupled receptors (FFAR2, FFAR3, GPR109a and Olfr78). SCFAs also modulate enzymatic activity and transcription factors including histone acetyltransferase and deacetylase, and the hypoxia-inducible factor (Corrêa-Oliveira et al. 2016; Kasubuchi et al. 2015).

The gut microbiota produces and regulates compounds with crucial local effects, but that also influence the function of distal organs and systems. In fact, metabolites generated by the gut microbiota have been shown to influence brain chemistry and behaviour independently from the autonomic nervous system, gastrointestinal neurotransmitters or inflammation through the microbiota-brain axis (Foster and McVey Neufeld 2013). This is a two-way communication pathway between the central nervous system and the intestine (Foster and McVey Neufeld 2013; Bercik et al. 2012). Butyrate has profound effects on mood and behaviour in mice (Schroeder et al. 2007). The microbial metabolism in the gut can also influence the level of neurotransmitters or hormones (Allen et al. 2013; O'Mahony et al. 2015). As examples, GABA and serotonin, related to anxiety, depression and mood states are influenced by the microbiota (Foster and McVey Neufeld 2013; Bercik et al. 2010). Inflammatory bowel diseases (IBD) are often accompanied by disorders of the nervous system such as irritability, anxiety, and depression (Hayley et al. 2016; Lee and Chua 2011). Moreover, dysbiosis can contribute to the development of psychiatric disorders in patients with intestinal symptoms (Bercik et al. 2010; Borre et al. 2014; Collins et al. 2012; Cryan and Dinan 2012; Dinan et al. 2014; Nemani et al. 2014).

The continuum human-microbiota represents a step of symbiosis along the genomic evolution. According to this perspective, the human genome would be a "hard drive" where is possible to observe integrated sequences of different origins as well as the integration processes. The human genome is consequently the result of the evolutionary history of the superorganism human-microbiota and their environmental interactions. This superorganism also includes mobile elements like plasmids, transposons, integrons, and bacteriophages that are called the 'mobilome' (Siefert 2009). These genetic elements constitute a genetic pool that fuel HGT and a sensitive response to environmental changes.

Main social and environmental changes along history caused variations in the human-microbes superorganism. These epidemiological transitions had and have today a profound impact on human health.

Epidemiological Transitions and Western Diseases

Two major epidemiological transitions resulted in dramatic changes to the human superorganism (McKeown 2009; Rook 2010). These shifts involved alterations in social behaviour and organization, which can be correlated with significant changes in the microbiota structure (Fig. 1). Two million years ago humans were nomadic hunter-gatherers (Palaeolithic period). With the establishment of populations, humans became sedentary. An agricultural revolution ensued with a profound shift in human lifestyle (Neolithic period, 10,000 years BC). In this period, new foods like cereals and dairy products appeared, and population density increased. It is reasonable to assume that dramatic changes occurred in the microbiota after the first epidemiological transition. However, this shift did not result in loss of microbial diversity because, until the modern era, more than 97% of the population still lived

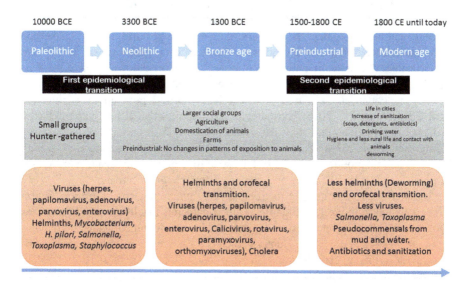

Fig. 1 Epidemiological transitions. The second epidemiological transition had great impact on human microbiota. Social and technological changes led to a depletion of microorganisms associated to human being.

in rural environments (Rook 2011). However, the new lifestyle increased faecal–oral transmission and prolonged contact with animals (Rook 2012). Viruses that were acquired previous to Neolithic have been maintained, including herpes, papilloma, adeno-, parvo-, some entero-, and perhaps hepatitis B virus (Rook 2012; Van Blerkom Linda 2003; Leal Éde and Zanotto 2000). It is possible that during this period characterized by a marked increase in population density, humans started suffering from "modern" diseases such as measles, influenza, and dengue (Rook 2012; Leal Éde and Zanotto 2000). Helminths, Mycobacteria, *Helicobacter pylori*, *Salmonella*, *Toxoplasma* and lactobacilli were acquired during this period (Rook 2011, 2012; Cremonini and Gasbarrini 2003).

The second remarkable shift in human lifestyle and dietary habits started with the Industrial Revolution less than 200 years ago, accompanied with a radical move, away from local and seasonal foods (McKeown 2009). People started to migrate into cities and the use of sanitization increased. This resulted in a diminished or delayed contact with microorganisms and viruses. With this second transition there were significant lifestyles changes. Sanitization was improved with the use of soap detergents, washed food, chlorinated water and, eventually, antibiotics. This transition was characterized by a significant decrease in orofecal transmission. Life in the city implied advances in housing with a wide use of construction materials like concrete and asphalt, and less soil material. Less contact with animals also contributed to de-worming. As a result, changes to the microbiota included decreased helminths, *H. pylori*, *Salmonella*, *Toxoplasma* and *Mycobacterium tuberculosis* (Fig. 1) (McKeown 2009; Rook 2009, 2012).

Changes in dietary habits were important factors in the alteration of the host–microbe symbiosis during epidemiological transitions since diet drives the generation of gut microbe-derived bioactive metabolites (Hunter 2008). According to recently published studies, the ancestral diet was based on high fiber intake and maintained a well-balanced microbiota. The microbiota was more structurally and functionally diverse with enhanced polysaccharide breakdown capacities. This highly diverse microbiota has been maintained in rural and remote populations from developing countries compared to urban industrialized populations (Crittenden et al. 2014; West et al. 2015; Turroni et al. 2016; Smits et al. 2017), with gut bacterial taxa or functions that may have disappeared due to cultural Westernization (Crittenden et al. 2014). Conversely, the Western diet is characterized by a high intake of animal products and sugars, the use of food preservatives, and a low intake of plant-based foods like fruits, vegetables, and whole grain cereals.

Researchers have compared microbiota of communities from Tanzania (Hadza tribe) with that from Western countries. Hunter-gatherers that live today as thousands years ago maintaining an ancestral diet, showed lower inflammation than individuals from the Western hemisphere. Hunter-gatherers have a more diverse enzymatic repertoire for utilizing carbohydrates in their microbiota than those from healthy American subjects (Schnorr et al. 2014). These studies made possible to identify organisms and functions that would have been abundant in the ancestral microbiota, but are diminished or absent from the modern city environment (Rook 2009, 2010). Specifically, the Hadza microbiota has higher functional capacity for utilization of

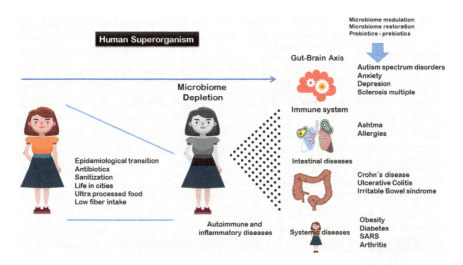

Fig. 2 Microbiota depletion. Human superorganism lost part of the microbiota along the last epidemiological transition. This depletion of microbiota is related to different emergent diseases with impact on gut-brain axis, immune system, gut and systemic disorders.

plant carbohydrates than the microbiota of Americans (Crittenden et al. 2014). The community consume tubers and baobab year-round. Conversely, the American microbiota with less consumption of fiber-rich diet have a greater mucin-utilization capacity than Hadzas (Turroni et al. 2016; Smits et al. 2017). These data are in agreement with other studies that indicate that the microbiota of people in urbanized regions is characteristic of a diet limited in plant-derived complex carbohydrates, with many bacterial species underrepresented or missing (Kau et al. 2011; Findley et al. 2016). Fiber-rich diets fuel the gut microbiota metabolism and maintain resident bacterial populations (Thorburn et al. 2014; De Filippo et al. 2010). The depletion of microbiota has been clearly intensified by social factors like industrialization and Western diets with consequences to equilibrium and homeostasis.

As a consequence of the disruption to the host-microbiota equilibrium, immunological, endocrine, metabolic and neurological alterations have arisen (Arrieta et al. 2014; Foster and McVey Neufeld 2013; Hayley et al. 2016; Rook 2009; Paun and Danska 2015; van der Meulen et al. 2016). Human diseases like inflammatory bowel diseases (IBD), irritable bowel syndrome (IBS), cancer, diabetes, asthma, allergies, obesity and metabolic syndrome appear to be related to the depleted microbiota (Fig. 2).

Obesity and metabolic syndrome have been recognized as epidemic diseases associated with Western diets and low microbial diversity. The altered microbiota in obese subjects has lower proportion of Bacteroidetes and more Firmicutes than healthy individuals (Remely et al. 2014; Ley et al. 2005). The obese microbiota can be inherited from parents with high fat diets and it has an increased capacity to harvest energy from the diet (Ley et al. 2005; Myles et al. 2013), producing significantly higher total body fat. Concomitantly, immunological alterations include infiltration of adipose tissue by macrophages, CD8+ T cells and CD4+ T cells, and high expression of inflammatory cytokines such as IL-6, IL-17, TNF- alpha and interferon-gamma (Kau et al. 2011). Additionally, obese individuals usually have high endotoxemia caused by LPS infiltration due to an increased intestinal permeability. A recent study showed that the diversity of the gut microbiota and the degree of methylation of the FFAR3 promoter region were significantly lower in obese and type 2 diabetic individuals compared to lean individuals. FFAR3 is a G protein couple receptor involved in the development and survival of IECs (Kasubuchi et al. 2015). Prebiotics and *Bifidobacterium* strains lowered the uptake of LPS from the gut lumen reinforcing tight junctions of epithelial cells (Al-Sheraji et al. 2015; Fukuda et al. 2011; Meyer and Stasse-Wolthuis 2009).

The genetic background necessary to develop these disorders (intrinsic factors) can be influenced by the metabolism of the gut microbiota (extrinsic factors) (Salvucci 2013, 2016; Proal et al. 2009). For example, multiple genetic risk variants associated with the type 1 (T1D) and type 2 (T2D) diabetes have been postulated. Most T1D-associated variants are in genes controlling immunity (e.g. HLA), whereas T2D-associated genes control transcription, adipo-cytokine signals and β cell cycle regulation. Naturally, the increased incidence of T1D and T2D is not due to only to the genetic background, but the result of gene interactions with dynamic environmental risk factors, particularly impactful in early childhood (Paun and Danska 2016). In the same way, an altered gut microbiota characterized by low diversity and resilience has been associated to T1D and T2D (Paun and Danska 2016).

Crohn's disease (CD) is a condition in which the lining of the gastrointestinal tract becomes inflamed, causing severe diarrhoea and abdominal pain (Reddy and Fried 2009). The incidence of CD has increased in Western countries. Both CD and ulcerative colitis (UC) are idiopathic pathologies of IBD. The manifestations of this disease include an aggressive cellular immune response, over expression of pro inflammatory mediators in different T lymphocyte subsets (Th1 and Th17, Th2), abnormal antigen presentation, and aberrant thymic education with a concomitant penetration of the intestinal barrier by luminal bacteria. Underlying inflammation triggers Crohn's disease and the luminal microbiota change. This process leads to nonstandard host-microbiota interactions that can aggravate the disease. In fact, CD patients present decreased bacterial diversity and richness (Imhann et al. 2016; Khanna and Raffals 2017). Moreover, reports agree that the phyla Bacteroidetes and Proteobacteria are increased, while the phylum Firmicutes is decreased in patients with UC. Also *Bifidobacterium*, *Roseburia*, *Faecalibacterium* and *Phascolarctobacterium* are in reduced abundance compared with healthy individuals (Imhann et al. 2016; Khanna and Raffals 2017) . Recently, it was reported that there are significant differences in the gut microbiota of healthy individuals who carried a high genetic risk for IBD, including a decrease in species of *Roseburia* known to be butyrate producers (Imhann et al. 2016).

The remnants of ancestral integration of the microbial DNA into the human genome can be detected today. *Helicobacter pylori* and other potentially pathogenic microorganisms, like *Toxoplasma*, which possibly only infects humans accidentally, can co-habit with the host with detrimental effects (Cremonini and Gasbarrini 2003; Marini et al. 2007). Likewise, herpes viruses like varicella-zoster remain dormant in

The Disappearing Microbiota: Diseases of the Western Civilization

the human nervous system after an infection in the cranial nerve ganglia and autonomic ganglia. Viruses can reactivate and cause neurologic conditions (Gilden et al. 2000). Schizophrenia has been associated with gene loci that include an ancestral cytomegalovirus (Børglum et al. 2014). All of the above listed organisms leave entire copies or fragments of their genomes, which remain lodged within us for the rest of our lives interacting with our nervous system (Kramer and Bressan 2015). Parasite co-habitation can drive evolutive changes of host organisms (Vannier-Santos and Lenzi 2011). With epidemiological transitions, there was an intensification in sanitization and de-worming of humans that might have led to immunological imbalances. Parasites could act as regulators of immune system helping to control excessive inflammatory responses (Helmby 2015). For example, the induction of suppressive regulatory T cells (Tregs) is a common mechanism to regulate inflammatory effects (Paun and Danska 2015). Interestingly, activation of Tregs appears to be a feature of both microbiota colonization (Gifford et al. 2008; Jamain et al. 2001) and helminth parasite infection (Medstrand and Mager 1998). *Bacteroides fragilis*, *Bifidobacterium infantis*, *Clostridium* spp., and *Lactobacillus* spp., as well nematode parasites, such as *Heligmosomoides polygyrus* and *Strongyloides ratti* can induce or suppress regulatory Tregs (Reynolds et al. 2015). Studies in animal models have demonstrated that intestinal helminth infections can inhibit the development of intestinal inflammation (Rook 2009; Reddy and Fried 2009; Reynolds et al. 2015). Likewise, pre-clinical assays have suggested a beneficial effect of helminth infections on inflammatory bowel conditions, allergies, asthma and multiple sclerosis (Reddy and Fried 2009; Correale and Farez 2011). Specifically, treatment with embryonated *Trichuris suis* eggs resulted in significant disease remission in patients with UC and CD (Reddy and Fried 2007).

Microbiota restoration by fecal microbiota transplantation (FMT) is another potential emerging therapeutic. This therapy has shown to be a successfully treatment in recurrent *Clostridium difficile* infection (CDI) (Gianotti and Moss 2017). However, CD like other autoimmune diseases, has a more complex pathogenesis and the success of FMT can be dependant of additional factors including the donor microbial profile, inflammatory burden, and the microbial diversity of the recipient (Gianotti and Moss 2017).

We can conclude from this section that two major epidemiological transitions (establishment of nomadic hunter-gatherers and industrial revolution), with the accompanying cultural and dietary shifts led to a significant reduction in microbiota diversity, with a concomitant loss of function that could be correlated to the increased occurrence of auto immune diseases. Restoration of the gut microbiota by FMT and helminth therapies are being considered as emergent treatments.

The Infant Microbiota and the Impact of Antibiotics

The natural drift of human-microbiota superorganism and the maintenance of its autopoiesis are observed in inheritance of the microbiota. This superorganism inherited epigenetic changes that are responsible of fine tuning along their development. This epigenetic landscape is involved in the adaptation to the environment including the interaction with the microbiota at different stages of development (Indrio et al. 2017).

The placental microbiome is composed of four dominant phyla: *Proteobacteria, Bacteroidetes, Actinobacteria* and *Firmicutes* (Zheng et al. 2017). Then, the mother's microbiota provides the newborn with a specific microbial inoculum at birth. Vaginal delivery supplies the baby with a bacterial composition resembling their mother's vaginal microbiota. *Lactobacillus, Prevotella*, and *Sneathia* are the predominant groups. Babies delivered by Cesarean section acquire bacteria that resembles those present on the skin, like *Staphylococcus, Corynebacterium*, and *Propionibacterium* (De Filippo et al. 2010; Koleva et al. 2015; Dominguez-Bello et al. 2010, 2016). Vaginal delivery may afford the neonate immediate access at birth to microbiota that allows maximal energy harvest during the incipient hours of life (Mueller et al. 2015). In fact, studies suggest a correlation between C-section delivery and the immune system due to the essential role of the maternal microbiota on the development of the perinatal immune system. Increased risk for asthma, allergies and obesity has been reported for C-section delivered infants (Ximenez and Torres 2017). Coincidentally, babies born via C-section have a different associated microbiome at least during the first 4 weeks of age. Full-term vaginally delivered infants present higher diversity compared to C-section babies (Ximenez and Torres 2017; von Mutius 2017). They also displayed an increased faecal abundance of *Firmicutes, Bacteroides* and *Bifidobacterium* and lower abundance of *Actinobacteria* compared to C-section babies (Ximenez and Torres 2017; Hill et al. 2017). However, the discrepancies between groups gradually disappear and are not observed by 6 months of age.

As elaborated in Part III, Chapter "Early Gut Microbiome, a Good Start in Nutrition and Growth May Have Lifelong Lasting Consequences", after birth, breastfeeding provides the baby with the microorganisms and immunological components to promote optimal growth. Moreover, breast milk also contains complex oligosaccharides (prebiotics) that promote the establishment of beneficial bacteria like *Bifidobacterium longum* (Indrio et al. 2017; Mueller et al. 2015). This is can be considered as the first example of how bacteria in food "feed" the gut microbiota.

The infant gut microbiota differ significantly from the adult microbiota. In fact its compositional structure and function are attained only after dynamic changes experienced during the first 3 years of life. *Staphylococcus, Streptococcus* and *Enterobacteria* are the first colonizers of the gut (Wampach et al. 2017; Voreades et al. 2014) . Then, these taxa are replaced by facultative anaerobic bacteria of the phyla Actinobacteria and Firmicutes (Turroni et al. 2012). At adulthood, 90% of microbes colonizing the gut are represented by only six phyla, Firmicutes, Bacteroides, Actinobacteria, Fusobacteria, Proteobacteria, and Verrucomicrobia (Arumugam et al. 2011). Other factors impact the infant gut microbiota including mom's diet and environmental factors. The degree of impact is more relevant early in life, when the gut microbiota has not yet been fully established.

Antibiotic treatment is a factor that can perturb the microbiota early in life and could be a missing link in autoimmunity disorders (Iizumi et al. 2017). Antibiotic use in children has become widespread. In fact, children in the US are prescribed a mean of three courses of antibiotic treatment before they are 2 years of age. This represents more than the double from European countries. A recent cross-national study has shown that South Korean children had the highest rate of antimicrobial prescriptions, with 3.41 prescribed courses per child-year during the first 2 years of life. Italy and Spain had a mean of 1 and 1.6, respectively, while Norway had only 0.5 courses per child-year (Youngster et al. 2017).

Antibiotic use, administered either orally or intravenously, reduces gut microbiota diversity (Iizumi et al. 2017) (Fig. 2). Repeated exposure to antibiotics during the first year of life caused a less stable microbial community, which lasted until the third year, and a decreased diversity (Maturana-Romesin and Mpodozis 2000). Also an epidemiological study showed that children that received antibiotics in the first 6 months of life had a significantly higher risk of being overweight at 7 years old (Trasande et al. 2013). Mouse studies have demonstrated the effects of antibiotics on microbiota. For example, penicillin G, V and vancomycin have been associated with increased weight, fat mass and insulin resistance (Iizumi et al. 2017). Likewise, azithromycin significantly increased weight gain risk. Other antibiotics including meropenem, cefotaxime and ticarcillin-clavulanate (Gibson et al. 2016) administered to preterm neonates affected severely intestinal species richness. Macrolides had a similar impact in children aged 2–7 years changing composition of the gut microbiome, decreased richness, and decreased abundance of *Bifidobacterium*, with increased levels of Proteobacteria (Enterobacteriaceae) and Bacteroidetes (Korpela et al. 2016). These changes persisted up to 2 years after macrolide treatment and were associated with increased risk of asthma and obesity (Korpela et al. 2016).

Administration of antibiotics to the mother during pregnancy may also affect the oral microbiota of the newborn. Maternal intrapartum antibiotic treatment is a key regulator of the initial neonatal oral microbiota. The oral microbiota of the infants was more similar to the oral microbiota than to the placenta or gut microbiota of the mother. Families belonging to *Proteobacteria* were abundant after antibiotics exposure of the mother while *Streptococcaceae*, *Gemellaceae* and *Lactobacillales* dominated in unexposed neonates (Gomez-Arango et al. 2017).

Finally, antibiotics have been shown to delay the maturation of microbiota, due to reduction of *Lachnospiraceae*, *Erysipelotrichaceae* and *Clostridiales*. These families are very sensitive to antibiotic exposure and this causes a reduction in production of butyrate and other SCFA. These reductions impact on infant immunity, signalling epithelial cell, colonic T regulatory cells and macrophages and maturation of the gut (Ximenez and Torres 2017). Therefore, *Lachnospiraceae* is useful as an indicator of microbiota maturation.

There is strong evidence that maternal exposure of antibiotics and administration of antibiotics in neonates alter and delay the maturation of microbiota. These changes have consequences at immunological level and it is associated to higher risk of diseases in adult life. Early antibiotic exposure may have other long term consequences related to higher risk to inflammatory and immune diseases. Also behaviour, anxiety, blood-brain-barrier integrity and brain cytokines expression could be related to antenatal and postnatal antibiotic exposure.

Epigenetics and the Gut Microbiota

Epigenetics refers to modifications of the genome that do not alter the DNA sequence but cause mitotically and meiotically heritable changes (Morgan and Whitelaw 2008). There is a wide variety of mechanisms that reduce, activate or inactivate genes and regulatory networks influencing early cellular differentiation, and creating new phenotypic traits during pregnancy and within the neonatal period (Indrio et al. 2017; Liu 2007). A number of antenatal and postnatal factors, including diet and composition of the microbiota, contribute to epigenetic changes that have an influence on lifelong health and disease by modifying inflammatory molecular pathways and the immune response (Indrio et al. 2017).

The main epigenetic mechanism is the methylation of cytosine residues in DNA, which results in remodelling of the chromatin structure and RNA-mediated regulation. These modifications may upregulate or downregulate gene expression according to the type of change and its position. A group of enzymes catalyze DNA methylation. It consists in methylation of cytosine residues followed by guanine or adenine and the consequent suppression of gene expression (Abdolmaleky et al. 2015).

Other epigenetic mechanisms are acetylation and phosphorylation. Different amino acids of histone tails can be methylated, acetylated or phosphorylated (Alam et al. 2017). Action of acetyltransferases results in acetylation of histone residues that increases accessibility of nucleosomal DNA to transcription factors, thus increasing the expression levels of corresponding genes (Alam et al. 2017).

Epigenetic modifications have key roles in the development of human organs, especially the central nervous system during the embryo-fetal, perinatal, and later stages of life (Alam et al. 2017; Jablonka and Raz 2009). These mechanisms are widely observed in nature acting in response to environmental factors and interact in a regulatory network involving more than 1000 microRNAs. Each microRNA can target hundreds of transcripts increasing cell adaptability in a tissue-specific manner. They demonstrate how environmental factors increase genomic flexibility, being maintained through generations (Table 1) (Alam et al. 2017).

Perturbations of the perinatal microbiota by specific practices (lack of skin contact, cesarean delivery, formula feeding, antibiotics) play a role in the susceptibility to late-onset diseases like obesity, diabetes, allergies, asthma, and other autoimmune disorders potentially by developing a particular genetic repertoire and modulating the immune development through epigenetic modifications (Salvucci 2016; Indrio et al. 2017; Jablonka and Raz 2009; Bossdorf et al. 2008). Individuals inherited from parents particular epigenetic changes (acetylation, methylation or phosphorylation of genes) that influence the adaptation of the newborn and are related to metabolism and immunological status (Myles et al. 2013).

Depletion of the gut microbiota and the related inflammatory, immune and neuroendocrine manifestations have shown to be linked by epigenetic changes (Table 1) (Myles et al. 2013; Indrio et al. 2017; Morgan and Whitelaw 2008; Liu 2007; Jablonka and Raz 2009). The fetal epigenetic program is influenced by diet and the

Table 1 Epigenetic effect persists in subsequent untreated generation of different organism

Organism	Effect and epigenetic mechanism	Reference
Arabidopsis thaliana	Short-wave radiation increases the somatic homologous recombination in a transgenic reporter gene	Molinier et al. (2006)
Rats	Exposure to glucocorticoids or a low-protein diet causes changes in the expression of liver enzymes, elevated blood pressure and endothelial dysfunction	Langley-Evans (2000) and Jensen Pena et al. (2012)
	Maternal choline supplementation improved development and functioning of the adult rat brain. DNA and histone methylation are mechanism implied	Davison et al. (2009)
	Maternal food restriction resulting in intrauterine growth restriction increased risk of obesity, insulin resistance and diabetes	Tosh et al. (2010) and Park et al. (2008)
Mice	Maternal supplementation with dietary methyl donors (folic acid, vitamin B12, choline, zinc, methionine, betaine) increased risk of allergic airway disease in offspring. decreased transcriptional activity by excessive methylation of Runx3 gene	Hollingsworth et al. (2008)
Yeast	Prion sup35 activates the expression of "silent" gene changing the fidelity in the translation process	Chernoff (2001) and Shorter and Lindquist (2006)
Drosophila melanogaster	Alteration of chromatin regulation. Decrease in heat shock protein HSP90 levels in response to environmental changes Stable heritable phenotypes up to 4 generations	Ruden and Lu (2008)
	Maternal undernutrition and increased risk of metabolic syndrome in adulthood. Methylation at the IGF2/H19 imprinting region	Hernández-Valero et al. (2013)
Humans (Human-microbiota superorganism)	*Lactobacilli* and *Bifidobacteria*. Inhibition of histone deacetylase. DNA methylation secondary to methyl donor production Modulation of local and systemic inflammation	Remely et al. (2014) and Indrio et al. (2017)
	Increased Firmicutes/Bacteroides ratio. DNA methylation in genes related to SCFA production and Toll-Like receptor	Indrio et al. (2017) and Sepulveda et al. (2010)
	Maternal vitamin D deficiency increase risk of preeclampsia development in humans and possible adverse pregnancy outcomes. Increased DNA methylation of CYP27B1, VDR and RXR genes	Anderson et al. (2015)
	Low maternal dietary intakes of long-chain polyunsaturated fatty acids. Vascular dysregulation, altered placentation and increased long-term risk of cardiovascular diseases. Aberrant DNA methylation patterns and alterations in the expression of angiogenic factor genes	Khot et al. (2015, 2017)
	Maternal high fat diet increase diabetes risk in their grandchildren. DNA methylation, histone modification and changes in microRNA	Myles et al. (2013) and Kaati et al. (2002)
	Maternal high-fat diet produce alteration in foetal chromatin structure and increase risk of non-alcoholic fatty liver disease. Histone deacetylase. SIRT1 involved. Regulation of hepatic Pon1	Suter et al. (2012), Strakovsky et al. (2014), and Chu et al. (2016)

Epigenetics mechanisms that connect risk factor and diseases in animal models and humans

microbiota along their development. Malnutrition or overnutrition during pregnancy cause negative effects on the offspring health at childhood and adulthood (Lee 2015; Alfaradhi et al. 2016; Roberts et al. 2015). For example, in a mouse study supplementation with folic acid, vitamin B12, methionine, zinc, betaine, and choline resulted in higher rates of allergic airway inflammation due to excessive methylation of the runt-related transcription factor 3 (RUNX3), a mediator of T-lymphocyte differentiation (Håberg et al. 2009) . Zinc status can exert a fundamental influence on the epigenome. Zinc deficiency during intrauterine life and childhood could contribute to the development of chronic inflammatory diseases by aberrant methylation (Tomat et al. 2011).

Microbiota has the ability to induce epigenetic changes in the human-microbiota superorganism (Salvucci 2013). Among the mechanisms through which intestinal bacteria can influence human health, epigenetic modifications are the most important. The epigenome influences the establishment of the microbiota but also microorganisms can introduce epigenetic changes in genes relevant to immunological, metabolic, and neurological development and functions. For instance, SCFAs that regulate gene expression by DNA methylation or histone modifications are crucial metabolites of microbiota (Indrio et al. 2017; McKenzie et al. 2017).

The host-microbial interactions that characterize human superorganism start before the early postnatal period. In fact, the microbiota found in the amniotic fluid starts to modulate the foetus epigenetically since its placental life (West et al. 2015; Zheng et al. 2017; Urushiyama et al. 2017). Moreover, diet, antibiotic exposure and other environmental factors influence the microbiota composition and their epigenetic changes. All these factors contribute the higher or lesser risk to develop allergies and inflammatory diseases (Indrio et al. 2017; Cremonini and Gasbarrini 2003; Schaub et al. 2009).

It is difficult to establish a causal relationship between the diversity or prevalence of certain species in microbiota and the epigenome. Still, there are some clues related to the ratios of Firmicutes/Bacteroides and distinct DNA methylation profiles. Ratios associated with metabolic disorders have also differences in methylation in genes related to obesity, metabolism, and inflammation (Kumar et al. 2014). In obesity, there are differences in methylation of the promoter of SCD5 that encodes an stearoyl-coenzyme A (CoA) desaturase, which has a key function in the catalysis of monounsaturated fatty acids from saturated fatty acids. The promoter was more methylated in the groups of individuals with higher Firmicutes/Bacteroides ratio (Kumar et al. 2014). In IBD there are epigenetic marks that define the visceral hypersensitivity and modulate stress-induced visceral pain. Altered microbiota profiles are concomitant (Indrio et al. 2017; Jeffery et al. 2012).

Epigenetic changes like aberrant DNA methylation, histone modifications and dysregulation of micro-RNAs are linked to the pathogenesis of mental disorders. Moreover, a number of psychiatric drugs modulates features of the epigenome, for instance, tubastatin can restore the reduction in tubulin acetylation observed in Rett syndrome (Abdolmaleky et al. 2015). Valproate, lithium, lamotrigine, haloperidol, clozapine, olanzapine and risperidone alter the expression of many miR-NAs (Abdolmaleky et al. 2015) . This implies that epigenetic modifications are a plausible alternative for treatment of mental disorders and, in consequence, modulation of gut microbiota can be a blank for therapies. The gut microbiota contributes to epigenetic fine-tuning confirming its role as an ontogenic missing link in mental illnesses. These changes are not only indirect effects mediated by metabolic by-products, it was observed that infection with some bacteria such as *Helicobacter pylori* is specifically linked to DNA methylation and may decrease expression of O6-methylguanine DNA methyltransferase (Sepulveda et al. 2010).

Maternal diet, lifestyle, mode of delivery, early nutrition and gut microbiota define an epigenome of newborn that potentially has a crucial role in the development of microbiota-related diseases. The mother (superorganism) transmits epigenetic changes to the foetus that interact with the microbiota and introducing changes at this level can be considered the missing link that could define the success of treatments like helminth therapy and FMT in the diseases related to the microbiota depletion, including mental disorders. The epigenetic heredity allows the child to be adapted to the environment that the mother has experienced. Many other antenatal and postnatal factors could distort that synchrony. The probiotic treatments and modulation of microbiota should take into account the ability of some bacteria to induce the epigenetic changes to re-establish the homeostasis. The role for microbiota-induced epigenetic modifications and their effects is an emerging research field that is in the initial stages and its development will contribute to a better understanding of the interrelationship host-microbiota superorganism.

Conclusions

Scientific and anecdotal evidence seem to suggest that overall loss of microbial diversity and the loss of specific bacterial groups associated with two historical epidemiological transitions could be potentially correlated with an over-reactive immune system and consequent increased occurrence of allergies and other autoimmune disorders (Salvucci 2013; Proal et al. 2009; Tlaskalová-Hogenová et al. 2004). The common background of immune, inflammatory or systemic imbalance point to treatments aiming to restore the gut microbiota through probiotics, prebiotics and diet are plausible treatments for these emergent diseases (van der Meulen et al. 2016).

References

Abdolmaleky, H. M., Zhou, J.-R., & Thiagalingam, S. (2015). An update on the epigenetics of psychotic diseases and autism. *Epigenomics, 7*(3), 427–449. Retrieved from https://www.ncbi.nlm.nih.gov/pubmed/26077430.

Alam, R., Abdolmaleky, H. M., & Zhou, J.-R. (2017). Microbiome, inflammation, epigenetic alterations, and mental diseases. *American Journal of Medical Genetics. Part B, Neuropsychiatric Genetics, 174*(6), 651–660.

Alfaradhi, M. Z., Kusinski, L. C., Fernandez-Twinn, D. S., Pantaleão, L. C., Carr, S. K., Ferland-McCollough, D., et al. (2016). Maternal obesity in pregnancy developmentally programs adipose tissue inflammation in young, lean male mice offspring. *Endocrinology, 157*(11), 4246–4256.

Allen, S. J., Watson, J. J., Shoemark, D. K., Barua, N. U., & Patel, N. K. (2013). GDNF, NGF and BDNF as therapeutic options for neurodegeneration. *Pharmacology & Therapeutics, 138*(2), 155–175.

Al-Sheraji, S. H., Amin, I., Azlan, A., Manap, M. Y., & Hassan, F. A. (2015). Effects of Bifidobacterium longum BB536 on lipid profile and histopathological changes in hypercholesterolaemic rats. *Beneficial Microbes, 6*(5), 661–668.

Anderson, C. M., Ralph, J. L., Johnson, L., Scheett, A., Wright, M. L., Taylor, J. Y., et al. (2015). First trimester vitamin D status and placental epigenomics in preeclampsia among Northern Plains primiparas. *Life Sciences, 129*, 10–15.

Arrieta, M.-C., Stiemsma, L. T., Amenyogbe, N., Brown, E. M., & Finlay, B. (2014). The intestinal microbiome in early life: Health and disease. *Frontiers in Immunology, 5*, 427. Retrieved from http://www.ncbi.nlm.nih.gov/pmc/articles/PMC4155789/.

Arumugam, M., Raes, J., Pelletier, E., Le Paslier, D., Yamada, T., Mende, D. R., et al. (2011). Enterotypes of the human gut microbiome. *Nature, 473*(7346), 174–180.

Bell, P. J. L. (2001). Viral eukaryogenesis: Was the ancestor of the nucleus a complex DNA virus? *Journal of Molecular Evolution, 53*(3), 251–256.

Belyi, V. A., Levine, A. J., & Skalka, A. M. (2010). Sequences from ancestral single-stranded DNA viruses in vertebrate genomes: The parvoviridae and circoviridae are more than 40 to 50 million years old. *Journal of Virology, 84*(23), 12458–12462.

Bercik, P., Verdu, E. F., Foster, J. A., Macri, J., Potter, M., Huang, X., et al. (2010). Chronic gastrointestinal inflammation induces anxiety-like behavior and alters central nervous system biochemistry in mice. *Gastroenterology, 139*(6), 2102–2112.e1.

Bercik, P., Collins, S. M., & Verdu, E. F. (2012). Microbes and the gut-brain axis. *Neurogastroenterology & Motility, 24*(5), 405–413.

Børglum, A. D., Demontis, D., Grove, J., Pallesen, J., Hollegaard, M. V., Pedersen, C. B., et al. (2014). Genome-wide study of association and interaction with maternal cytomegalovirus infection suggests new schizophrenia loci. *Molecular Psychiatry, 19*(3), 325–333.

Borre, Y. E., O'Keeffe, G. W., Clarke, G., Stanton, C., Dinan, T. G., & Cryan, J. F. (2014). Microbiota and neurodevelopmental windows: Implications for brain disorders. *Trends in Molecular Medicine, 20*(9), 509–518.

Bossdorf, O., Richards, C. L., & Pigliucci, M. (2008). Epigenetics for ecologists. *Ecology Letters, 11*(2), 106–115.

Brucker, R. M., & Bordenstein, S. R. (2012). Speciation by symbiosis. *Trends in Ecology & Evolution, 27*(8), 443–451.

Casjens, S. (2003). Prophages and bacterial genomics: What have we learned so far? *Molecular Microbiology, 49*(2), 277–300.

Chernoff, Y. O. (2001). Mutation processes at the protein level: Is Lamarck back? *Mutation Research, 488*(1), 39–64.

Chu, D. M., Antony, K. M., Ma, J., Prince, A. L., Showalter, L., Moller, M., et al. (2016). The early infant gut microbiome varies in association with a maternal high-fat diet. *Genome Medicine, 8*(1), 77.

Collins, S. M., Surette, M., & Bercik, P. (2012). The interplay between the intestinal microbiota and the brain. *Nature Reviews. Microbiology, 10*(11), 735–742.

Correale, J., & Farez, M. F. (2011). The impact of parasite infections on the course of multiple sclerosis. *Journal of Neuroimmunology, 233*(1–2), 6–11.

Corrêa-Oliveira, R., Fachi, J. L., Vieira, A., Sato, F. T., & Vinolo, M. A. R. (2016). Regulation of immune cell function by short-chain fatty acids. *Clinical & Translational Immunology, 5*(4), e73.

Craig, J. P., Bekal, S., Niblack, T., Domier, L., & Lambert, K. N. (2009). Evidence for horizontally transferred genes involved in the biosynthesis of vitamin B1, B5, and B7 in Heterodera glycines. *Journal of Nematology, 41*(4), 281–290.

Cremonini, F., & Gasbarrini, A. (2003). Atopy, Helicobacter pylori and the hygiene hypothesis. *European Journal of Gastroenterology & Hepatology, 15*(6), 635–636.

Crittenden, A. N., Henry, A. G., Mabulla, A., Consolandi, C., Peano, C., Luiselli, D., et al. (2014). Gut microbiome of the Hadza hunter-gatherers. *Nature Communications, 5*, 3654.

Cryan, J. F., & Dinan, T. G. (2012). Mind-altering microorganisms: The impact of the gut microbiota on brain and behaviour. *Nature Reviews. Neuroscience, 13*(10), 701–712.

Davison, J. M., Mellott, T. J., Kovacheva, V. P., & Blusztajn, J. K. (2009). Gestational choline supply regulates methylation of histone H3, expression of histone methyltransferases G9a (Kmt1c) and Suv39h1 (Kmt1a), and DNA methylation of their genes in rat fetal liver and brain. *The Journal of Biological Chemistry, 284*(4), 1982–1989.

De Filippo, C., Cavalieri, D., Di Paola, M., Ramazzotti, M., Poullet, J. B., Massart, S., et al. (2010). Impact of diet in shaping gut microbiota revealed by a comparative study in children from Europe and rural Africa. *Proceedings of the National Academy of Sciences of the United States of America, 107*(33), 14691–14696.

Deschamps, P., Zivanovic, Y., Moreira, D., Rodriguez-Valera, F., & López-García, P. (2014). Pangenome evidence for extensive interdomain horizontal transfer affecting lineage core and shell genes in uncultured planktonic thaumarchaeota and euryarchaeota. *Genome Biology and Evolution, 6*(7), 1549–1563.

Dinan, T. G., Borre, Y. E., & Cryan, J. F. (2014). Genomics of schizophrenia: Time to consider the gut microbiome? *Molecular Psychiatry, 19*(12), 1252–1257.

Ding, D., & Lipshitz, H. D. (1994). Spatially regulated expression of retrovirus-like transposons during Drosophila melanogaster embryogenesis. *Genetical Research, 64*(03), 167–181.

Dominguez-Bello, M. G., Costello, E. K., Contreras, M., Magris, M., Hidalgo, G., Fierer, N., et al. (2010). Delivery mode shapes the acquisition and structure of the initial microbiota across multiple body habitats in newborns. *Proceedings of the National Academy of Sciences of the United States of America, 107*(26), 11971–11975.

Dominguez-Bello, M. G., De Jesus-Laboy, K. M., Shen, N., Cox, L. M., Amir, A., Gonzalez, A., et al. (2016). Partial restoration of the microbiota of cesarean-born infants via vaginal microbial transfer. *Nature Medicine, 22*(3), 250–253.

Evans, J. M., Morris, L. S., & Marchesi, J. (2013). The gut microbiome: The role of a virtual organ in the endocrinology of the host. *The Journal of Endocrinology, 218*(8), R37–R47.

Feschotte, C., & Gilbert, C. (2012). Endogenous viruses: Insights into viral evolution and impact on host biology. *Nature Reviews. Genetics, 13*(4), 283–296.

Findley, K., Williams, D. R., Grice, E. A., & Bonham, V. L. (2016). Health disparities and the microbiome. *Trends in Microbiology, 24*(11), 847–850.

Forterre, P. (2006). The origin of viruses and their possible roles in major evolutionary transitions. *Virus Research, 117*(1), 5–16.

Foster, J. A., & McVey Neufeld, K.-A. (2013). Gut-brain axis: How the microbiome influences anxiety and depression. *Trends in Neurosciences, 36*(5), 305–312.

Fuchsman, C. A., Collins, R. E., Rocap, G., & Brazelton, W. J. (2017). Effect of the environment on horizontal gene transfer between bacteria and archaea. *PeerJ, 5*, e3865. Retrieved from https://www.ncbi.nlm.nih.gov/pmc/articles/PMC5624296/.

Fukuda, S., Toh, H., Hase, K., Oshima, K., Nakanishi, Y., Yoshimura, K., et al. (2011). Bifidobacteria can protect from enteropathogenic infection through production of acetate. *Nature, 469*(7331), 543–547.

Georgiades, K., & Raoult, D. (2011). The rhizome of Reclinomonas americana, Homo sapiens, Pediculus humanus and Saccharomyces cerevisiae mitochondria. *Biology Direct, 6*, 55.

Georgiades, K., & Raoult, D. (2012). How microbiology helps define the rhizome of life. *Frontiers in Cellular and Infection Microbiology, 2*, 60.

Gianotti, R. J., & Moss, A. C. (2017). Fecal microbiota transplantation: From Clostridium difficile to inflammatory bowel disease. *Gastroenterología y Hepatología, 13*(4), 209–213.

Gibson, M. K., Wang, B., Ahmadi, S., Burnham, C.-A. D., Tarr, P. I., Warner, B. B., et al. (2016). Developmental dynamics of the preterm infant gut microbiota and antibiotic resistome. *Nature Microbiology, 1*, 16024.

Gifford, R. J., Katzourakis, A., Tristem, M., Pybus, O. G., Winters, M., & Shafer, R. W. (2008). A transitional endogenous lentivirus from the genome of a basal primate and implications for lentivirus evolution. *Proceedings of the National Academy of Sciences of the United States of America, 105*(51), 20362–20367.

Gilbert, S. F., McDonald, E., Boyle, N., Buttino, N., Gyi, L., Mai, M., et al. (2010). Symbiosis as a source of selectable epigenetic variation: Taking the heat for the big guy. *Philosophical Transactions of the Royal Society of London. Series B, Biological Sciences, 365*(1540), 671–678.

Gilden, D. H., Kleinschmidt-DeMasters, B. K., LaGuardia, J. J., Mahalingam, R., & Cohrs, R. J. (2000). Neurologic complications of the reactivation of Varicella–Zoster Virus. *The New England Journal of Medicine, 342*(9), 635–645.

Gomez-Arango, L. F., Barrett, H. L., McIntyre, H. D., Callaway, L. K., Morrison, M., & Dekker, N. M. (2017). Antibiotic treatment at delivery shapes the initial oral microbiome in neonates. *Scientific Reports, 7*, 43481.

Håberg, S. E., London, S. J., Stigum, H., Nafstad, P., & Nystad, W. (2009). Folic acid supplements in pregnancy and early childhood respiratory health. *Archives of Disease in Childhood, 94*(3), 180–184.

Hayley, S., Audet, M.-C., & Anisman, H. (2016). Inflammation and the microbiome: Implications for depressive disorders. *Current Opinion in Pharmacology, 29*, 42–46.

Helmby, H. (2015). Human helminth therapy to treat inflammatory disorders—Where do we stand? *BMC Immunology, 16*, 12. Retrieved from https://www.ncbi.nlm.nih.gov/pmc/articles/PMC4374592/.

Hernández-Valero, M. A., Rother, J., Gorlov, I., Frazier, M., & Gorlova, O. Y. (2013). Interplay between polymorphisms and methylation in the H19/IGF2 gene region may contribute to obesity in Mexican-American children. *Journal of Developmental Origins of Health and Disease, 4*(6), 499–506.

Hill, C. J., Lynch, D. B., Murphy, K., Ulaszewska, M., Jeffery, I. B., O'Shea, C. A., et al. (2017). Evolution of gut microbiota composition from birth to 24 weeks in the INFANTMET Cohort. *Microbiome, 5*(1), 4.

Hollingsworth, J. W., Maruoka, S., Boon, K., Garantziotis, S., Li, Z., Tomfohr, J., et al. (2008). In utero supplementation with methyl donors enhances allergic airway disease in mice. *The Journal of Clinical Investigation, 118*(10), 3462–3469.

Holmes, E. C. (2011). The evolution of endogenous viral elements. *Cell Host & Microbe, 10*(4), 368–377.

Hotopp, J. C. (2011). Horizontal gene transfer between bacteria and animals. *Trends in Genetics, 27*(4), 157–163.

Hunter, P. (2008). We are what we eat. The link between diet, evolution and non-genetic inheritance. *EMBO Reports, 9*(5), 413–415.

Iizumi, T., Battaglia, T., Ruiz, V., & Perez Perez, G. I. (2017). Gut microbiome and antibiotics. *Archives of Medical Research, 48*(8), 727–734. Retrieved from https://www.sciencedirect.com/science/article/pii/S0188440917302333.

Imhann, F., Vila, A. V., Bonder, M. J., Fu, J., Gevers, D., Visschedijk, M. C., et al. (2016). Interplay of host genetics and gut microbiota underlying the onset and clinical presentation of inflammatory bowel disease. *Gut, 67*(1), 108–119.

Indrio, F., Martini, S., Francavilla, R., Corvaglia, L., Cristofori, F., Mastrolia, S. A., et al. (2017). Epigenetic matters: The link between early nutrition, microbiome, and long-term health development. *Frontiers in Pediatrics, 5*, 178. Retrieved from http://journal.frontiersin.org/article/10.3389/fped.2017.00178/full?&utm_source=Email_to_rerev_&utm_medium=Email&utm_content=T1_11.5e4_reviewer&utm_campaign=Email_publication&journalName=Frontiers_in_Pediatrics&id=286554.

Jablonka, E., & Raz, G. (2009). Transgenerational epigenetic inheritance: Prevalence, mechanisms, and implications for the study of heredity and evolution. *The Quarterly Review of Biology, 84*(2), 131–176.

Jamain, S., Girondot, M., Leroy, P., Clergue, M., Quach, H., Fellous, M., et al. (2001). Transduction of the human gene FAM8A1 by endogenous retrovirus during primate evolution. *Genomics, 78*(1–2), 38–45.

Jeffery, I. B., O'Toole, P. W., Öhman, L., Claesson, M. J., Deane, J., Quigley, E. M. M., et al. (2012). An irritable bowel syndrome subtype defined by species-specific alterations in faecal microbiota. *Gut, 61*(7), 997–1006.

Jensen Pena, C., Monk, C., & Champagne, F. A. (2012). Epigenetic effects of prenatal stress on 11beta-hydroxysteroid dehydrogenase-2 in the placenta and fetal brain. *PLoS One, 7*(6), e39791.

Kaati, G., Bygren, L. O., & Edvinsson, S. (2002). Cardiovascular and diabetes mortality determined by nutrition during parents' and grandparents' slow growth period. *European Journal of Human Genetics, 10*(11), 682–688.

Kasubuchi, M., Hasegawa, S., Hiramatsu, T., Ichimura, A., & Kimura, I. (2015). Dietary gut microbial metabolites, short-chain fatty acids, and host metabolic regulation. *Nutrients, 7*(4), 2839–2849.

Katzourakis, A., & Gifford, R. J. (2010). Endogenous viral elements in animal genomes. *PLoS Genetics, 6*(11), e1001191.

Kau, A. L., Ahern, P. P., Griffin, N. W., Goodman, A. L., & Gordon, J. I. (2011). Human nutrition, the gut microbiome and the immune system. *Nature, 474*(7351), 327–336.

Khanna, S., & Raffals, L. E. (2017). The microbiome in Crohn's disease: Role in pathogenesis and role of microbiome replacement therapies. *Gastroenterology Clinics of North America, 46*(3), 481–492.

Khot, V., Chavan-Gautam, P., & Joshi, S. (2015). Proposing interactions between maternal phospholipids and the one carbon cycle: A novel mechanism influencing the risk for cardiovascular diseases in the offspring in later life. *Life Sciences, 129*, 16–21.

Khot, V. V., Chavan-Gautam, P., Mehendale, S., & Joshi, S. R. (2017). Variable methylation potential in preterm placenta: Implication for epigenetic programming of the offspring. *Reproductive sciences (Thousand Oaks, Calif.), 24*(6), 891–901.

Kidwell, M. G., & Lisch, D. R. (2000). Transposable elements and host genome evolution. *Trends in Ecology & Evolution, 15*(3), 95–99.

Koleva, P. T., Bridgman, S. L., & Kozyrskyj, A. L. (2015). The infant gut microbiome: Evidence for obesity risk and dietary intervention. *Nutrients, 7*(4), 2237–2260.

Korpela, K., Salonen, A., Virta, L. J., Kekkonen, R. A., Forslund, K., Bork, P., et al. (2016). Intestinal microbiome is related to lifetime antibiotic use in Finnish pre-school children. *Nature Communications, 7*, 10410.

Kozo-Polyansky, B. M., & Margulis, L. (2010). *Symbiogenesis: A new principle of evolution* (198p). Cambridge, MA: Paperbackshop UK Import.

Kramer, P., & Bressan, P. (2015). Humans as superorganisms: How microbes, viruses, imprinted genes, and other selfish entities shape our behavior. *Perspectives on Psychological Science, 10*(4), 464–481.

Kumar, H., Lund, R., Laiho, A., Lundelin, K., Ley, R. E., Isolauri, E., et al. (2014). Gut microbiota as an epigenetic regulator: Pilot study based on whole-genome methylation analysis. *mBio, 5*(6), e02113–e02114.

Langley-Evans, S. C. (2000). Critical differences between two low protein diet protocols in the programming of hypertension in the rat. *International Journal of Food Sciences and Nutrition, 51*(1), 11–17.

Leal Éde, S., & Zanotto, P. M. (2000). Viral diseases and human evolution. *Memórias do Instituto Oswaldo Cruz, 95*, 193–200.

Lederberg, J., & McCray, A. T. (2001). 'Ome sweet 'omics—A genealogical treasury of words. *The Scientist., 15*(7), 8.

Lee, H.-S. (2015). Impact of Maternal Diet on the Epigenome during In Utero Life and the Developmental Programming of Diseases in Childhood and Adulthood. *Nutrients, 7*(11), 9492–9507.

Lee, Y. Y., & Chua, A. S. B. (2011). Influence of Gut Microbes on the Brain-Gut Axis (Gut 2011;60:307-317). *Journal of Neurogastroenterology and Motility, 17*(4), 427–429.

Ley, R. E., Backhed, F., Turnbaugh, P., Lozupone, C. A., Knight, R. D., & Gordon, J. I. (2005). Obesity alters gut microbial ecology. *Proceedings of the National Academy of Sciences of the United States of America, 102*(31), 11070–11075.

Liu, Y. (2007). Like father like son. A fresh review of the inheritance of acquired characteristics. *EMBO Reports, 8*(9), 798–803.

Mallet, F., Bouton, O., Prudhomme, S., Cheynet, V., Oriol, G., Bonnaud, B., et al. (2004). The endogenous retroviral locus ERVWE1 is a bona fide gene involved in hominoid placental physiology. *Proceedings of the National Academy of Sciences of the United States of America, 101*(6), 1731–1736.

Margulis, L. (2010). Symbiogenesis. A new principle of evolution rediscovery of Boris Mikhaylovich Kozo-Polyansky (1890–1957). *Paleontological Journal, 44*(12), 1525–1539.

Margulis, L., & Sagan, D. (2003). *Acquiring genomes: A theory of the origin of species* (1st ed., 256p). Princeton, NJ: Basic Books.

Marini, E., Maldonado-Contreras, A. L., Cabras, S., Hidalgo, G., Buffa, R., Marin, A., et al. (2007). Helicobacter pylori and intestinal parasites are not detrimental to the nutritional status of Amerindians. *The American Journal of Tropical Medicine and Hygiene, 76*(3), 534–540.

Maturana-Romesin, H., & Mpodozis, J. (2000). The origin of species by means of natural drift. *Revista Chilena de Historia Natural, 73*(2), 261–310. Retrieved from https://www.biodiversitylibrary.org/part/116245.

McDonald, J. F., Matyunina, L. V., Wilson, S., Jordan, I. K., Bowen, N. J., & Miller, W. J. (1997). LTR retrotransposons and the evolution of eukaryotic enhancers. *Genetica, 100*(1–3), 3–13.

McKenzie, C., Tan, J., Macia, L., & Mackay, C. R. (2017). The nutrition-gut microbiome-physiology axis and allergic diseases. *Immunological Reviews, 278*(1), 277–295.

McKeown, R. E. (2009). The epidemiologic transition: Changing patterns of mortality and population dynamics. *American Journal of Lifestyle Medicine, 3*(1 Suppl), 19S–26S.

Medstrand, P., & Mager, D. L. (1998). Human-specific integrations of the HERV-K endogenous retrovirus family. *Journal of Virology, 72*(12), 9782–9787.

Merhej, V., & Raoult, D. (2012). Rhizome of life, catastrophes, sequence exchanges, gene creations, and giant viruses: How microbial genomics challenges Darwin. *Frontiers in Cellular and Infection Microbiology, 2*, 113. Retrieved from http://www.ncbi.nlm.nih.gov/pmc/articles/PMC3428605/.

Meyer, D., & Stasse-Wolthuis, M. (2009). The bifidogenic effect of inulin and oligofructose and its consequences for gut health. *European Journal of Clinical Nutrition, 63*(11), 1277–1289.

Mi, S., Lee, X., Li, X., Veldman, G. M., Finnerty, H., Racie, L., et al. (2000). Syncytin is a captive retroviral envelope protein involved in human placental morphogenesis. *Nature, 403*(6771), 785–789.

Molinier, J., Ries, G., Zipfel, C., & Hohn, B. (2006). Transgeneration memory of stress in plants. *Nature, 442*(7106), 1046–1049.

Morgan, D. K., & Whitelaw, E. (2008). The case for transgenerational epigenetic inheritance in humans. *Mammalian Genome, 19*(6), 394–397.

Mueller, N. T., Bakacs, E., Combellick, J., Grigoryan, Z., & Dominguez-Bello, M. G. (2015). The infant microbiome development: Mom matters. *Trends in Molecular Medicine, 21*(2), 109–117.

Myles, I. A., Fontecilla, N. M., Janelsins, B. M., Vithayathil, P. J., Segre, J. A., & Datta, S. K. (2013). Parental dietary fat intake alters offspring microbiome and immunity. *Journal of Immunology (Baltimore, MD: 1950), 191*(6), 3200–3209.

Nemani, K., Hosseini Ghomi, R., McCormick, B., & Fan, X. (2014). Schizophrenia and the gut-brain axis. *Progress in Neuro-Psychopharmacology & Biological Psychiatry, 56C*, 155–160.

O'Mahony, S. M., Clarke, G., Borre, Y. E., Dinan, T. G., & Cryan, J. F. (2015). Serotonin, tryptophan metabolism and the brain-gut-microbiome axis. *Behavioural Brain Research, 277*, 32–48.

Parfrey, L. W., Walters, W. A., & Knight, R. (2011). Microbial eukaryotes in the human microbiome: ecology, evolution, and future directions. *Frontiers in Microbiology, 2*, 153. Retrieved from https://www.ncbi.nlm.nih.gov/pmc/articles/PMC3135866/.

Park, J. H., Stoffers, D. A., Nicholls, R. D., & Simmons, R. A. (2008). Development of type 2 diabetes following intrauterine growth retardation in rats is associated with progressive epigenetic silencing of Pdx1. *The Journal of Clinical Investigation, 118*(6), 2316–2324.

Paun, A., & Danska, J. S. (2015). Immuno-ecology: How the microbiome regulates tolerance and autoimmunity. *Current Opinion in Immunology, 37*, 34–39.

Paun, A., & Danska, J. S. (2016). Modulation of type 1 and type 2 diabetes risk by the intestinal microbiome. *Pediatric Diabetes, 17*(7), 469–477.

Proal, A. D., Albert, P. J., & Marshall, T. (2009). Autoimmune disease in the era of the metagenome. *Autoimmunity Reviews, 8*(8), 677–681.

Ramulu, H. G., Raoult, D., & Pontarotti, P. (2012). The rhizome of life: What about metazoa? *Frontiers in Cellular and Infection Microbiology, 2*, 50.

Reddy, A., & Fried, B. (2007). The use of *Trichuris suis* and other helminth therapies to treat Crohn's disease. *Parasitology Research, 100*(5), 921–927.

Reddy, A., & Fried, B. (2009). An update on the use of helminths to treat Crohn's and other autoimmunune diseases. *Parasitology Research, 104*(2), 217–221.

Remely, M., Aumueller, E., Jahn, D., Hippe, B., Brath, H., & Haslberger, A. G. (2014). Microbiota and epigenetic regulation of inflammatory mediators in type 2 diabetes and obesity. *Beneficial Microbes, 5*(1), 33–43.

Reynolds, L. A., Finlay, B. B., & Maizels, R. M. (2015). Cohabitation in the intestine: Interactions among helminth parasites, bacterial microbiota, and host immunity. *Journal of Immunology (Baltimore, MD: 1950), 195*(9), 4059–4066.

Roberts, V. H. J., Frias, A. E., & Grove, K. L. (2015). Impact of maternal obesity on fetal programming of cardiovascular disease. *Physiology, 30*(3), 224–231.

Rook, G. A. (2009). Review series on helminths, immune modulation and the hygiene hypothesis: The broader implications of the hygiene hypothesis. *Immunology, 126*(1), 3–11.

Rook, G. A. (2010). 99th dahlem conference on infection, inflammation and chronic inflammatory disorders: Darwinian medicine and the 'hygiene' or 'old friends' hypothesis. *Clinical and Experimental Immunology, 160*(1), 70–79.

Rook, G. A. W. (2011). Hygiene and other early childhood influences on the subsequent function of the immune system. *Digestive Diseases (Basel, Switzerland), 29*(2), 144–153.

Rook, G. A. W. (2012). Hygiene hypothesis and autoimmune diseases. *Clinical Reviews in Allergy and Immunology, 42*(1), 5–15.

Ruden, D. M., & Lu, X. (2008). Hsp90 affecting chromatin remodeling might explain transgenerational epigenetic inheritance in Drosophila. *Current Genomics, 9*(7), 500–508.

Salvucci, E. (2013). Crohn's disease within the hologenome paradigm. In *Crohn's disease: Classification, diagnosis and treatment options* (pp. 19–32). New York: Nova Publishers.

Salvucci, E. (2016). Microbiome, holobiont and the net of life. *Critical Reviews in Microbiology, 42*(3), 485–494.

Schaub, B., Liu, J., Höppler, S., Schleich, I., Huehn, J., Olek, S., et al. (2009). Maternal farm exposure modulates neonatal immune mechanisms through regulatory T cells. *Journal of Allergy and Clinical Immunology, 123*(4), 774–782.e5.

Schnorr, S. L., Candela, M., Rampelli, S., Centanni, M., Consolandi, C., Basaglia, G., et al. (2014). Gut microbiome of the Hadza hunter-gatherers. *Nature Communications, 5*, 3654. Retrieved from http://www.ncbi.nlm.nih.gov/pmc/articles/PMC3996546/.

Schroeder, F. A., Lin, C. L., Crusio, W. E., & Akbarian, S. (2007). Antidepressant-like effects of the histone deacetylase inhibitor, sodium butyrate, in the mouse. *Biological Psychiatry, 62*(1), 55–64.

Sender, R., Fuchs, S., & Milo, R. (2016). Revised estimates for the number of human and bacteria cells in the body. *PLoS Biology, 14*(8), e1002533.

Sepulveda, A. R., Yao, Y., Yan, W., Park, D. I., Kim, J. J., Gooding, W., et al. (2010). CpG methylation and reduced expression of O6-methylguanine DNA methyltransferase is associated with Helicobacter pylori infection. *Gastroenterology, 138*(5), 1836–1844.e4.

Shorter, J., & Lindquist, S. (2006). Destruction or potentiation of different prions catalyzed by similar Hsp104 remodeling activities. *Molecular Cell, 23*(3), 425–438.

Siefert, J. L. (2009). Defining the mobilome. *Methods in Molecular Biology, 532*, 13–27.

Smits, S. A., Leach, J., Sonnenburg, E. D., Gonzalez, C. G., Lichtman, J. S., Reid, G., et al. (2017). Seasonal cycling in the gut microbiome of the Hadza hunter-gatherers of Tanzania. *Science, 357*(6353), 802–806.

Strakovsky, R. S., Zhang, X., Zhou, D., & Pan, Y.-X. (2014). The regulation of hepatic Pon1 by a maternal high-fat diet is gender specific and may occur through promoter histone modifications in neonatal rats. *The Journal of Nutritional Biochemistry, 25*(2), 170–176.

Suter, M. A., Chen, A., Burdine, M. S., Choudhury, M., Harris, R. A., Lane, R. H., et al. (2012). A maternal high-fat diet modulates fetal SIRT1 histone and protein deacetylase activity in nonhuman primates. *The FASEB Journal, 26*(12), 5106–5114.

Symbiosis as a Source of Evolutionary Innovation. MIT Press. 2017. Retrieved from https://mitpress.mit.edu/books/symbiosis-source-evolutionary-innovation

Thorburn, A. N., Macia, L., & Mackay, C. R. (2014). Diet, metabolites, and "western-lifestyle" inflammatory diseases. *Immunity, 40*(6), 833–842.

Tlaskalová-Hogenová, H., Stepánková, R., Hudcovic, T., Tucková, L., Cukrowska, B., Lodinová-Zádníková, R., et al. (2004). Commensal bacteria (normal microflora), mucosal immunity and chronic inflammatory and autoimmune diseases. *Immunology Letters, 93*(2–3), 97–108.

Tomat, A. L., Costa, M., de Los, Á., & Arranz, C. T. (2011). Zinc restriction during different periods of life: Influence in renal and cardiovascular diseases. *Nutrition (Burbank, Los Angeles County, Calif.), 27*(4), 392–398.

Tosh, D. N., Fu, Q., Callaway, C. W., McKnight, R. A., McMillen, I. C., Ross, M. G., et al. (2010). Epigenetics of programmed obesity: Alteration in IUGR rat hepatic IGF1 mRNA expression and histone structure in rapid vs. delayed postnatal catch-up growth. *American Journal of Physiology-Gastrointestinal and Liver Physiology, 299*(5), G1023–G1029.

Trasande, L., Blustein, J., Liu, M., Corwin, E., Cox, L. M., & Blaser, M. J. (2013). Infant antibiotic exposures and early-life body mass. *International Journal of Obesity, 37*(1), 16.

Turroni, F., Peano, C., Pass, D. A., Foroni, E., Severgnini, M., Claesson, M. J., et al. (2012). Diversity of bifidobacteria within the infant gut microbiota. *PLoS One, 7*(5), e36957.

Turroni, S., Fiori, J., Rampelli, S., Schnorr, S. L., Consolandi, C., Barone, M., et al. (2016). Fecal metabolome of the Hadza hunter-gatherers: A host-microbiome integrative view. *Scientific Reports, 6*, 32826.

Urushiyama, D., Suda, W., Ohnishi, E., Araki, R., Kiyoshima, C., Kurakazu, M., et al. (2017). Microbiome profile of the amniotic fluid as a predictive biomarker of perinatal outcome. *Scientific Reports, 7*(1), 12171.

Van Blerkom Linda, M. (2003). Role of viruses in human evolution. *American Journal of Physical Anthropology, 122*(37), 14–46.

van der Meulen, T. A., Harmsen, H., Bootsma, H., Spijkervet, F., Kroese, F., & Vissink, A. (2016). The microbiome-systemic diseases connection. *Oral Diseases, 22*(8), 719–734.

Vannier-Santos, M. A., & Lenzi, H. L. (2011). Parasites or cohabitants: Cruel omnipresent usurpers or creative "eminences grises"? *Journal of Parasitology Research, 2011*, 1–19.

Varela, F. G., Maturana, H. R., & Uribe, R. (1974). Autopoiesis: The organization of living systems, its characterization and a model. *Currents in Modern Biology, 5*(4), 187–196.

Villarreal, L. P., & DeFilippis, V. R. (2000). A hypothesis for DNA viruses as the origin of eukaryotic replication proteins. *Journal of Virology, 74*(15), 7079–7084.

von Mutius, E. (2017). The shape of the microbiome in early life. *Nature Medicine, 23*(3), 274–275.

Voreades, N., Kozil, A., & Weir, T. L. (2014). Diet and the development of the human intestinal microbiome. *Frontiers in Microbiology, 5*, 494. Retrieved from https://www.ncbi.nlm.nih.gov/pmc/articles/PMC4170138/.

Wallin, I. E. (1927). *Symbionticism and the origin of species* (208p). Baltimore: Williams & Wilkins Company. Retrieved from http://archive.org/details/symbionticismori00wall

Wampach, L., Heintz-Buschart, A., Hogan, A., Muller, E. E. L., Narayanasamy, S., Laczny, C. C., et al. (2017). Colonization and succession within the human gut microbiome by archaea, bacteria, and microeukaryotes during the first year of life. *Frontiers in Microbiology, 8*, 738. Retrieved from https://www.ncbi.nlm.nih.gov/pmc/articles/PMC5411419/.

West, C. E., Jenmalm, M. C., & Prescott, S. L. (2015). The gut microbiota and its role in the development of allergic disease: A wider perspective. *Clinical & Experimental Allergy, 45*(1), 43–53.

Ximenez, C., & Torres, J. (2017). Development of the microbiota in infants and its role in maturation of the gut mucosa and the immune system. *Arch Med Res, 48*(8), 666–680. Retrieved from https://www.sciencedirect.com/science/article/pii/S0188440917302369.

Youngster, I., Avorn, J., Belleudi, V., Cantarutti, A., Díez-Domingo, J., Kirchmayer, U., et al. (2017). Antibiotic use in children—A cross-national analysis of 6 countries. *The Journal of Pediatrics, 182*(Supplement C), 239–244.e1.

Zelený, M. (1981). *Autopoiesis, a theory of living organizations* (342p). New York: North Holland.

Zheng, J., Xiao, X., Zhang, Q., Mao, L., Yu, M., Xu, J., et al. (2017). The placental microbiota is altered among subjects with gestational diabetes mellitus: A pilot study. *Frontiers in Physiology, 8*, 675.

Conclusions: What Is Next for the Healthy Human-Microbe "Holobiome"?

M. Andrea Azcarate-Peril

> *It will be well for the unscientific reader to understand distinctly that Professor Metchnikoff does not offer a cure for old age. Old age is not a disease and cannot be cured; it is an accumulation of changes which begin during earliest youth and continue throughout the entire life of the individual. To overcome old age, either the process or the result of the normal life of man would have to be radically changed, and there seems little prospect of our ever being able to overthrow the natural course of individual development. On the other hand, we may reasonable hope, by improving the health of the individual, to prolong life.*
> *Introduction by P. Chalmers Mitchell to the American Edition of the Prolongation of Life—Optimistic Studies by Elie Metchnikoff (1910)*

Abstract Our gut microbiota is composed of an assortment of (semi)permanent intestinal inhabitants that co-exist with temporary microbes. Although it is hard to keep up with the fast advancing field of microbiome research, this book bid to provide, first, a look into key players of the gut microbiota found in food from breast milk to dairy, meats and vegetables, to a revision of methods to generate efficacious probiotics. Then, we approached the oral microbiota as the host-microbe ecosystem that selects which bacteria will progress into the gastrointestinal tract, highlighting the importance of oral health for the overall health of the host. Finally, the third part of the book focused on age and the gut microbiome, how the consumption of foods with high microbial loads may lead to temporary changes in microbial diversity, to end on what is commonly known now as "the diseases of the Western civilization" and the potential to modulate the gut microbiota for a more resilient microbial population. Hence, what do we know now that we did not know 10 years ago?

M. A. Azcarate-Peril (✉)
Division of Gastroenterology and Hepatology, Department of Medicine, University of North Carolina at Chapel Hill, Chapel Hill, NC, USA
e-mail: azcarate@med.unc.edu

© Springer Nature Switzerland AG 2019
M. A. Azcarate-Peril et al. (eds.), *How Fermented Foods Feed a Healthy Gut Microbiota*, https://doi.org/10.1007/978-3-030-28737-5_15

Keywords Holobiome · Responders and non-responders · Healthy gut microbiota · Microbiome diversity

Who Are the Gut Bacterial Key Players and Where Do We Get Them from?

The most equilibrated and complete source of nutrition for the human infant and its associated microbiota is, without argument, breast milk. As pointed out by Dr. Collado and collaborators in the first chapter of this book, all national and international organizations focused on health, infancy, paediatrics, nutrition and epidemiology recommend exclusive breast-feeding during at least the first 6 months of life, and breastfeeding supplemented with solid foods up to year two or beyond (Horta and Victora 2013). Although exciting new research has demonstrated the contribution of other environmental influences (like mode of delivery) to the healthy establishment of baby's gut microbiota and beyond, breast milk is the main source of health-associated microorganisms (bifidobacteria) that will dominate the infants' gut during the first year of life, plus the prebiotic human milk oligosaccharides (HMOs) intended to support the establishment of beneficial bacteria in the gastrointestinal tract. Later, solid foods will expose baby's gut to a whole new world of microbes.

Dr. Requena in the chapter "Fermented Dairy Products", speculates that "because humans have consumed fermented foods since ancient times, the human gastrointestinal tract adapted to a constant supply of live bacteria on a nearly daily basis. In fact, many of the microbial species found in fermented foods are either identical to or share physiological traits with species known to promote gastrointestinal health". She emphasizes the exploration not only of the microbiota in fermented dairy products, but also the bacterial bioactive components that enrich those products, being one example the yeast *S. cerevisiae*, used as starter in kefir or koumiss, which is a folate producer. Another example of bacterial metabolites in fermented dairy products is gamma-aminobutyric acid (GABA), a major inhibitory neurotransmitter in the adult mammalian brain, generated by lactic acid bacteria that encode glutamate decarboxylases, which in the bacteria, act as a resistance mechanism against acid stress. As nicely described by the chapters "Fermented Dairy Products", "Meat and Meat Products", and "Fermented Vegetables as Vectors for Relocation of Microbial Diversity from the Environment to the Human Gut", the indigenous microbiota associated to dairy, meats, and vegetables is diverse and normally established in their original habitat in symbiotic or commensal relationships. For example, species of *Bacillus* help cucumber plants to scavenge and fix nitrogen converting gas into a solid and usable form. *Bacillus*, Enterobacteriaceae and lactic acid bacteria also assist with phosphate solubilization through the production of organic acids or phosphatases. Furthermore, Enterobacteriaceae and *Pseudomonas* can produce auxin, a plant growth hormone, and siderophores to chelate iron (Khalaf and Raizada 2016). The plant-associated microbiota is essential for the host, to the point where

Conclusions: What Is Next for the Healthy Human-Microbe "Holobiome"? 351

scientists have speculated that it may be transmitted by seeds and conserved for future generations to secure the symbiotic relationship between plants and their microbiomes. Important points are made by Dr. Perez-Diaz in the chapter "Fermented Vegetables as Vectors for Relocation of Microbial Diversity from the Environment to the Human Gut". For example, that starter cultures in vegetable fermentations eliminate or reduce biodiversity selecting for those microbes that can tolerate a number of stresses associated with the specific habitat. Lowering salt content and adding a more diverse starter could increase diversity of the end product; however, is this really desirable at the risk of shortening shelf life by increasing the risk of spoilage or adverse effects? In other words, how can we improve current food processing practices for a better gut microbiome health, preserving nutritional properties and enhancing healthy bacteria? Dr. Bruno-Barcena in the chapter "Production and Conservation of Starter Cultures: From "Backslopping" to Controlled Fermentations" takes us back to the origin of fermentation, which extended the life of dairy, meats and vegetables by bio-transforming them into new, enriched products. Trial and error practices led then to artisanal fermentations, which were aggressively restricted when humanity declared the war on microbes. The rise on chronic diseases as well as consumer preferences of new generations are leading us back to traditional, *but controlled*, fermentation practices seeking the equilibrium between food nutrition/taste and safety.

Oral Health Is Essential for Gut Health

The oral microbiome is the second most diverse ecosystem in the human body, after the colon. The mouth can (and should) be considered a complex macro ecosystem in which teeth, gingiva, tongue and the other oral surfaces act as differentiated microecosystems, each with a characteristic microbiome. The most diverse of these microecosystems is the non-shedding surface of teeth, which provide the conditions for establishment of biofilms. Oral biofilms can be composed of over 700 species, where streptococcal species become the first colonizers able to bind tooth surfaces and promote arrival of secondary colonizers like *Actinomyces naeslundii* and *Fusobacterium nucleatum*. Although the superficial epithelial surfaces of the mouth are continually shedding, the oral mucosa is constantly populated with microorganisms; however, the cellular turnover and the efficiency of the washing action of saliva result in a limited microbial colonization and low diversity communities. In following with the theme of this book, saliva is the most relevant ecosystem of the oral habitat as it provides the vehicle for the delivery of nutrients and microbes into the gastrointestinal tract. Although it has no indigenous microbiota, whole saliva can contain up to 10^8 colony forming units per mL of cultivable bacteria, most of which are also found in the oral mucosa. Saliva has important roles in biofilm formation and colonization of the oral mucosa. It also delivers antimicrobial molecules like secretory IgA antibodies, lactoferrin, lysozyme, salivary peroxidase, cationic peptides, proline-rich proteins, defensins, mucins and the salivary agglutinin GP340.

Salivary glands secrete an average of 1–1.5 L in healthy human individuals leading us to conclude that, in addition to shaping the composition of the oral microbiome, saliva and its microbial passengers influence establishment and adult composition of the gut microbiome.

As with the gut, birth is the single, most important event in shaping the oral microbiota. In fact, as detailed by Drs. Ribeiro and Roland in the chapter "Introduction to the Oral Cavity", there are significant differences in the oral microbiome composition of babies delivered by C-section compared to vaginally delivered infants at 3 months of age. Breastfeeding will then maintain an oral microbiota rich in lactobacilli, which will change with eruption and loss of primary teeth as well as with dietary changes. Healthy establishment and stabilization of the microbiota around year 3 of life leads to a symbiotic balance, where established commensals prevent pathogen colonization while contributing to overall host health. Unfavorable dietary and/or oral hygiene habits lead to a dysbiotic, and hence vulnerable, community, eventually resulting in prevalent diseases including dental caries and periodontal diseases like gingivitis and periodontitis.

Links between oral diseases and systemic chronic diseases including cardiovascular disease, stroke, abnormal pregnancy outcomes, diabetes, aspiration pneumonia, cancers and Alzheimer's disease, have been identified in a number of studies. In fact, research studies connected maternal periodontal disease with adverse pregnancy outcomes, including preterm delivery, preeclampsia, and low birth weight. Interventional studies however failed to demonstrate that treatment of periodontitis reduced the incidence of preterm birth or low birth weight. It is clear that the concept of dysbiosis of the oral microbiota and systemic effects is new; therefore, further studies will provide a better understanding of how this imbalance can be related to human diseases.

The connection between oral health and diet was recognized at least in the 1920s. The omics tools available today, like sequencing of microbial DNA, demonstrated how drastically changes in human dietary patterns impacted diversity of the oral microbiome, paralleling the effect on the gut microbiota. Sequencing data from calcified dental plaque of Mesolithic to medieval humans showed that the oral microbiome of individuals who lived in early farming communities were much less diverse than those of hunter-gatherers, harboring more groups linked to periodontal diseases (Fig. 1).

Complex hunter-gatherers as well as humans that turned to agriculture for sustenance, and later to modern starch- and sugar-rich diets, went through dramatic changes in the composition of their oral microbiome. It is interesting, however, that Pleistocene hunter-gatherers from North Africa had a very high prevalence of caries (51.2% of teeth in adult dentitions), comparable to modern industrialized populations with a diet high in refined sugars and processed cereals, due to their reliance on wild plants rich in fermentable carbohydrates and changes in food processing, which caused an early change toward a disease-associated oral microbiota (Humphrey et al. 2014). We can conclude then that, despite a current diet rich in simple carbohydrates, hygiene and dentistry practices counteract an environment conducive to caries, leading however to an oral macro ecosystem that is different from that of our ancestors whose diet was rich in fibrous foods.

Conclusions: What Is Next for the Healthy Human-Microbe "Holobiome"? 353

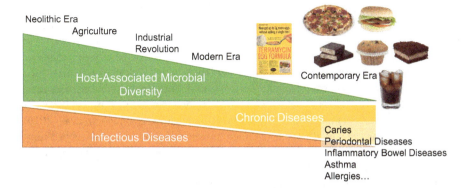

Fig. 1 Dramatic shifts introduced into the human culture and lifestyle led to loss of diversity of the host-associated microbiome. The loss of diversity was exacerbated in the second half of the twentieth century by consumption of the Western diet, characterized by low fiber and high levels of refined carbohydrates. While antibiotic use and sanitation practices resulted in a reduction of the prevalence of infectious diseases, they also led to the increased prevalence of chronic and autoimmune conditions

Are Diversity and Resilience the Key to a Healthy Gut Microbiota?

Proper establishment of beneficial and commensal microbes supports nutrient absorption, pathogen exclusion and maturation of baby's immune and nervous system. Interruption of this natural process has lifelong consequences as highlighted by recent research and throughout this book. As remarked by Dr. Thompson in the chapter "Early Gut Microbiome: A Good Start in Nutrition and Growth May Have Lifelong Lasting Consequences", "since initial colonization may shape final adult colonization, disturbances during this period may alter growth, contribute to the development of immune disorders, such as allergy and infectious disease, and metabolic conditions, like obesity and diabetes, and, potentially, shape brain development, with effects on developmental disorders and cognitive function".

Once microbial communities are established (around 3 years of age), the gut microbiome of adults is host-specific and unique to each individual. Conversely, functionality is well conserved across subjects. The evolutive meaning of this is under debate. Clearly, environmental factors impose selective pressures on the host which in turn affect its microbiota. Thus, selecting for a function or a genetic trait rather than specific taxa is advantageous for the host. However, recent evolutionary theories may need to be reformulated or expanded to consider adaptation of the holobiont (the host plus its associated microbiota) as a unit (Huitzil et al. 2018).

Dr. Zoetendal in chapter the "We Are What We Eat": How Diet Impacts the Gut Microbiota in Adulthood" details how long-term diets can shape gut microbial composition in contrast to short-term dietary changes, which have in general a minor impact on microbial composition, but may change overall activity patterns. In the adult microbiome, adequate support of diversity imparts the resilience to buffer against dysbiosis, transient changes in intestinal permeability, inflammation, pre-disposition to illness and infection and psychological imbalances. The most important factor with a clear impact on the gut microbiome of healthy individuals are fiber and, within fibers, those considered prebiotics. This has been recognized by numerous studies comparing our current Western diet (high in sugar and fat, but low in fiber) with fiber-rich diets. The observations that long-term adaptation to a Western diet over generations may have resulted in progressive loss of gut microbial species, and that diversity probably cannot be fully recovered by the reintroduction of traditional diets (Sonnenburg et al. 2016) suggest that improvement of our current microbiomes will most likely take generations of conscious, healthy eating. More importantly, the consequences of dietary long-terms patterns directly affect how individuals age.

Extensive research has demonstrated that the aging gut is characterized by a compromised barrier integrity, which is thought to allow the translocation of bacteria and bacterial metabolites into circulation from the gut lumen, and chronic, low-level inflammation or 'inflammaging'. As highlighted by Dr. Ja, the aging gut microbiota is characterized by a declining diversity and the over representation of bacterial taxa that could be considered pathogenic (like Enterobacteria) while butyrate producers are in reduced numbers. Although the general features of the aging gut holobiome have been confirmed by several studies, a causal relationship is not clear. Different lifestyles and life events like changes in diet, medications, frailty and residency status, have such an impact that the emerging features of the host-microbiota system are difficult to compile. Although research implies that life-long dietary patterns and lifestyle condition gut aging, the human life span limits longitudinal studies, which can only be undertaken in animal models like *Caenorhabditis elegans.*

Studies of the gut microbiome at every age have identified critical life events or choices that will have a long-term impact (for example, breastfeeding), which define essential windows of modulation. As nature exemplifies in breast milk, beneficial bacteria plus prebiotics, will greatly contribute to a healthy microbiota in infants. We can speculate that adult and elderly populations benefit from this approach as increasing number of studies show the effects of probiotics on composition, functionality, and disease or disease prevention, even if the need for probiotic colonization of the gut is under debate. Similarly, prebiotics and fiber-rich diets have known beneficial effects.

The logical follow up question to this book is, what actions can we take at each age and life stage to establish and maintain a healthy gut microbiota? Figure 2 presents a very simplified view of the characteristics of the gut microbiome at different ages. One commonality of the healthy microbiome, as emphasized by Dr. Salvucci in the chapter "The Disappearing Microbiota: Diseases of the Western Civilization", is the biodiversity status of the microbiota that will ensure the optimal functionality of the system at every age, therefore, should we attempt to increase gut microbiota diversity? If so, how do we promote *healthy* diversity in the gut microbiota?

As biologists, we have to acknowledge variations that complicate a clear categorization of individuals (although attempts to do this have been carried out, for example by classifying specific microbiome configurations as enterotypes (Arumugam et al. 2011; Wu et al. 2011). The fluidity of these differences makes essential to

Conclusions: What Is Next for the Healthy Human-Microbe "Holobiome"?

Characteristic of the Gut Microbiome	Infant	Adult	Elder
Vulnerability	High	Low	High
Stability	Low	High	Low
Diversity	Low	High	Low(er)
Characteristic taxa	*Bifidobacterium*, Bacteroidetes	Ruminococcaceae, Lachnospiraceae, and Bacteroidaceae	Facultative anaerobes, streptococci, staphylococci, enterococci, and enterobacteria
Life events of relevance	Birth, Breastfeeding, Disease	Disease	Disease
Microbiome Objective	To establish a healthy gut microbiota through breastfeeding	To maintain a healthy gut microbiota through high-fiber, probiotics diet	To prevent loss of diversity through high-fiber, probiotics diet

Fig. 2 A simplified summary of the characteristics of the gut microbiota in infancy, adulthood and aging

identify treatment-specific signature microbes that can predict host physiological responses and efficacy. For example, a high percent of lactose intolerant individuals responded to treatment with lactose-free prebiotics, which act via modulation of the gut microbiota increasing lactose-metabolizing bacteria (Azcarate-Peril et al. 2017); however, some individuals did not. Elucidating the factors that contribute to the responder and non-responder phenotypes is the first step to increasing efficacy of modulatory treatments of the gut microbiota.

Overall, the association between hosts and their microbiotas and the intersectionality between systems is complex and fascinating. It is also very fragile as demonstrated by the impacts that "hits" like disease conditions, antibiotic treatments, and lifestyle can have in one system (the gut) that will then cascade into others (the host's overall health). Although immense progress has been made, the study of such systems and their connections is still in its infancy. Advancement will depend on education of the consumer on the best approaches to maintain health as well as policy makers to regulate food industries and labeling.

References

Arumugam, M., Raes, J., Pelletier, E., Le Paslier, D., Yamada, T., Mende, D. R., Fernandes, G. R., Tap, J., Bruls, T., Batto, J. M., Bertalan, M., Borruel, N., Casellas, F., Fernandez, L., Gautier, L., Hansen, T., Hattori, M., Hayashi, T., Kleerebezem, M., Kurokawa, K., Leclerc, M., Levenez, F., Manichanh, C., Nielsen, H. B., Nielsen, T., Pons, N., Poulain, J., Qin, J., Sicheritz-Ponten, T., Tims, S., Torrents, D., Ugarte, E., Zoetendal, E. G., Wang, J., Guarner, F., Pedersen, O., de Vos, W. M., Brunak, S., Dore, J., Antolin, M., Artiguenave, F., Blottiere, H. M., Almeida, M., Brechot, C., Cara, C., Chervaux, C., Cultrone, A., Delorme, C., Denariaz, G., Dervyn, R., Foerstner, K. U., Friss, C., van de Guchte, M., Guedon, E., Haimet, F., Huber, W., van Hylckama, J., Vlieg, A., Jamet, C., Juste, G., Kaci, J., Knol, O., Lakhdari, S., Layec,

K., Le Roux, E., Maguin, A., Merieux, R., Melo Minardi, C., M'Rini, J., Muller, R., Oozeer, J., Parkhill, P., Renault, M., Rescigno, N., Sanchez, S., Sunagawa, A., Torrejon, K., Turner, G., Vandemeulebrouck, E., Varela, Y., Winogradsky, G., Zeller, J., Weissenbach, S. D. E., & Bork, P. (2011). Enterotypes of the human gut microbiome. *Nature, 473*(7346), 174–180.

Azcarate-Peril, M. A., Ritter, A. J., Savaiano, D., Monteagudo-Mera, A., Anderson, C., Magness, S. T., & Klaenhammer, T. R. (2017). Impact of short-chain galactooligosaccharides on the gut microbiome of lactose-intolerant individuals. *Proceedings of the National Academy of Sciences of the United States of America, 114*(3), E367–E375.

Elie Metchnikoff (1910). The Prolongation of Life: Optimistic Studies, New York & London, G.P. Putnam's Sons.

Huitzil, S., Sandoval-Motta, S., Frank, A., & Aldana, M. (2018). Modeling the role of the microbiome in evolution. *Frontiers in Physiology, 9*, 1836.

Horta, B. L., Victora, C. G. (2013). Long-term effects of breastfeeding: a systematic review. World Health Organization. ISBN: 978 92 4 150530 7

Humphrey, L. T., De Groote, I., Morales, J., Barton, N., Collcutt, S., Bronk Ramsey, C., & Bouzouggar, A. (2014). Earliest evidence for caries and exploitation of starchy plant foods in Pleistocene hunter-gatherers from Morocco. *Proceedings of the National Academy of Sciences of the United States of America, 111*(3), 954–959.

Khalaf, E. M., & Raizada, M. N. (2016). Taxonomic and functional diversity of cultured seed associated microbes of the cucurbit family. *BMC Microbiology, 16*(1), 131.

Sonnenburg, E. D., Smits, S. A., Tikhonov, M., Higginbottom, S. K., Wingreen, N. S., & Sonnenburg, J. L. (2016). Diet-induced extinctions in the gut microbiota compound over generations. *Nature, 529*(7585), 212–215.

Wu, G. D., Chen, J., Hoffmann, C., Bittinger, K., Chen, Y. Y., Keilbaugh, S. A., Bewtra, M., Knights, D., Walters, W. A., Knight, R., Sinha, R., Gilroy, E., Gupta, K., Baldassano, R., Nessel, L., Li, H., Bushman, F. D., & Lewis, J. D. (2011). Linking long-term dietary patterns with gut microbial enterotypes. *Science, 334*(6052), 105–108.

Index

A
ACE-inhibitory activity, 44
Acetate, 263
Aciduric, 177, 179, 185
Acinetobacter, 175
Actinobacteria, 245, 250, 261
Actinobaculum, 175
Actinomyces, 173, 175–177, 185
Adult microbiome
  carbohydrates, 271, 272
  dietary modulation, 265, 266, 271
  functional capacity, 260
  individual and niche-specific, 260, 261
  long-term dietary patterns, 260
  microbial diversity, stability and resilience,
      262, 263
  microbial metabolites, 263–265
Aging microbiome
  adulthood, 288
  age-related changes, 286
  bacteria/bacterial products, 286
  community dwellers, 289
  copper cell region, 294
  diet and environmental factors, 288
  dietary interventions, 297
  diversity and stability, 288
  dysbiotic ecology, 297
  ecosystem, 288
  ELDERMET project, 289, 297
  enterobacteria, 288
  fecal metagenomes, 287
  fecal microbiota composition, 289
  healthy microbiota, 287
  host-commensal physiology, 293
  host–microbe homeostasis, 285, 294, 295
  immune system development, 286

  inflammatory response, 294
  intestinal barrier and immune system, 286
  intestinal homeostasis, 297
  invertebrate and vertebrate model
      organisms, 286
  JAK/Stat pathway, 294
  longitudinal studies, 289
  luminal microbes, 293–294
  mammalian and fly intestinal biology, 293
  microbe-driven pathologies, 286
  mucosal and systemic immune tissues,
      286–287
  passive *vs.* active microbial effects, 295
  physiological changes, 286
  solid foods, 287
Akkermansia, 293
Akkermansia muciniphila, 243, 262
Alzheimer's disease, 48, 352
Anaerobic gut-associated microbes, 6
Angiotensin converting enzyme (ACE), 43
Animal-based diet, 73
Animalcules, 155
Animal-derived microbiota, 59
Antibiotic-resistant staphylococci, 20
Antibiotic therapy, 21
Antibiotic treatments, 355
Antibiotics
  autoimmunity disorders, 334
  gut microbiota diversity, 335
  maturation of microbiota, 335
  penicillin G,V and vancomycin, 335
  pregnancy, 335
Antihypertensive peptides, 43
Antimicrobial compounds, 13
Antimicrobial substances, 196, 197
Antimutagenic, 44

© Springer Nature Switzerland AG 2019
M. A. Azcarate-Peril et al. (eds.), *How Fermented Foods Feed a Healthy Gut Microbiota*, https://doi.org/10.1007/978-3-030-28737-5

358

Index

Antioxidant, 44
Arginine, 200
Artisanal products, 128

B
'Back slopping' culture, 92, 127–129, 134
Bacterial colonization process, 12
Bacteriocins, 196
Bacteriophage, 101, 102, 133
Bacteroidaceae, 288
Bacteroides, 243, 273
Bacteroides fragilis, 246
Bacteroidetes, 45, 242, 249, 261
β-galactosidases (β-Gal), 314
Bifidobacteria, 5, 48, 74, 194, 196
Bifidobacterium, 4, 48, 185, 244–246, 311, 313, 314, 317
Bifidobacterium animalis, 296
Bifidobacterium infantis, 246
Bifidogenic factors, 4
Bile acids, 266
Bile salt hydrolases (BSH), 272
Bioactive compounds, 49
  acidic environment, stomach, 41
  bioactive peptides, 43, 44
  EPS, 43
  fermentation of milk, 41
  GABA, 42
  human health, 41
  vitamins, 41, 42
Bioactive peptides, 43, 44
Biologically active substances, 20
Bioprotective strains, 75, 76
Bleeding on probing (BOP), 211
Blood-brain barrier, 263
Bowel syndrome (IBS) symptoms, 260
Branched-chain fatty acids (BCFAs), 264
Breast cancer, 17
Breast-fed infants, 8, 11–14, 16, 19
Breastfeeding, 352
Breastmilk, 246
Breast tissue biopsies, 11
Bulky vegetables, 92
Butyrate, 15, 263

C
Cabbage, 94
Caenorhabditis elegans, 298, 354
Calpis, 43
Candidate probiotic strains, 73, 74
Carbohydrates, 263, 271, 272
Carcasses, 59
Cardiometabolic diseases, 46, 47

Caries prevention, 199, 200, 205, 224
Carnobacteria, 69
Cemento-enamel junction, 179
Checkerboard DNA-DNA hybridization technique, 156
Cheese, 36, 38, 39, 42–45, 48
Chemokines, 15
Chilling temperatures, 60
Chronic diseases, 351
Chronic intestinal inflammation, 46
Clostridium difficile infection (CDI), 333
Coagulase-negative staphylococci (CNS), 15, 16
Cold chain application, 60
Colony-forming units (CFU), 67
Colorectal cancer (CRC), 260
Community and system-level interactive optimization (CASINO), 271
Complex hunter-gatherers, 352
Conjugated linoleic acid (CLA), 38
Control halitosis, 220
Cooked meat products, 68, 69
Corynebacteria, 19
Crohn's disease (CD), 332
Cucumber fermentation, 98, 99, 101–106, 108, 109, 113
Cucurbits rhizoplane, 93
Cultivable bacteria, 5
Culture-based studies testing
  bifidobacteria, 5
  contaminants, 4
  cultivable bacteria, 5
  human-milk strains, 5
  infant infections, 4
  limitations of culture-dependent methods, 5
  microbial load in human milk, 5
  mother-to-infant bacterial transfer, 5
  viruses, 5
Culture-dependent methods, 5, 12
Culture-independent methods, 12, 61, 72
Culture-independent molecular techniques
  application, 6
  bacterial cell wall, 6
  DNA, 6, 7
  human milk/breast tissue, 7
  lactobacilli, bifidobacteria and strict anaerobes, 8
  limitations and bias, 6
  metagenomic analysis, 8
  microorganisms, 7
  non-critical analysis, 7
  nucleic acids, 6
Culture production, 130–132, 134
Cystatins, 146

D

Debaryomyces hansenii, 70
Demineralization, 173, 174, 176, 185
Dental caries, 173–178
    deciduous/primary teeth, 198
    demineralization, 198
    etio-pathogenesis, 224
    management, 200, 201, 204, 205
    pathogenesis, prevention and treatment,
        199, 200
    severity, 198
Dentin, 173, 174, 176
Deoxycholic acid, 272
Dietary fiber, 36
Diet-microbiota interactions, 262
Disease-associated oral microbiota, 352
Double-blind placebo prospective study, 39
Dry-cured (fermented) meat products, 69–71
Dysbiosis process, 18

E

Early childhood caries (ECC), 177
Early gut microbiome
    adult-like pattern, 247
    allergy and infectious disease, 241
    bacterial communities, 240
    birth type, 242–244
    complementary feeding and maturation,
        247, 248
    complex ecosystem, 239
    development of diseases, 240
    early infant feeding, 244–246
    environmental and dietary exposures, 239
    environmental bacteria, 240
    implications
        breastfeeding, 249
        dietary interventions/supplementation,
          251
        epidemiological studies, 249
        food/supplements, 251
        high fat/carbohydrate diets, 251
        infancy and early childhood, 249
        maternal and child health, 251
        maternal bacteria, 249
        metabolic and inflammatory pathways,
          251
        metabolic processes, 251
        obese and non-obese adults, 250
        vaginally-born infants, 249
    infant feeding, 244
    metabolic function, 247
    microbes, 240
    obesity and chronic disease, 239

organ systems, 240
    predominant microbial groups, 240
    prenatal development, 241
    risk factors, 250
Endodontic infections, 178, 179
Enterobacteriaceae, 93, 244
Enterococci, 67
Entero-mammary pathway, 22
Epidemiological transitions
    animals contact, 330
    behaviour and organization, 329
    dietary habits, 330
    hunter-gatherers, 330
    Industrial Revolution, 330
    sanitization, 330
    study, 331
    western diet, 330
Epigenetics
    acetylation and phosphorylation, 336
    allergies and inflammatory diseases, 338
    DNA methylation profiles, 338
    DNA sequence, 336
    factors, 336
    fetal epigenetic program, 336
    Firmicutes/Bacteroides ratio, 338
    genes, 338
    Helicobacter pylori, 339
    human-microbiota superorganism, 338
    methylation of cytosine residues, 336
    microRNA, 336
    modifications, 336
    perturbations, 336
    probiotic treatments and modulation of
        microbiota, 339
    Rett syndrome, 338
    risk factor, 337
    T-lymphocyte differentiation, 338
Epstein–Barr virus, 164
Erythritol, 200
Etio-pathogenesis, 219
Etiopathogenesis, lactational mastitis, 17–19
Eukaryotic antimicrobial peptides, 20
European Food Safety Authority (EFSA), 13,
    194
Evolus, 43
Exopolysaccharides (EPS), 43, 176

F

Facility-specific microbial consortia, 60
Faecalibacterium prausnitzii, 287
Fat, 272, 273
Fecal microbiota transplantation (FMT), 262,
    318, 333

Index

Fermentable oligosaccharides, disaccharides, monosaccharides and polyols (FODMAP) diets, 260
Fermentation
  artisanal practices, 127
  "boiling without heat", 126
  food, 129
  industrial, 134
  microbial, 132, 134
  niche-specific strains, 129
Fermented dairy
  bioactive compounds (*see* Bioactive compounds)
  cheese, 38
  GIT, 39, 40
  health promoting effects (*see* Health promoting effects)
  kefir, 37
  LAB (*see* Lactic acid bacteria (LAB))
  lactose hydrolysis, 40, 41
  metabolites, 36
  microorganisms, 38, 39
  milk components, 35, 36
  milk preservation, 36
  probiotics, 47–49
  yogurt, 36, 37
Fermented vegetables
  acids, 104
  artisanal preparations, table olives, 105
  augmenting biodiversity
    food consumption guidelines, 111
    human body, 109
    human gut microbiome, 110
    lactobacilli, 111
    Lb. plantarum, 111, 112
    microbial diversity, 110
  back slopping, 92
  bacterial consortia, 114
  bulky vegetables, 92
  cucumber, 106
  Enterobacteriaceae, 105
  era of controlled fermentations, 106
  exclusion of air, 106
  fresh vegetables (*see* Fresh vegetables)
  human gut microbiome (*see* Human gut microbiome)
  indigenous microbiota, 92
  industrial production of, 106
  microbiota of natural fermentations (*see* Natural fermentation of vegetables)
  processing parameters, 106–109
  products, 106
  sauerkraut, 92
  sodium chloride, 105
  spoilage (*see* Spoilage)
  starter culture, 105
  vessels, 104
Firmicutes, 72, 248, 249, 261, 272
Fluorescence in situ hybridization (FISH), 175
Food and Agriculture Organization (FAO), 195
Food and Drug Administration (FDA), 13
Food biotransformation, 128
Food preservation, 126, 127, 134
Freeze-drying, 133
Freezing temperatures, 60
Fresh vegetables
  adhesion of bacteria, 94
  Bacillus species, 93
  bacteriome, 94
  biotic and abiotic factors, 93
  cabbage, 94
  cucurbits rhizoplane, 93
  gram-negative bacteria, 94
  LAB, 96
  lactobacilli, 96
  lettuce core microbiota, 94
  microbes and plants, 93
  MRS agar plates, 96
Fructo-oligosaccharides (FOS), 314
Fusobacteria, 179

G

Galacto-oligosaccharides (GOS), 13, 314
Gamma-aminobutyric acid (GABA), 38, 42, 350
Gardnerella, 242
Gastrointestinal tract (GIT), 38–40
Generally Recognised As Safe (GRAS), 13, 317
Genetic mutations, 40
Genome sequencing, 48
Geographical links, 69
Gingival crevice, 143
Gingival crevicular fluid (GCF), 179, 198, 211
Gingivitis, 179, 180, 204–212, 218, 224
Glucosyl transferases, 143
Glyoxylate cycle, 184
Gram-negative bacteria, 94, 97
Gram-positive bacteria, 97
Granulicatella, 175
Gut microbiome, 183, 184, 354
  colon, 313–315
  effective probiotics, 311–313
  methods and vehicles, 316–317
  modulators, 317–318
  origin and evolution, 309–311

Index 361

Gut microbiota, 354
  acid stress, 350
  beneficial and commensal microbes, 353
  breastfeeding, 350
  dietary changes, 353
  environmental factors, 353
  fiber-rich diets, 354
  gastrointestinal tract, 350
  meat microbiome
    animal-based diet, 73
    bacterial species, 72
    bifidobacteria, 74
    bioprotective strains, 75, 76
    candidate probiotic strains, 73, 74
    consumption of foods, 72
    culture-independent methods, 72
    fermented meats, 73
    health status, 72
    LAB, 73, 75, 76
    lactobacilli, 74
    presence and abundance, 76
    probiotic properties, 74
    short-time dietary, 72
  modulation, 44, 45
  nutrition, 350
  physiological traits, 350

H
Hafnia, 67
Halitosis, 218–220, 224
  prebiotic treatment, 220
  probiotic treatment, 221–223
Hamburger formulations, 68
Hampering, 16
Health promoting effects
  cardiometabolic diseases, 46, 47
  inflammation, 46
  in vitro and in animal models, 44
  modulation of gut microbiota, 44, 45
  prevention of infection, 45
Healthier insulin profile, 46
Healthy aging, 285, 297
Herbal components, 68
Highly sensitive CRP (hsCRP), 315
High-oxygen MAP red meats, 61, 66, 67
High-protein foods, 58
High-throughput DNA sequencing technology, 39
Histatins, 146
HIV-1 breast milk transmission, 14
HIV-inhibitory activity, 13
Homeostasis, 262
Horizontal gene transfer (HGT), 327

Host defenses, 197, 198
Host–microbiota system
  bacterial cell-wall product, 292
  bioinformatics analysis, 292
  cell culture-based tests, 293
  fecal microbiota, 290
  germ-free mice, 292
  humans and model organisms, 291
  inflammation status, 291
  intercellular spaces, 292
  intestinal barrier function, 292
  intestinal microbes, 290
  mucus layer, 291
  nutrition, older adults, 290
  physiologic systems, 290
  TNF, 292
Human breast milk
  bifidogenic factors, 4
  composition, 4
  infant nutrition, 4
  lactational mastitis (*see* Lactational mastitis)
  microbial diversity (*see* Microbiomes in human milk)
  mother-to-infant transfer of bacteria, 12
  probiotics, 12–16
  short- and long-term health-promoting effects, 4
Human cytomegalovirus, 164
Human gut microbiome
  concept of nutrition, 112
  consumption, un-pasteurized fermented vegetables, 113
  energy, 113
  fibers, 113
  GI tract, 112
  lactic acid, 113, 114
  pathogens, 113
Human-microbes superorganism
  complex models, 327
  cross-feeding interactions, 328
  C-section delivery and immune system, 334
  cultural and dietary shifts, 333
  eukaryotes, 327
  G protein coupled receptors, 328
  GABA and serotonin, 328
  gut microbiota, 328
  health status, 328
  HGT, 327
  microbiota, 327
  mitochondria and chloroplasts, 326
  'ontogenic drift', 326
  phyla, 334

362 Index

Human-microbes superorganism (*cont.*)
  symbiogenesis, 326
  system evolution, 326
  vaginal delivery, 334
Human milk oligosaccharides (HMOs), 8, 19,
    245, 314, 350
Human neutrophil peptides 1-3 (HNP1-3), 197
Human omnivore diet, 57
Human Oral Microbe Identification with Next
    Generation Sequencing
    (HOMINGS), 163
Human oral taxon (HOT), 160
Human papilloma virus (HPV), 164
Human systems, 43
Hydrogen, 264
Hydrogen sulfide, 265
Hygienic industrial practices, 38
Hypothesis, 262

I
Immunomodulating, 44
In vitro meat production systems, 58
Industrial fermentation, 134
Industrial-scale technologies, 43
Infant gut microbiota, 334
Infant nutrition, 4, 12
Inflammation, 46
Inflammatory bowel diseases (IBD), 328, 331
International Scientific Association for
    Probiotics and Prebiotics (ISAPP),
    310
Intestinal transit time (ITT), 265
Intraoral halitosis, 219
Inulin (IN), 194
Irritable bowel syndrome (IBS), 272

J
JAK/Stat pathway, 294

K
Kefir, 36, 37, 39, 42, 43
Klebsiella, 175

L
Lachnospiraceae, 288, 289
Lactational mastitis
  ecological niches, 16
  etiological diagnosis, 17
  etiopathogenesis, 17–19
  hampering, 16

  mammary gland, 16, 17
  microbiological analysis, 17
  molecular microbiology techniques, 17
  predisposing factors, 19–20
  probiotics, 20–22
  undesired premature weaning, 16
Lactic acid bacteria (LAB), 36, 38–44, 46, 61,
    69, 70, 73, 75, 76, 93, 94, 96–98,
    101, 104–106, 109–111, 113, 114
Lactobacillus, 48, 242–245, 311, 314, 317,
    318
Lactobacillus acidophilus, 246
Lactobacillus curvatus, 69
Lactobacillus plantarum, 296
Lactobacillus reuteri, 223
Lactobacillus salivarius CECT 5713, 14
Lactobacillus salivarius WB21, 222, 223
Lactose fermentation, 40, 41
Leuconostocs, 69
Life stages, 19
Lithocholic acid, 272
Live microorganisms, 44
Long-term dietary patterns, 266
Low diversity, 332

M
Malpositioned teeth, 206
Maltotetrose, 147
Maltotriose, 147
Mammary gland, 17
Mammary-associated microbiome, 19
Marination, 67
Maternal dysbiosis, 243
Matrix metalloproteinases (MMP), 198
Matrix-assisted laser desorption/ionization
    time-of-flight mass spectrometry
    (MALDI-TOF MS), 17
Meat fermented products
  consumption, 58
  cooked, 68, 69
  dry-cured, 69–71
  herbal components, 68
  popularity, 58
  preservatives, 68
  ready-to-eat foods, 58
  salting, 68
  smoked, 71, 72
  types, 58, 68
Meat matrix, 58
Meat microbiome
  gut microbiota (*see* Gut microbiota)
  and meat products, 68–72
  microbial ecosystem, 58–60

poultry, 66–68
raw meats, 60–65
red meats, 61–66
Meat-processing plants, 59
Mesenteric lymph nodes, 246
Metagenetic analysis, 66
Metagenomic analysis, 8
Metagenomics, 48
Metaproteomic and metatranscriptomic
    analyses, 48
Methods to study the oral microbiome, 157,
    158
Microbial communities, 353
Microbial diversity, 67
Microbial ecosystem, 58–60
Microbiological analysis, 17
Microbiomes in human milk
    culture-based studies testing, 4–6
    culture-independent studies, 6–8
    factors, 8–10
    HMOs, 8
    milk samples, 10
    origin of bacteria, 10, 11
    quantitative/qualitative composition, 8
Microbiota, 350
Microbiota modulation, 310, 318, 319
Microorganisms
    fermented dairy, 38, 39
Milk-oriented microbiota (MOM), 245
Milk preservation, 36
Modified-atmosphere packaging (MAP), 60
Molecular microbiology techniques, 17
Mucin-stimulating, 44
Muscle myoglobin, 61

N

National Health and Nutrition Examination
    Survey (NHANES), 46
Natural fermentation of vegetables
    bacteriophages, 101
    cabbage´and sauerkraut, 101
    commercial cucumber, 99
    cucumbers, 98
    gram-positive and gram-negative bacteria,
        97
    Korean cabbage kimchi, 98
    LAB, 97
    molds, 101
    olive, 101
    process, 97
    salting, 97, 98
    Spanish-style table olive, 98
    viruses, 101

watery kimchi, 98
    yeast, 99, 101
Necrotizing enterocolitis (NEC), 251
Next-generation sequencing, 61
Non-bacterial oral microbiome, 163, 164
Non-negligible vehicles, 58
Non-starter lactobacilli (NSLAB), 38
Normal healthy gut microbiota, 45

O

Oligofructose (OF), 194
Omnivore dietary setup, 72
Open-ended molecular diagnostic, 157, 158
Operational taxonomic units (OTUs), 60, 184
Opioid effects, 44
Oral administration, 10, 11
Oral biofilms, 157, 158, 160, 164, 165, 351
Oral cavity
    antibiotics, 143
    complexity, 144
    ecological niches, 141
    environmental conditions, 142
    exocrine secretions, 143
    gut microbiome, 142
    infant, 143
    microbial DNA sequencing, 142
    microbiome, 142
    normal microbiota of mouth, 142
    oral environment (see Oral environment)
    saliva and gingival crevicular fluid, 144
    systemic chronic diseases, 142
Oral environment, 141
    definition, 144
    saliva (see Saliva microbiome)
    soft tissues, 150
    teeth (see Teeth microbiome)
Oral halitosis, 218–220
Oral health, 156, 159, 160, 162
    breastfeeding, 352
    characteristic microbiome, 351
    and diet, 352
    and systemic chronic diseases, 352
Oral microbiome, 351, 352
    characterization, 156
    colon, 142
    composition, 143, 148, 155
    cultivable methods, 156
    culture-independent methods, 156
    diagnostic techniques, 157
    gut microbiome, 142
    homeostasis, 147
    host-protection mechanisms, 143
    infant's life, 143

Oral microbiome (*cont.*)
 microbial diversity, 164, 165
 multispecies communities, 157
 non-bacterial species, 163, 164
 open-ended diagnostic techniques, 166
 open-ended sequencing techniques, 156
 oral and broader systemic health, 142
 oral biofilm diversity, 157, 158
 oral cavity, 162, 163
 oral health, 159–162
 oral mucosal surfaces, 144
 OTUs, 161
 16S rRNA gene, 156
 subgingival biofilm samples, 162
Oral microbiome dysbiosis
 acidogenic and aciduric species, 185
 aciduric, 185
 cemento-enamel junction, 179
 dental caries, 173–178
 dietary characteristics, 171
 dietary composition, 172
 endodontic infections, 178, 179
 GCF, 179
 gingivitis, 179, 180
 gut microbiome, 183, 184
 healthy individuals, 172
 non-shedding surfaces, 172
 oral cavity, 172, 173
 periodontal diseases, 182, 183, 185
 periodontitis, 180–182
 protected surfaces, 173
 saccharolytic, acidogenic and aciduric
     populations, 179
 smooth free surfaces of tooth, 173
 Streptococcus, 172
 supragingival plaque, 179
 systemic consequences, 184, 185
Oral prebiotics, 194, 195
Oral probiotics, 195–198, 204
Oral tissue microbiome, 144
Origin of bacteria, 10, 11
Osteoporosis, 206
2-Oxoglutarate ferredoxin oxidoreductase, 184
Oxymyoglobin development, 66

P
Pasteurization, 127, 134
Pathobionts, 182, 186
Peri-implant diseases, 205, 206, 214, 217, 218
Peri-implantitis, 157
Periodontal diseases, 172, 173, 180, 182, 183,
     185, 205, 206
Periodontitis, 157, 160, 162, 180–182
 treatment, 212, 214

Peripartum antibiotherapy, 20
Peyer's patches, 10
Pharyngeal cavities, 145
Photodynamic therapy (PDT), 217
Planktonic cells, 130
Plant-associated microbiota, 350
Plaque-induced gingivitis, 207, 208, 211, 212
Polymerase chain reaction (PCR), 157, 199
Porphyromonas, 175, 181, 183
Porphyromonas gingivalis, 298
Poultry microbiome, 59–61, 66–68
Prebiotic therapy, 200, 201
Prebiotics, 47
 Bifidobacterium species, 314
 definition, 310
 β-Gal enzymes, 314
 GOS, 315
 health and disease, 315, 316
 lactases, 314
 and synbiotics, 311
Prevotella, 175–177, 179, 181–183, 242, 265
Pro- and anti-inflammatory cytokines, 15
Probiotic properties, 74
Probiotics
 animal models, 317
 antimicrobial compounds, 13
 bacterial species, 13
 Bifidobacterium breve BBG-001, 313
 breastfeeding, 14
 butyrate, 15
 category, 311
 CNS, 15, 16
 criterion, 311
 definition, 310
 delivery matrix, 316
 dietary supplementation, 295
 disease prevention, 313
 double-blinded controlled study, 13
 efficacy, 316
 fermented dairy, 47–49
 fermented foods, 295
 functional, 316
 gastrointestinal function, 15
 gut microbiota, 313
 health and disease, 315, 316
 HIV-1 breast milk transmission, 14
 HIV-inhibitory activity, 13
 host–microbe homeostasis, 296
 human mastitis
   antibiotics, 20, 21
   antimicrobials, 22
   development, 20
   entero-mammary pathway, 22
   L. gasseri CECT 5714, 20, 21
   L. salivarius CECT 5713, 20, 21

Index

L. *salivarius* PS2, 21, 22
  microbiological, biochemical/
    immunological biomarkers, 21
  NMR characterization, 21
  potential mechanisms, 22
  staphylococcal/streptococcal counts, 21
  supplementation, 21
  TCA cycle, 21
immobilization and encapsulation, 316
immune response of healthy humans, 15
immune-stimulating activity, 14
immunomodulatory effect, 14, 295
infant gut, 13
intestinal microbial species, 295
L. *fermentum* CECT5716, 15
L. *salivarius* CECT 5713, 14
lactic acid bacteria, 13
Lactobacillus genera, 311
maturation of infant immune system, 14
methods and vehicles, 317
milk component, 12
nutrient absorption, 15
study, 313
synbiotics, 311
Th1 cytokines, 14
WCFS1 strain, 296
ZO-1, 15
Probiotic therapy, 201, 204, 205
Professionally administered plaque removal
  (PAPR), 217
Prokaryote, 10
Propionate, 263
Propionibacterium, 273
Protein, 273
Proteobacteria, 261, 272, 293
Pseudomonas, 175
Psychobiotics, 48

Q

Qualified Presumption of Safety (QPS), 13
Quantitative/qualitative composition, 8

R

Randomized clinical trials (RCTs), 212
Raw meat microbiome, 60–65
Raynaud's disease, 17
Red meat microbiome, 61–66
Regulatory T cells (Tregs), 333
Remineralization, 174, 184, 185
16S ribosomal RNA (16S rRNA), 156, 158,
  160, 161, 164
Ruminococcaceae, 288
Runt-related transcription factor 3 (RUNX3), 338

S

Saccharolytic, 179, 185
Saliva microbiome
  acquired enamel pellicle, 145
  antimicrobial influences, 148
  dental biofilms, 148
  electrolytes, 145
  exocrine secretions, 148
  hydroxyapatite surfaces, 146
  hypotonic nature, 145
  lingual lipases, 147
  microbial profile, 147
  microbiome development, 147
  oligosaccharides, 147
  oral cavity, 144, 147
  pH range, 146
  salivary constituents, 148
  salivary flow, 145–147
  salivary glands, 145
  tooth integrity, 146
  water, 145
Sanger sequencing, 156–159
Sauerkraut, 92, 99, 101, 103, 105, 106
Scaling and root planing (SRP), 212
Selectin, 19
Shelf-life of red meats, 61
Short-chain fatty acids (SCFAs), 194, 245,
  263, 272, 328
Short-term intervention studies, 266
Single nucleotide polymorphism (SNP), 20
Slaughterhouses, 59
Slaughtering process, 66
Smoked meat products, 71, 72
Spoilage
  anaerobiosis, 102
  extreme acidic pH, 103
  high organic acid concentrations, 103
  industrial vegetable fermentations, 102
  LAB, 104
  olive and cucumber fermentations, 103
  oxidative yeast, 103
  pH, 102
  Pichia manshurica and *Issatchenkia*
    occidentalis, 103
  pink sauerkraut, 103
  Propionibacterium spp., 103
  residual sugars and viable microbes, 102
  softening of vegetables, 103
  undesirable microbes, 102
Staphylococci, 76
Starter cultures
  agricultural development, 126
  backslopping culture methods, 127–129
  culture production, 130–132
  fermentation practices, 126

Starter cultures (*cont.*)
  food preservation techniques, 134
  food storage practices, 126
  foods, 126
  germophobic interpretation, 127
  knowledge-based hygiene practices, 127
  monocultures, 129, 130
  non-pathogenic microbes, 127
  pasteurization methods, 127
  physical separation, 134
  preservation, 132, 133
  vigorous and stable, 134
Statherin, 146
Streptococcus, 172, 175–178, 185
Streptococcus salivarius K12, 221, 222
Stress-induced dysfunction, 15
Structural food matrix, fermented milk, 39
Subgingival biofilm, 206
Subgingival sulcus, 149
Succinate dehydrogenase, 184
Sulphidogenic bacteria, 265
Superorganisms, 260
Supragingival biofilm, 160, 206
Symbiogenetic theory, 326
Synbiotics, 311

T
Teeth microbiome
  approximal surfaces, 149
  dental plaque, 149
  early colonizers, 149
  hydroxyapatite mineral, 148
  initial colonizers, 148
  lingual surfaces, 149
  pathogenic microbial communities, 150
  pioneer streptococci, 148
  subgingival sulcus, 149
Th1 cytokines, 14
Thermal treatment, 69
Thermotolerant bacteria, 69
Traditional diet, 260, 266

Traditional fermentation
  monocultures, 129, 130
Treatment-specific signature microbes, 355
Tricarboxylic acid (TCA) cycle, 21
Type 2 diabetes, 47

U
Ulcerative colitis (UC), 262, 332

V
Vacuum-packed meat variants, 66
Vaginal delivery, 242
Veillonella, 173, 175, 183, 185
Verrucomicrobia, 261
Viridans streptococci, 17
Vitamins, 41, 42
Volatile sulfur compounds (VSCs), 219

W
Weissella cibaria, 223
Western diet, 260, 262, 264–266, 274, 330,
  331
Western diseases
  CD, 332
  diabetes, 332
  helminth infections, 333
  obesity and metabolic syndrome, 331
  schizophrenia, 333
  Tregs, 333
  UC, 332

Y
Yeast, 99, 101
Yogurt, 36, 37, 39–44, 46, 47

Z
Zonula occludens-1 (ZO-1), 15